Scanning Tunneling Microscopy and Related Methods

NATO ASI Series

Advanced Science Institutes Series

A Series presenting the results of activities sponsored by the NATO Science Committee, which aims at the dissemination of advanced scientific and technological knowledge, with a view to strengthening links between scientific communities.

The Series is published by an international board of publishers in conjunction with the NATO Scientific Affairs Division

A	**Life Sciences**	Plenum Publishing Corporation
B	**Physics**	London and New York
C	**Mathematical and Physical Sciences**	Kluwer Academic Publishers Dordrecht, Boston and London
D	**Behavioural and Social Sciences**	
E	**Applied Sciences**	
F	**Computer and Systems Sciences**	Springer-Verlag
G	**Ecological Sciences**	Berlin, Heidelberg, New York, London,
H	**Cell Biology**	Paris and Tokyo

Series E: Applied Sciences - Vol. 184

Scanning Tunneling Microscopy and Related Methods

edited by

R. J. Behm
Institut für Kristallographie und Mineralogie,
Universität München, München, F.R.G.

N. Garcia
Department of Physics,
Universidad Autonoma de Madrid, Madrid, Spain

and

H. Rohrer
IBM Research Division,
Zurich Research Laboratory, Ruschlikon, Switzerland

Kluwer Academic Publishers

Dordrecht / Boston / London

Published in cooperation with NATO Scientific Affairs Division

Proceedings of the NATO Advanced Study Institute on
Basic Concepts and Applications of Scanning Tunneling Microscopy
Erice, Italy
April 17–29, 1989

Library of Congress Cataloging in Publication Data

```
NATO Advanced Study Institute on Basic Concept and Applications of
  Scanning Tunneling Microscopy (1989 : Erice, Italy)
    Scanning tunneling microscopy and related methods : Proceedings of
  the NATO Advanced Study Institute on Basic Concepts and Applications
  of Scanning Tunneling Microscopy, Erice, Italy, April 17-29, 1989 /
  edited by R.J. Behm, N. Garcia, H. Rohrer.
       p.   cm. -- (NATO ASI series. Series E, Applied sciences ; no.
  184)
    Includes index.
    ISBN 0-7923-0861-1 (alk. paper)
    1. Scanning tunneling microscopy--Congresses.  2. Surfaces
  (Physics)--Congresses.   I. Behm, R. J.   II. Garcia, N. (Nicolás)
  III. Rohrer, Hermann.   IV. Series.
  QC173.4.S94N379  1989
  502'.8'2--dc20                                             90-40494
```

ISBN 0–7923–0861–1

Published by Kluwer Academic Publishers,
P.O. Box 17, 3300 AA Dordrecht, The Netherlands.

Kluwer Academic Publishers incorporates the publishing programmes of
D. Reidel, Martinus Nijhoff, Dr W. Junk and MTP Press.

Sold and distributed in the U.S.A. and Canada
by Kluwer Academic Publishers,
101 Philip Drive, Norwell, MA 02061, U.S.A.

In all other countries, sold and distributed
by Kluwer Academic Publishers Group,
P.O. Box 322, 3300 AH Dordrecht, The Netherlands.

Printed on acid-free paper

All Rights Reserved
© 1990 by Kluwer Academic Publishers
No part of the material protected by this copyright notice may be reproduced or utilized in any form or by any means, electronic or mechanical, including photocopying, recording or by any information storage and retrieval system, without written permission from the copyright owner.

Printed in the Netherlands

TABLE OF CONTENTS

Preface (H.Rohrer)   ix

I. Methods

1) Scanning Tunneling Microscopy - methods and variations (H.Rohrer)   1

II. Theory

2) A brief introduction to tunneling theory (C.R.Leavens and G.C.Aers)   27

3) Tunneling times for one-dimensional barriers (C.R.Leavens and G.C.Aers)   59

4) Theory of Scanning Tunneling Microscopy and Spectroscopy (J.Tersoff)   77

5) Theory of tunneling from transition metal tips (G.Doyen, E.Koetter, J.Barth and D.Drakova)   97

6) Tip-surface interactions (S.Ciraci)   113

7) On the quantized conductance of small contacts (L.Escapa and N.Garcia)   143

8) Adiabatic evolution and resonant tunneling through a one-dimensional constriction (E.Tekman and S.Ciraci)   157

9) What do we mean by 'work function' (R.G.Forbes)   163

III. Applications of STM at Solid State Surfaces

10) Scanning Tunneling Microscopy: Metal surfaces, adsorption and surface reactions (R.J.Behm)   173

11) Scanning Tunneling Microscopy: Semiconductor surfaces, adsorption and epitaxy (R.M.Feenstra)   211

12) Spectroscopy using conduction electrons 241
    (H.van Kempen)

13) Scanning Tunneling Optical Microscopy (STOM) 269
    of silver nanostructures
    (R.Berndt, A.Baratoff and J.K.Gimzewski)

14) Surface modification with the STM and the AFM 281
    (C.F.Quate)

IV. Liquid-Solid Interface

15) Scanning Probe Microscopy of liquid-solid 299
    interfaces
    (P.K.Hansma, R.Sonnenfeld, J.Schneir, O.Marti,
    S.A.C.Gould, C.B.Prater, A.L.Weissenhorn,
    B.Drake, H.Hansma, G.Slough, W.W.McNairy and
    R.V.Coleman)

16) In-situ Scanning Tunneling Microscopy in 315
    electrochemistry
    (H.Siegenthaler and R.Christoph)

V. Applications of STM at Organic and Biological Materials

17) Imaging and conductivity of biological and 335
    organic material
    (G.Travaglini, M.Amrein, B.Michel and H.Gross)

18) Study of the biocompatibility of surgical 349
    implant materials at the atomic and molecular
    level using Scanning Tunneling Microscopy
    (R.Emch, X.Clivaz, C.Taylor-Denes, P.Vaudaux,
    D.Lew and P.Descouts)

19) Naked DNA helicity observed by Scanning Tunneling 359
    Microscopy
    (A.Cricenti, S.Selci, A.C.Felici, R.Generosi,
    E.Gori, W.Djaczenko and G.Chiarotti)

20) Applications of Scanning Tunneling Microscopy 367
    to layered materials, organic charge transfer
    complexes and conductive polymers
    (S.N.Magonov and H.-J.Cantow)

21) Electron tunneling through a molecule 377
    (Ch.Joachim and P.Sautet)

22) Electronic transport in disordered organic 391
    chains (R.Garcia and N.Garcia)

VI. Electron and Ion Point Sources

23) Electron and ion point sources, properties 399
    and applications (H.W.Fink)

24) Field electron emission from atomic-size 409
    microtips
    (J.J.Saenz, N.Garcia, Vu Thien Binh and
    H.De Raedt)

VII. Force Microscopy

25) Force Microscopy 443
    (H.Heinzelmann, E.Meyer, H.Rudin and H.J.Güntherodt)

26) Electret-condensor-microphone used as a very 469
    sensitive force sensor
    (E.Schreck, J.Knittel and K.Dransfeld)

VIII. Optical and Acoustic Microscopy

27) Resolution and contrast generation in Scanning 475
    Near-Field Optical Microscopy (U.C.Fischer)

28) Scanning Tunneling Optical Microscopy 497
    (D.Courjon)

29) Scanning Near-Field Acoustic Microscopy 507
    (P.Güthner, E.Schreck, K.Dransfeld and
    U.C.Fischer)

Index 515

PREFACE

Some eight years ago, in March of 1981, the first successful experiments of vacuum tunneling in an STM configuration were carried out and later topographic images with mono-atomic steps on CaIrSn$_4$ were obtained within three months. Four years later, in July of 1985, about fifty STMers, or practically the entire STM community, gathered for the first time at a workshop of the IBM Europe Institute in Oberlech, Austria. The workshop was a cross between a school and a research conference: everybody learned like a student but practically everybody also had an opportunity to lecture. The main interest was focussed on such instrumental questions as vibrations, speed and tip stability, and on surface topographies, some of them with atomic resolution.

Within the past four years, the scene has completely changed. At the present NATO Advanced Study Institute fourteen main lecturers taught an audience of about 140 participants of which more than half were graduate students in one of the respective fields. From the variety of topics in this issue it is evident that not only the number of scientists in the field has grown but also their competence as well as the depth and breadth of the field itself. Atomic resolution is obtained routinely in surface science applications, spectroscopy has become indispensable, STMs are operated at cryogenic temperatures and in insulating and electrolytic liquids, and a large family of instruments has been built, and about ten kinds are even commercially available.

Real space imaging with atomic resolution and in various environments is an attractive microscopy capability, but equally significant is the general acceptance of distance and displacement monitoring and control with subangstrom precision. This cleared the way for a variety of new local probe methods which had not even been thought of in Oberlech or were mentioned if at all as exotic curiosities. Force microscopy has since been invented and has even developed into a field with a wide range of variations. Other techniques such as ballistic electron emission microscopy, thermal profiling, and ion flow microscopy have emerged and show interesting potential. The importance of the single-atom tip as a point source for low-energy electrons and ions goes far beyond its original purpose of a well-defined tunneling tip for STM. It again opens a new area of "microscopic" thinking. Finally, the tip as an active tool for surface modifications has begun to receive the attention it deserves. The "Fifth International STM Conference on Scanning Tunneling Microscopy" in 1990 will also be Nano I, the "First International Conference on Nanometer Science and Technology".

Crucial to this exciting progress was the openness which exists in the STM community, as openness is a key for any scientific development. I hope this openness continues. Then, I am sure that four years from now the scene will have progressed as much as it has in the past four-year period.

I would like to thank all the lecturers for their conscientious and excellent presentations, the special lecturers for presenting their newest achievements, and the students for their incessant and unwavering attention. On behalf of all participants I express our gratitude to our hosts at the Ettore Majorana Center for their generosity, efficient organization and comprehensive assistance and for having accepted such an unusually large number of us in the first place, and also to the NATO Science Committee for financial support.

H. Rohrer

SCANNING TUNNELING MICROSCOPY — METHODS AND VARIATIONS

H. Rohrer
IBM Research Division
Zurich Research Laboratory
CH-8803 Rüschlikon
Switzerland

ABSTRACT. In these two lectures, I give a short introduction to local probe methods, pointing out certain aspects which very often do not receive proper attention, and present some of the interesting methods not represented in the main lectures. The various problems addressed should in no way discourage anybody from solving them, and doing it right.

1. Local Probe Methods

There are two major methods to perform experiments on a single microscopically small object or part thereof. In the first method, a lens system connects the microscopic object with the world of the macroscopic observer or experimenter. This is the classical microscopy with its variations, such as scanning techniques, where the lateral resolution is limited by the wavelength and/or aberrations of the lens system. For the vertical resolution other criteria apply: in optical techniques, such as contrast microscopy or ellipsometry, angstrom or even subangstrom resolution is attainable, provided homogeneity extends over lateral distances greater than the wavelength. For electron microscopy, depth resolution is usually poorer than the lateral resolution.

In the second method, the microscopic part of the object under investigation is addressed by a small probe in close proximity to the object. This is the local probe method, which probes a local property of the object or produces a local, reversible or irreversible modification via some interaction between probe and object. It is a natural and conceptionally the simplest way to perform a local experiment and, together with the scanning capability of the probe, to image a microscopic object. A well-known local probe technique is point contact spectroscopy, but without scanning capability [1,2]. Another very old and practical example is the medical doctor's stethoscope, which is not really microscopic, but the "experiment" is done on a scale much smaller than the wavelength of the sound emitted by the heart. But it remained for scanning tunneling microscopy to truly pioneer the technique for imaging and making local modifications on a nanometer scale.

In the following, I shall concentrate on imaging. Local modifications are treated in the lectures by Quate. In the local probe method, imaging is performed by scanning a probe over a fixed sample, while in force microscopy the probe is usually kept fixed and the sample is scanned. The drives for probe or sample are generally built from piezoelectric components. Tripods [3] or tube [4] drives cover fields of view from a few thousand angstrom up to several microns; larger scan areas can be covered by bimorphs [5]. The piezo-electric materials used are ceramics and it was not *a priori* clear that their elongations or contractions were continuous on an angstrom or subangstrom level. So far, no discontinuity in their piezo-electric response is reported. However, hysteresis effects can be appreciable [6]. For a summary on instrumental issues, see Ref. 7.

The major instrumental problems are the vibration isolation of the local probe-object distance and the coarse positioning of the probe within the range of the piezo drives. Vibration isolation is achieved by either protecting the instrument as a whole from its environment [8,9] or by making the lowest eigenfrequency of the mechanical path between object and local probe much higher than any mechanical vibrations of the environment. Various coarse-positioning methods have been used since the "louse" [8]. Many of them are variations of reductions by lever systems [10,11], often combined with some kind of differential spring and screw systems. Convenient, remote coarse positioning over macroscopic distances in three dimensions, as done in the bioscope [12], *and* at high mechanical eigenfrequencies for fast imaging are challenges for future versatile instrumentation. An interesting approach in this direction are the micromechanical instruments [13,14] or "beetle" type instruments [15].

For detailed and comprehensive information on local probe techniques, in particular scanning tunneling microscopy (STM) and atomic force microscopy (AFM), I refer the reader to conference proceedings [16-21], reviews [22-25], and of course to the lectures of this NATO Meeting. The latter are referred to as "lecture by" followed by the author's name and appear in the NATO ASI series volume "Basic Concepts and Applications of Scanning Tunneling Microscopy (STM) and Related Techniques" (Kluwer Academic Publishers).

Finally, there are semi-local methods in which microscopic and macroscopic lengths are involved, such as field ion microscopy (FIM), projection microscopy and mask lithography. For the first two, see the lectures by Fink.

1.1. RESOLUTION

The vertical resolution is limited mainly by the stability of the probe-object distance and we will not discuss it further. We also assume stable conditions, i.e. that the probe and object do not change during the measurement.

The crucial elements for the lateral resolution of local probe methods are the size of the probe, the distance between probe and object and its control, and the distance dependence and lateral variation of the interaction under consideration. Therefore, the resolution depends on the local sample properties, so it is not possible to give simple general resolution criteria. This problem is addressed in the lectures by Tersoff. Nevertheless, we can get a feeling for the resolution by considering the width L of the interaction filament, which contains $1 - 1/\eta$ of the total interaction intensity, as shown in Fig. 1. For various distance dependences, $f(d)$, we obtain:

$$L \simeq 2(R+s)\eta, \qquad f(d) = \text{const} \qquad (1a)$$

$$L \simeq 2\sqrt{2}\ \beta(\eta)\sqrt{(s+R)s} \qquad f(d) \propto d^{-n},\ n \geq 2 \text{ and} \qquad (1b)$$
$$\beta(\eta) \text{ is of order 1 and varies slowly with } \eta$$

$$L \simeq 2\sqrt{2}\ \sqrt{\ln \eta}\ \sqrt{(s+R)/\kappa} \qquad f(d) \propto \exp-(\kappa d). \qquad (1c)$$

The dependence of Eq. (1c) on length is the same as that obtained for STM [26,27]. The value of η depends on the ratio of intensity change due to the object and that due to noise. Stoll [28] obtains

$$a = \pi \ln^{-1}(h/\delta)\sqrt{(s+R)/\kappa}, \qquad (2)$$

where a is the minimum period of a corrugation of height h which can be detected and δ the noise in terms of distance fluctuations.

The square-root dependence of the resolution on certain lengths is also encountered in some of the semilocal methods. The two examples shown in Fig. 2 are projection microscopy and mask lithography. In projection microscopy, the distance between light source and object is microscopic and the object-image distance

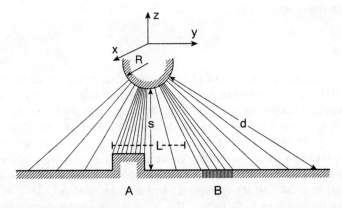

Figure 1. "Interaction" $f(d)$ between a spherical tip of radius R and the surface of an object at distance s with a geometrical feature at A and an interaction inhomogeneity at B. L is the width of the interaction filament containing $1 - 1/\eta$ of the total intensity in the absence of features A and B.

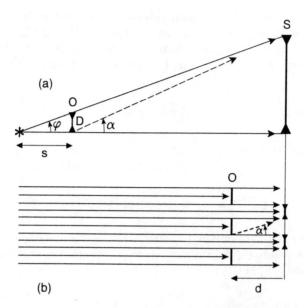

Figure 2. Resolution of (a) projection microscopy and (b) mask lithography. In both cases the object O is projected onto screen I. Solid lines: undiffracted beams; dashed lines: first-order diffracted beams.

macroscopic; in mask lithography this is reversed. The resolution limit for projection microscopy is obtained by requiring that the illumination angle $\varphi = D/s$ be larger than the angle of the first-order diffracted beam, $\alpha = \lambda/D$, where λ is the wavelength. This gives $D > \sqrt{s\lambda}$. In addition, the light-source size must be much smaller than D. The resolution for mask lithography is analogous, with s the distance between mask and resist.

2. Interactions

Figure 3 summarizes the various local probe methods invented or implemented in the wake of STM (see the lectures by Behm, Feenstra, and Tersoff). Interactions used for imaging include forces, see Fig. 3(b) and the lectures by Güntherodt; optical near fields, see Fig. 3(c) and the lectures by Fischer and Courjon; capacitance, Fig. 3(d) [29]; thermal sensor, Fig. 3(c) [29], and ion flow, see Fig. 3(f) and the lecture by Hansma. The resolution limits can vary appreciably according to the property probed and the smallest practicable probe size. Atomic resolution is generally achieved in STM and its variations, and also in AFM on many surfaces with strong in-plane bonds. In both cases, the apex atom of the probe tip serves as local probe. A magnetic force sensor, on the other hand, contains several thousands of atomic magnetic moments in order to yield a measurable force and to

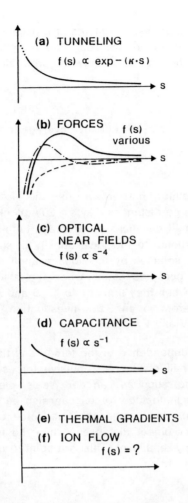

Figure 3. "Interactions." Illustration courtesy of D. Pohl.

have a fixed direction of its total magnetic moment. At present, this limits the magnetic force resolution to a few tens of nm. For optical near fields, capacitance and ion flow, the minimal probe size is both a matter of practical fabrication as well as of reasonably measurable intensity, since e.g. optical near fields decrease with the fourth power of the probe size.

2.1. COMMENTS

(a) Tunneling: The tunneling current is a measure of the overlap of the electronic wave functions of tip and sample in the gap between them, i.e. the tip wave functions probe the object wave functions. In the approximation of Tersoff and Hamann

[26], the tip probes the local density of states at a distance $R+s$ from the surface (see the lecture by Tersoff and references cited therein). The tunneling current depends exponentially on the separation between probe tip and sample,

$$I = B \, \exp(-\kappa s). \tag{3}$$

The decay length $1/\kappa$ is related to the effective average tunnel barrier height $\bar{\phi}$,

$$\kappa = \sqrt{2m/\hbar} \, \sqrt{\bar{\phi}} = A\sqrt{\bar{\phi}}. \tag{4}$$

In the case of a vacuum barrier, m is the free electron mass and $A = 1.025 \, \text{Å}^{-1} \, \text{eV}^{-1/2}$. At large distances, say $s \geq 20$ Å, $\bar{\phi}$ closely corresponds to the work function, while at small distances the average tunnel barrier decreases with decreasing s. B is proportional to the local density of states involved in the tunneling process, which are selected by the voltage applied between probe tip and object, i.e. the sample's occupied states when the electrons tunnel out of the sample, the empty states when they tunnel into it. B and κ are the local electronic sample properties of interest in an STM investigation. For metallic densities, $I \simeq 1$ nA for $s \simeq 6$ Å, $\bar{\phi} \simeq 5$V and $V = 1$ Volt.

(b) Forces: The distance dependence of the forces and their magnitudes can vary considerably. At very small distances, repulsive forces dominate, at large distances attractive forces do, such as van der Waals, electrostatic and attractive magnetostatic forces. Magnetostatic forces can also be repulsive. Because the attractive forces are usually long-range, the total force on the probe tip can be attractive at medium probe-object distances, while the force at the apex of the probe tip is already strongly repulsive. This is a point to be considered, especially when working with soft objects.

3. Operating Modes

Local probe methods use two interactions between the local probe and the object: the control interaction for control of the probe position and the probing interaction to perform the local experiment, be it a measurement or a modification. In the following, the control of the sensor position concerns only the sensor-object distance, the lateral sensor displacement usually follows a standard raster scan pattern. We take the x-direction as the fast scan direction and the y-direction as the slow scan displacement. For object modifications, a vector scan operation is also possible. So far, two operating modes have been applied for the vertical control of the probe position: the constant interaction mode and the constant height or z-position mode. Position control is thus performed with the distance dependence of the control interaction, while the experiment addresses a local property via the probing interaction. The control interaction should be as homogeneous as possible across the object area under consideration, whereas the probing interactions should sense the slightest heterogeneities, which is the very purpose of a local probe method in the first place. It lies in the nature of the local probe method that the same type of interaction is most conveniently used for both control and probing. This is a matter of the interaction sensing and measurement as well as of the signal-to-noise ratio given by the signal of the features to be detected by the probing interactions to the noise caused by position fluctuations of the local probe. However, different interactions can also be used for control and measurement, some cases of which will be mentioned below. The central problem is to separate position control and local experiment or, in other words, to distinguish between structural and interaction features as shown in Fig. 4. In one class of approaches, the property of interest is obtained from a series of experiments under different control conditions, which contain the property with different weights. In another class, the control is intermittently disabled such that the z-position of the local probe remains unchanged when performing a local experiment.

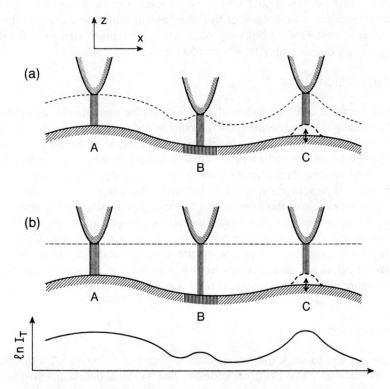

Figure 4. (a) Constant interaction and (b) constant height mode. Around A, the interaction is homogeneous, at B it is stronger, and at C the object surface responds to it. The curve at the bottom of (b) gives the measured tunnel current in the constant height mode.

3.1. CONSTANT INTERACTION MODE: TOPOGRAPHY

Tracing the contours of constant interaction intensity with the probe tip is so far the easiest and most widely used control and imaging mode. It produces a "constant interaction surface". In this mode, the deviation from a set interaction intensity is corrected by constantly adjusting the separation of tip and sample via a feedback system. In STM, this is called the "constant current mode" [see Fig. 4(a)]. Control and probing interactions are the same. In region A, the interaction is homogeneous and the tip trace corresponds to the surface topography − up to resolution broadening. In B, an inhomogeneity in the interaction changes the tip trace accordingly. In C, the local property is affected by the proximity of the tip. Shown is the effect of an elastic response of the object surface to an attractive force between tip and object, which induces a reversible change of the surface geometry. Effects of possible tip surface interactions on the imaging should always be given due consideration. A more detailed account is given in the lectures by Ciraci.

The imaging speed is limited by the motion of the probe tip in vertical or z-direction. This motion should only have frequency components smaller than the mechanical resonance frequency of the mechanical path connecting probe tip and sample. For a resonance frequency of 5 kHz and a pixel size of 1 Å, the scan speed should be considerably less than 5000 Å/sec.

3.2 CONSTANT HEIGHT MODE

The imaging speed can be increased substantially by keeping the z-position of the probe tip constant and recording the change in intensity [22] [see Fig. 4(b)]. In practice, the time constant of the feedback loop is made sufficiently long or, if the relative z-position of tip and sample remains stable in time, the feedback loop is disabled. The limiting factor is then the scan frequency. This is called "constant height mode". It has to be applied to relatively flat parts of a sample due to the limited dynamic range of the control electronics. For instance, four orders of magnitude in the tunnel current at a tunnel barrier height of 2 eV allow maximal height differences of about 6 Å. Constant height images give a qualitative topography; for conversion to a quantitative topography the detailed distance dependence of the interaction is required, which is not known *a priori*. In spite of these two limitations, the constant height mode should be very useful for studying dynamic aspects.

3.3. TRACKING MODE

This mode is used to track the lateral displacement of a feature. The displacement can be instrumental, like drift, or intrinsic, like surface diffusion. In the general case two control interactions are needed: one for the vertical position and one sensing the feature to be tracked. In the case of structural features, the same interaction can be used. Pohl and Möller [30] have used circular lateral tip motion to find and track an indentation in the object surface with the method of steepest descent.

3.4. ABSOLUTE PROBE-OBJECT DISTANCE

Of prime importance in microscopy are the lateral and, for topographic structures, also the vertical sizes of specific features. They are obtained from the calibrated displacements of the probe. However, the absolute probe-object distance, s, is required when carrying out deconvolution procedures, since the lateral resolution depends s. In other experiments, such as those aimed at the general distance dependence of an interaction, s is even a primary parameter.

In one approach, contact between probe and object surface is taken as $s = 0$. This seems straightforward and quite natural. In practice, however, contact is a rather delicate and not well understood problem. For force microscopy, contact can be defined as the closest point of zero force. However, there might be metastable positions in the probe-object force interaction, e.g. physisorption − if there is a zero force at all, e.g. magnetic repulsive forces. Other practical problems include possible hysteresis effects caused by attractive forces, and partial compensation in the experimental force measurement of repulsive forces at the apex of the probe tip by long-range attractive forces such as those mentioned above. In tunneling, contact is usually associated with a discontinuity in the current vs. distance relation [31]. Since such a discontinuity is not accounted for by theory, it is associated with a mechanical instability due to rearrangements or displacements of probe and object atoms in the contact area. The discontinuity at contact can vary appreciably, from less than a factor of two in the current on clean metal surfaces [31] up to two orders of magnitude in an electrolytic environment [32]. Moreover, the predicted flattening of the $I(s)$ curve in the two-angstrom region before contact is not always observed [32,33]. Thus, contact is in practice a somewhat fuzzy concept − at least experimentally on a subangstrom scale.

A second approach uses an extrapolation from a region where the distance dependence of the interaction is believed to be well known. In tunneling, one method uses the field emission or Gundlach resonances [8,34,35] or the fact that in the field emission regime, $I \propto \exp(-\phi^{3/2}/E)$, where $E = V/s$, as long as the radius of curvature of the tunnel tip R is large compared to s.

Still another estimate is based on the Tersoff-Hamann model and compares a measured surface corrugation amplitude with a calculated one.

4. Imaging Modes

Most of this chapter deals with imaging modes which involve tunneling as control interaction and/or probing interaction. Many names given to the various modes were adapted from other traditional experimental procedures. It is not my intention to invent a more systematic nomenclature and I therefore retain this established usage. Some people narrow microscopy down to structural features; I shall use microscopy in a much broader sense. By microscopy I mean assembling an image of a specific property obtained from many local experiments. Actually, nanoscopy would be a more appropriate name.

4.1. TOPOGRAPHY

This imaging mode was described as operating or control mode in Sects. 3.1 and 3.2. An image closest to the topography or geometrical structure is obtained as a constant interaction surface, provided the interaction itself and the response of the object to it are homogeneous. This is usually the case for clean metallic surfaces as shown in Fig. 5 for the 1×2 reconstruction of Au(110) [36]. However, not the topography is shown, but an image of $\partial I/\partial x$ on a constant current surface. It was obtained by modulating the x-position of the tunneling tip by $\delta x \simeq 0.1$ Å at a frequency much larger than the response of the feedback loop. This method has been proposed by Abrahams *et al.* [37]. Since the lock-in technique used to measure

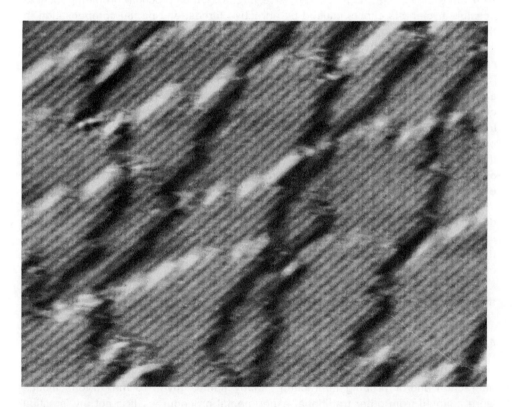

Figure 5. $\partial I/\partial x$ image on a constant current surface exhibiting terraces of 1×2 reconstructed Au(110). Image courtesy of J. K. Gimzewski.

$\partial I/\partial x$ is an averaging technique, structural features as seen through $\partial I/\partial x$ on a constant current surface $I(x,y)$ appear more clearly than in the constant current surface or topography. In the example given, the reconstruction is hardly recognizable in the topography. But even the restoration of a topography from the $\partial I/\partial z$ image requires $1/f$ filtering, as shown by Stoll [38], before the 2×1 reconstruction and step lines become clearly visible. Note that the lock-in technique averages only on a time scale shorter than the integration time and does not quench the noise at low frequency.

On electronically inhomogeneous surfaces, the central task is to separate the inhomogeneity effects, like those at B and C in Fig. 4, from geometric features. In some cases, such as STM of clean metal surfaces or AFM in the repulsive mode of uniformly hard surfaces, the constant interaction surface reflects the surface structure. In many others, however, the variations in the interaction dominate the image contrast. They reflect local properties such as electronic structure, chemical composition and magnetic polarization − to name just a few − which in their own right are as much of interest as the surface structure or topography. A variety of modes aim at imaging these different specific properties. Generally, however, there is no single, distinct imaging mode for a specific property and the separation of the various properties contained in any type of image is by no means trivial.

In scanning tunneling microscopy, it is the local electronic properties contained in B and κ of Eq. (3), which make the interaction inhomogeneous. The imaging methods concerned with B and κ are scanning tunneling spectroscopy with its variations and work-function profiles.

4.2. SPECTROSCOPY

Spectroscopy is concerned with the local density of electronic states, which appears in the prefactor B of Eq. (3). The electronic states of interest are selected by the voltage applied between tip and sample. A certain "coarse selection" is made by the energy dependence of the tunneling probability as shown in Fig. 6.

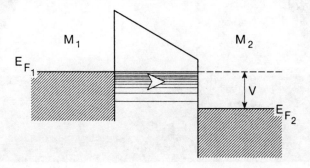

Figure 6. Tunneling transmission is highest between occupied states at E_{F_1} and the empty states at $E_{F_2} + V$.

The simplest type of spectroscopy consists of taking two constant current images at reversed polarity [39]. We then obtain an image each of predominantly occupied and empty states, respectively; in the case of GaAs, As appears in the image of the occupied states, Ga in those of the empty states. In a second mode, images taken at different voltages are compared, where the voltages are chosen such that specific electronic features are particularly evident in the corresponding constant current images [40]. An elegant variation of this approach is the current imaging tunneling spectroscopy (CITS) [41]. There, the z-position of the tip is controlled in the constant current mode at a set current and voltage I_0 and V_0, respectively. The control is disabled intermittently to record $I-V$ curves at every pixel. Besides the constant current image at V_0, a series of images of $I-I_0$ at $V-V_0$ are obtained, which should exhibit the relevant electronic features. The art of the method consists of choosing V_0 such that the constant current image is electronically as homogeneous as possible. Finally the classical way of performing tunneling spectroscopy entails measuring dI/dV (or better $d(\ln I)/d(\ln V)$ for normalizing purposes) simultaneously with the topographic mode so as to obtain the local density of states at an energy $E_F \pm V$, depending on polarity, and at a distance from the surface given by the set current and the set voltage [22,42]. For problems arising in semiconductors at small voltages, see the lecture by Feenstra. Structurally induced electronic inhomogeneities can mix structural features into the dI/dV images [22] which then resemble an inverted topography. Spectroscopy proved to be a powerful tool in the imaging of semiconductor surfaces. Two more recent, very elegant applications of spectroscopy deal with coulomb blocking in single-electron tunneling [43] and with the superconducting vortices in 2 H-NbSe$_2$ [44], which is shown in Fig. 7.

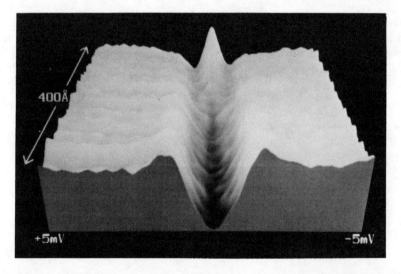

Figure 7. dI/dV vs. V characteristic of 2 H-NbSe$_3$ taken on a line from the superconducting edge of the vortex (front) to its center (back). From [44].

4.3. WORK FUNCTION PROFILES

Another set of STM experiments is concerned with the inverse decay length or the tunnel barrier height, the other local electronic property of Eq. (3). The interpretation of such experiments is often subject to considerable confusion and it appears worthwhile to comment more extensively on this aspect. The obvious experimental methods consist in measuring I vs. s at every point, in analogy to CITS in spectroscopy, or $d\ln I/ds$ on a constant current surface in analogy to classical spectroscopy. In the framework of Eq. (3), this then directly yields $\bar{\phi}$, and for this reason the corresponding images are referred to as work function profiles. This, however, is quite misleading. First, what we measure is I as a function of the voltage applied to the piezo drive. Even assuming that the z-piezo elongates proportionally to the applied voltage does not imply that the tunnel gap s also varies accordingly. This is discussed in Sect. 4.7. Secondly, even given that we can determine I vs. s, its connection with the local electronic property and, even more so, with the work function might be quite involved.

The work function, φ_∞, is by definition the energy needed to remove an electron to infinity as shown in Fig. 8(a). In practice, the work function is taken as $\varphi(z)$ at some large distance L averaged laterally over distances of order L. It is therefore not a local but a global material property. The tunnel barrier height, on the other hand, is a local property but depends also on the tip-sample separation s, as evident from Fig. 8(b). At separations $s > 2$ Å, the s-dependence is dominated by image forces [45], at smaller distances by exchange and correlation effects [46]. Approximating $\phi(s, z)$ by a square barrier $\bar{\phi}(s)$, we obtain for the image force range

$$\bar{\phi}(s) = \bar{\phi}_\infty - \alpha/s, \tag{5}$$

where $\alpha \sim 10$ eV Å. Note that $\bar{\phi}_\infty$ is a local property; it should not be considered the true value of ϕ and thus φ at $s = \infty$, which is an averaged global property. The extrapolation to $s = \infty$ is a type of normalizing procedure, which eliminates the distance dependence from a distance-dependent local property, but retains its local character. On a homogeneous surface, however, $\bar{\phi}_\infty = \varphi_\infty$. At this point, it is appropriate to return to Eqs. (3) and (4). Actually, the exponent there, $\beta' = -A\sqrt{\bar{\phi}}\, s$, where

$$\bar{\phi} = \frac{1}{s}\int_0^s \phi(s, z)dz, \tag{6}$$

should be replaced [47] by

$$\beta = -\int_0^s \sqrt{\phi(s, z)}\, dz, \tag{7}$$

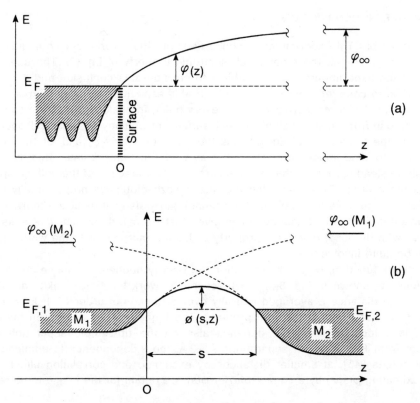

Figure 8. (a) Electron potential $\varphi(z)$ along a line perpendicular to the surface. φ_∞ is the work function. (b) Formation of the tunnel barrier $\phi(s,z)$ between two metals with work functions $\varphi_\infty(M_1)$ and $\varphi_\infty(M_2)$, respectively.

where we have set the prefactor $A = 1$. We then obtain

$$-d\ln I/ds = \sqrt{\phi(s,s)} + \int_0^s \left(d\sqrt{\phi(s,z)}/ds\right)dz, \tag{8}$$

which simplifies to

$$\sqrt{\phi(s)} + s\, d\left(\sqrt{\phi(s)}\right)/ds \tag{9}$$

for a rectangular barrier. The exponents β' and β differ only in $\sigma(\phi)/\overline{\phi}^2$, where σ is the variance. Thus Eq. (3) is a good approximation for the total current, but not necessarily for $d(\ln I)/ds$.

Let us now revert to the image force reduction of the tunnel barrier. Inserting Eq. (5) into (8), we obtain

$$-d \ln I/ds = \sqrt{\bar{\phi}_\infty}\left(1 + \frac{3}{8}(\alpha/s\bar{\phi}_\infty)^2\right), \tag{10}$$

i.e. the distance dependence enters in second order only.

Whether we use a more complicated barrier or the simple one of Eq. (5), the important result is that for a leading $1/s$ decrease of the tunnel barrier, $d \ln I/ds$ yields a local electronic property which is independent of the experimental geometry to second order in $\alpha/s\bar{\phi}_\infty$. This is generally confirmed by experiments down to very small values of s [31-33], where exchange and correlation effects become dominant [46]. Although the image force reduction of the tunnel barrier only slightly affects the value of $d \ln I/ds$, it increases the tunnel current by $\exp(\alpha/2) \simeq 150$, independent of distance.

For the trapezoidal barrier of Fig. 9(a), which is a model barrier for tunneling between two different metals, $d \ln I/ds \simeq (\phi_1 + \phi_2)/2 = \bar{\phi}$. However, for a composite barrier of Fig. 9(b), a model barrier for e.g. tunneling through a nonconducting object O on a conducting substrate, $\bar{\phi} \simeq \phi_1$, but $d \ln I/ds = (\phi_1 + \phi_2)/2$.

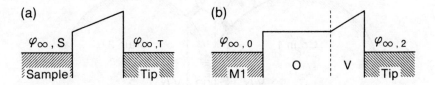

Figure 9. Idealized tunnel barriers (a) between two metals with different work functions. (b) Composite barrier with oxide layer, O, and vacuum gap V.

4.4. SCANNING TUNNELING POTENTIOMETRY

The object of scanning tunneling potentiometry is the potential distribution in a current-carrying structure. The method was pioneered by Muralt and Pohl [48]. The principle is shown in Fig. 10. The potential V_T of the probe tip at the position x of the current carrying structure S is adjusted such that the tunneling current $I = 0$. Hence $V(x) = V_T$. In one approach, the tip position is controlled by an ac tunnel current while the dc component is compensated such that $V_T = V(x)$ [48]; in the other [49], the dc current controls the tip position and $V(x)$ is determined from $I = 0$ of the I-V characteristic. In practice, however, it is not as simple as it sounds. In the application to semiconductor heterostructures a major problem arises from the fact that I can be below a measurable size over a voltage range up to several volts and that the I-V characteristic is asymmetric. The voltage at which $I = 0$ can therefore not be determined accurately. On rough surfaces, such as granular structures, a discontinuity in the potential distribution can be mimicked by a jump of the tunnel current from A to B in the tip as shown in Fig. 11.

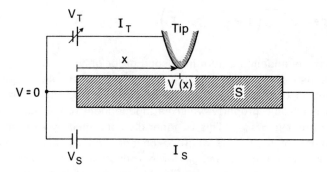

Figure 10. Principle of scanning tunneling potentiometry.

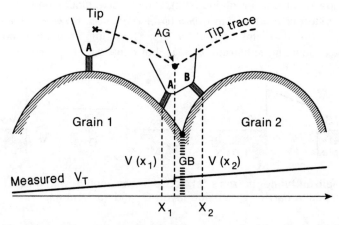

Figure 11. Apparent voltage drop at an apparent grain boundary AG in a current-carrying granular structure. The true grain boundary is at GB with no voltage drop across it.

4.5. BALLISTIC ELECTRON EMISSION MICROSCOPY (BEEM)

This most elegant and powerful method developed by Kaiser and Bell is a subsurface method [50]. It has been successfully applied to Au-Si, Au-GaAs and Au-AlGaAs interfaces. Tunneling is surface-sensitive and detects bulk properties only when they manifest themselves at the surface. BEEM on the other hand is sensitive to properties at interfaces which lie too far below the surface – 100 Å is a good number – to be seen at the surface. The principle is shown in Fig. 12(a) for a metal-semiconductor interface. Electrons tunneling from the tip to the surface of the metal layer traverse the latter ballistically. For a layer thickness of 100 Å, typically half of them reach the metal-semiconductor interface. Those with less energy than the conduction band edge of the semiconductor are reflected, the others are collected at the far side of the semiconductor [see Fig. 12(b)]. The collector

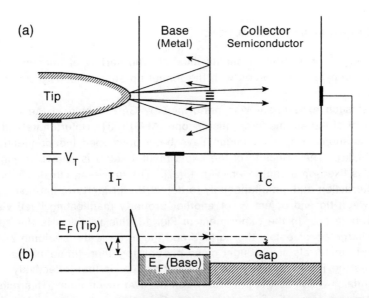

Figure 12. (a) Schematic of BEEM with the tip as electron injector; (b) corresponding energy diagram. Electrons reaching the base-collector (metal-semiconductor) interface ballistically can be reflected because their energy is insufficient [see (b)] or because of perpendicular momentum mismatch [see (a)].

current, I_c, then reflects the semiconductor band edges at the interface, the relevant property of semiconductor-metal interfaces. Spatial resolution might at first glance be a problem since the probe tip is relatively far from the interface, and even more so since the electrons traverse most of the distance ballistically with an angular distribution of halfwidth of ten to fifteen degrees. However, due to conservation of perpendicular momentum when traversing the interface, only metal electrons with very small perpendicular momentum can penetrate the semiconductor, the others are reflected [see Fig. 12(a)]. In practice, a resolution of about 10 Å is achieved.

4.6. IMAGING OF DYNAMIC PHENOMENA

This type of imaging has not yet received much attention. The first experiment dealt with surface diffusion of oxygen on Ni(110) [51]. An atom diffusing across the tunnel area gives rise to a specific spike in the tunnel current. In another experiment, the properties of electron traps in SiO_2 were investigated [52]. An electron trap in the tunneling area causes the tunnel current to switch between two values: a large current when the trap is empty and a small current when it is occupied. This is a type of coulomb blockade, which is discussed in more detail in another context in the lecture by van Kempen.

4.7. IMAGING WITH MIXED INTERACTIONS

The following three examples should illustrate the variety of imaging possibilities. All involve tunneling and forces with the tunnel tip simultaneously serving as force probe.

The first such experiment was performed by Gerd Binnig in the course of the development of the atomic force microscope (AFM) [53]. Unfortunately, it is unpublished. Variations of this technique have been developed independently by other groups [54,55]. The principle of the experimental setup is shown in Fig. 13. The tunnel tip is fixed on a cantilever [Fig. 13(a)]. The tunneling current is used as the control interaction and simultaneously to detect the property of interest, namely the force between tip and object – or another property manifesting itself via its force on the probing tip. In the experiment in Fig. 13, this property is the spatial variation of charge on or below the surface and the force is the coulomb attraction. If two different materials are brought close enough together, tunneling between them equilibrates the Fermi levels. The tip-object system is then effectively a capacitor with potential $\phi = \phi_T - \phi_O$ applied between the two electrodes. This establishes a static, attractive force $F \simeq \frac{1}{2}C\phi^2/s$, where C is the capacitance of the tip-object system. A dc voltage V between tip and object changes this force to $F \simeq \frac{1}{2}C(\phi + V)^2/s$. An ac voltage $\delta V = \delta V_0 \sin(\omega t)$, superimposed on V produces a direct ac component in the tunneling current, $\delta I_V = (\partial I/\partial V)\delta V$, and an indirect one, $\delta I_s = (\partial I/\partial s)\delta s$, due to the vibration δs of the tip-object distance induced by the

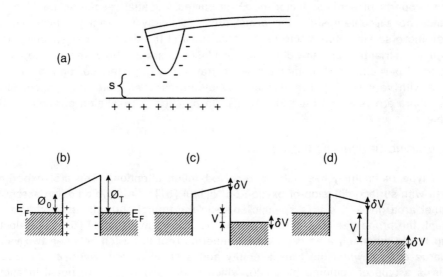

Figure 13. Schematic of electrostatic forces between tunnel tip and object. Electrons have flown from the object to the tip in order to equilibrate the Fermi levels, shown in (a) and (b). The static force in (c) and (d) is the same, but the dynamic force is in phase and out of phase, respectively, with δV.

force $\delta F \simeq \frac{1}{2} C(2(\phi + V) + \delta V)\delta V/s$. For a cantilever with good Q and a frequency ω close enough to the cantilever resonance frequency, δI_s dominates. For the situations shown in Figs. 13(c) and 13(d), the vibration amplitudes are the same, but δI_s is in-phase and out-of-phase, respectively, with δV. Hence a phase image gives the regions with $\phi > -V$ and $\phi < -V$. A lateral resolution of better than 1 nm was obtained [53]. For $V = -\phi = \phi_O - \phi_T$, the cantilever vibrates at 2ω since $\delta F \propto \delta V^2$. At this voltage the electrostatic force between tip and sample is zero, which can be important when imaging soft biological material. This is the situation of "field-less" tunneling.

A setup similar to that of Fig. 13(a) but with the sample instead of the tip on the cantilever is used by U. Dürig and his co-workers to study the adhesive forces between metal surfaces [33]. The tunnel current is used for both control and measurement of the tip-sample distance. The z-position of the tip is modulated by δs at a frequency ω, the quantities measured being $I(s)$ and $\delta I(\omega)$, from which the adhesive forces are derived.

A third set of experiments with mixed interactions concerns the local elastic response of the object surface to the tip-object force interaction. The experimental situation is similar to that used by Dürig et al. mentioned above except that the soft sample itself provides the spring action of the cantilever as shown in Fig. 14(a). For a small sample compliance, a, the width of the tunnel gap is always stable. For a tip position z_p, the width of the tunnel gap is s_A, as shown in Fig. 14(b). However, for a large sample compliance, b, the gap is only stable for tip positions $z > z_{p''}$ and $z < z_{p'}$. For $z_{p'} < z < z_{p''}$, we have a stable and a metastable gap of width s_A and s_C as shown for the tip position z_p. At a fixed tip position z_p, the gap width can jump randomly between s_A and s_C. For imaging soft materials, it is therefore important to

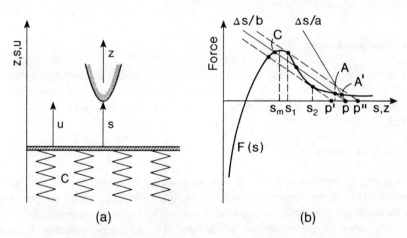

Figure 14. Elastic response of object to tip-object forces. (a) Denotations, with the effective sample compliance c. (b) Equilibrium tunnel gap widths for a tip-object interaction force $F(s)$ and two different sample compliances a and b.

choose appropriate imaging conditions which avoid instabilities. The instabilities can, however, also be used to perform "instability" imaging with atomic resolution [56]. Finally, no stable gap of width $s_1 < s < s_2$ exists.

The elastic response to the tip-surface force interaction can have drastic effects on the "work function profiles". The measured response of the tunnel current to a tip modulation δz is

$$\mathrm{d} \ln I/\mathrm{d}z = \mathrm{d} \ln I/\mathrm{d}s \, (1 + c \, \partial F/\partial s)^{-1}, \qquad (11)$$

where c is the effective compliance. The measured signal, $\mathrm{d} \ln I/\mathrm{d}z$ is enhanced or reduced according to the sign of $\partial F/\partial s$, where $\partial F/\partial s < 0$ for increasing attractive forces with decreasing gap width, $s > s_m$ in Fig. 14(b), and $\partial F/\partial s > 0$ for $s < s_m$. The latter case has been proposed by Pethica et al. [57] as the main factor for the extremely small values of $\mathrm{d} \ln I/\mathrm{d}z$ usually obtained at ambient imaging, with some particle between tip and surface acting as force mediator. The same mechanism — with or without additional force mediator — is also responsible for the giant corrugations observed on graphite and other materials of similar electronic structures [58,59].

Strong enhancements of $\mathrm{d} \ln I/\mathrm{d}z$ by attractive forces have, on the other hand, been observed on biological material [60,61]. Usually, the $\mathrm{d} \ln I/\mathrm{d}z$ image shows a much richer structure than the topography. Values of $\mathrm{d} \ln I/\mathrm{d}z$ considerably in excess of 2 Å^{-1}, which translates into an effective tunnel barrier height of 4 eV, have to be attributed to such enhancement effects. The images of $\mathrm{d} \ln I/\mathrm{d}z$ are therefore a reflection of local force interactions and compliances rather than "work function" profiles. They might well be as relevant as "work function" profiles, in particular for biological materials.

5. Other Imaging Modes

In all the imaging, the central problems are (i) the separation of topographic features, interaction inhomogeneities, and local response to a probe-surface interaction, and (ii) the size of the probe tip. Let me conclude with some illustrative examples.

5.1. IMAGING OF MAGNETIC FORCES

Force imaging is treated in detail in the lectures by Güntherodt. Here I would like to mention a recent experiment of Schönenberger et al. [62] as an elegant example of how to separate magnetic forces from topography by force microscopy.

Generally, short-range repulsive forces are used to image surface structures by force microscopy. This contact method can achieve atomic resolution on hard materials or on surfaces where the atoms are laterally fixed by strong bonds. The forces are controlled and monitored by the deflection of the cantilever carrying the probe tip. The long-range forces, on the other hand, are usually probed dynamically by the vibrations of the cantilever. These are noncontact methods. The measurements yield force derivatives: $\partial F/\partial s$ from the shift of resonance frequency, and higher-order derivatives from the entire frequency spectrum of the vibrating cantilever [33].

The method of Schönenberger et al. is based on the fact that magnetic forces due to a magnetization pattern vary slowly with distance at distances which are small or comparable to the dimension of the magnetic probe [63]. They introduce an additional force, the coulomb force, obtained by applying a voltage between tip and sample, as control interaction, see also Sect. 4.7. The coulomb force diverges at small distances with $\sim s^{-2}$. The magnetic force F_m, on the other hand, saturates at $F_{m,0}$ (see Fig. 15); we approximate the slow variation of the magnetic force with distance by $dF_m/ds \lesssim F_m(s)/\ell$, where ℓ is the lateral extension of a magnetic domain. One can then always find a distance s_0 such that $dF_c/ds \gg dF_m/ds$, even if $F_c(s_0)$ might be smaller than $F_m(s_0)$. Without magnetic forces, $dF/ds = dF_c/ds = $ const is essentially a topography. The magnetic forces induce only a small deviation, $\delta s/s_0 = \frac{1}{6}(F_m/F_c)(s_0/\ell)$. The lateral variation of the force on a constant dF/ds surface, however, is dominated by the lateral variation of F_m. The total force on a $dF/ds = $ const surface is

$$F(s) = F_c(s) + F_m(s) = F_c(s_0) + F_m(s)\left(1 + \frac{1}{3}(s_0/\ell)\right). \tag{12}$$

Since $F_c(s_0) = $ const, the force image reflects the magnetic force up to a correction of order s_0/ℓ.

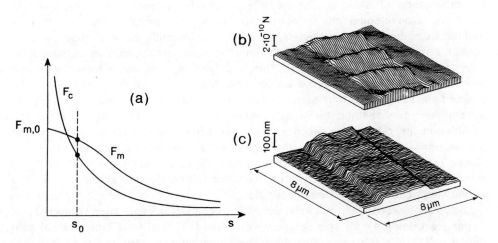

Figure 15. (a) Distance dependence of magnetic and coulomb forces. (b) and (c) give the magnetic force image and topography (as a $dF/ds = $ const image), respectively, at a stripe of alternatively up and down-oriented magnetic bits.

5.2. CAPACITANCE IMAGING AND THERMAL PROFILING

The work in this area as well as a great deal of the SXM applications were pioneered mainly at IBM's Yorktown [64-66] and Almaden Research Laboratories [67,68], the main motivating force being to a great extent the development of SXM applications in a technological environment. A summary of SXM techniques appears in the Proceedings of the Fourth STM Conference [21].

It is interesting to note how many variations of an STM tip can be used to probe various interactions, thus providing an ultimately small interaction sensor.

As a capacitance probe, it allows capacitance imaging in the 10 nm range. The capacitance is of the order of atofarads (10^{-18} Farad). It can be measured to an accuracy of about 10^{-22} Farad either with a classical capacitance-measuring circuit or via the coulomb force (see Sect. 4.7).

For the thermal profiles, initially a separate, miniaturized thermocouple was employed as a temperature sensor. The resolution was limited to the micron range due to the size of the sensor. It was realized later [69], however, that a tunneling junction is itself already a thermocouple, one of atomic size. This "tunneling thermometer" opens very interesting prospects in temperature sensing.

5.3. SECONDARY PARTICLES

Finally, another imaging method uses secondary particles in analogy to the secondary electrons which produce the image in scanning electron microscopy (SEM). Imaging by secondary particles generated from local-probe primary-particle beams was suggested nearly two decades ago in the context of the topografiner [70,71]. Owing to the limited resolution of the topografiner, it was not pursued further. With the emergence of STM, secondary-particle imaging is receiving more attention again and is developing into an interesting and powerful imaging method. Photons emitted from the tunneling area of an STM have recently been studied by Gimzewski et al. [72] and imaging with subnanometer resolution was achieved on silver (see lecture by Gimzewski). This is a type of local inverse photoemission method, except that there is no energy restriction of the primary particle — the injected electron — and thus also to some extent of the emitted photons. It also is a complementary method to inelastic tunneling spectroscopy [73] as far as the radiative inelastic tunneling processes are concerned. The main advantage of the photon emission method is that the signal is not lost against the background of elastic tunneling processes. The latter limits inelastic tunneling spectroscopy to relatively sharp features in energy and requires extremely stable tunnel gaps — a serious restriction for an STM configuration.

Local-probe secondary-electron detection or imaging has been reported for Auger electrons [74], for spin-polarized electrons [75], and for a conventional configuration [76]. In the case of relatively high energetic primary electrons ($\lesssim 1$ keV) required for Auger electron spectroscopy, the probe-sample separation must be large to keep the electric fields at the tip at an a acceptable level. This impairs the resolution. In the case of low-energy primary and secondary electrons, the electric fields in the escape path of the secondary electrons prevent the escape of the very-low-energy secondary electrons, and also complicate the energy analysis of the secondary electrons. In spite of these difficulties, local-probe secondary-electron imaging will develop into a viable method.

A most interesting alternative approach that avoids the restrictions of and problems with local-probe secondary electrons are point-source beams − electrons and ions − pioneered by Fink [76,77] and discussed in the lectures by Fink and Sáenz. The extraordinary properties of point-source beams indeed open new and most exciting possibilities [78,79].

6. Prospects

The prospects of science and technology reside with the scientists themselves, their ideas, visions and, of course, their work. This is all there is to be said about the prospects of the local probe methods.

7. References

[1] Yanson, I. K. (1974) Soviet Phys. JEPT 39, 506.
[2] Jansen, A. G. N., van Gelder, A. P., and Wyder, P. (1980) J. Phys. C 13, 6073; Duif, A. N., Jansen, A. G. N., and Wyder, P. (1989) J. Phys. Condens. Matter 1, 3157.
[3] Binnig, G., Rohrer, H., Gerber, Ch., and Weibel, E. (1982) Phys. Rev. Lett. 49, 57.
[4] Binnig, G. and Smith, D. P. E. (1986) Rev. Sci. Instrum. 57, 1688.
[5] Pohl, D. W. (1984) Appl. Phys. Lett. 44, 653.
[6] Vieira, S. (1986) IBM J. Res. Develop. 30, 553.
[7] Kuk, Y. and Silverman, P. J. (1989) Rev. Sci. Instrum. 60, 165.
[8] Binnig, G. and Rohrer H. (1982) Helv. Phys. Acta 55, 726; for coarse positioning based on friction, see Pohl, D. W. (1987) Rev. Sci. Instrum. 58, 54; Renner, Ch., Niedermann, Ph., Kent, A. D., and Fischer, Ø. (1989) Rev. Sci. Instrum. (in press).
[9] Gerber, C., Binnig, G., Fuchs, H., Marti, O., and Rohrer, H. (1986) Rev. Sci. Instrum. 57, 221.
[10] Kaiser, W. J. and Jaklevic, R. C. (1987) Surface Sci. 181, 55.
[11] Demuth, J. E., Hamers, R. J., Tromp, R. M., and Welland, M. E. (1986) IBM J. Res. Develop. 30, 396.
[12] Michel, B. and Travaglini, G. (1988) J. Microscopy 152, Pt. 3, 681.
[13] Pohl, D. W. (1986) Europ. Patent Nr. 0247219.
[14] Albrecht T. R., Akamine, S., Zdeblick, M. J., and Quate, C. F. (1989) in Ref. 21.
[15] Besocke, K. H., Teske, M., and Frohn, J. (1987) Surface Sci. 181, 145.
[16] Proceedings of the STM Workshop in Oberlech, Austria, IBM Europe Institute (1985), published in IBM. J. Res. Develop. 30, Nos. 4 and 5 (1986).
[17] Proceedings First International Conference on STM, Santiago de Compostela, Spain (1986), published in Surface Science 181, Nos. 1/2.

[18] Proceedings Second International Conference on STM, Oxnard, CA, USA (1987), published in J. Vac. Sci. Technol. A 6, No. 2 (1988).
[19] Proceedings of the Adriatico Research Conference on STM, Trieste, Italy (1987), published in Physica Scripta 38 (1988).
[20] Proceedings Third International Conference on STM, Oxford, England (1988), published in J. Microscopy 125, Pt. 1-3.
[21] Proceedings Fourth International Conference on STM, Oarai, Japan (1989), to be published in J. Vac. Sci. Technol. A (1990).
For a descriptive summary of the Erice NATO Meeting, see di Capua, M. S., and Marti, O. (1989) Europ. Sci. Notes Information Bulletin ESNIB 89-09, 49 (U.S. Navy, Office of Naval Research, European Office, London, UK)
[22] Binnig, G. and Rohrer, H. (1986) IBM J. Res. Develop. 30, 355.
[23] Hansma, P. and Tersoff, J. (1987) J. Appl. Phys. 61, R1.
[24] Golovchenko, J. A. (1986) Science 232, 4.
[25] Behm, J. and Hössler, W. (1986) Physics and Chemistry of Solid Surfaces, Vol. VI, Springer-Verlag, Berlin, p. 361.
[26] Tersoff, J. and Hamann, D. R. (1983) Phys. Rev. Lett. 50, 1998.
[27] Stoll, E., Baratoff, A., Selloni, A., and Carnevali, P. (1984) J. Phys. C 17, 3073.
[28] Stoll, E. (1984) Surface Sci. 143, L411.
[29] Wickramasinghe, H. K. (1989) in Ref. 21.
[30] Pohl, D. and Möller, R. (1988) Rev. Sci. Instrum. 59, 840.
[31] Gimzewski, J. K. and Möller, R. (1987) Phys. Rev. B 36, 1284.
[32] Christoph, R., Siegentaler, H., and Rohrer, H. (to be published).
[33] Dürig, U., Züger, O., and Pohl, D. W. (1988) J. Microscopy 152, 259.
[34] Binnig, G., Frank, K. H., Fuchs, H., García, N., Reihl, B., Rohrer, H., Salvan, F., and Williams, A. R. (1985) Phys. Rev. Lett. 55, 991.
[35] Becker, R. S., Golovchenko, J. A., and Swartzentruber, B. S. (1985) Phys. Rev. Lett. 55, 987.
[36] I thank J. K. Gimzewski for providing Fig. 5 before publication.
[37] Abraham, D. W., Williams, C. C., and Wickramasinghe, H. K. (1988) Appl. Phys. Lett. 53, 1503.
[38] Stoll, E. P. (1989) IBM J. Res. Develop. (in press).
[39] Feenstra, R. M. and Fein, A. P. (1985) Phys. Rev. B 32, 1394.
[40] Wolkow, R. and Avouris, Ph. (1988) Phys. Rev. Lett. 60, 1049.
[41] Hamers, R. J., Tromp, R. M., and Demuth, J. E. (1986) Phys. Rev. Lett. 56, 1972.
[42] Becker, R. S. and Golovchenko, J. A. (1985) Phys. Rev. Lett. 55, 2032; Lang, N. D. (1986) Phys. Rev. B 34, 5947; Ihm, J. (1988) Physica Scripta 38, 269.
[43] Wilkins, R., Ben-Jacob, E., and Jaklevic, R. C. (1989) Phys. Rev. Lett. 63, 801.
[44] Hess, H. R., Robinson, R. B., Dynes, R. C., Volles, J. M., and Waszczak, J. V. (1989) Phys. Rev. Lett. 62, 214, and in Ref. 21.
[45] Binnig, G., García, N., Rohrer, H., Soler, J. M., and Flores, F. (1984) Phys. Rev. B 30, 4816.
[46] Lang, N. D. (1987) Phys. Rev. B 36, 8173.
[47] Simmons, J. (1983) J. Appl. Phys. 34, 1793.
[48] Muralt, P. and Pohl, D. (1986) Appl. Phys. Lett. 48, 514.
[49] Kirtley, J. R., Washburn, S., and Brady, M. J. (1988) Phys. Rev. Lett. 60, 1546.
[50] Kaiser, W. J. and Bell, L. D. (1988) Phys. Rev. Lett. 60, 1406.
[51] Binnig, G., Fuchs, H., and Stoll, E. (1986) Surface Sci. 169, L235.
[52] Koch, R. H. and Hamers, R. J. (1987) Surface Sci. 181, 333.
[53] Binnig, G. (unpublished).

[54] Stern, J. E., Terris, B. D., Mamin, H. J., and Rugar, D. (1988) Appl. Phys. Lett. 53, 2717; Terris, B. D., Stern, J. E., Rugar, D., and Mamin, H. J. (1989) in Ref. 21.
[55] Martin, Y., Abraham, D. W., and Wickramasinghe, H. K. (1988) Appl. Phys. Lett. 52, 1103.
[56] Binnig, G. (1987) Physica Scripta T19, 53.
[57] Coombs, J. H. and Pethica, J. B. (1986) in Ref. 16, p. 455.
[58] Soler, J. M., Baró, A. M., García, N., and Rohrer, H. (1986) Phys. Rev. Lett. 57, 444.
[59] Mamin, H. J., Ganz, E., Abraham, D. W., Thomson, R. E., and Clarke, J. (1986) Phys. Rev. B 34, 9015.
[60] Travaglini, G., Rohrer, H., Amrein, M., and Gross, H. (1987) Surface Sci. 181, 380; Travaglini, G., Rohrer, H., Stoll, E., Amrein, M., Stasiak, A., Sogo, J., and Gross, H. (1988) Physica Scripta 38, 309.
[61] Selci, S., Cricenti, A., Felici, A. C., Generosi, R., Gori, E., Djaczenko, W., and Chiarotti, G. (1989) in Ref. 21.
[62] Schönenberger, Ch., Alvarado, S., Lambert, S., and Sanders, I. (to be published).
[63] Sáenz, J. J., García, N., Grütter, P., Meyer, E., Heinzelmann, H., Wiesendanger, R., Rosenthaler, L., Hidber, H. R., and Güntherodt, H.-J. (1987) J. Appl. Phys. 62, 4293.
[64] Martin, Y., Abraham, D. W., and Wickramasinghe, H. K. (1988) Appl. Phys. Lett. 52, 1103; Hobbs, P. C. D., Abraham, D. W., and Wickramasinghe, H. K. (1989) Appl. Phys. Lett. (in press).
[65] Williams, C. C. and Wickramasinghe, H. K. (1986) Appl. Phys. Lett. 49, 1587; idem, in B. R. McAvoy (ed.), 1986 Ultrasonics Symposium Proceedings, IEEE Cat. No. 0090-5607/86, p. 393, and idem (1988) Proc. SPIE 897, 129.
[66] Amer, N. M., Skumanich, A., and Rippel, D. (1986) in Proc. 18th Int'l Conf. Physics of Semiconductors, World Scientific, Singapore, 1987, Vol. 1, p. 53; Meyer, G. and Amer, N. M. (1988) Appl. Phys. Lett. 53, 1045.
[67] Mate, C. M., McClelland, G. M., Erlandsson, R., and Chiang, S. (1987) Phys. Rev. Lett. 59, 1942.
[68] Mate, C. M., Lorenz, M. R., and Novotny, V. J. (1989) J. Chem. Phys. 90, 7550.
[69] Weaver, J. M. R., Walpita, L. M., and Wickramasinghe, H. K. (1989) Nature (in press).
[70] Young, R., Ward, J., and Scire, F. (1972) Rev. Sci. Instrum. 43, 999.
[71] Young, R. (1971) Physics Today, November, 42.
[72] Gimzewski, J. K., Sass, J. K., Schlittler, R. R., and Schott, J. (1989) Europhys. Lett. 8, 435.
[73] See, e.g., Paul K. Hansma (ed.), Tunneling Spectroscopy, Plenum Press, New York, London, 1982.
[74] Reihl, B. and Gimzewski, J. K. (1987) Surface Sci. 189/190, 36.
[75] Allenspach, R. and Bischof, A. (1989) Appl. Phys. Lett. 54, 587.
[76] Fink, H.-W. (1988) Physica Scripta 38, 260.
[77] Fink, H.-W. (1986) IBM J. Res. Develop. 30, 460.
[78] Stocker, W., Fink, H.-W., and Morin, R. (1989) Ultramicroscopy (in press).
[79] Serena, P. A., Escapa, L., Sáenz, J. J., García, N., and Rohrer, H. (1988) J. Microscopy 152, Pt. 1, 43; de Raedt, H., García, N., Sáenz, J. J. (1989) Phys. Rev. Lett. (in press); Chang, T. H. P., Kern, D. P., McCord, M. A. (1989) J. Vac. Sci. Technol. (in press).

A BRIEF INTRODUCTION TO TUNNELING THEORY

C.R. LEAVENS AND G.C. AERS
Division of Physics,
National Research Council of Canada,
Ottawa, Ontario K1A 0R6
Canada

ABSTRACT. In this set of two lectures the emphasis is on the tunneling of electrons between two semi-infinite electrodes with parallel, planar surfaces. Although directly relevant to scanning tunneling microscopy only in the "blunt tip" regime, it is hoped that they will provide a useful background for the other lectures. The topics include the stationary-state scattering and transfer Hamiltonian approaches to elastic and inelastic tunneling, resonant tunneling in semiconductor heterostructures, electron tunneling spectroscopy of single-particle electronic excitations and electron-phonon interactions in (conventional) superconductors and of the electron-phonon interaction in normal metals, inelastic electron tunneling spectroscopy of molecular vibrations, and single-electron charging effects. A third lecture is devoted to tunneling times.

1. Introduction

In classical mechanics kinetic energy is a non-negative quantity and hence any region in which the potential energy $V(\vec{r})$ of a particle would be greater than its total energy E is completely inaccessible to that particle. In quantum mechanics the solution $\Psi(\vec{r})$ of the Schrödinger equation $[-(\hbar^2/2m)\nabla^2 + V(\vec{r})]\Psi(\vec{r}) = E\Psi(\vec{r})$ is not necessarily equal to zero wherever $V(\vec{r}) > E$, provided $V(\vec{r})$ is not infinite. Hence, there is a non-zero probability that an incident particle can "tunnel" through a finite region in which $V(\vec{r}) > E$. In fact, under very special circumstances the particle tunnels through such a region with unit probability.

These introductory lectures are for the most part concerned with the tunneling of electrons through a barrier $V(\vec{r}) = V(z)\Theta(z)\Theta(d-z)$, such as that shown in Fig. 1, extending from z=0 to z=d and varying spatially only in the z direction. It will be assumed that the individual electrodes occupying the half-spaces z<0 and z>d are regions of constant potential energy. For the stationary-state elastic scattering problem with a beam of electrons of wave-vector $\vec{k}\equiv(k_z,\vec{k}_t)$ and conserved energy E incident on the barrier from the left, the wavefunction $\Psi_{\vec{k}}(\vec{r})$ is equal to

$$\exp(i\vec{k}_t\cdot\vec{r})\,[\exp(ik_zz) + R_{L\to R}\exp(-ik_zz)] \quad (z<0) \quad ,$$

$$\exp(i\vec{k}_t\cdot\vec{r})\,T_{L\to R}\exp(ik_z'z) \quad (z>d) \quad ,$$

outside the barrier. Conservation of transverse wavevector, i.e. $\vec{k}_t' = \vec{k}_t$ has been assumed. Calculation of the all important reflection and transmission probability amplitudes, $R_{L \to R}$ and $T_{L \to R}$ respectively, presents no difficulties for the (separable) one-dimensional tunneling problems considered here. Explicit expressions are readily available for rectangular and trapezoidal barriers and numerical integration of the time-independent Schrödinger equation across more realistic barriers to obtain $R_{L \to R}$ and $T_{L \to R}$ requires very little computer time. However, useful insight is often obtained from the simplest WKB approximation for the transition probability

$$|T_{WKB}(E_z)|^2 = \exp\left[-2 \int_{z_\ell}^{z_r} dz \kappa(z)\right] \qquad (\kappa(z) = \hbar^{-1}\sqrt{2m[V(z)-E_z]} \quad) \qquad (1)$$

where $E_z \equiv \hbar^2 k_z^2 / 2m$ is the energy associated with the tunneling direction and z_ℓ and z_r are the classical turning points of Fig. 1 where $\kappa(z)$ is zero. Provided that $V(z)$ is sufficiently smooth and that E_z is not too close to the top of the barrier this simple expression usually gives a reliable order-of-magnitude description of the dependence of $|T|^2$ on the various barrier parameters. However, there are situations in which the approximation of replacing the pre-exponential factor by unity leads to serious error (e.g., for a rectangular barrier the exact result for $|T(E_z=0)|^2$ is zero). Unless explicitly stated otherwise it has been assumed in the following that $|T|^2$ has been obtained by accurate solution of the Schrödinger equation.

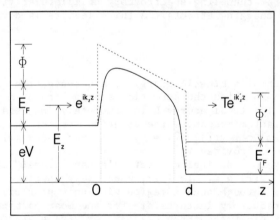

Figure 1. Schematic potential energy profile, $V(\vec{r})=V(z)$, for a planar metal-vacuum-metal tunnel junction. E_z is the component of the incident energy associated with motion in the tunneling direction, V is the applied voltage, E_F and Φ denote Fermi energy and work function respectively. The trapezoidal barrier (-----) has been lowered by a static multiple-image potential. The vertical dotted lines denote the locations of the left and right classical turning points z_ℓ and z_r where $V(z)=E_z$.

In section 2 an expression for the elastic current density J is derived for free-electron electrodes using the stationary-state scattering approach and extended to include band-structure effects within the effective-mass approximation. The applied voltage (V) and temperature (T) dependences of J(V,T) are studied for a metal-insulator-metal junction and for a semiconductor double-barrier quantum well structure exhibiting resonant tunneling. As a prelude to the discussions of self-energy effects in the electrodes and inelastic scattering of tunneling electrons in the barrier region the transfer-Hamiltonian model is considered briefly in section 3. The tunneling spectroscopy of the single-particle electronic excitations and electron-phonon interactions in superconducting electrodes and of the electron-phonon interactions in normal-metal electrodes is discussed in section 4. In section 5 the inelastic tunneling spectroscopy of molecules in the barrier is considered. Single-electron tunneling is discussed very briefly in section 6.

2. Stationary-State Scattering Approach

2.1 THE ELASTIC TUNNELING CURRENT

Consider a free electron with wavevector $\vec{k} \equiv (k_z, \vec{k}_t)$ and energy $E = E_z + E_t = (\hbar^2 k_z^2/2m + eV) + \hbar^2 k_t^2/2m$ incident on the barrier from the left (i.e. $k_z > 0$). Conservation of energy E for elastic scattering and transverse momentum $\vec{k}_t$ uniquely specifies the corresponding state $\vec{k}'$ of the right-hand electrode into which the incident electron can tunnel elastically: $\vec{k}'_t = \vec{k}_t$; $k'_z = \hbar^{-1}[2m(E_z + E'_F - E_F)]^{\frac{1}{2}}$. Those states $\vec{k}'$ for which $E_z < E_F - E'_F$ are localized at the $z=d$ interface (k'_z is pure imaginary) and do not contribute to the tunneling current if the thickness of the right electrode is much greater than $|k'_z|^{-1}$. The z component of the incident current density associated with the state $\vec{k}$ is $e\hbar k_z/m$. The corresponding transmitted current density is $|T_{L \to R}(E_z)|^2 e\hbar k'_z/m$ where $|T_{L \to R}(E_z)|^2$ is the left-to-right barrier transmission probability. Taking into account the probability that the state $\vec{k}$ of the left electrode is initially occupied and the state $\vec{k}'$ of the right electrode is initially unoccupied at temperature T one readily obtains the left-going current density:

$$J_{L \to R}(V,T) = 2 \int \frac{d\vec{k}}{(2\pi)^3} \Theta(k_z) \Theta(E_z - E_F + E'_F) |T_{L \to R}(E_z)|^2 (e\hbar k'_z/m) f(E-eV)[1-f(E)] \quad ,$$

where f is the Fermi function $f(E) = [1+\exp((E-\mu)/k_B T)]^{-1}$ with μ the chemical potential, and the factor of 2 accounts for spin. Using

$$\int d\vec{k} \Theta(k_z) = \int d\vec{k}_t \int_0^\infty dk_z = \frac{2\pi m}{\hbar^3} \int_0^\infty dE_t \int_0^\infty dE_z \left(\frac{m}{\hbar k_z}\right)$$

it immediately follows that

$$J_{L\to R}(V,T) = \frac{4\pi em}{(2\pi\hbar)^3} \int_{E_o}^{\infty} dE_z \frac{k_z'}{k_z} |T_{L\to R}(E_z)|^2 \int_0^{\infty} dE_t f(E_z+E_t-eV)[1-f(E_z+E_t)]$$

where $E_o \equiv \text{Max}[eV, E_F-E_F']$. Since $|T_{L\to R}(E_z)|^2(k_z'/k_z) = |T_{R\to L}(E_z)|^2(k_z/k_z')$ the expression for $J_{R\to L}(V,T)$ is obtained from that for $J_{L\to R}(V,T)$ simply by replacing the thermal factors by $f(E_z+E_t)[1-f(E_z+E_t-eV)]$. The final expression[1,2] for the total current density $J(V,T) \equiv J_{L\to R}(V,T) - J_{R\to L}(V,T)$ then follows after integration over E_t:

$$J(V,T) = \frac{4\pi emk_BT}{(2\pi\hbar)^3} \int_{E_o}^{\infty} dE_z \frac{k_z'}{k_z} |T_{L\to R}(E_z)|^2$$

$$\{\ln[1+\exp(\frac{\mu+eV-E_z}{k_BT})] - \ln[1+\exp(\frac{\mu-E_z}{k_BT})]\} \qquad . \quad (2)$$

For semiconductor heterojunctions subject to a slowly varying potential $U(\vec{r})$ in addition to the underlying periodic potential in each layer, the effect of the latter can often be included in the effective-mass approximation.[3] In each layer the wavefunction $\Psi(\vec{r})$ is expanded in terms of the complete set of Bloch functions $\psi_{n\vec{k}}(\vec{r})$ of eigenenergy $E_n(\vec{k})$ for the corresponding perfectly periodic bulk system:

$$\Psi(\vec{r}) = \sum_{n\vec{k}} \Phi_n(\vec{k}) \psi_{n\vec{k}}(\vec{r}) \qquad ,$$

$$\psi_{n\vec{k}}(\vec{r}) = e^{i\vec{k}\cdot\vec{r}} u_{n\vec{k}}(\vec{r}) \quad (u_{n\vec{k}}(\vec{r}) \text{ is periodic}) \qquad .$$

Here the wave vector $\vec{k}$ is restricted to the first Brillouin zone and n is a band index. It can be shown[3] that the envelope function

$$F(\vec{r}) = (2\pi)^{-3/2} \int d\vec{k}\, \Phi_n(\vec{k}) e^{i\vec{k}\cdot\vec{r}}$$

satisfies the Schrödinger-like equation (with effective mass m^*)

$$[-\frac{\hbar^2}{2m^*}\nabla^2 + U(\vec{r})]F(\vec{r}) = [E-E_n(\vec{k}=0)]F(\vec{r})$$

provided that (1) $U(\vec{r})$ varies sufficiently slowly on the scale of the lattice period; (2) $|<n\vec{k}|U(\vec{r})|n'\vec{k}'>| \ll |E_n(\vec{k}=0)-E_{n'}(\vec{k}=0)|$ ($n \neq n'$); (3) The states of interest are in the neighbourhood of a single, well separated, critical point of the band structure so that $E_n(\vec{k}) \cong E_n(\vec{k}=0) + \hbar^2k^2/2m^*$. The usual effective mass boundary conditions are continuity of $F(z)$ and of $m^*(z)^{-1}\partial F(z)/\partial z$. Continuity of $\Psi(\vec{r}) \cong u_{n0}(\vec{r})F(\vec{r})$ (assuming that the $\vec{k}$ dependence of $u_{n\vec{k}}(\vec{r})$ is weak for $\vec{k}\cong 0$) requires that u_{n0} be approximately constant throughout the heterostructure.

For the heterostructure junctions considered below in section 2.3 it is a reasonable approximation to ignore the z dependence of $m^*(z)$, in which case the only change required in Eq. (2) is to replace m everywhere by m^*. If m^* in the barrier differs from m^* in either of the electrodes then the transition probability depends on E_t as well as E_z and the integral over E_t cannot in general be evaluated analytically.

2.2 NORMAL-METAL ELECTRODES

The voltage and temperature dependences of the elastic tunneling current are considered for normal-metal electrodes.

Fig. 2 shows the dependence of $J(V,T=0)$ on V for the electrode parameters $E_F = E_F' = 5.5$ eV, $\Phi = 4.6$ eV and $\Phi' = 5.4$ eV (modelling tungsten and gold respectively) with a (vacuum) gap d of 10 Å. The initial linear dependence steepens rapidly for V>0.5 volts. In the Fowler-Nordheim regime (eV>Φ') reflection from the discontinuity in the trapezoidal barrier at z=d leads to strong interference at energies $E_z > \Phi'$ in the region between the right-hand classical turning point z_r and the interface at z=d. The resulting Gundlach[4] oscillations in the current density still persist, although considerably diminished in amplitude, when the sharp corners of the trapezoidal barrier are rounded by the inclusion of a static multiple-image potential (see Fig. 1). Mendez[5] has shown that the oscillations are much more prominent if there is a significant difference in effective-mass between electrode and barrier regions. The sensitivity of the Gundlach oscillations to details of the barrier potential has been emphasized by Becker et al.[6]. This is illustrated in Fig. 3 which shows the constant-current conductance, $[\partial J(V',T=0)/\partial V']_{d=d(V)}|_{V'=V}$, and distance-voltage characteristic, $d(V)$, calculated with and without the image potential. For each value of bias V the gap d was fixed during the partial differentiation at the value $d(V)$ determined by requiring that $J(V,T=0)|_{d=d(V)}$ be equal to 5 pA Å^{-2}. The Gundlach oscillations in the distance-voltage characteristic die out rapidly with increasing voltage, particularly when the image potential is included. Numerical calculations indicate[7] that, for the planar geometry, $d(V) = c_0 + c_1 V$ at very high voltages and that extrapolation of this relation to d=0 leads to a voltage intercept $V_0 = -c_0/c_1$ that is very insensitive to the value of the constant current density J (see Fig. 4). Hence, in principle, the local vacuum gap between a planar electrode and a sufficiently blunt STM tip could be estimated from the linearly extrapolated point of intersection of several high voltage d-V characteristics with different constant values of J.

Since metallic Fermi energies and barrier heights are very much greater than $k_B T$ in usual experimental situations the temperature dependence of the tunneling current is usually very small.

Fig. 5 shows the voltage dependence of $T^{-2}[J(V,T)-J(V,0)]/J(V,0)$ with $J(V,T)$ given by Eq. (2) at T=200 and 1000 K for the barrier parameters of Fig. 2. The (normalized) temperature dependent part of $J(V,T)$ is obviously very small and proportional to T^2 to a good approximation over the temperature range considered. The dependence on applied voltage is quite complex.[1]

2.3 RESONANT TUNNELING

A decade after the first demonstration of resonant tunneling in double-barrier AlGaAs heterostructures by Chang, Esaki and Tsu[8], the work of Sollner and collaborators[9,10] sparked an explosion of interest in the field. One of the reasons for this interest is that these structures exhibit negative differential resistance (NDR) and can therefore be used as the basis for high speed devices such as high frequency oscillators.

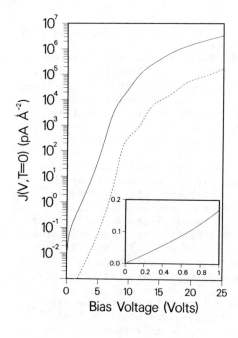

Figure 2. Voltage dependence of the zero temperature current density for a planar metal-vacuum-metal junction with $E_F=E_F'=5.5$eV, $\Phi=4.6$eV and $\Phi'=5.4$eV, calculated with (——) and without (-----) a static multiple-image contribution to the barrier potential.

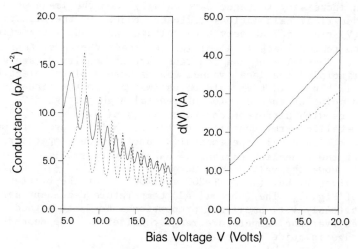

Figure 3. Voltage dependence of the conductance at constant current (left) and the corresponding distance-voltage characteristic (right) in the Fowler-Nordheim regime for the junction of Fig. 2.

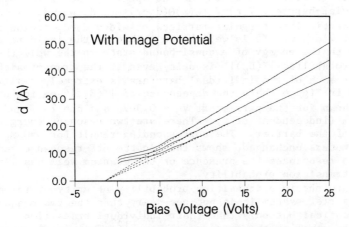

Figure 4. Distance-voltage characteristics for a metal-vacuum-metal junction with $E_F=E_F'=5.5\text{eV}$, $\Phi=5.4\text{eV}$ and $\Phi'=4.6\text{eV}$ at constant current densities of 0.5, 5.0 and 50 pA Å^{-2} (top to bottom). The dashed lines are linear extrapolations from high voltages.

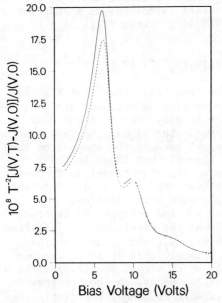

Figure 5. Voltage dependence of the (normalized) temperature-dependent part of the current density for the metal-vacuum-metal tunnel junction of Fig. 2 (including a static multiple-image contribution to the barrier potential) at T=200K (-----) and T=1000K (———).

For the simplest example of resonant tunneling consider the stationary-state scattering problem for the special case of the symmetric double-rectangular-barrier, $V(z)=V_0[\Theta(z)\Theta(d_\ell-z)+\Theta(z-d_\ell-w)\Theta(d_\ell+w+d_r-z)]$ with the two identical rectangular barriers of width $d_\ell=d_r$ enclosing a quantum well of width w. It is well known[11] that if the incident energy E_z coincides with the energy of a quasi-bound state in the well then the transmission probability $|T(E_z)|^2$ is unity even if the transmission probability for each of the individual barriers is extremely small. This is illustrated in Fig. 6 where the dependence of $|T(E_z)|^2$ on incident energy E_z is shown for the parameters $V_0 = 0.3$eV, $d_\ell = d_r = 75$ Å and w=50 Å with $m^* = 0.067$m (independent of z). There are two resonant energies below the top of the barrier. The corresponding result for w=0 Å, with all other parameters unchanged, shows that in the off-resonance regime between the two resonances the presence of the quantum well has little effect on the transition probability.

In order to obtain a transition probability of unity at resonance for a double-barrier system it is not necessary that the two opaque barriers be identical but only that their individual transition probabilities be equal at the resonance energy.[12] This is illustrated in Fig. 7. The curve, actually three coincident curves, was generated as follows: for a double rectangular barrier with V_ℓ, d_ℓ and w fixed at 0.30eV, 75Å and 50Å respectively, V_r was fixed at 0.20, 0.30 and 0.40eV successively; for each value of V_r, the transition probability $|T|^2$ for the lowest resonance as well as the corresponding single-barrier transition probabilities $|T_L|^2$ and $|T_R|^2$ were calculated as a function of d_r. It is clear from the figure that $|T(E_r)|^2 = 1$ when $|T_R(E_r)|^2 = |T_L(E_r)|^2$. Büttiker's expression[12],

$$|T(E_r)|^2 = 4|T_R(E_r)|^2|T_L(E_r)|^2/[|T_R(E_r)|^2 + |T_L(E_r)|^2]^2$$

($|T_L|^2, |T_R|^2 \ll 1$), accurately describes the numerical results of Fig. 7.

In an AlGaAs double-barrier heterojunction an undoped GaAs layer is sandwiched between two undoped $Al_xGa_{1-x}As$ layers. The adjacent GaAs electrode layers are heavily n (or p) doped. The $Al_xGa_{1-x}As$ layers act as barriers to the flow of electrons (holes) because the conduction band minima (valence band maxima) are higher (lower) than those in the adjacent GaAs regions. A simple picture of resonant tunneling in these structures was given by Luryi[13] and is illustrated in Fig. 8. Consider at T=0 a simple model of such a heterostructure with identical rectangular barriers in the absence of an applied voltage and ignoring charging and depletion effects. Assume that the Fermi energy E_F is less than the lowest resonance energy at zero bias $E_r(V=0)$. For eV>0 the resonance energy shifts upwards by approximately eV/2 if it is assumed that the potential energy drops uniformly by eV across the undoped region of the structure, i.e. $E_r(V) \cong E_r(0) + eV/2$. Hence, the most energetic electrons of the left electrode are first able to tunnel resonantly when $eV = 2(E_r(0)-E_F)$ for the lowest quasi-bound state. Resonant tunneling via this state continues with further increase in eV until the bottom of the left-hand conduction band coincides with the resonant level, i.e.

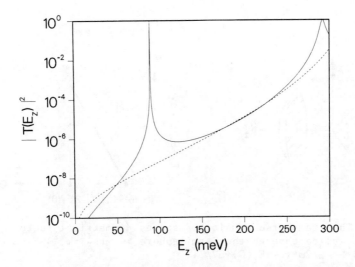

Figure 6. Energy dependence of the transmission probability for the symmetric double-rectangular-barrier of height $V_0=300$ meV, individual barrier widths $d_\ell=d_r=75$ Å and well width $w=50$ Å (———) and $w=0$ Å (-----).

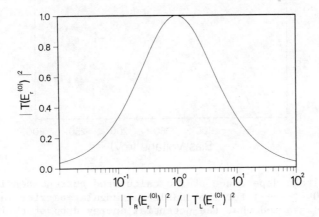

Figure 7. Transmission probability at the first resonance, $E=E_r^{(0)}$, of the double-rectangular-barrier $V_\ell\Theta(z)\Theta(d_\ell-z)+V_r\Theta(z-d_\ell-w)\Theta(d_\ell+w+d_r-z)$ as function of the ratio of the transmission probabilities of the isolated right and left rectangular barriers $V_\ell\Theta(z)\Theta(d_\ell-z)$ and $V_r\Theta(z)\Theta(d_r-z)$ respectively.

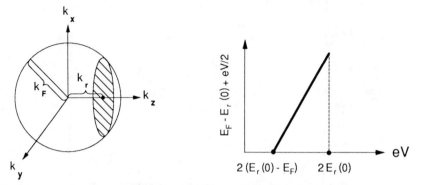

Figure 8. The transverse density of states geometrical factor. The area of the $k_z = k_r$ slice through the Fermi sphere is $\pi(k_F^2 - k_r^2) \propto E_F - E_r(0) + eV/2$ where $E_r(V) = \hbar^2 k_r^2/2m^* + eV \cong E_r(0) + eV/2$.

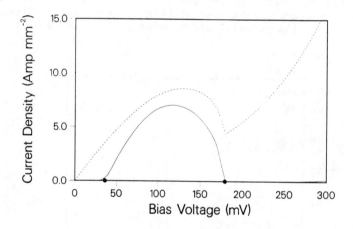

Figure 9. Voltage dependence of the calculated current density at T=0 (───) and 300K (-----) for the double-rectangular-barrier (at eV=0) of Fig. 6. It is assumed that the potential energy drop of eV is linear and confined to the undoped barrier region of width $d = d_\ell + w + d_r$. The zero temperature Fermi energy of the electrodes is 71.3 meV corresponding to $n = 1.5 \times 10^{18}$ cm^{-3} and an effective mass m* of 0.067m. The dots (●) at $eV = 2(E_r^{(0)}(0) - E_F)$ and $2E_r^{(0)}(0)$ denote the endpoints of the geometrical factor of Fig. 8.

until $eV = 2E_r(0)$. For $2(E_r(0)-E_F) \leqslant eV \leqslant 2E_r(0)$ an important factor in determining the resonant contribution to the tunneling current is the number of occupied states of the left electrode for which $E_z = E_r(V)$, i.e. for which $E_r(V) + E_t \leqslant E_F$. This is proportional to the area $\pi(k_F^2 - k_r^2) = 2\pi m^* \hbar^{-2}(E_F - E_r(V) + eV)$ of that circular cross section of the Fermi sphere for which $E_z = \hbar^2 k_r^2/2m^* + eV = E_r(V)$. Hence the voltage dependence of the tunneling current at T=0 contains

$$\Theta[eV - 2(E_r(0) - E_F)]\Theta(2E_r(0) - eV)(E_F - E_r(0) + \frac{eV}{2})$$

as an important factor. The negative differential resistance portion of the I-V curve associated with the discontinuous step decrease to zero at $eV = 2E_r(0)$ is rounded by the rapidly decreasing value of $|T(E_r)|^2$ as the bottom of the left-hand conduction band approaches $E_r(V)$. Fig. 9 shows the T=0 tunneling current calculated with Eq. (2) for the double-barrier heterostructure of Fig. 6 assuming that the potential energy drop of eV is linear across the double barrier. Also shown is the current at T=300K. Thermal smearing allows tunneling through the resonant level to contribute to the current at bias voltages below $2(E_r(0)-E_F)$.

Because of the long lifetime associated with the quasi-bound states of the double-barrier, resonant tunneling leads to an accumulation of charge in the well region which distorts the potential energy profile of the structure by the addition of an electrostatic term. Hence the steady-state tunneling current should be calculated by simultaneous solution of the Schrödinger and Poisson equations. The choice of boundary conditions for such self-consistent problems is important and may lead to very different results appropriate to different kinds of devices. For example, Figure 10 shows the self-consistent current voltage (I-V) characteristic calculated for a structure with the same parameters as used in Figure 9 except for the addition of 50 Å buffer regions on either side of the barriers. The higher current for the non self-consistent calculation in Figure 10, compared to the self-consistent result or to the unbuffered system of Figure 9 is principally due to the artificially large linear voltage drop across the left-hand buffer region as shown in Figure 11. The self-consistent method used here was similar to that of Cahay et al.[14] in which the carriers are assumed to be in thermal equilibrium only outside the extended doped regions on either side of the barriers. Scattering is assumed to be completely absent throughout the device and the model would therefore be expected to be appropriate to situations where the scattering mean-free-path is very long.[15] The redistribution of charge in the neighbourhood of the double barrier shifts the potential profile upwards around the barrier regions, as shown in Figure 11 and hence moves the NDR region to higher voltage. In systems for which the scattering mean-free-path is very short compared to the device size it may be more appropriate to use a model in which the injected carriers are in thermal equilibrium everywhere outside the barriers.[15,16]

Another phenomenon exhibited both experimentally and theoretically is hysteresis in the I-V characteristic. The presence of charge in the well delays the onset of the NDR when the voltage is being ramped upwards. If the voltage is reduced from a high value, on the other hand,

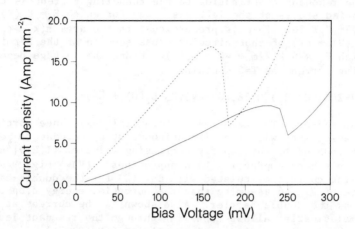

Figure 10. Voltage dependence of the current density at T=300 K for non self-consistent (-----) and self-consistent (———) calculations for the double-rectangular-barrier structure used in Figure 9 but with 50Å undoped buffer layers outside the barrier regions.

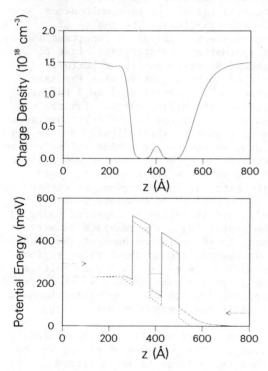

Figure 11. Self-consistent charge density (top panel) and potential energy profile (bottom panel) for the structure used in Figure 10 at a bias voltage of 230 mV. In the lower panel the dashed curve is the non self-consistent potential used for the calculation of the dashed curve of figure 10. The dotted line represents the lowest resonance energy at this bias and the arrows indicate the chemical potentials in the contact regions.

there is little charge in the well and the switchover to resonant
tunneling will occur at a slightly lower voltage. Self-consistent
calculations performed with zero and high charge starting conditions
exhibit a similar hysteresis effect.[15]

The above models do not give a good account of the important
peak-to-valley ratio in the NDR region due to effects such as interface
roughness[17], phonon interactions[18] and Γ-X mixing at the interfaces.[19]
Furthermore the above models all assume 'coherent' tunneling whereby the
tunneling electron retains 'phase memory' while crossing the device.
However, if the barriers are wide enough or high enough the electron may
spend sufficient time in the well region to be scattered before escaping.
In this case the electron is commonly pictured as tunneling
'sequentially'.[13] Although Wiel and Vinter[20] showed that the current
calculated for this process is mathematically equal to that for the
coherent process, Büttiker[12] pointed out that this equivalence is not
meaningful unless a specific scattering mechanism is included
consistently in the calculation. Obviously the time spent in the well is
of considerable interest in attempting to understand the behaviour of
these devices.

In the next section Price's derivation[21] of the relation between
the lifetime of a quasi-bound state and the width $2\Delta E$ of the
corresponding resonance is sketched as an application of the transfer
matrix approach.

2.4 RELATION BETWEEN LINE WIDTH AND LIFETIME

Price[21] has shown that for a symmetric double-barrier the full width $2\Delta E$
of a resonant peak in the transition probability $|T(E_z)|^2$ is related to
the lifetime τ of the corresponding quasi-bound state by $\tau = \hbar/2\Delta E$. In
his approach, a given stationary-state wavefunction is described by

$$\Psi_k(z) = A_{\ell(r)} e^{ikz} + B_{\ell(r)} e^{-ikz} \equiv (A_{\ell(r)}, B_{\ell(r)}) \tag{3}$$

in the region to the left (right) of the barrier. These are related by a
2 x 2 transfer matrix $\widetilde{M}$:

$$(A_r, B_r) = (A_\ell, B_\ell) \widetilde{M} \tag{4}$$

For a system that is symmetric in z, $\widetilde{M}$ is of the form

$$\widetilde{M} = \begin{pmatrix} P & U \\ -U & Q \end{pmatrix} \quad (\det \widetilde{M} = 1), \tag{5}$$

where P, Q and U are in general complex.

First consider the resonant peak in $|T(E_z)|^2$ for the scattering
problem where E_z and $k \equiv \hbar^{-1}\sqrt{2mE_z}$ are, of course, real quantities. For k
real, $\exp(ikz) = [\exp(-ikz)]^*$ and it readily follows from (3) to (5) that
$P=Q^*$ and $U=-U^*=iW$ with W real. Now, the special case $(A_\ell, B_\ell) = (0,1) = \exp(-ikz)$ corresponds to $(A_r, B_r) = (-iW, Q) = -iW\exp(ikz) + Q\exp(-ikz)$.
For k>0 this describes the scattering of a beam of electrons incident
from the right with transmission and reflection probability amplitudes
$T=Q^{-1}$ and $R=-iWQ^{-1}$ respectively. Conservation of current requires that

$|T|^2 + |R|^2 = 1$, i.e. $|Q|^2 = 1 + W^2$. Requiring that $|T(E_z)|^2$ have the Breit-Wigner resonance form, i.e.

$$|T(E_z)|^2 = \frac{1}{1+W(E_z)^2} = \frac{1}{1+[(E_z-E_r)/\Delta E]^2},$$

at least for E_z within a half-width ΔE of E_r, implies that

$$W(E_r) = 0 \text{ and } |dW(E)/dE|_{E_r} = 1/\Delta E .$$

Now consider the decay of the quasi-bound state. For real E_z and k it is impossible to construct a stationary-state wavefunction $\propto$ $\exp(ik|z|)$ for z outside the symmetric barrier. For complex E_z, Eq. (5) for the transfer matrix of a symmetrical barrier can be rewritten as

$$\overleftrightarrow{M} = \begin{pmatrix} P & iW \\ -iW & Q \end{pmatrix}$$

with $W=U/i$ in general complex. The choice $(A_\ell, B_\ell) = (0,1) = \exp(-ikz)$ and $(A_r, B_r) = (1,0) = \exp(ikz)$ to describe the asymptotic behaviour of a symmetric wavefunction $\Psi_k(z)$ localized in the well requires that $Q(E_z)=0$ and $W(E_z) = i$. Since $W(E_r)=0$, the assumption that $W(E_z)=i$ is satisfied for a complex energy $E_z = E_r - i\varepsilon$, with ε real and small, implies that $W(E_r - i\varepsilon) = i \cong -i\varepsilon(dW(E_z)/dE_z)_{E_r}$. Hence, $\varepsilon = -[(dW(E_z)/dE_z)_{E_r}]^{-1} = \Delta E$, using Eq. (6) assuming that $\varepsilon > 0$. For a wavefunction $\Psi_k(z)$ of eigenenergy $E_z = E_r - i\varepsilon$, the time dependence of $|\Psi_k(z)|^2$ for any z is given by $|\exp[-i(E_r - i\varepsilon)t/\hbar]|^2 = \exp(-2\varepsilon t/\hbar)$. Hence the lifetime of the quasi-bound state is given by $\tau = \hbar/2\varepsilon = \hbar/2\Delta E$ where $2\Delta E$ is the full-width of a sharp Breit-Wigner resonance in $|T(E_z)|^2$. This has been confirmed by the time-dependent wave-packet calculations of Collins et al.[22]

3. Transfer Hamiltonian Approach

The basic ingredients of the stationary-state approach considered in section 2 are the scattering solutions of the time-independent Schrödinger equation for the entire junction. There is no restriction in this method to states for which the coupling between the electrodes is weak. In this section we consider a first-order perturbation theory approach to the calculation of the tunneling current based on the assumption that the electrodes are very weakly coupled systems (i.e. $|T(E_z)|^2 << 1$ for the important range of E_z).[23,24,1] Although limited to opaque barriers this approach has been successfully applied in many areas, most notably superconducting tunneling[23,24,1,2,25], inelastic tunneling spectroscopy[26-29] and, more recently, scanning tunneling microscopy[30,31].

The following derivation[1] of the basic results of Bardeen's transfer Hamiltonian approach[23] has been simplified by considering a one-electron rather than a many-electron Hamiltonian.

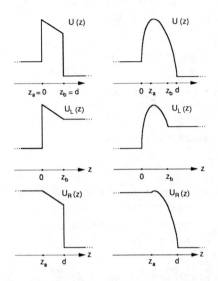

Figure 12. Potential energy profiles for the entire junction, U(z), and for the uncoupled left, $U_L(z)$, and right, $U_R(z)$, electrodes for a trapezoidal (left) and rounded (right) barrier.

Consider a single electron with Hamiltonian $H \equiv -(\hbar^2/2m)\nabla^2 + U(z)$ where $U(z)$ contains a barrier $V(z)\Theta(z)\Theta(d-z)$ that is opaque ($|T|^2 \ll 1$) for the electron energies of interest. It is desired to calculate the transition matrix element $M_{\ell r}$ for the transfer of an electron from an eigenstate $\Psi_\ell(\vec{r})$ of H in which the probability density $|\Psi_\ell(\vec{r})|^2$ is strongly localized in the left (L) electrode ($z \leqslant 0$) to an eigenstate $\Psi_r(\vec{r})$ of H strongly localized in the right (R) electrode ($z \geqslant d$). In the transfer Hamiltonian model such localized eigenstates of the entire junction are assumed sufficiently well approximated over the spatial ranges of interest (see Fig. 12) by eigenstates of the corresponding uncoupled electrodes, i.e. $\Psi_\ell(\vec{r})$ by $\Psi_\ell^{(L)}(\vec{r})$ for $z \leqslant z_b$ and $\Psi_r(\vec{r})$ by $\Psi_r^{(R)}(\vec{r})$ for $z \geqslant z_a$. For all z,

$$H_L \Psi_\ell^{(L)}(\vec{r}) = E_\ell^{(L)} \Psi_\ell^{(L)}(\vec{r}), \quad H_R \Psi_r^{(R)}(\vec{r}) = E_r^{(R)} \Psi_r^{(R)}(\vec{r}),$$

$$H_L = \Theta(z_b - z)H + \Theta(z - z_b)\left[-\frac{\hbar^2}{2m}\nabla^2 + U(z_b)\right],$$

$$H_R = \Theta(z - z_a)H + \Theta(z_a - z)\left[-\frac{\hbar^2}{2m}\nabla^2 + U(z_a)\right].$$

z_a and z_b are located within the barrier with $z_a \leqslant z_b$. For the trapezoidal barrier shown on the left side of Fig. 12 a natural choice is $z_a = 0$ and $z_b = d$; for the sloping barrier on the right there is no obvious choice, illustrating the lack of uniqueness in the procedure. Together, the L→R and R→L scattering states for the Hamiltonian H form a complete basis set of orthogonal wavefunctions for the entire junction. It is assumed that the basis set of almost orthogonal wavefunctions obtained by combining the L→R scattering states for H_L and the R→L scattering states for H_R can adequately represent the wavefunction $\Psi(\vec{r},t)$

for the entire barrier. To calculate $M_{\ell r}$ the eigenstates of H_L and H_R are used to construct a trial time-dependent wavefunction

$$\Psi(\vec{r},t) = a_\ell(t)\Psi_\ell^{(L)}(\vec{r})e^{-iE_\ell^{(L)}t/\hbar} + \sum_r b_r(t)\Psi_r^{(R)}(\vec{r})e^{-iE_r^{(R)}t/\hbar} \quad (7)$$

with $a_\ell(0)=1$ and $b_r(0)=0$. Substitution of Eq. (7) in the time-dependent Schrödinger equation $H\Psi(\vec{r},t)=i\hbar\partial\Psi(\vec{r},t)/\partial t$ and application of first-order perturbation theory to the calculation of $db_r(t)/dt=(i\hbar)^{-1}M_{\ell r}\exp[i(E_r^{(R)}-E_\ell^{(L)})t/\hbar]$ immediately gives

$$M_{\ell r} = \int d\vec{r}\Psi_r^{(R)}(\vec{r})^*(H-E_\ell^{(L)})\Psi_\ell^{(L)}(\vec{r}).$$

Since $H=H_L$ for $z \leqslant z_b$ the lower limit on the integral over z can be replaced by any $z_0 \leqslant z_b$. Similarly, since $H=H_R$ for $z \geqslant z_a$ the quantity $\Psi_\ell^{(L)}(\vec{r})(H-E_r^{(R)})\Psi_r^{(R)}(\vec{r})^*$ is zero for $z \geqslant z_a$ (recall that $z_a \leqslant z_b$). Hence for energy conserving final states ($E_r^{(R)} = E_\ell^{(L)}$)

$$M_{\ell r} = \int_{-\infty}^{\infty}dx\int_{-\infty}^{\infty}dy\int_{z_0}^{\infty}dz[\Psi_r^{(R)}(\vec{r})^*H\Psi_\ell^{(L)}(\vec{r})-\Psi_\ell^{(L)}(\vec{r})H\Psi_r^{(R)}(\vec{r})^*]$$

$$= -\frac{\hbar^2}{2m}\int_{-\infty}^{\infty}dx\int_{-\infty}^{\infty}dy\int_{z_0}^{\infty}dz\;[\Psi_r^{(R)}(\vec{r})^*\nabla^2\Psi_\ell^{(L)}(\vec{r})-\Psi_\ell^{(L)}(\vec{r})\nabla^2\Psi_r^{(R)}(\vec{r})^*]$$

for ($z_a \leqslant z_0 \leqslant z_b$). Integrating by parts using the fact that

$$\Psi_\ell^{(L)}(\vec{r}) \propto \exp[-\hbar^{-1}(2m(U(z_b)-E_{\ell z}^{(L)}))^{\frac{1}{2}}z] \qquad (z > z_b)$$

finally gives

$$M_{\ell r} = -\frac{\hbar^2}{2m}\int_{-\infty}^{\infty}dx\int_{-\infty}^{\infty}dy\left[\Psi_r^{(R)}(\vec{r})^*\frac{\partial\Psi_\ell^{(L)}(\vec{r})}{\partial z} - \Psi_\ell^{(L)}(\vec{r})\frac{\partial\Psi_r^{(R)}(\vec{r})^*}{\partial z}\right]_{z=z_0}. \quad (z_a \leqslant z_0 \leqslant z_b) \quad (8)$$

$M_{\ell,r}$ is independent of z_0 in the range $z_a \leqslant z_0 \leqslant z_b$ since $U_L(z)=U_R(z)=U(z)$ for $z_a \leqslant z \leqslant z_b$.

The transition probability per unit time from an occupied initial state ℓ of the left electrode to a continuum of empty states r of the right electrode is

$$P_\ell = \frac{2\pi}{\hbar}\sum_r |M_{\ell r}|^2 \delta(E_r^{(R)}-E_\ell^{(L)}) \quad .$$

Hence the tunneling current density $J(V,T)$ is given by[23,24]

$$J(V,T) = 2e\left(\frac{2\pi}{\hbar}\right)\sum_{\ell,r}|M_{\ell r}|^2[f(E_\ell^{(L)}-eV)-f(E_r^{(R)})]\delta(E_\ell^{(L)}-E_r^{(R)}) \quad . \quad (9)$$

This equation was the starting point for Tersoff and Hamann's[30] and Lang's[31] theories of STM imaging.

For spectroscopy of the energy gap and phonons in conventional superconductors, the energy ranges of interest, O(meV) and O(10meV) respectively, are very small and the energy dependence of $|M_{\ell r}|^2$ is usually ignored. For a superconductor-insulator-superconductor (S-I-S) junction the single-particle tunneling current density is then given by

$$J(V,T) = 2e(\frac{2\pi}{\hbar})|\overline{M}|^2 \int_{-\infty}^{\infty} dE \; N_S^{(L)}(E-eV) \; N_S^{(R)}(E) [f(E-eV)-f(E)]$$

where $N_S(E)$ is the density of states for single-particle excitations out of the superconducting ground state. If one electrode is normal (N) the appropriate density of states factor N_S is replaced by N_N.

4. Tunneling Spectroscopy

4.1. GAP SPECTROSCOPY OF CONVENTIONAL SUPERCONDUCTORS

In the BCS theory of superconductivity,[32-34,2] the difference between the phonon-mediated electron-electron pairing interaction and the screened Coulomb repulsion is modelled by

$$-V_{\vec{k},\vec{k}'} = -V \; \Theta(\Omega_D - |\varepsilon_{\vec{k}}|) \Theta(\Omega_D - |\varepsilon_{\vec{k}'}|)$$

where Ω_D is a characteristic phonon energy and $\varepsilon_{\vec{k}}$ is the electronic (band-structure) energy measured from the chemical potential. At T=0, in the ground state all of the electrons are condensed into $(\vec{k}\uparrow, -\vec{k}\downarrow)$ pairs with $v_{\vec{k}}^2 = (1-\varepsilon_{\vec{k}}/E_{\vec{k}})/2$ the probability that the state $(\vec{k}\uparrow, -\vec{k}\downarrow)$ is occupied. The quantity $E_{\vec{k}} = (\varepsilon_{\vec{k}}^2 + \Delta^2)^{\frac{1}{2}}$ is the energy associated with a quasiparticle excitation out of the BCS ground state. The density of states for such quasiparticle excitations is

$$N_S(E) = N_N(0) \; E/\sqrt{E^2 - \Delta^2}$$

where it has been assumed that the ε dependence of the normal density of states $N_N(\varepsilon)$ is negligible for $|\varepsilon| \lesssim \Omega_D$.
The transfer Hamiltonian approach[23,24] confirms Giaever's phenomenological expression[35] for the low bias quasiparticle tunneling current between two superconducting electrodes,

$$I(V,T) \propto \int_{-\infty}^{\infty} dE \; N_S^{(L)}(E-eV) \; N_S^{(R)}(E) \; [f(E-eV) - f(E)] \quad .$$

Fig. 13 shows $N_S^{(L)}(E-eV) \; f(E-eV)$ and $N_S^{(R)}(E) \; f(E)$ in the vicinity of the chemical potential for finite T at eV=0, $\Delta_R - \Delta_L$ and $\Delta_R + \Delta_L$. When eV = $\Delta_R - \Delta_L$ the peak in the density of thermally occupied states immediately above the energy gap of the left electrode faces the peak in the density of unoccupied states immediately above the energy gap of the right electrode and there is a well defined local maximum in the quasiparticle tunneling current. When eV = $\Delta_R + \Delta_L$ the peak in the density of occupied states immediately below the energy gap of the left electrode gains access to the large density of unoccupied states above the energy gap of the right electrode leading to an abrupt increase in current with further increase in bias. These two features of the finite temperature I-V characteristic allow accurate determination of both Δ_L and Δ_R.[1,2,25]

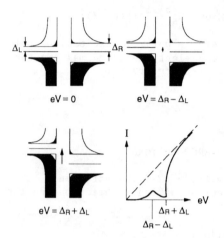

Figure 13. Determination of the BCS gaps Δ_L and Δ_R from the finite temperature I-V characteristics of a S-I-S tunnel junction. The alignment of the densities of occupied (shaded) and unoccupied (unshaded) single particle states, $N_s f$ and $N_s(1-f)$ respectively, are shown for eV=0, $\Delta_R - \Delta_L$ and $\Delta_R + \Delta_L$. The chemical potentials of the electrodes are indicated by the horizontal lines at mid-gap.

4.2. PHONON SPECTROSCOPY OF CONVENTIONAL SUPERCONDUCTORS

Since the pairing interaction is instantaneous in the BCS model of a superconductor it does not provide a theoretical framework for a spectroscopy of the electron-phonon interaction. Such a framework is provided by an extension of the BCS theory that accurately takes into account the frequency (ω) dependence of the phonon-mediated electron-electron pairing interaction[36]. This self-consistent mean-field theory leads to the Eliashberg equations[36], a set of coupled non-linear equations for the complex gap function $\Delta(\omega,T)$ and renormalization function $Z(\omega,T)$ that are believed accurate to $\sim 1\%$ (i.e. $O(\Omega_D/E_F)$). If the energy dependence of the normal-state band structure density of states is negligible within several Ω_D of the Fermi energy then all the essential information about the electron-phonon interaction needed as input to these equations is contained in the spectral function.

$$\alpha^2 F_{\vec{k}}(\Omega) \equiv \int_{S_F} \frac{dS_{\vec{k}'}}{\hbar v_{\vec{k}'}} \sum_\lambda |g_{\vec{k},\vec{k}',\lambda}|^2 \delta(\Omega - \Omega_{\vec{k}'-\vec{k},\lambda}) \qquad (10).$$

Here $g_{\vec{k},\vec{k}',\lambda}$ is the electron-phonon matrix element for the process in which an electron in state $\vec{k}$ is scattered to state $\vec{k}'$ with the emission or absorption of a phonon of polarization λ, wave vector $\vec{k}'-\vec{k}$ and frequency $\Omega_{\vec{k}'-\vec{k},\lambda}$. If impurity scattering is strong enough, as is usually the case, to average out any anisotropy in the superconducting gap function $\Delta_{\vec{k}}(\omega,T)$, then the Fermi surface averaged spectral function

$$\alpha^2 F(\Omega) = \int_{S_F} \frac{dS_{\vec{k}}}{\hbar v_{\vec{k}}} \alpha^2 F_{\vec{k}}(\Omega) / \int_{S_F} \frac{dS_{\vec{k}}}{\hbar v_{\vec{k}}}$$

is the relevant quantity. The repulsive Coulomb interaction usually has negligible frequency dependence over the range of interest and is consequently replaced by the Coulomb pseudopotential $\mu^*(\omega_c)\Theta(\omega_c-|\omega|)$ with ω_c typically 5 to 10 times Ω_D.

Given the electron-phonon spectral function and Coulomb pseudopotential, the Eliashberg equations can be solved numerically to obtain the complex, frequency and temperature dependent gap function $\Delta(\omega,T)$ and renormalization function $Z(\omega,T)$, the latter for both the superconducting and normal states. Knowledge of $\Delta(\omega,T)$ and $Z(\omega,T)$ for a given superconductor is sufficient for straightforward calculation of its thermodynamic properties. The quantity $\lambda(T) \equiv \text{Re}[Z_N(\omega=0,T)]-1$ is an important normal-state parameter, specifying the renormalization of many electronic properties by the electron-phonon interaction. To date, however, the Eliashberg equations have been most successfully applied, following the classic pioneering work of McMillan and Rowell[25,2], in the extraction of the electron-phonon spectral function and Coulomb pseudopotential from tunneling measurements on both S-I-N and S-I-S junctions. Besides producing a wealth of microscopic information on the electron-phonon interaction in s-p superconductors, McMillan and Rowell obtained convincing evidence for the correctness of the Eliashberg theory to the anticipated few % level.[36]

The density of states for the single-particle excitations of an Eliashberg superconductor is given by[37,25,34,2]

$$N_S(\omega) = N_N(0)\text{Re}\left\{\frac{|\omega|}{\sqrt{\omega^2-\Delta^2(\omega,T)}}\right\}, \tag{11}$$

where it has been assumed that the normal-metal density of states $N_N(\omega)$ has negligible dependence on ω in the immediate vicinity ($\pm$ several Ω_D) of the Fermi energy ($\omega=0$). In the BCS theory the gap function is real and independent of energy except for cutoffs at $\pm\Omega_D$. In this case, for all $T<T_c$, there are no single-particle states between $-\Delta(T)$ and $+\Delta(T)$. This is only true at T=0 in the Eliashberg theory because $\text{Im}\Delta(\omega,T)$ is in general non-zero for $\omega>0$ and $0 < T < T_c$. The analogue of the BCS energy gap $2\Delta(T)$ is the quantity $2\Delta_0(T)$ where $\Delta_0(T) = \text{Re}[\Delta(\omega=\Delta_0(T),T)]$.

The transfer Hamiltonian result for the single-particle tunneling current for an S-I-S junction is

$$I(V,T) = \sigma_{NN} \int_{-\infty}^{\infty} d\omega \, \rho_L(\omega-eV)\rho_R(\omega)[f(\omega-eV)-f(\omega)], \tag{12}$$

where $\rho(\omega) \equiv N_S(\omega)/N_N(0)$ and σ_{NN} is the zero bias conductance when both electrodes are in the normal state.

If the left metal is in the normal state the tunneling current is given by Eq. (12) with $\rho_L(\omega-eV)$ replaced by 1. For such a S-I-N junction a particularly important quantity is the normalized conductance[25,2]

$$\tilde{\sigma}(V,T) \equiv \frac{\sigma(V,T)}{\sigma_{NN}} = \int_{-\infty}^{\infty} d\omega \, \rho_R(\omega) \left[\frac{\partial f(\omega-eV)}{\partial eV} \right]$$

Since, at T=0, $\partial f(\omega-eV)/\partial eV = \delta(\omega-eV)$ it follows that

$$\tilde{\sigma}(V,T=0) = \rho(eV,T=0) \quad . \tag{13}$$

Hence at sufficiently low temperatures measurement of the normalized conductance directly gives the normalized density of states for single-particle excitations out of the superconducting ground state. Moreover, as shown by Galkin, D'Yachenko and Svistunov[38], knowledge of $\tilde{\sigma}(V,T=0)$ also yields the imaginary part of $1/\sqrt{\omega^2-\Delta^2(\omega,0)}$ via the dispersion relation

$$\text{Im} \left\{ \frac{1}{\sqrt{\omega^2-\Delta^2(\omega,0)}} \right\} = \frac{2}{\pi} \int_{\Delta_0(0)}^{\infty} d\omega' \, \text{Re} \left\{ \frac{\omega'}{\sqrt{\omega'^2-\Delta^2(\omega',0)}} \right\} \frac{1}{\omega^2-\omega'^2} \quad .$$

With the real and imaginary parts of $\omega/\sqrt{\omega^2-\Delta^2(\omega,0)}$ thus determined the complex gap function $\Delta(\omega,0)$ follows immediately. The electron-phonon spectral function $\alpha^2 F(\Omega)$ can then be extracted from the Eliashberg equations by straightforward numerical solution of a linear integral equation. The imaginary part of this equation does not involve the Coulomb pseudopotential which is readily obtained, once $\alpha^2 F(\Omega)$ is known, by fitting to the experimental gap edge $\Delta_0(0)$. The inversion procedure of Galkin et al. provides an elegant alternative to the double iteration scheme of McMillan and Rowell[25] in which the gap function is iterated to convergence for each estimate of $\alpha^2 F(\Omega)$.

The remarkable success of superconducting tunneling spectroscopy is well documented in the book by Wolf[2] and can largely be attributed to the high accuracy with which the Eliashberg theory treats the electron-phonon interaction, i.e. to the existence of the small parameter Ω_D/E_F. Another important factor is the large coherence length of most conventional superconductors which lessens the relative influence of the strong degradation of bulk properties near the metal-insulator interface. For high temperature superconductors the analogue of Ω_D/E_F, if it exists, may be of order unity and the coherence length is very much smaller than for conventional superconductors. In the same context, it is important to recall the assumptions made in deriving equation (12) for the tunneling current, in particular that the normal density of states and the transfer matrix element are independent of energy over the range of interest and that the latter is the same in the superconducting and normal states.

4.3 NORMAL STATE TUNNELING PHONON SPECTROSCOPY

Two decades ago Hermann and Schmid[39] studied the effect of electron-phonon interactions in the electrodes on the elastic N-I-N tunneling current between normal metals. Their starting point was the transfer Hamiltonian expression

$$I(V) = (4\pi e/\hbar)N_L(0)N_R(0) \qquad (14)$$

$$\int_{-\infty}^{\infty}d\omega\int_{-\infty}^{\infty}d\varepsilon_L\int_{-\infty}^{\infty}d\varepsilon_R|T(\varepsilon_L,\varepsilon_R)|^2 A_R(\varepsilon_R,\omega)A_L(\varepsilon_L,\omega-eV)[f(\omega-eV)-f(\omega)],$$

where ε denotes the band-energy measured from the Fermi level μ and

$$A(\varepsilon,\omega) = \frac{\pi^{-1}|\text{Im}[\sum(\varepsilon,\omega)]|}{\{\varepsilon-\omega + \text{Re}[\sum(\varepsilon,\omega)]\}^2 + \{\text{Im}[\sum(\varepsilon,\omega)]\}^2} \qquad (15)$$

is the electron spectral weight function with $\sum(\varepsilon,\omega)$ the electron self-energy in the isotropic approximation. $A(\varepsilon,\omega)$ is the probability that an electron with wave vector $\vec{k}$ and bare energy $\varepsilon_{\vec{k}}$ has renormalized energy ω in the presence of electron-phonon interactions. In the absence of such many-body interactions $A(\varepsilon,\omega) = \delta(\omega-\varepsilon)$. If one assumes that $\sum(\varepsilon,\omega) = \sum(0,\omega) \equiv \sum(\omega)$ and $|T(\varepsilon_L,\varepsilon_R)|^2 = |T(0,0)|^2 \equiv |T|^2$ it follows from Eq. (14) that $I(V) \propto V$ for low bias. Hermann and Schmid removed the latter assumption, replacing $|T|^2$ by $|T|^2\{1 + \eta(\varepsilon_L + \varepsilon_R)/\mu\}$ in (14), and showed that to linear order in μ^{-1} the odd part of the conductance, $\sigma_{odd}(V) \equiv [\sigma(V)-\sigma(-V)]/2$, provides a signature of electron self-energy effects in the electrodes:

$$\sigma_{odd}(V) = -\frac{\eta}{\mu}\sigma_0\{\text{Re}[\sum_R(eV)] + \text{Re}[\sum_L(-eV)]\} \quad (T=0), \qquad (16)$$

with σ_0 the conductance in the absence of such self-energy effects.

For the electron-phonon interaction the imaginary part of the electron self-energy at T=0 is given by[33,34]

$$\text{Im}[\sum{}^{(0)}(\omega)] = -\pi\int_0^{|\omega|}d\Omega\alpha^2 F(\Omega). \qquad (17)$$

The superscript (0) affixed to $\sum(\omega)$ indicates that the energy dependence of both $N(\varepsilon)$ and $\alpha^2 F(\Omega;\varepsilon,\varepsilon')$ has been ignored. Svistunov et al.[40] used a dispersion relation to obtain $\text{Re}[\sum{}^{(0)}(\omega)]$ from (17) and then applied (16) with the implicit assumption that $\sum(\omega) = \sum{}^{(0)}(\omega)$ to derive

$$\alpha^2 F_R(\Omega) = \frac{2\mu\Omega}{\eta\sigma_0\pi^2}\int_0^{\infty}dV\frac{d\sigma_{odd}(V)/dV}{\Omega^2 - (eV)^2} \qquad (18)$$

for the ideal situation in which the self-energy effects are negligibly small in the left-hand electrode, at least in the region of interest $0 \leq \Omega \leq \Omega_R^{Max}$ (where Ω^{Max} is the maximum phonon energy). It has been assumed that the linear background arising from barrier asymmetry has been subtracted from $d\sigma_{odd}(V)/dV$.

For Pb (driven normal by the application of a magnetic field) Svistunov et al. used (18) to invert their normal-metal tunneling data. As shown in Fig. 14, when the magnitude was adjusted using η as a fitting parameter, they obtained an electron-phonon spectral function $\alpha^2 F(\Omega)$ in promising agreement with that obtained by McMillan and Rowell using the superconducting tunneling technique[25]. The measurements of Svistunov et al. were carried out at 2.3K and the differences between the two experimental spectra are largely due to thermal smearing. This is indicated by the dashed curve in the figure which was obtained as follows: the spectral function of McMillan and Rowell was used to calculate $\Sigma^{(0)}(\omega)$ and then, using (14), $d\sigma_{odd}(V)/dV$ both for T = 2.3K. The latter quantity was then used in the T=0 inversion formula, Eq. (18), to obtain the thermally smeared spectral function shown.

The results obtained by Svistunov et al. provide experimental confirmation of (16) and (18). However, Appelbaum and Brinkman[41] have argued that the renormalized energy $\omega=\varepsilon + \mathrm{Re}[\Sigma(\varepsilon,\omega)]$ rather than the bare band-structure energy ε should determine the exponential behaviour of the transfer Hamiltonian matrix element, in which case $\sigma_{odd}(V) = 0$. Davis[42] subsequently confirmed this rigorously for a simple model. Fortunately, equations (16) and (18) can be recovered from the weak ε dependence of the electron self-energy $\Sigma(\varepsilon,\omega)$ if it is assumed that, for ε and ε' within a few Ω^{Max} of the Fermi level,

$$N(\varepsilon) \cong N(0) (1 + \beta\varepsilon/\mu) \qquad (19)$$

$$\alpha^2 F(\Omega;\varepsilon,\varepsilon') \cong \alpha^2 F(\Omega)[1 + \gamma(\varepsilon + \varepsilon')/\mu] \qquad (20)$$

for the individual electrodes ($\varepsilon = \varepsilon_L$ or ε_R) with β and γ of order unity (at most).[43] If the calculation of Hermann and Schmid is repeated with the above changes one obtains to linear order in μ^{-1}

$$\sigma_{odd}(V) = -\sigma_0 \left\{ \frac{\gamma_R}{\mu_R} \mathrm{Re}[\Sigma_R^{(0)}(eV)] - \frac{\gamma_L}{\mu_L} \mathrm{Re}[\Sigma_L^{(0)}(eV)] \right\} \qquad (21)$$

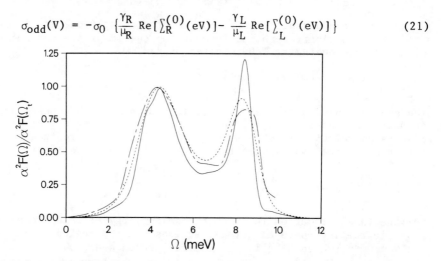

Figure 14. Comparison of the electron-phonon spectral function $\alpha^2 F(\Omega)$ of Pb obtained by the superconducting tunneling technique of McMillan and Rowell[25] (———) with that obtained by the normal-state tunneling technique of Svistunov et al.[40] at T=2.3K (— — — —). Application of the T=0K inversion procedure of Svistunov et al. to tunneling "data" calculated for T=2.3K starting from the spectral function of McMillan and Rowell produces a thermally smeared spectral function (-----). The three curves are normalized to unity at the transverse peak frequency Ω_t.

in place of Eq. (16). It is clear that the two results are equivalent as regards extraction of the frequency dependence of $\alpha^2F(\Omega)$ because neither of the quantities η or γ are accurately known and must be regarded as fitting parameters.

Mitrovic[44] recently calculated $\alpha^2F(\Omega;\varepsilon,\varepsilon')$ for a nearly-free-electron metal assuming that $\sum_\lambda |g_{\vec{k}',\vec{k},\lambda}|^2 \delta(\Omega-\Omega_{\vec{k}'-\vec{k},\lambda})$ is a function only of $\vec{k}'-\vec{k}$. He found that, for $|\varepsilon|/\mu$ and $|\varepsilon'|/\mu \ll 1$ Eq. (20) for $\alpha^2F(\Omega;\varepsilon,\varepsilon')$ holds with $\gamma=-\frac{1}{2}$, consistent with the assumption that γ be at most of order unity. The values of the parameter η obtained for Pb and $Pb_{70}Bi_{30}$ by Svistunov et al. on fitting Eq. (18) for $\alpha^2F(\Omega)$ to the superconducting tunneling spectroscopy results are -0.5 to -0.7 and -0.5 respectively in good agreement with Mitrovic's estimate for the corresponding fitting parameter, γ, of the modified Hermann-Schmid theory.

According to (16) and (21), σ_{odd} (V) should be zero for a symmetrical N-I-N tunnel junction. Hence, it is puzzling that McMillan et al.[45] found $\sigma_{odd}(V) \propto Re\Sigma_{Pb}(V)$ for Pb-I-Pb, after subtraction of a linear background term.

5. Inelastic Electron Tunneling Spectroscopy

The field of inelastic electron tunneling spectroscopy (IETS)[46,47,48] began when Jaklevic and Lambe[49] identified the positions of peaks in their d^2I/dV^2 tunneling spectra with vibrational energies of molecular impurities or dopants in the barrier. It is an extremely sensitive probe of the vibrational spectra of organic molecules deliberately introduced at monolayer or submonolayer coverage at a barrier interface.

Consider a N-I-N tunnel junction at T=0. In an elastic (el) tunneling process for eV > 0, an electron of energy E with $E_F+eV \geq E \geq E_F$ is transferred from an occupied state of the left electrode to an unoccupied state of the right electrode. For normal metals such processes give the dominant contribution, $J_{el}(V,T=0)$, to the tunneling current density, a contribution that varies smoothly with increasing eV. If a tunneling electron has sufficient energy to create an excitation of the barrier with energy $\hbar\omega_n$ this leads to an additional tunneling channel provided that the scattered electron with energy $E-\hbar\omega_n$ can find an empty state of that energy in the right electrode. Hence the threshold for this inelastic (inel) contribution to the current is $eV = \hbar\omega_n$. Assuming that this contribution varies smoothly above threshold as $[\alpha(eV-\hbar\omega_n) + \beta(eV-\hbar\omega_n)^2+...]\Theta(eV-\hbar\omega_n)$ it follows that

$$\frac{dJ(V,T=0)}{deV} = \frac{dJ_{el}(V,T=0)}{deV} + \alpha\Theta(eV-\hbar\omega_n) ,$$

$$\frac{d^2J(V,T=0)}{deV^2} = \frac{d^2J_{el}(V,T=0)}{deV^2} + \alpha\delta(eV-\hbar\omega_n) .$$

Repeating the argument for eV<0 shows that the inelastic contributions to $dJ(V,T=0)/dV$ and $d^2J(V,T=0)/dV^2$ are even and odd, respectively, in V to the extent that one can ignore the dependence of α on polarity. Such a dependence can arise[50], for example, if the excitation process is strongly localized at the left side of the barrier. In this case α for $eV>0$ is in general different than α for $eV < 0$ because in the former case $\kappa(z) \propto \sqrt{V(z)+\hbar\omega_n - E_z}$ over most of the barrier for the exciting electron while $\kappa(z) \propto \sqrt{V(z)-E_z}$ in the latter case. Hence, in this example, the above symmetry rules are accurately obeyed only if $\hbar\omega_n <<< V(z)-E_z$. These symmetry rules are an important signature of inelastic contributions to the tunneling current because they are the reverse of the symmetry rules for self-energy contributions to the elastic current.

The discontinuous upward steps in dJ/dV and delta function spikes in d^2J/dV^2 are of course broadened by the intrinsic width of the barrier excitation due to its finite lifetime. In addition there is broadening arising from thermal blurring of the Fermi surfaces of the two metals. As regards the latter, the widely used transfer Hamiltonian expression for the inelastic contribution to the tunneling current is

$$J_{inel}(V,T) = 2e\left(\frac{2\pi}{\hbar}\right) \sum_{\ell rn} |M_{\ell r;n}|^2 f(E_\ell^{(L)}-eV)[1-f(E_r^{(R)})]\delta(E_\ell^{(L)}-E_r^{(R)}-\hbar\omega_n) \quad (22)$$

It has been assumed that the barrier excitation is infinitely long-lived and that the left-going contribution to the inelastic current is negligible at the temperatures and voltages of interest. In addition, the inelastic channel associated with the absorption of a barrier excitation by an electron tunneling from left to right has not been included (there is no voltage threshold and hence no sharp spectral features for this process). To determine the amount of thermal smearing of the delta function peaks of $d^2J_{inel}(V,T=0)/dV^2$ at finite T, Lambe and Jaklevic[27] assumed that only the energy dependence of the Fermi functions is important and evaluated the resulting thermal broadening function:

$$\frac{d^2}{deV^2}\left(\int_{-\infty}^{\infty} dE f(E-eV)[1-f(E-\hbar\omega_n)]\right)$$

$$= \frac{e^u}{k_BT}\left[\frac{(u-2)e^u + (u+2)}{(e^u-1)^3}\right] \quad u \equiv (eV-\hbar\omega_n)/k_BT .$$

This bell-shaped function has a peak at $eV = \hbar\omega_n$ of height $(6k_BT)^{-1}$ and a full-width at half-maximum of $5.44\,k_BT$.

In practice a finite modulation technique is usually employed to obtain d^2I/dV^2. For example, suppose that a voltage $V + V_\omega \cos\omega t$ is applied and the second harmonic

$$I_{2\omega}(eV) \equiv (2\omega/\pi) \int_0^{\pi/\omega} I(eV + eV_\omega \cos\omega t)\cos 2\omega t\, dt$$

of the current measured. Changing the integration variable to $\varepsilon = eV_\omega \cos\omega t$ and integrating by parts twice shows that

$$\lim_{V_\omega \to 0} \left(\frac{2}{V_\omega}\right)^2 I_{2\omega}(eV) = \frac{d^2I(eV)}{dV^2} ,$$

while for finite V_ω

$$\left(\frac{2}{eV_\omega}\right)^2 I_{2\omega}(eV) = \int_{-eV_\omega}^{eV_\omega} d\varepsilon\, C_2(\varepsilon)\, \frac{d^2 I(eV+\varepsilon)}{d(eV+\varepsilon)^2} .$$

The modulation voltage broadening function

$$C_2(\varepsilon) \equiv (8/3\pi eV_\omega)[1-(\varepsilon/eV_\omega)^2]^{3/2} \Theta(eV_\omega-\varepsilon)\Theta(eV_\omega+\varepsilon)$$

of Klein et al.[51] has a peak at $\varepsilon=0$ of height $(8/3\pi eV_\omega)$ and a full-width at half-maximum of $1.22\ eV_\omega$. When dI/dV is determined by the modulation technique from the first harmonic the corresponding broadening function

$$C_1(\varepsilon) \equiv (2/\pi eV_\omega)[1-(\varepsilon/eV_\omega)^2]^{\frac{1}{2}} \Theta(eV_\omega-\varepsilon)\Theta(eV_\omega+\varepsilon)$$

has a peak at $\varepsilon=0$ of height $(2/\pi eV_\omega)$ and full-width at half-maximum of $1.73\ eV_\omega$.

The excitations routinely detected by inelastic electron tunneling spectroscopy include phonons characteristic of the electrodes (presumably emitted in the immediate vicinity of the electrode-barrier interfaces), phonons characteristic of the barrier material, and internal vibrations of molecular impurities or molecules deliberately introduced into the barrier region. These are now discussed briefly in turn.

When phonon structure characteristic of either electrode is observed in the elastic, i.e. even (in V), part of d^2I/dV^2 it is attributed to the virtual emission and absorption of phonons in that electrode as described by the real part of the electronic self-energy. When such structure is seen in the inelastic, i.e. odd, part of d^2I/dV^2 it is attributed to the emission of real (electrode) phonons very close to the electrode-barrier interface. (The emission of real phonons in the interior of the electrode contribute to the imaginary part of the electronic self-energy and within the approximations of the theory of section 4.3 do not contribute to the tunneling current.) Clearly there is some ambiguity in this description as to where the electrodes end and the barrier begins. Moreover, since the elastic and inelastic channels for an incident electron of initial energy E are not orthogonal the tunneling current should contain an interference term which according to the theory of Ivanchenko[52] contributes to the even part of d^2I/dV^2. Unfortunately, the inversion scheme of Svistunov et al.[40] does not allow for this interference term and may not be of general applicability, although it certainly worked very well for Pb.

The phonon structure in the tunneling current is very weak for normal metal electrodes, corresponding to a relative change in dI/dV of order 10^{-2} at most.[40,45,51] There are situations in which the phonon thresholds in the conductance are very prominent because the elastic background is greatly suppressed. Consider, for example, a perfect single-crystal Ge p-n tunnel junction at T=0 with the applied bias such that it is energetically possible for an electron near a conduction band minimum at $\vec{k}=(1,1,1)\pi/a$ (a is the lattice constant) on the n side to tunnel through the depletion layer barrier to an unoccupied state near the maximum of the valence band at $\vec{k}=0$ on the p side. Since the entire junction is a single crystal and the zone boundary $\vec{q}=(1,1,1)\pi/a$ phonons have finite energy, this elastic tunneling transition is forbidden by conservation of crystal momentum. It is not until the threshold

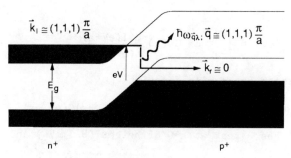

Figure 15. Phonon-assisted tunneling in an indirect gap p-n tunnel junction.

condition for inelastic tunneling with the emission of the lowest energy $\vec{q} = (111)\pi/a$ phonon is satisfied that an appreciable current flows with a very marked jump in the conductance[53].

The pioneering theoretical work on the tunneling spectroscopy of molecular vibrations was done by Scalapino and Marcus[26] and Lambe and Jaklevic[27]. Here the more recent approach of Kirtley, Scalpino and Hansma (KSH)[28] is briefly summarized.

Consider a layer of uniformly distributed, non-interacting molecules in the insulating barrier of a planar tunnel junction. For the purpose of calculating its interaction with a tunneling electron, KSH regard a molecule as a group of N atoms each with a point partial charge arising from unequal sharing of bond electrons:

$$V_{el-mol}(\vec{r}) = -\sum_{j=1}^{N} \frac{e^2 z_j}{|\vec{r}-\vec{R}_j|}, \qquad (23)$$

where $\vec{r}$ and $\vec{R}_j$ denote the positions of the tunneling electron and jth partial charge respectively. Expanding the interaction to linear order in the displacements $\delta\vec{R}_j$ from the equilibrium positions $\vec{R}_j^{(0)}$ and keeping only the contribution from the nth vibrational mode of frequency ω_n gives

$$V_{el-mol}^{(n)}(\vec{r}) = -\sum_{j=1}^{N} e^2 z_j^{(n)} \delta\vec{R}_j^{(n)} \cdot \frac{\partial}{\partial \vec{R}_j} \left(\frac{1}{|\vec{r}-\vec{R}_j|} \right)_{\vec{R}_j = \vec{R}_j^0}.$$

When the multiple images of the partial charges are included this becomes

$$V_{el-mol}^{(n)}(\vec{r}) =$$

$$-\sum_{i=-\infty}^{+\infty} \sum_{j=1}^{N} \frac{e^2 z_j^{(n)} \delta\vec{R}_j^{(n)}}{\varepsilon(\omega_n)} \cdot \frac{\partial}{\partial \vec{R}_j} \left(\frac{1}{|\vec{r}-\vec{R}_j-2id\hat{z}|} - \frac{1}{|\vec{r}-\vec{R}_j-2(id-R_{jz})\hat{z}|} \right)_{\vec{R}_j=\vec{R}_j^0} \qquad (24)$$

where $\varepsilon(\omega_n)$ is the dielectric constant assumed for the combined oxide plus molecular layer barrier (the metal electrodes are assumed to be perfect conductors). This interaction does not allow for any molecular polarizability.

KSH use the transfer Hamiltonian approach with WKB wavefunctions $\psi_\ell^{(L)}(\vec{r})$ and $\psi_r^{(R)}(\vec{r})$ for the uncoupled electrodes (assuming a barrier of constant height) as a basis for the expansion of the time-dependent wavefunction $\Psi(\vec{r},t)$ of Eq. (7). The resulting matrix element describing the inelastic tunneling of an electron from the state $\psi_\ell^{(L)}(\vec{r})$ of energy $E_\ell^{(L)}$ to the state $\psi_r^{(R)}(\vec{r})$ of energy $E_r^{(R)} = E_\ell^{(L)} - \hbar\omega_n$, leaving the excited molecule in its nth vibrational level, is

$$M_{\ell r;n} = \int d\vec{r}\, \psi_r^{(R)}(\vec{r})^* V_{el-mol}^{(n)}(\vec{r}) \psi_\ell^{(L)}(\vec{r}). \qquad (25)$$

The integration is confined to the barrier because of electronic screening in the electrodes. The inelastic tunneling current density is obtained by substituting $M_{\ell r;n}$ into Eq. (22) and multiplying by the number of molecules per unit area. The resulting expression has been used by KSH to calculate integrated IETS intensities and the ratio of intensities for opposite voltage polarities. These are in good agreement with experiment. Since the three dimensional nature of the electron-molecule interaction is included in the theory the tunneling electron can change its transverse momentum leading to a matrix element that varies spatially in the plane of the barrier. This leads to a breakdown of selection rules which are based on spatial homogeneity. Hence the theory predicts that even in the absence of molecular polarizability, which it does not include, Raman active modes can be of comparable intensity to infrared modes in IETS spectra and that modes forbidden to both optical techniques may be detected by IETS. The orientation rule of Scalapino and Marcus,[26] that the IETS intensity is zero for a vibrational mode associated with a dipole moment oscillating in the plane of the barrier, is weakened slightly in the theory of Kirtley, Scalapino and Hansma. However, it is still sufficiently strong that orientational information can be inferred from intensity data for molecules of high symmetry at the electrode-barrier interface.[54]

6. Single-Electron Charging Effects

The discussion so far has centred on tunnel junctions of macroscopic extent in the x-y plane. The pioneering experimental and theoretical work on mesoscopic tunnel junctions was carried out two decades ago by Giaever and Zeller[55] and Lambe and Jaklevic[56] who investigated the effects of a layer of well-separated small metal particles (average diameter less than 100Å) embedded in an oxide barrier between two metallic electrodes. Giaever and Zeller correctly attributed the suppression of the tunneling current at low temperature and voltages to the non-negligible electrostatic charging energy associated with the transfer of a single electron to a particle with an ultrasmall capacitance (for a capacitance C of 10^{-17}F the charging energy, $e^2/2C$, is about 8 meV). In the last few years there has been a resurgence of interest, both experimental[55-64] and theoretical[65-72], in the properties of mesoscopic tunnel junctions. The following brief introduction to single-electron charging effects is based on the semiclassical approach of Ben-Jacob and coworkers[69-73] and relies heavily on references 71 and 72.

Consider two tunnel junctions in series driven by an ideal dc voltage source (zero internal resistance). The usual transfer

Hamiltonian expression for the rate of transfer of electrons across, say, the right-hand barrier from the middle (M) to right (R) electrode is

$$\ell_R(V_R) \equiv J_{M \to R}(V_R, T)/e$$

$$= \frac{4\pi}{\hbar} \overline{|M_{mr}|^2} \int_{-\infty}^{\infty} dE \, N^{(M)}(E-eV_R) \, N^{(R)}(E) \, f(E-eV_R)[1-f(E)]$$

where V_R and $V_L = V_{dc} - V_R$ are the voltage changes across the right and left barriers respectively. It has been assumed that the transfer matrix element M_{mr} is independent of energy over the energy range of interest. The density of states factors N apply to either normal metal (N_N) or superconducting (N_S) electrodes. There are similar expressions for $r_R(V_R) \equiv J_{R \to M}(V_R,T)/e$, $\ell_L(V_L) \equiv J_{L \to M}(V_L,T)/e$, and $r_L(V_L) \equiv J_{M \to L}(V_L,T)/e$. For the macroscopic tunnel junctions considered so far it has been assumed that the difference in chemical potentials between adjacent electrodes is unaffected by the transfer of an electron across the intervening barriers. Moreover, eV_L and eV_R are usually assumed independent of the amount of excess charge $\delta Q = ne$ stored between the two junctions. Strictly speaking[71,72], however,

$$V_L(\delta Q) = V_{dc} \frac{C_R}{C_L + C_R} - \frac{\delta Q}{C_L + C_R} \quad , \tag{26}$$

$$V_R(\delta Q) = V_{dc} \frac{C_L}{C_L + C_R} + \frac{\delta Q}{C_L + C_R} \quad , \tag{27}$$

where C_L and C_R are the capacitances of the individual junctions. For mesoscopic junctions at sufficiently low voltages and temperatures ($eV_{dc}, k_B T \lesssim |e\delta Q|/(C_L + C_R)$) the terms $\pm \delta Q/(C_L + C_R)$ can no longer be assumed negligible. In this case, for the calculation of, say, $\ell_L(V_L)$ the appropriate difference $\mu_L - \mu_M$ in chemical potentials between the left and middle electrodes is not just $eV_L(0)$ nor $eV_L(\delta Q)$ but

$$\mu_L - \mu_M = \int_{\delta Q}^{\delta Q + e} V_L(Q) dQ = eV_L(\delta Q) - \frac{e^2}{2(C_L + C_R)} \quad . \tag{28}$$

At T=0 for normal-metal electrodes there will be empty states available in the middle electrode to an electron tunneling from the left only if $\mu_L > \mu_M$. Hence, at low temperatures the charging energy $e^2/2(C_L + C_R)$ leads to a suppression of electron tunneling from the left to the middle electrode when $eV_L(\delta Q) < e^2/2(C_L + C_R)$. If the electrodes are identical BCS superconductors this becomes $eV_L(\delta Q) < 2\Delta(T) + e^2/2(C_L + C_R)$ because of the energy gap $2\Delta(T)$ in the single-particle density of states.

To calculate the I-V characteristic of the system under discussion Mullen and coworkers[71,70] used the semiclassical approach in which it is assumed that V_{dc} and the classical variable $\delta Q = ne$, the amount of excess charge on the middle electrode, completely determine the behaviour of the tunneling current. The distribution function $\rho(n,t)$ gives the probability that the middle electrode contains n excess electrons at time t. Since the single-election transfers described by $\ell_L(V_L)$ and $r_R(V_R)$

($r_L(V_L)$ and $\ell_R(V_R)$) increase (decrease) n by unity, the master equation for $\rho(n,t)$ is

$$\frac{\partial \rho(n,t)}{\partial t} = [\ell_L(n-1)+r_R(n-1)]\rho(n-1,t)+[r_L(n+1)+\ell_R(n+1)]\rho(n+1,t)$$
$$- [\ell_L(n)+r_L(n)+\ell_R(n)+r_R(n)]\rho(n,t) \quad , \tag{29}$$

where the tunneling rates have been expressed in terms of n (and implicitly V_{dc}), rather than V_L or V_R using $V_{dc} = V_L + V_R$ and ne = $C_R V_R - C_L V_L$. It has been assumed that the simultaneous transfer of more than one electron through a given barrier has negligible probability (for the case of superconducting electrodes this requires that the Josephson tunneling of Cooper pairs is suppressed). The tunneling current is obtained from the stationary-state distribution $\rho(n)$ by

$$I(V_{dc}) = \sum_n I(n)\rho(n) \quad , \quad I(n) \equiv e[\ell_L(n)-r_L(n)] \quad .$$

Mullen et al. showed by differentiating the tunneling rates with respect to V_{dc} that steps in the current are possible whenever eV_{dc} takes on one of the values

$$eV_X = \pm \frac{(C_L + C_R)}{C_X} 2\Delta(T) + (n \pm \tfrac{1}{2}) \frac{e^2}{C_X} \qquad (X=L,R) \quad .$$

For the special case of identical junctions ($C_L = C_R = C$) this becomes

$$eV = \pm 4\Delta(T) + (n \pm \tfrac{1}{2})e^2/C \quad .$$

The magnitude of the steps depends on the system parameters in a complicated way. For the case of identical superconducting electrodes the step heights decrease with $|n|$ while for the case of identical normal metal electrodes the steps are not discernable although there is a voltage offset of $|e|/2C$. For normal metal electrodes with, say, junction resistances and capacitances satisfying $R_L \ll R_R$ and $C_L < C_R$ there are well defined steps at $V_{dc} = (n \pm \tfrac{1}{2})e^2/C_R$ of height $\Delta I = |e|/R_R(C_L+C_R)$, except for the first which has height $\Delta I/2$, and flat plateaus of width $|e|/C_R$. This structure disappears when $k_B T \gtrsim e^2/2C_R$.

These and related mesoscopic quantum effects, including the interesting dynamic behaviour of a single ultralow capacitance junction driven by a direct-current source, will be discussed at greater length by Prof. García and Prof. van Kempen.

Acknowledgement

We profited from useful discussions of resonant tunneling with D. Landheer and H.C. Liu.

References

1. C.B. Duke, Tunneling in Solids (Academic, New York, 1969).
2. E.L. Wolf, Principles of Electron Tunneling Spectroscopy (Oxford University Press, New York, 1985).
3. M. Altarelli, in Heterojunctions and Semiconductor Superlattices (Springer, Berlin, 1986), edited by G. Allan, G. Bastard, N. Boccara, M. Lannoo and M. Voos.

4. K.H. Gundlach, Solid St. Electronics 9, 949 (1966).
5. E.E. Mendez, in Physics and Applications of Quantum Wells and Superlattices (Plenum, New York, 1988) edited by E.E. Mendez and K. von Klitzing.
6. R.S. Becker, J.A. Golovchenko and B.S. Swartzentruber, Phys. Rev. Lett. 55, 987 (1985).
7. G.C. Aers and C.R. Leavens, Solid St. Commun. 60, 427 (1986).
8. L.L. Chang, L. Esaki and R. Tsu, Appl. Phys. Lett. 24, 593 (1974).
9. T.C.L.G. Sollner, W.D. Goodhue, P.E. Tannenwald, C.D. Parker and D.D. Peck, Appl. Phys. Lett 43, 588 (1983).
10. E.R. Brown, W.D. Goodhue and T.C.L.G. Sollner, J. Appl. Phys. 64, 1519 (1988).
11. R. Tsu and L. Esaki, Appl. Phys. Lett. 22, 562 (1973).
12. M. Büttiker, IBM J. Res. Develop. 32, 63 (1988).
13. S. Luryi, Appl. Phys. Lett. 47, 490 (1985).
14. M. Cahay, M. McLennan, S. Datta and M.S. Lundstrom, Appl. Phys. Lett. 50, 612 (1987). M.J. McLennan, Master's Thesis, Purdue University, May 1987.
15. D. Landheer and G.C. Aers, in press.
16. H. Ohnishi, T. Indta, S. Muto, N. Yokoyama and A. Shibatomi, Appl. Phys. Lett. 49, 1248 (1986).
17. H.C. Liu and D.D. Coon, J. Appl. Phys. 64, 6785 (1988).
18. N.S. Wingreen, K.W. Jacobsen and J.W. Wilkins, Phys. Rev. Lett. 61, 1396 (1988).
19. H.C. Liu. Appl. Phys. Lett. 51, 1019 (1987).
20. T. Wiel and B. Vinter, Appl. Phys. Lett. 50, 1281 (1987).
21. P. J. Price, Phys. Rev. B 38, 1994 (1988).
22. S. Collins, D. Lowe and J.R. Barker, J. Phys. C 20, 6233 (1987).
23. J. Bardeen, Phys. Rev. Lett. 6, 57 (1961).
24. M.H. Cohen, L.M. Falicov and J.C. Phillips, Phys. Rev. Lett. 8, 316 (1962).
25. W.L. McMillan and J.M. Rowell, in Superconductivity (Marcel-Dekker, New York, 1969), edited by R.D. Parks.
26. D.J. Scalapino and S.M. Marcus, Phys. Rev. Lett. 18, 459 (1967).
27. J. Lambe and R.C. Jaklevic, Phys. Rev. 165, 821 (1968).
28. J. Kirtley, D.J. Scalapino and P.K. Hansma, Phys. Rev. B 14, 3177 (1976).
29. J. Kirtley and J.T. Hall, Phys. Rev. B 22, 848 (1980).
30. J. Tersoff and D.R. Hamann, Phys. Rev. B 31, 805 (1985).
31. N.D. Lang, Phys. Rev. Lett. 55, 230 (1985).
32. J. Bardeen, L.N. Cooper and J.R. Schrieffer, Phys. Rev. 106, 162 (1957): 108, 1175 (1957).
33. J.R. Schrieffer, Theory of Superconductivity (Benjamin, New York, 1964).
34. D.J. Scalapino, in Superconductivity (Marcel-Dekker, New York, 1969), edited by R.D. Parks.
35. I. Giaever, Phys. Rev. Lett. 5, 147, 464 (1960).
36. G.M. Eliashberg, Sov. Phys. JETP 11, 696 (1960).
37. D.J. Scalapino, J.R. Schrieffer and J.W. Wilkins, Phys. Rev. 148, 263 (1966).
38. A.A. Galkin, A.I. D'yachenko and V.M. Svistunov, Soc. Phys. JETP 39, 1115 (1974).

39. H. Hermann and A. Schmid, Zeitschrift für Physik 211, 313 (1968).
40. V.M. Svistunov, M.A. Beloglovskii, O.I. Chernysh, A.I. Khachaturov, and A.P. Kvachev, Sov. Phys. JETP 57, 1038 (1983).
41. J.A. Appelbaum and W.F. Brinkman, Phys. Rev. 183, 553 (1969).
42. L.C. Davis, Phys. Rev. 187, 1177 (1969).
43. C.R. Leavens, Solid State Commun. 55, 13 (1985).
44. B. Mitrovic, to be published.
45. J.M. Rowell, W.L. McMillan and W.L. Feldmann, Phys. Rev. 180, 658 (1969).
46. E. Burstein and S. Lundqvist (ed.), Tunneling Phenomena in Solids (Plenum, New York, 1969).
47. P.K. Hansma (ed.), Tunneling Spectroscopy: Capabilities, Applications and New Techniques, (Plenum, New York, 1982).
48. T. Wolfram (ed.), Inelastic Electron Tunneling Spectroscopy, (Springer-Verlag, Berlin, 1978).
49. R.C. Jaklevic and J. Lambe, Phys. Rev. Lett. 17, 1139 (1966).
50. I.K. Yanson, N.I. Bogatina, B.I. Verkin and O.I. Shklyarevski, Sov. Phys. JETP 35, 540 (1972).
51. J. Klein, A Léger, M. Belin, D. Défourneau and M.J.L. Sangster, Phys. Rev. B 7, 2336 (1973).
52. M.A. Belogolovskii, Yu. M. Ivanchenko and Yu. V. Medvedev, Sov. Phys. Solid State 17, 1937 (1976).
53. R.T. Payne, Phys. Rev. 139, A570 (1965).
54. J. Kirtley and J.T. Hall, Phys. Rev. B 22, 848 (1980).
55. I. Giaever and H.R. Zeller, Phys. Rev. Lett. 20, 1504 (1968).
56. J. Lambe and R. Jaklevic, Phys. Rev. Lett. 22, 1371 (1969).
57. M.S. Raven, Phys. Rev. B 29, 6218 (1984).
58. R.E. Cavicchi and R.H. Silsbee, Phys. Rev. Lett. 52, 1453 (1984).
59. L.S. Kuzmin and K.K. Likharev, JETP Lett. 45, 495 (1987).
60. T.A. Fulton and G.J. Dolan, Phys. Rev. Lett. 59, 109 (1987).
61. J.B. Barner and S.T. Ruggiero, Phys. Rev. Lett. 59, 807 (1987).
62. R.E. Cavicchi and R.H. Silsbee, Phys. Rev. B 37, 706 (1988); Phys. Rev. B 38, 6407 (1988).
63. P.J.M. van Bentum, H. van Kempen, L.E.C. van de Leemput and P.A.A. Teunissen, Phys. Rev. Lett. 60, 369 (1988).
64. P.J.M. van Bentum, R.T.M. Smokers and H. van Kempen, Phys. Rev. Lett. 60, 2543 (1988).
65. D.V. Averin and K.K. Likharev in SQUID'85 (w. de Gruyter, Berlin, 1985), edited by H. Lübbig and H.D. Hahlbohm.
66. D.V. Averin and K.K. Likharev, J. Low Temp. Phys. 62, 345 (1986).
67. F. Guinea and G. Schön, Europhys. Lett. 1, 585 (1986).
68. K.K. Likharev, IBM J. Res. Develop. 32, 144 (1988).
69. E. Ben-Jacob and Y. Gefen, Phys. Lett. A 108, 289 (1985).
70. E. Ben-Jacob, E. Mottola and G. Schön, Phys. Rev. Lett. 51, 2064 (1983).
71. K. Mullen, E. Ben-Jacob, R.C. Jaklevic, and Z. Schuss, Phys. Rev. B 37, 98 (1988).
72. K. Mullen, E. Ben-Jacob, and S. Ruggiero, Phys. Rev. B 38, 5150 (1988).
73. E. Ben-Jacob, Y. Gefen, K. Mullen, and Z. Schuss, Phys. Rev. B 37, 7400 (1988).

TUNNELING TIMES FOR ONE-DIMENSIONAL BARRIERS

C.R. LEAVENS AND G.C. AERS
Division of Physics,
National Research Council of Canada,
Ottawa, Ontario K1A 0R6
Canada

ABSTRACT. The desire to minimize the response time of tunneling devices has led to a renewed interest in an unresolved fundamental question: "How long does an incident electron of energy E take, on average, to tunnel through a classically inaccessible region?" Since there is considerable controversy and confusion surrounding this problem, several recent approaches that illustrate the main points of dispute are discussed in some detail.

1. Introduction

Consider a monoenergetic beam of electrons with wavevector $\vec{k}=(0,0,k>0)$ incident from the left on the one-dimensional barrier $V(\vec{r})=V(z)\Theta(z)\Theta(d-z)$. In the absence of inelastic scattering and absorption there are actually three electron-barrier interaction times of interest. The dwell time, $\tau_D(k;0,d)$, is the average time spent by an incident electron of energy $E(k) \equiv \hbar^2 k^2/2m$ in the barrier region $0 \leq z \leq d$ regardless of whether it is ultimately transmitted or reflected; the transmission time, $\tau_T(k;0,d)$, is the corresponding average time if it is finally transmitted and the reflection time, $\tau_R(k;0,d)$, the average time if it is finally reflected. Literal interpretation of these precise definitions requires that τ_D, τ_T and τ_R satisfy

$$\tau_D(k;0,d) = |T(k)|^2 \tau_T(k;0,d) + |R(k)|^2 \tau_R(k;0,d) \qquad (1)$$

if they are to be consistent with each other. If the above definitions are extended to an arbitrary region $z_1 \leq z \leq z_2$ within the barrier then the more general relation

$$\tau_D(k;z_1,z_2) = |T(k)|^2 \tau_T(k;z_1,z_2) + |R(k)|^2 \tau_R(k;z_1,z_2) \qquad (2)$$

must be satisfied. In (1) and (2), $|T(k)|^2$ and $|R(k)|^2$ are the transmission and reflection probabilities for the entire barrier. Another condition that must be satisfied by each of the local times if they are to be useful quantities, is,

$$\tau_X(k;z_1,z_3) = \tau_X(k;z_1,z_2) + \tau_X(k;z_2,z_3) \quad (X = D,T,R) \qquad (3)$$

with $z_1 \leq z_2 \leq z_3$.

The question of whether or not the quantities τ_T and τ_R must be real is a controversial one. However, if the times τ_X are real then the additional requirement

$$\tau_X(k;z_1,z_2) > 0 \quad (X = D,T,R) \tag{4}$$

for $z_2 > z_1$ must be met. Another requirement for real local transmission times is that the corresponding local average 'speed' be non-negative and not exceed the speed of light c, i.e.

$$0 \leq v_T(z) \equiv \left[\frac{\partial \tau_T(k;0,z)}{\partial z}\right]^{-1} \leq c . \tag{5}$$

A small selection of the many approaches [1-33] to determining tunneling times are discussed in the following sections and the criteria (1) to (5) applied where appropriate. Section 2 is devoted to a discussion of Büttiker's expression[10] for the dwell time. The Feynman path integral approach of Sokolovski and Baskin[15] is considered in 3, the Larmor clock approach of Baz'[5], Rybachenko[6] and Büttiker[10] in 4, the wavepacket centroid-extrapolation method of Hauge, Falck and Fjeldly[17] in 5, and the Larmor clock analysis of Huang et al.[22] in 6. A brief summary follows in section 7.

2. The Dwell Time

Büttiker[10] identified the dwell time $\tau_D(k;0,d)$ with the average number of electrons within the barrier region $0 \leq z \leq d$ divided by the average number incident on the barrier per unit time, i.e.

$$\tau_D(k;0,d) \equiv \int_0^d |\Psi_k(z)|^2 dz / (\hbar k/m) \quad (k > 0). \tag{6}$$

Here $\Psi_k(z)$ is the steady-state scattering solution of the time-independent Schrödinger equation for incident probability current density $j_I(k) = \hbar k/m$. The right-hand-side of (6) is obviously a real non-negative quantity. Eq. (6) has been criticized on the grounds that it does not properly take into account quantum mechanical interference effects. It is now shown that Büttiker's expression is in fact correct.[33] Consider the quantity

$$\int_{-\infty}^{\infty} dt \int_0^d dz\, \Psi^*(z,t)\Psi(z,t) \tag{7}$$

where $\Psi(z,t)$ is a wavepacket solution of the time-dependent Schrödinger equation constructed from the solutions $\Psi_k(z)$ of the corresponding stationary-state scattering problem:

$$\Psi(z,t) = \int_{-\infty}^{\infty} \frac{dk}{2\pi} \Phi(k)\Psi_k(z)e^{-i\hbar k^2 t/2m} . \tag{8}$$

$\Phi(k)$ is the Fourier transform of $\Psi(z,t=0)$. If equation (8) for $\Psi(z,t)$ and the corresponding one for $\Psi^*(z,t)$, with integration variable k' in place of k, is substituted into (7) integration of the resulting time-dependent factor $\exp[i\hbar(k'^2-k^2)t/2m]$ over t gives $2\pi\delta[\hbar(k'^2-k^2)/2m]$ = $2\pi(m/\hbar|k|)[\delta(k'-k)+\delta(k'+k)]$. Integration over k' then gives two terms, one involving $|\Phi(k)|^2$ and the other $\Phi^*(-k)\Phi(k)$. If one considers only those wavepackets for which $|\Phi(k)|$ is negligibly small for $k<0$ then the term involving $\Phi^*(-k)\Phi(k)$ can be dropped and the final result for (7) is

$$\int_{-\infty}^{\infty} dt \int_0^d dz |\Psi(z,t)|^2 = \int_0^{\infty} \frac{dk}{2\pi} |\Phi(k)|^2 \tau_D(k;0,d) \quad , \tag{9}$$

with $\tau_D(k;0,d)$ given by (6). The left-hand-side of this equation is an exact expression for the average time spent in the barrier region $0 \leq z \leq d$ by an electron with wave function $\Psi(z,t)$. The quantity $dk|\Phi(k)|^2/2\pi$ is the probability that an electron described by the wavefunction $\Psi(z,t)$ has wave number between k and k + dk. Now, equation (6) for $\tau_D(k;0,d)$ applies to the stationary-state scattering problem and hence has no connection whatsoever to the Fourier transform $\Phi(k)$ of the initial wavefunction $\Psi(z,t=0)$ for the time-dependent wavepacket approach. Hence, for all k>0, equation (6) must be an exact expression for the dwell time as precisely defined in the introduction. The dwell time $\tau_D(k;0,d)$ for the entire barrier and all of the discussion of this section is trivially generalized to an arbitrary region $z_1 \leq z \leq z_2$ within the barrier on replacing 0 by z_1 and d by z_2 with $0 \leq z_1 \leq z_2 \leq d$.

The following sections are concerned with the more difficult question of whether or not it is possible to calculate transmission and reflection times, $\tau_T(k;0,d)$ and $\tau_R(k;0,d)$, which are completely consistent with the dwell time, $\tau_D(k;0,d)$, in the sense of satisfying equation (1). τ_T and τ_R must, of course, be consistent with each other, i.e. they must be determined by precisely parallel calculations within a given approach.

3. Feynman Path Integral Approach

Sokolovski and Baskin[15] began with the classical result

$$t_{0,d}^{cl} = \int_{t_a}^{t_b} dt \Theta(z(t))\Theta(d-z(t)) \tag{10}$$

for the time spent by a particle in the region $0 \leq z \leq d$ during the time interval $t_a \leq t \leq t_b$. Here z(t) is the classical path connecting (z_a,t_a) and (z_b,t_b) followed by a particle moving in the potential V(z). They used Feynman's path integral technique[34] to extend the classical expression into the quantum mechanical regime. The mean time <t(0,d)> spent in the region $0 \leq z \leq d$ by a quantum particle is obtained by averaging the classical functional $t_{0,d}^{cl}[z(.)]$ over all paths z(.) connecting (z_a,t_a) and (z_b,t_b):

$$\langle t(0,d)\rangle \equiv \overline{t_{0,d}^{cl}} \equiv g^{-1} \int_{z_a,t_a}^{z_b,t_b} Dz(\cdot) t_{0,d}^{cl}[z(\cdot)] \exp(iS[z(\cdot)]/\hbar) \qquad (11)$$

$$g \equiv \int_{z_a,t_a}^{z_b,t_b} Dz(\cdot) \exp(iS[z(\cdot)]/\hbar) \qquad (12)$$

where

$$S[z(\cdot)] = \int_{t_a}^{t_b} dt \left(\frac{m}{2}\dot{z}(\cdot)^2 - V(z(\cdot))\right) \qquad (13)$$

is the action. Let the potential $V(z)$ be changed by a very small amount $\Delta V(z) = \Delta V \Theta(z) \Theta(d-z)$ with the ΔV on the right independent of z. Since g is a functional of $V(z)$, the first order in ΔV change, Δg, in g can be calculated in two ways. In terms of the functional derivative $\delta g[V(z)]/\delta V(z)$ the result is

$$\Delta g = \Delta V \int_0^d dz \, \frac{\delta g[V(z)]}{\delta V(z)} \quad . \qquad (14)$$

Alternatively, expansion of the exponential in equation (12) to first order in ΔV gives

$$\Delta g = -\frac{i}{\hbar} \Delta V \int_{z_a,t_a}^{z_b,t_b} Dz(\cdot) \left(\int_{t_a}^{t_b} dt \Theta(z(\cdot)) \Theta(d-z(\cdot))\right) \exp(\frac{i}{\hbar} S[z(\cdot)]) . \qquad (15)$$

Equating (14) and (15) and noting that (15) is proportional to $\langle t(0,d)\rangle$, as defined by (11), immediately gives

$$\langle t(0,d)\rangle = i\hbar \int_0^d \frac{\delta \ln g[V(z)]}{\delta V(z)} dz \quad . \qquad (16)$$

For the one dimensional tunneling problem of interest here Sokolovski and Baskin specified the energy of the incident particle by letting $z_a \to -\infty$ and $t_b - t_a \to \infty$ with $|z_a|/(t_b - t_a) = v = \hbar k/m$. For transmission $z_b > d$ and for reflection $z_b < 0$ with z_b finite. Straightforward application of the method of stationary phase to the evaluation of the propagator

$$g(z_b,z_a;t_b,t_a) = \int_{-\infty}^{\infty} dk \Psi_k(z_b) \Psi_k^*(z_a) \exp(-\frac{i\hbar k^2}{2m}(t_b - t_a)) \qquad (17)$$

finally gives the Sokolovski-Baskin transmission and reflection times

$$\tau_T^{SB}(k;0,d) \equiv \langle t(0,d)\rangle_T = i\hbar \int_0^d \frac{\delta \ln T[V(z)] dz}{\delta V(z)} \qquad (18a)$$

$$\tau_R^{SB}(k;0,d) \equiv \langle t(0,d)\rangle_R = i\hbar \int_0^d \frac{\delta \ln R[V(z)] dz}{\delta V(z)} \qquad (18b)$$

which are, in general, complex quantities.

Sokolovski and Baskin also proved that their complex transmission and reflection times satisfy equation (1) and hence are consistent with the dwell time $\tau_D(k;0,d)$. Since the only factor in g, i.e. T or R, to survive the logarithmic functional derivative in (16) is independent of time they conclude that the energy-time uncertainty relation applies only to the time of arrival at or departure from the barrier but not to the mean time spent in the barrier region. They also point out that their generalization of $t_{0,d}^{cl}$ to the quantal regime is not unique except in the semiclassical limit. However, if one asserts that Büttiker's expression, equation (6), uniquely specifies the dwell time then application of the consistency relation (1) may eliminate other possibilities.

It should be noted that the Sokolovski-Baskin analysis is readily generalized to the region $z_1 \leq z \leq z_2$ within the barrier simply by replacing 0 by z_1 and d by z_2. It is clear from the resulting generalization of equations (18a) and (18b) that the local Sokolovski-Baskin transmission and reflection times are additive in the sense of equation (3).

The Sokolovski-Baskin approach can be generalized to obtain information on the distribution of transmission and reflection times.[35] For example, consideration of the second order in ΔV change in g leads to

$$<t(0,d)^2>_T = \frac{(i\hbar)^2}{T} \int_0^d dz' \int_0^d dz'' \frac{\delta^2 T[V(z)]}{\delta V(z')\delta V(z'')} . \quad (19)$$

A measure of the relative width of the distribution $P_T(t_{0,d})$ of complex transmission times is

$$\Delta \tau_T / \tau_T \equiv$$

$$\left\{ \frac{[\text{Re}<t(0,d)^2>_T - \text{Re}(<t(0,d)>_T)^2]^2 + [\text{Im}<t(0,d)^2>_T - \text{Im}(<t(0,d)>_T)^2]^2}{[\text{Re}<t(0,d)>_T]^2 + [\text{Im}<t(0,d)>_T]^2} \right\}^{\frac{1}{2}} . \quad (20)$$

Results for $\Delta \tau_T / \tau_T$ are presented in the next section.

4. The Local Larmor Clock

Two decades ago Baz' introduced the Larmor clock approach for the calculation of collision times.[5] He investigated the scattering of a spherically symmetric monoenergetic beam of spin $\frac{1}{2}$ particles incident on a central potential $V(r)$ of finite range $r=r_0$. To determine the mean time $<t(0,R_0)>$ spent by a particle within a distance $R_0 \geq r_0$ of the scattering centre at $r=0$ he imagined that an infinitesimal uniform magnetic field $\vec{B} = B\Theta(R_0-r)\hat{x}$ was present in the spherical region $r \leq R_0$ and that the particles in the incident beam were completely spin-polarized in the y direction. He defined $<t(0,R_0)>$ as $<\psi>/\omega_L$ where $<\psi>$ is the average precession angle in the y-z plane and ω_L the Larmor frequency. He also showed that $<t(0,R_0)^2> = <t(0,R_0)>^2$ and hence that the distribution of collision times $t(0,R_0)$ is a delta function. This "remarkable result" implies that all the particles in the incident beam spend precisely the same length of time in the spherical shell

$R_0 \leqslant r \leqslant R_0 + \Delta R_0$ for any $R_0 \gg r_0$ and $\Delta R_0 > 0$.

Rybachenko[6] applied the idea of Baz' to derive mean Larmor precession transmission and reflection times $\tau_{zT}(k;0,d)$ and $\tau_{zR}(k;0,d)$ for particles incident on a one-dimensional barrier $V(z)\Theta(z)\Theta(d-z)$. Büttiker[10], for the special case of a rectangular barrier, included the spin-dependent Zeeman shift of the effective barrier height and introduced the additional times $\tau_{xT}(k;0,d)$ and $\tau_{xR}(k;0,d)$ associated with the resulting spin-polarization of the transmitted and reflected beams in the field direction. He identified the transmission and reflection times $\tau_T(k;0,d)$ and $\tau_R(k;0,d)$ as defined in the introduction, with the quantities $[\tau_{xT}^2(k;0,d) + \tau_{zT}^2(k;0,d)]^{\frac{1}{2}}$ and $[\tau_{xR}^2(k;0,d) + \tau_{zR}^2(k;0,d)]^{\frac{1}{2}}$ respectively.

Generalization of Büttiker's analysis[31,32] to any region $z_1 \leqslant z \leqslant z_2$ within an arbitrary barrier $V(z)\Theta(z)\Theta(d-z)$ requires confining the infinitesimal uniform magnetic field to the region of interest, i.e.

$$\vec{B} = B\Theta(z-z_1)\Theta(z_2-z)\hat{x} \qquad (0 \leqslant z_1 \leqslant z \leqslant z_2 \leqslant d). \qquad (21)$$

Since the magnetic field changes the effective barrier potential in the region $z_1 \leqslant z \leqslant z_2$ by $\mp\hbar\omega_L/2$ for electrons with $S_x = \pm\hbar/2$ there is, in general, differential transmission of spin-up ($S_x = +\hbar/2$) and spin-down ($S_x = -\hbar/2$) electrons. Hence the transmitted and reflected beams associated with an incident monoenergetic beam completely spin-polarized in the y direction ($\langle S_x \rangle_I = 0$, $\langle S_y \rangle_I = +\hbar/2$, $\langle S_z \rangle_I = 0$) are, in general, spin-polarized in the field direction:

$$\langle S_x \rangle_T = \frac{\hbar}{2} \frac{|T+|^2 - |T-|^2}{|T+|^2 + |T-|^2}; \quad \langle S_x \rangle_R = \frac{\hbar}{2} \frac{|R+|^2 - |R-|^2}{|R+|^2 + |R-|^2}. \qquad (22a)$$

Larmor precession leds to spin-polarization of the transmitted and reflected beams in the tunneling direction:

$$\langle S_z \rangle_T = i\frac{\hbar}{2} \frac{(T+)(T-)^* - (T+)^*(T-)}{|T+|^2 + |T-|^2}; \quad \langle S_z \rangle_R = i\frac{\hbar}{2} \frac{(R+)(R-)^* - (R+)^*(R-)}{|R+|^2 + |R-|^2}. \qquad (22b)$$

In (22a) and (22b) T± and R± are the transmission and reflection probability amplitudes for spin-up (+) and spin-down (-) electrons.

$$\langle \vec{S} \rangle \equiv \tilde{\Psi}^+(z)\vec{\sigma}\tilde{\Psi}(z)/\tilde{\Psi}^+(z)\tilde{\Psi}(z)$$

is the local average spin per electron ($\tilde{\Psi}$ is the two component Pauli spinor wavefunction and $\vec{\sigma} \equiv (\sigma_x, \sigma_y, \sigma_z)$ with σ_x, σ_y and σ_z the 2 × 2 Pauli matrices). In the limit of vanishingly small field these quantities are used to define local characteristic electron-barrier interaction times for transmitted and reflected electrons:

$$\tau_{xT(R)}(k;z_1,z_2) \equiv \lim_{\omega_L \to 0} (\hbar\omega_L/2)^{-1} \langle S_x \rangle_{T(R)}, \qquad (23a)$$

$$\tau_{zT(R)}(k;z_1,z_2) \equiv -\lim_{\omega_L \to 0} (\hbar\omega_L/2)^{-1} \langle S_z \rangle_{T(R)}. \qquad (23b)$$

Expansion of T± and R± about their zero field values to linear order in ω_L leads upon substitution of 22a(b) into 23a(b) to explicit expressions for the local times. To obtain compact expressions it is convenient to introduce the auxiliary barrier potential

$$\tilde{V}(z) \equiv V(z) + \Delta V\ \Theta(z-z_1)\Theta(z_2-z) \tag{24}$$

with ΔV independent of z and the complex transmission and reflection times

$$\tau_T^V(k;z_1,z_2) \equiv i\hbar\ \left(\frac{\partial \ln T}{\partial \Delta V}\right)\bigg|_{\Delta V=0} , \tag{25a}$$

$$\tau_R^V(k;z_1,z_2) \equiv i\hbar\ \left(\frac{\partial \ln R}{\partial \Delta V}\right)\bigg|_{\Delta V=0} . \tag{25b}$$

In terms of these complex times

$$\tau_{xT(R)}(k;z_1,z_2) = -\ \mathrm{Im}[\tau_{T(R)}^V(k;z_1,z_2)] , \tag{26a}$$

$$\tau_{zT(R)}(k;z_1,z_2) = \mathrm{Re}[\tau_{T(R)}^V(k;z_1,z_2)] . \tag{26b}$$

It is easy to prove, using the properties of functional derivatives, that

$$\tau_{T(R)}^V(k;z_1,z_2) = \tau_{T(R)}^{SB}(k;z_1,z_2) . \tag{27}$$

Hence the transmission and reflection times τ_T^V and τ_R^V satisfy equations (1) to (3). Since they are complex quantities (4) and (5) cannot, in general, be applied.

The (generalized) Büttiker-Landauer expressions [9-11] for the transmission and reflection times are

$$\tau_{T(R)}^{BL}(k;z_1,z_2) = |\tau_{T(R)}^V(k;z_1,z_2)| . \tag{28}$$

These quantities have the desirable properties of being real and non-negative but they do not, in general, satisfy equations (1) to (3). Because they are not additive in the sense of equation (3) it is possible to have $|\tau_T^V(k;z_1,z_3>z_2)| < |\tau_T^V(k;z_1,z_2)|$.

Since $\tau_D(k;z_1,z_2)$ is real, the real parts of $\tau_T^V(k;z_1,z_2)$ and $\tau_R^V(k;z_1,z_2)$ obviously satisfy equations (1) to (3). However, there are strong grounds for not identifying $\mathrm{Re}[\tau_T^V]$ with the actual transmission time τ_T. For example, $\mathrm{Re}[\tau_T^V]$ is virtually independent of barrier thickness d for an opaque ($\kappa d \gg 1$) rectangular barrier.[10] Hence, for sufficiently large d the mean transmission speed for the entire barrier, $v_T = d/\mathrm{Re}[\tau_T^V(k;0,d)]$, is greater than the speed of light c. For the typical metal electrode parameters $k=\kappa=1$ Å^{-1} this occurs for $d \gtrsim 260$ Å, far beyond the practical tunneling regime (d=O(10Å)). The situation is much worse, however, if one considers the local transmission speed, $v_T(z) = (\partial \text{Re}[\tau_T^V(k;0,z)]/\partial z)^{-1}$. The dependence of $\text{Re}[\tau_T^V(k;0,z)]$ on z is extremely small in the middle of an opaque rectangular barrier (see Fig. 3, below). Consequently, for $k = \kappa = 1$ Å^{-1}, $v_T(z=d/2) > c$ for $d \gtrsim 6$Å.[32] These difficulties do not disappear when the calculations are repeated with the Dirac equation.[36]

Analytic expressions are readily obtained for the rectangular barrier $V_0\Theta(z)\Theta(d-z)$.[32] Fig. 1 shows the dependence of the real and imaginary parts of $\tau_T^V(k;0,d)$ on incident energy E for V_0= 10eV and d = 2,4 and 8 Å. In the limit that E approaches zero $\text{Re}[\tau_T^V(k;0,d)] \propto E^2$ while $\text{Im}[\tau_T^V(k;0,d)] \propto E^0$. As noted above $\text{Re}[\tau_T^V(k;0,d)]$ shows negligible dependence on d in the deep tunneling regime $\kappa d >> 1$. It is $-\text{Im}[\tau_T^V(k;0,d)]$ that shows a more "reasonable" dependence on d.

Fig. 2 shows for the barrier parameters of Fig. 1, the quantity $\Delta\tau_T/\tau_T$ defined by equation (20) as a function of incident energy E. It is seen that deep in the tunneling regime this measure of the relative width of the distribution of complex transmission times decreases with increasing opacity of the barrier, i.e. both with increasing d and decreasing E. Above the barrier $\Delta\tau_T/\tau_T$ can be large for a thick barrier when E is close to V_0. After smoothing of the oscillations there is a general decrease with increasing E so that for $E >> V_0$, where the oscillatory behaviour of |T| is no longer important, $\Delta\tau_T/\tau_T$ is again very small.

The transition probability amplitude T is zero for the potential barrier $V(z) = V_0\Theta(z)\Theta(a-z) + V_\infty\Theta(z-a)\Theta(d-z)$ with $V_\infty = \infty$. Hence this barrier provides a one dimensional analogue of the three-dimensional scattering problem considered by Baz'.[5] The incident and reflected plane waves exp(ikz) and Rexp(-ikz) of the former correspond to the incoming and outgoing spherical waves $r^{-1}\exp(-ikr)$ and $-Sr^{-1}\exp(ikr)$ of the latter. Since $|R| = 1$, it follows immediately from equation (18b) that $\text{Im}<t(0,d)>_R=0$. A little more effort leads to the one-dimensional analogue of the 'remarkable result' due to Baz', namely that $\text{Re}<t(0,d)^2>_R = [\text{Re}<t(0,d)>_R]^2$. Hence, if we completely ignore the complex nature of t(0,d), the distribution of reflection times is a delta function, i.e. $P_R(t(0,d)) = \delta(t(0,d)-<t(0,d)>_R)$. This result is remarkable in the sense that it is what one would expect in the classical regime. However, consideration of the complex nature of t(0,d) restores the expected quantum mechancial indefiniteness: the relative width of $P_R(t(0,d))$ then involves comparison of $\text{Re}<t(0,d)>_R$ and $\text{Im}<t(0,d)>_R$ with $\text{Re}[<t(0,d)^2>_R]^{\frac{1}{2}}$ and $\text{Im}[<t(0,d)^2>_R]^{\frac{1}{2}}$, respectively, not with $[\text{Re}<t(0,d)^2>_R]^{\frac{1}{2}}$ and $[\text{Im}<t(0,d)^2>_R]^{\frac{1}{2}}$. Since $\text{Im}<t(0,d)^2>_R \neq 0$ it follows that in general $\text{Re}<t(0,d)>_R \neq \text{Re}[<t(0,d)^2>_R^{\frac{1}{2}}]$ and $P_R(t(0,d))$ is not a delta function.

Returning to the (isolated) rectangular barrier, Fig. 3 shows the real and imaginary parts of $\tau_T^V(k;0,z)$ as a function of z for $0 \leq z \leq d$ with d = 10 Å, V_0= 10 eV and E = 2.5, 5.0 and 7.5 eV. The results for $\text{Re}[\tau_T^V(k;0,z)] = \tau_{zT}(k;0,z)$ show that for an opaque rectangular barrier Larmor precession is an edge effect: the transverse magnetic field inside the barrier has negligible effect on $<S_z>_T$ unless it extends to

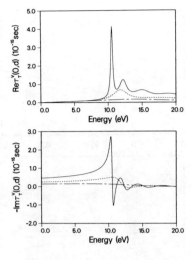

Figure 1. The energy dependence of the real and imaginary parts of the Larmor clock transmission time $\tau_T^V(k;0,d)$ for the rectangular barrier $V_0\Theta(z)\Theta(d-z)$ with V_0 = 10eV and d = 2 (— — — —), 4(— — — —), and 8Å (————).

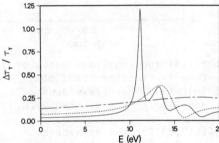

Figure 2. The relative width $\Delta\tau_T/\tau_T$ for the transmission time distribution $P_T(t_{0,d})$ as a function of incident energy for the rectangular barrier $V_0\Theta(z)\Theta(d-z)$ with V_0=10eV and d = 2 (— — — —), 4 (——————), and 8Å (————).

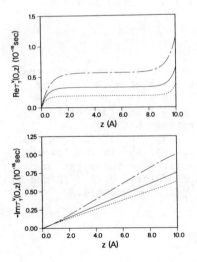

Figure 3. The real and imaginary parts of $\tau_T^V(k;0,z)$ as a function of z for the rectangular barrier $V_0\Theta(z)\Theta(d-z)$ with V_0=10eV, d=10Å and incident energy E=2.5 (——————), 5.0 (————) and 7.5 eV (— — — —) The corresponding tunneling decay lengths are κ^{-1}= 0.71, 0.87 and 1.23Å respectively.

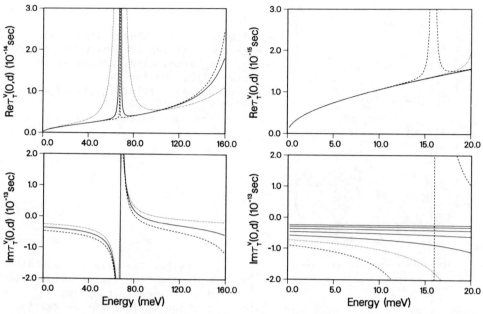

Figure 4. The energy dependence of the real and imaginary parts of $\tau_T^V(k;0,d)$ for the symmetric double-rectangular-barrier (DRB) of height $V_o=0.16$eV, well width =50Å and individual barrier widths $d_1=d_2=a=$ 50 (......), 100 (———), and 150Å (------). $m^*=0.067m$.

Figure 5. The energy dependence of $\tau_T^V(k;0,d)$ for the symmetric DRB with $V_o=0.16$eV, $d_1=d_2=100$ Å, and $w=0, 25, 50, 75, 100$ Å (———), 125 Å (......) and 150 Å (------). $m^*=0.067m$. In the bottom panel the results for $w=0$ to 100 Å are ordered in terms of increasing w from top to bottom.

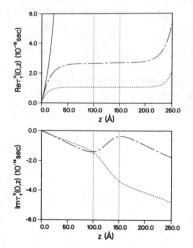

Figure 6. The z dependence of the real and imaginary parts of $\tau_T^V(k;0,z)$ for the symmetric DRB with $V_o=0.16$eV, $d_1=d_2=100$ Å, $w=50$ Å and for three incident energies: $E=E_r^{(0)} \approx 0.06769$eV (———), $E=E_r^{(0)}/2$ (------), $E=3E_r^{(0)}/2$ (......). $E_r^{(0)}$ is the lowest resonance energy. The tunneling decay length κ^{-1} is 21 Å for $E = E_r^{(0)}/2$ and 31 Å for $E = 3E_r^{(0)}/2$. $m^*=0.067m$.

within a tunneling decay length κ^{-1} of either interface at $z=0$ or $z=d$. (For reflection, it is only the interface at $z=0$ that matters). This is entirely consistent with the fact that $\text{Re}[\tau_T^V(k;0,d)]$ is virtually independent of d when $\kappa d \gg 1$.

It is instructive to apply the Larmor clock approach to the symmetric double-rectangular-barrier (DRB)

$$V_0[\Theta(z)\Theta(a-z)+\Theta(z-b)\Theta(b+a-z)] \tag{29}$$

of individual barrier widths $d_1=d_2=a$, well width $w=b-a$ and total width $d=b+a$. As shown in Fig. 4 and Fig. 5, many of the qualitative features noted above for the single-rectangular-barrier (SRB) apply to the DRB well off resonance, even though the barrier parameters are very different. $\text{Re}[\tau_T^V(k;0,d)]$ is proportional to $E^{\frac{1}{2}}$ as E goes to zero and is virtually independent of the individual barrier and well widths; $\text{Im}[\tau_T^V(k;0,d)]$ remains finite as E approaches zero and shows a more "reasonable" dependence on $d_1=d_2$, w and d. Fig. 6 shows that $\text{Re}[\tau_T^V(k;0,z)] \equiv \tau_{zT}(k;0,z)$ changes very slowly with z except when z is within a tunneling decay length κ^{-1} of the outside barrier edges at $z=0$ and $z=d$.

For a symmetric DRB exactly on resonance $|T|^2=1$ and $\text{Im}[\tau_T^V(k;z_1,z_2)]=0$. This special case is important for three reasons: (1) $\tau_T^V(k;z_1,z_2)$ can be studied for a non-trivial system without the complication of it being complex; (2) comparison with the real and imaginary parts of the same quantity for the corresponding (isolated) SRB actually sheds light on that complication; (3) it provides examples of local average transmission speeds $v_T(z)$ very much in excess of the speed of light.

Focusing on the well region $a \leqslant z \leqslant b$, Fig. 7 shows the local Larmor transmission time $\tau_T^V(k;a,z)$ and inverse speed $v_T^{-1}(z)$ for $E=E_r^{(0)}$, the lowest resonance energy of the DRB with $V_0=0.16\text{eV}$, $d_1=d_2=100$ Å and $w=50$ Å (assuming an effective mass m^* of $0.067m$). $\tau_T^V(k;a,z)$ increases most rapidly and $v_T^{-1}(z)$ is a maximum when z is in the centre of the well where the probability of finding a trapped electron is largest. The trapped electron makes very many excursions into the barrier regions before its eventual escape through the second barrier. This is evident in Fig. 8 which shows the local Larmor transmission time $\tau_T^V(k;0,z)$ for the first barrier. Thinking in terms of Feynman trajectories, one might expect important contributions to the mean time spent in the first barrier from two very different types of trajectories: 1) those in which the incident particle tunnels completely through the first barrier into the well region and 2) those in which the particle oscillating back and forth in the well repeatedly tunnels back into the first barrier. Despite the small average penetration depth, $(2\kappa)^{-1}$, the latter type of trajectory completely dominates in Fig. 8 because of the very large average number of oscillations before escape. In order to see the contribution to $\tau_T^V(k;0,z)$ from the former type of trajectory it is necessary to take advantage of the exponential fall-off with $a-z$ of the latter type and to focus on that part of the first barrier furthest from the well, i.e. $0 \leqslant z \ll a$. Intuitively, one might also expect that $\tau_T^V(k;0,0 \leqslant z \ll a)$ is closely related to the same quantity for the SRB $V(z) = V_0\Theta(z)\Theta(a-z)$ at the same incident energy. However, $\tau_T^V(k;0,z)$ for the single barrier is a complex quantity. In Fig. 9, the real quantities $\tau_T^V(k;0,z)$ and $v_T^{-1}(z)$ for the resonant symmetric DRB of Fig. 7 and Fig. 8 are compared at very small z with the real and imaginary parts of the same quantities for the corresponding SRB. A similar comparison is made in Fig. 10 for a different set of DRB parameters. In both cases, for very small z the real transmission time $\tau_T^V(k;0,z)$ for the symmetric DRB at resonance merges with $-\text{Im}[\tau_T^V(k;0,z)]$ for the corresponding SRB rather than with $\text{Re}[\tau_T^V(k;0,z)]$. This supports Büttiker and Landauer's claim that for an opaque rectangular barrier it is τ_{xT} (i.e. $-\text{Im}[\tau_T^V]$) rather than τ_{zT} (i.e. $\text{Re}[\tau_T^V]$) that makes the dominant contribution to the actual transmission time. However, the results do not support their identification of $[\tau_{xT}^2 + \tau_{zT}^2]^{\frac{1}{2}}$ (i.e. $|\tau_T^V|$) with τ_T.

Fig. 11 shows the local Larmor clock transmission time and inverse speed for the well region of a symmetric DRB when the incident energy coincides with $E_r^{(1)}$, the resonant level with one quasi-node in the well. $\tau_T^V(k;a,a\leqslant z\leqslant b)$ increases most slowly with z in the centre of the well where the tunneling probability of finding a particle is a minimum. Because the tunneling current is finite the wavefunction does not have an actual node anywhere and hence $v_T^{-1}(z)$ is never zero. However, for the case shown, in the immediate vicinity of the point at which $|\Psi_k(z)|^2$ is a minimum $v_T^{-1}(z)$ is so small that $v_T(z)$ is much greater than the speed of light. Fortunately, it is not difficult to prove that $v_T(z) \leqslant c$ for a symmetric double-barrier at resonance when the calculation is carried out with the Dirac equation instead of the Schrödinger equation.[36]

An interesting question is "How much time, on average, does an incident electron of energy E that is finally reflected spend in the region $z_1 \leqslant z \leqslant z_2$ on the far side $(z_1 > a)$ of an isolated rectangular barrier $V_0\Theta(z)\Theta(a-z)$?" Intuitively one might think that the result must be zero. The Larmor clock approach leads to a complex reflection time $\tau_R^V(k;z_1,z_2)$, the (oscillatory) real and imaginary parts of which are exponentially small ($\propto \exp[-2\kappa a]$), but finite, for an opaque barrier. Does this signal a breakdown of the Larmor clock approach or provide a simple example of quantum non-locality? This question is currently under investigation.

5. The Wave Packet Approach Of Hauge, Falck, And Fjeldly

Recently Hauge, Falck and Fjeldly[17] proposed "the proper basis for the calculation of transmission and reflection times for wave packets scattered off arbitrary tunneling structures in one dimension". Central to this approach are their exact results for the limiting time dependences of the centroids $\bar{z}(t)$ of the incident, transmitted, and reflected wave packets:

$$\bar{z}_I(t) = z_0 + \frac{\hbar}{m} <k>_I t \qquad (t \to 0) \quad , \qquad (30a)$$

$$\bar{z}_T(t) = z_0 + \frac{\hbar}{m} <k>_T t - <d\phi_T/dk>_T \qquad (t \to \infty) \quad , \qquad (30b)$$

$$\bar{z}_R(t) = -z_0 - \frac{\hbar}{m} <k>_R t + <d\phi_R/dk>_R \qquad (t \to \infty) \quad , \qquad (30c)$$

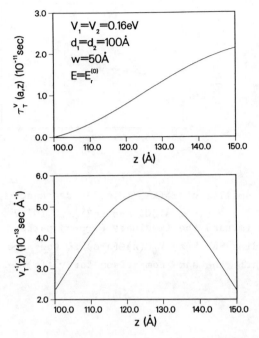
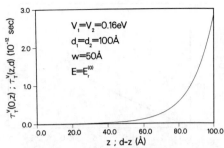

Figure 8. The z dependence of $\tau_T^V(k_r^{(0)};0,z)$ at resonance for the first barrier $0 \leqslant z \leqslant a$ of the symmetric DRB with $V_0=0.16$eV, $d_1=d_2=100$Å and $w=50$Å. The incident energy is $E_r^{(0)}$ and $m^*=0.067m$.

Figure 7. The z dependence of $\tau_T^V(k_r^{(0)};a,z)$ and $v_T^{-1}(z)$ at resonance for the well region $a \leqslant z \leqslant b$ of the symmetric DRB with $V_0=0.16$eV, $d_1=d_2=100$Å and $w=50$Å. The incident energy is $E_r^{(0)} \cong 0.06769$eV. $m^*=0.067m$.

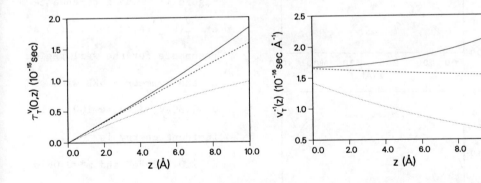

Figure 9. Left: a comparison for small z of $\tau_T^V(k_r^{(0)};0,z)$ at resonance for the symmetric DRB of Fig. 8 (———) with the real (......) and (minus) the imaginary (------) parts of $\tau_T^V(k=k_r^{(0)};0,z)$ for the corresponding SRB $V(z)=V_0\Theta(z)\Theta(a-z)$ at the same incident energy (i.e. $E=E_r^{(0)}$). Right: the same comparison for $v_T^{-1}(z)$.

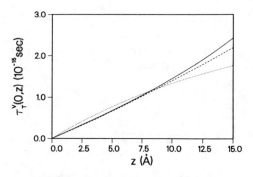
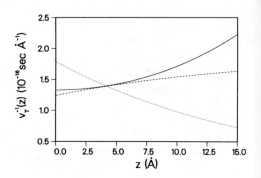

Figure 10. Left: a comparison for small z of $\tau_T^V(k_r^{(1)};0,z)$ at resonance for the symmetric DRB with $V_0=0.16$eV, $d_1 = d_2 = w = 100$Å and $E=E_r^{(1)} \cong 0.10797$ eV (———) with the real (·····) and (minus) the imaginary (-----) parts of $\tau_T^V(k=k_r^{(1)};0,z)$ for the corresponding SRB $V(z)=V_0\Theta(z)\Theta(a-z)$ at the same incident energy (i.e. $E=E_r^{(1)}$). Right: the same comparison for $v_T^{-1}(z)$.

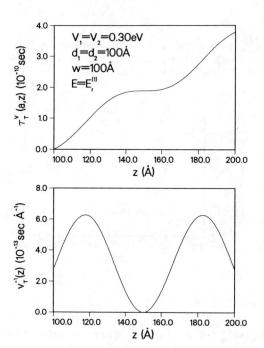

Figure 11. The z dependence of $\tau_T^V(k_r^{(1)};a,z)$ and $v_T^{-1}(z)$ at resonance for the well region of the symmetric DRB with $V_0=0.30$eV, $d_1=d_2=w=100$Å. The incident energy is $E_r^{(1)} \cong 0.13259$eV and $m^*=0.067m$.

respectively, where $z_0 \equiv \bar{z}_I(t=0)$, and ϕ_T and ϕ_R are the phases of the transmission and reflection probability amplitudes ($T \equiv |T|\exp i\phi_T$, $R \equiv |R|\exp i\phi_R$). The angular brackets denote

$$\langle \ldots \rangle \equiv \int_{-\infty}^{+\infty} \frac{dk}{2\pi} \ldots |\Phi(k)|^2 \quad ,$$

$$\langle \ldots \rangle_I \equiv \langle \ldots 1 \rangle / \langle 1 \rangle = \langle \ldots \rangle \quad ,$$

$$\langle \ldots \rangle_T \equiv \langle \ldots |T|^2 \rangle / \langle |T|^2 \rangle \quad , \quad (31)$$

$$\langle \ldots \rangle_R \equiv \langle \ldots |R|^2 \rangle / \langle |R|^2 \rangle \quad ,$$

with $\Phi(k) \equiv |\Phi(k)|\exp(i\xi(k))$ the Fourier transform of the wavefunction describing the wave packet at t=0. It is assumed that $|\Psi(z,t\leqslant 0)|^2$ is negligible for $z \geqslant 0$. The expressions of Hauge et al. have been simplified here by assuming that $d\xi(k)/dk = -z_0$, as is the case for the much used minimum-uncertainty-product wavepacket, so that the cross correlation[17] $\langle (k_0-k)d\xi/dk \rangle$ between k and z is zero. Hauge et al. extrapolated Eq. (30a) forward in time to determine the time $t(0)_I$ at which the centroid of the incident packet reaches the left edge of the barrier at z=0. They extrapolated (30b) and (30c) backwards in time to determine the times $t(d)_T$ and $t(0)_R$, respectively, at which the centroids of the transmitted and reflected packets leave the barrier region. They then identified the time differences $t(d)_T-t(0)_I$ and $t(0)_R-t(0)_I$ with the transmission and reflection times for scattering of wave packets off potential barriers:

$$\tau_T^{HFF}(0,d) \equiv t(d)_T - t(0)_I = \frac{m}{\hbar} \left[\frac{d - z_0 + \langle d\phi_T/dk \rangle_T}{\langle k \rangle_T} + \frac{z_0}{\langle k \rangle_I} \right], \quad (32a)$$

$$\tau_R^{HFF}(0,d) \equiv t(0)_R - t(0)_I = \frac{m}{\hbar} \left[\frac{-z_0 + \langle d\phi_R/dk \rangle_R}{\langle k \rangle_R} + \frac{z_0}{\langle k \rangle_I} \right], \quad (32b)$$

assuming that $d\xi(k)/dk = -z_0$. These equations are only implicit in Ref. 17 because early in the derivation the authors expanded all k-dependent quantities, except $|\Phi(k)|^2$ which was assumed to be sharply peaked near $k_0 \equiv \langle k \rangle_I$, about k_0 to second order in $(k-k_0)$. An attractive feature of the approach of Hauge et al. is that $\tau_T^{HFF}(0,d)$ and $\tau_R^{HFF}(0,d)$ can be obtained without solving the time-dependent Schrödinger equation resulting in an enormous saving in computer time. It should be noted that $\tau_T^{HFF}(0,d) < 0$ when $z_0 < -k_0[d+\langle d\phi_T/dk \rangle_T]/[\langle k \rangle_T - k_0]$. Hauge et al. showed that in the limit that the width, Δk, of $|\Phi(k)|^2$ goes to zero the classical phase times of Bohm[1] and Wigner[2] are obtained:

$$\lim_{\Delta k/k_0 \to 0} \tau_T^{HFF}(0,d) = \tau_T^{\phi}(k_0;0,d) \equiv \frac{m}{\hbar k_0} [d + (d\phi_T/dk)_{k=k_0}], \quad (33a)$$

$$\lim_{\Delta k/k_o \to 0} \tau_R^{HFF}(0,d) = \tau_R^\phi(k_o;0,d) \equiv \frac{m}{\hbar k_o}(d\phi_R/dk)_{k=k_o} \quad . \quad (33b)$$

This is consistent with the conclusion of Collins, Lowe, and Barker[25] who calculated the time evolution of minimum-uncertainty-product wave packets by explicit solution of the time-dependent Schrödinger equation. From their numerical solutions they accurately determined the time at which the peak in the transmitted packet left the barrier region but determined when the peak in the incident packet reached the barrier by linear extrapolation of the asymptotic trajectory.

Hauge et al.[17] obtained an important exact relation between Büttiker's expression, Eq. (6), for the dwell time and the phase times:

$$\tau_D(k;,0,d) = |T(k)|^2 \tau_T^\phi(k;0,d) + |R(k)|^2 \tau_R^\phi(k;0,d) + (m/\hbar k^2)|R(k)|\sin(\phi_R). \quad (34)$$

Comparison of this equation with Eq. (1) leads to the conclusion[33] that the classic phase times $\tau_T^\phi(k;0,d)$ and $\tau_R^\phi(k;0,d)$ cannot, in general, be identified with the long sought for transmission and reflection times defined in the introduction. The method of Hauge et al. fails because the centroid of the incident wave packet is extrapolated through the region of strong quantum mechanical interference between the incident and reflected components of $\Psi(z,t)$.

6. The Larmor Clock Approach Of Huang Et Al.

Huang et al.[22] repeated Büttiker's stationary-state analysis of the Larmor clock but allowed the infinitesimal uniform magnetic field to extend throughout all of space rather than confining it to the barrier region $0 \leq z \leq d$. Central to the analysis are the two z-dependent angles $\psi(z)$ and $\theta(z)$ specifying the orientation of the local average spin value $\langle \vec{S}(z) \rangle$. $\psi(z)$ is associated with spin-precession in the plane perpendicular to the magnetic field and $\theta(z)$ with rotation of $\langle \vec{S}(z) \rangle$ into the field direction.

For the spatially very extended but finite wavepackets that one approximates by planes waves in the stationary-state scattering approach, the angle ψ_I can be fixed and the angles ψ_T and ψ_R measured asymptotically where there is no interference between the incident and reflected components of the wavepacket (similarly for θ_I, θ_T and θ_R). When the magnetic field is confined to the barrier region the angles ψ and θ calculated for each of the incident, transmitted, and reflected components of the stationary-state wavefunction are independent of z for z outside the barrier. Hence one can use the asymptotic measurements to determine the Larmor clock times $\tau_{zT}(k;0,d)$ and $\tau_{xT}(k;0,d)$ without any knowledge of the precise distances of the points of measurement from the nearest barrier edges. On the other hand, when the magnetic field extends from $z = -\infty$ to $z = +\infty$ the angles ψ and θ calculated for each of the three components (I,T, and R) of the wavefunction depend on z for all z. Hence, the asymptotic measurements are useless unless one knows precisely (on the scale of the barrier width d) the distances to the edges of the barrier. Hence, Huang et al. calculate the Larmor clock precession and spin-rotation transmission times directly from the differences $[\phi_T(z=d)-\phi_I(z=0)]$ and $[\theta_T(z=d)-\theta_I(z=0)]$ respectively. In this case, $\phi_I(z=0)$ and $\theta_I(z=0)$ are required precisely in the region where interference between the incident and reflected components of the wavepacket cannot be ignored no matter what the extent of the packet. Because of this interference $\phi(z=0)$ and $\theta(z=0)$ cannot be partitioned between the incident and reflected components of the packet. Hence, it is not surprising that this prescription for evaluating $\tau_{zT}(k;0,d)$ leads to the phase time $v_T^\phi(k;0,d)$.

7. Concluding Remarks

Of the three tunneling times τ_D, τ_T and τ_R precisely defined above it is only the dwell time τ_D for which a satisfactory result[10] has been derived and there still remains the question of whether or not viable alternatives to Büttiker's result exist. None of the real transmission and reflection times τ_T and τ_R discussed above satisfy all of the criteria set out in Section 2. There are many possible responses to this conclusion, including the following: 1) Further effort will eventually yield real transmission and reflection times, perhaps some function of Büttiker's $\tau_{xT(R)}$ and $\tau_{zT(R)}$, that will satisfy all of the above criteria. 2) The criteria, not all of which are universally accepted, should be relaxed. 3) The precise definitions of τ_D, τ_T and τ_R should be modified or not interpreted so rigidly. 4) Quantum mechanical transmission and reflection times are in fact complex quantities, in which case a crucial question is whether the Sokolovski-Baskin (or the equivalent complex Larmor clock) times are unique in satisfying criteria (1) to (3). (Are there any other criteria to be satisfied by complex times?) 5) Transmission and reflection times are meaningless concepts within quantum mechanics because $|\Psi(z,t)|^2$, with z inside the barrier, cannot be separated into two parts one describing particles that will be transmitted and the other those that will be reflected, except for the special cases of perfect transmission or reflection. 6) There is a basic flaw in the time-dependent Schrödinger equation because, according to Peres[37], the operator $\partial/\partial t$ is not operationally well defined. (Does the Larmor clock actually circumvent the difficulties inherent in Peres' quantum clocks or is this an artifact of the treatment of the magnetic field as a fixed externally imposed quantity?)

REFERENCES

1. D. Bohm, Quantum Theory (Prentice-Hall, New York, 1951).
2. E.P. Wigner, Phys. Rev. 98, 145 (1955).
3. F.T. Smith, Phys. Rev. 118, 349 (1960).
4. T.E. Hartman, J. Appl. Phys. 33, 3427 (1962).
5. A.I. Baz', Sov. J. Nucl. Phys. 4, 182 (1967), 5, 161 (1967).
6. V.F. Rybachenko, Sov. J. Nucl. Phys. 5, 635 (1967).
7. M. Jonson, Solid State Commun. 33, 743 (1980).
8. K.W.H. Stevens, J. Phys. C 16, 3649 (1983).
9. M. Büttiker and R. Landauer, Phys. Rev. Lett. 49, 1739 (1982).
10. M. Büttiker, Phys. Rev. B 27, 6178 (1983).
11. M. Büttiker and R. Landauer, IBM J. Res. Develop. 30, 451 (1986).
12. E. Pollak and W.H. Miller, Phys. Rev. Lett. 53, 115 (1984).
13. E. Pollak, J. Chem. Phys. 83, 1111 (1985).
14. P. Hedegård, Phys. Rev. B 35, 533 (1987).
15. D. Sokolovski and L.M. Baskin, Phys. Rev. A 36, 4604 (1987).
16. D. Sokolovski and P. Hänggi, Europhys. Lett. 7, 7 (1988).
17. E.H. Hauge, J.P. Falck and T.A. Fjeldly, Phys. Rev. B 36, 4203 (1987).
18. J.P. Falck and E.H. Hauge, Phys. Rev. B 38, 3287 (1988).
19. N. Teranishi, A.M. Kriman and D.K. Ferry, Superlattices and Microstructures 3, 509 (1987).
20. Z. Kotler and A. Nitzan, J. Chem. Phys. 88, 3871 (1988).
21. W. Jaworski and D.M. Wardlaw, Phys. Rev. A 37, 2843 (1988).
22. Z. Huang, P.H. Cutler, T.E. Feuchtwang, R.H. Good, Jr., E. Kazes, H.Q. Nguyen, and S.K. Park, Journal de Physique C6, 17 (1988).
23. A.P. Jauho and M. Nieto, Superlattices and Microstructures 2, 401 (1986).
24. A. Tagliacozzo, Nuovo Cimento 10D, 363 (1988).
25. S. Collins, D. Lowe and J.R. Barker, J. Phys. C 20, 6213 (1987), 6233 (1987).
26. A.B. Nasser, Phys. Rev. A 38, 683 (1988).
27. W.R. Frensley, Phys. Rev. B 36, 1570 ('87).
28. W. Jaworski and D.M. Wardlaw, Phys. Rev. A 38, 5404 (88).
29. A.P. Jauho and M. Jonson, preprint.
30. J.A. Stovneng and E.H. Hauge, preprint
31. C.R. Leavens and G.C. Aers, Solid State Commun. 63, 1101 (1987).
32. C.R. Leavens and G.C. Aers, Solid State Commun. 67, 1135 (1988).
33. C.R. Leavens and G.C. Aers, Phys. Rev. B 39, 1202 (1989).
34. R.P. Feynman and A.R. Hibbs, Quantum Mechanics and Path Integrals (New York, McGraw-Hill, 1965).
35. C.R. Leavens, Solid State Commun. 68, 13 (1988); 68, ii (1988).
36. C.R. Leavens and G.C. Aers, submitted to Phys. Rev. B.
37. A. Peres, Am. J. Phys. 48, 553 (1980).

THEORY OF SCANNING TUNNELING MICROSCOPY AND SPECTROSCOPY

J. TERSOFF
IBM Research Division, T. J. Watson Research Center
Yorktown Heights, NY 10598 USA

1. Introduction

In the few years which have passed since its invention[1] by Binnig, Rohrer, and coworkers, scanning tunneling microscopy (STM) has established itself as a remarkably powerful and versatile tool for studying surfaces. This paper reviews the present theoretical understanding of STM, with emphasis on the interpretation of atomic-resolution STM images. Some familiarity with the basic principles of STM is assumed here; the basic ideas and instrumentation have been described in detail elsewhere.[1]

The primary task in any theoretical analysis of STM is to understand how the tunneling current J varies with voltage V and tip position $\vec{r}_t$, i.e. to determine the function $J(\vec{r}_t, V)$. Most (though not all) issues in the theory of STM reduce to determining this function. Some general aspects of the tunneling problem, and of the determination of $J(\vec{r}_t, V)$, are discussed in Section 2.

A function of two variables is already a bit complicated to consider, so it is convenient for discussion to consider separately the dependence of current on tip position at constant voltage, $J(\vec{r}_t)|_V$, and the dependence of current on voltage at fixed tip position, $J(V)|_{\vec{r}_t}$. These two functions determine the (voltage-dependent) STM image, and the (position-dependent) tunneling spectrum, respectively. The first of these, the STM image, is the subject of Section 3. The second, spectroscopy, is discussed in Section 4.

Because STM is, above all, a microscopy, Section 3 is the longest and most important section here. It includes sub-sections on (1) modes of imaging, (2) imaging of metals, (3) imaging of semiconductors, (4) imaging band-edge states, and (5) spatial resolution of STM.

Finally, Section 5 introduces an issue which is crucial for STM, but which is outside the scope of the tunneling problem. This is the effect of mechanical interactions between tip and surface on the operation of STM. Some concluding remarks are offered in Section 6.

2. General Tunneling Theory

2.1 Non-perturbative Treatment

The tunneling problem may be viewed as a special case of the more general problem, of the current-voltage characteristic of an interface. Unfortunately, the latter problem is extremely difficult in general. Therefore, it is desirable to take advantage of the fact that, in tunneling, the coupling between surface and tip is weak. Such a perturbative approach is discussed in Section 2.2 below.

Nevertheless, it is sometimes necessary to treat the problem non-perturbatively, in particular when the tip approaches very close to the surface. For example, Gimzewski and Möller[2] measured the resistance between tip and sample as the tip is brought down to the surface. They found that the resistance decreased exponentially with decreasing distance, as expected from simple arguments, until at very small distances the resistance reached a plateau, presumably where the tip and surface are touching. At this point, the interaction between surface and tip certainly cannot be assumed to be weak.

To understand these results, Lang[3] calculated the resistance between a model surface (jellium), and a model tip (a single atom adsorbed on jellium). (Jellium is simply a metal in which the positive ions are replaced by a uniform positive background.) The simplification of using jellium made it feasible to calculate the resistance "exactly", i.e. non-perturbatively, so the results are valid even when the surface and tip are in contact. The results describe the experiment very well, showing that the plateau corresponds to the point where the resistance is dominated by a one-atom contact, rather than a vacuum gap.

For real systems, however, the matching of wavefunctions across the interface becomes extremely difficult. Even for an atomically perfect planar interface, such a calculation represents a *tour de force*.[4] For the STM problem, non-perturbative calculations have not proven feasible to date except by using models which ignore the atomic character of the electrodes. Thus, while some early calculations[5] for STM were non-perturbative, this approach has now generally been abandoned except for cases where the tip and surface interact strongly.

2.2 Tunneling-Hamiltonian Treatment

Under typical tunneling conditions, the interaction between the two electrodes is sufficiently weak that it may be treated in first order perturbation theory, giving

$$J = \frac{2\pi e}{\hbar} \sum_{\mu,\nu} \{f(E_\mu)[1 - f(E_\nu)] - f(E_\nu)[1 - f(E_\mu)]\} |M_{\mu\nu}|^2 \delta(E_\nu + V - E_\mu), \quad (1)$$

where $f(E)$ is the Fermi function, V is the applied voltage (in units of energy, i.e. eV), $M_{\mu\nu}$ is the tunneling matrix element between states ψ_μ and ψ_ν of the respective electrodes, and E_μ is the energy of ψ_μ. For most purposes, the Fermi functions can be replaced by their zero-temperature values, i.e. unit step functions, in which case one of the two terms in braces becomes zero. In the limit of small voltage, this expression further simplies to

$$J = \frac{2\pi}{\hbar} e^2 V \sum_{\mu,\nu} |M_{\mu\nu}|^2 \delta(E_\mu - E_F)\delta(E_\nu - E_F). \tag{2}$$

These results are quite simple. The real difficulty is in evaluating the tunneling matrix elements. Bardeen[6] showed that, under certain assumptions, (regarding which see Section 2.4 below), the tunneling matrix element could be expressed as

$$M_{\mu\nu} = \frac{\hbar^2}{2m} \int d\vec{S} \cdot (\psi_\mu^* \vec{\nabla} \psi_\nu - \psi_\nu \vec{\nabla} \psi_\mu^*), \tag{3}$$

where the integral is over any surface lying entirely within the barrier region. If we choose a plane for the surface of integration, and neglect the variation of the potential in the region of integration, then the surface wavefunction at this plane can be conveniently expanded in the generalized planewave form

$$\psi = \int d\vec{q}\, a_q \exp[-(\kappa^2 + |\vec{q}|^2)^{1/2} z] \exp(i\vec{q}\cdot\vec{x}), \tag{4}$$

where z is measured from a convenient origin at the surface, $\kappa = \hbar^{-1}(2m\phi)^{1/2}$, and ϕ is the "local workfunction" (or, more precisely, the potential in the region of interest, relative to the Fermi level). A similar expansion applies for the other electrode, replacing a_q with b_q, z with $z_t - z$, and $\vec{x}$ with $\vec{x} - \vec{x}_t$. Here $\vec{x}_t$ and z_t are the lateral and vertical components of the position of the tip. Then, substituting these wavefunctions into (3), one obtains

$$M_{\mu\nu} = -\frac{4\pi^2\hbar^2}{m} \int d\vec{q}\, a_q b_q^* (\kappa^2 + |\vec{q}|^2)^{1/2} \exp[-(\kappa^2 + |\vec{q}|^2)^{1/2} z_t] \exp(i\vec{q}\cdot\vec{x}_t). \tag{5}$$

Thus, given the wavefunctions of the surface and tip separately, i.e. a_q and b_q, one has a reasonably simple expression for the matrix element and tunneling current. Note that $\phi \sim 4$ eV, so evaluating κ and its role in the matrix element (5), one finds that the current decreases by roughly an order of magnitude for each angstrom increase in the surface-tip separation.

2.3 Modeling the Tip

In order to calculate the tunneling current, and hence the STM image or spectrum, it is first necessary to have explicitly the wavefunctions of the surface and tip. Unfortunately, the actual atomic structure of the tip is generally not known.[7] Even if it were known, the very low symmetry would probably make accurate calculation of the tip wavefunctions infeasible.

One must therefore adopt a reasonable but somewhat arbitrary model for the tip. The most realistic model which has been used to date is that of Lang,[8] an atom adsorbed on jellium. However, to facilitate the treatment of real surfaces, an even more severe approximation for the tip is convenient. The simplest model which has been used for the tip is that of Tersoff and Hamann.[9] Because this model leads to a particularly simple interpretation of the STM image, it is worth describing in a little detail.

To motivate the simplest possible model for the tip, Ref. 9 considered what would be the ideal STM. First, one wants the maximum possible resolution, and therefore the smallest possible tip. Second, one wants to measure the properties of the bare surface, not of the more complex interacting system of surface and tip. Therefore, the ideal STM tip would consist of a mathematical point source of current. In that case, Equation (2) reduces to[9]

$$J \propto \sum_v |\psi_v(\vec{r}_t)|^2 \delta(E_v - E_F)$$
$$\equiv \rho(\vec{r}_t, E_F) \tag{6}$$

Thus the ideal STM would simply measure $\rho(\vec{r}_t, E_F)$ of the bare surface. This is a familiar quantity, being simply the local density of states at E_F, i.e. the charge density from states at the Fermi level, at the position of the tip. Thus within this model, STM has quite a simple interpretation.

It is important to see how far this interpretation can be applied for more realistic models of the tip. Ref. 9 showed that (6) remains valid, regardless of tip size, so long at the tunneling matrix elements can be adequately approximately by those for an s-wave tip wavefunction. The tip position r_t must then be interpreted as the center of the tip, i.e. the origin of the s-wave which best approximates the tip wavefunction.

Even for a realistic model of a one-atom tip, Lang verified[8] that the image corresponds quite closely to a contour of constant $\rho(\vec{r}_t, E_F)$, confirming the applicability of (6). Of course, the s-wave model must break down when there is appreciable tunneling to several tip atoms at once. However, in view of the reduced resolution, and the lesser accuracy which is consequently acceptable for the theory in such a case, (6) is probably still an adequate approximation, if r_t is interpreted[9] as an effective center of curvature of the tip.

2.4 Some Technical Issues

There are several points which can be extremely important in specific measurements or calculations, but which are not sufficiently fundamental to include under the other major headings of this paper. These miscellaneous issues are addressed in this Section below: the basis set and potential used in calculations of tunneling current; the validity of the Bardeen formula for the tunneling matrix elements; and the role of tip shape.

2.4.1 *Basis Set for Wavefunction Calculations.* In most calculations of electronic structure, one avoids the complexity of solving the Schrödinger equation explicitly as a differential equation, by working in a fixed finite basis set, and transforming the problem to one of diagonalizing the Hamiltonian matrix. Unfortunately, most such methods, though well adapted to problems of total energy and bandstructure, are quite inappropriate for the quantitative calculation of the STM current.

For STM, it is the tails of the wavefunctions at relatively large distances from the nuclei, perhaps 3-10 Å, which are important. These tails contribute little to the energy, so most methods treat them rather inaccurately in order to simplify the calculation.

In a planewave basis, though there is no problem in principle, in practice it is never feasible to accurately describe the wavefunction tails at large distances, because of the enormous number of basis functions required. Methods using a local-orbital basis often employ Gaussian orbitals, which behave unphysically at large distances. Even with Slater-type local orbitals, the exponential decay does not depend on the energy of the wavefunction, as it should. Thus these methods may be useful in qualitative comparisons with STM images, based on the calculated densities very near the surface; but the calculated behavior at larger distances is unreliable.

Use of such local orbitals as a basis can lead to particularly severe errors in a compound system, such as GaAs. There, the different atoms are described by orbitals with different decay rates. As a result, at sufficiently large distances, the atom whose basis function has the slower decay will always dominate the image. In contrast to this, the true wavefunction has a single decay rate for both Ga and As components, and either atom can dominate the image, depending on the tunneling voltage.[9] Similar problems can be anticipated for virtually any adsorbate system.

Thus it seems that the only methods which permit accurate calculation of STM images or currents at realistic distances, are those which explicitly integrate Schrödinger's equation to obtain the wavefunctions outside the surface. The only such method currently being used for calculations of real materials appears to be the linearized augmented planewave method (LAPW).[9] (Even this method involves a matrix formulation; but the basis set is energy-dependent, and is obtained by direct integration with the true potential in the vacuum region.)

The method used by Lang,[3,8] although confined at present to strictly free-electron substrates, also treats the wavefunctions correctly. Alternatively, if the potential is modeled as piecewise constant, one can match the exact solutions in each region to construct the wavefunction,[5] although obviously such a model is very restrictive.

2.4.2 *Sensitivity to Errors.* There are several approximations discussed below, which are routinely made in the calculation of STM images or spectra, and which introduce errors whose magnitude is not well known. It is therefore worth first making a general point here, concerning the importance (or unimportance) of such errors.

An approximation which leads to an error of, say, one order of magnitude in the current J, might be considered devastating at first glance. However, the net affect of such an error is that the tip is inferred to be about 1 Å closer or farther from the surface than it really is. This will typically have rather little effect on the image, and even that effect will be quite similar to that of having a slightly sharper or duller tip.[9] Such an error is thus not so important with respect to observable properties. In fact, any theoretical treatment of STM relies heavily on this insensitivity, since the exponential dependence on barrier height makes high absolute accuracy in the calculation of J unachievable in general.

The errors which are important, therefore, are those which show a strong dependence on tip position or on wavefunction. Fortunately, there is little reason to suppose that the sources of error discussed below show such a stong dependence. (In contrast, use of a poor basis set to describe the wavefunction, as discussed above, leads to errors which may depend strongly on the tip position and the particular wavefunction involved, and so lead to unacceptable distortions at realistically large distances.)

2.4.3 *Potential in the Barrier Region.* Besides the issue of the basis set, discussed above, one possible problem is the potential used in calculating the wavefunctions. Most solid-state calculations at present use the local density approximation (LDA) for correlation and exchange. This approximation does not give the long-ranged image potential correctly. If the error is small everywhere compared to the workfunction, then this problem is unimportant. But since the true potential is known only in the limit of large distance, there is no accurate way at present to quantitatively evaluate the errors involved. At least, we may assume that the error does not depend sensitively on the lateral tip position, since the tip is essentially a constant distance from the surface.

When the tip and surface are brought close together, their potentials will overlap. This effect will lead to errors if one uses the wavefunctions calculated for the bare tip and surface. Again, because of the constant distance between tip and surface, we may expect that the resulting error will be a constant multiplicative factor for the current, and therefore relatively unimportant as discussed in Section 2.4.2 above.

Finally, for voltages which are not very small compared to the workfunction, one really needs to know the wavefunction in the presence of the applied electric field, as discussed below.

2.4.4 *Accuracy of the Bardeen Formula.* There are also questions regarding the accuracy of the Bardeen formula (3) for the matrix elements. The derivation[6] assumes that the potential in the barrier is identical to that outside the bare surface and tip. In practice this is never exactly true, but once again, even if the absolute error is large, there is no reason to believe that it leads to significant errors in images or spectra.

Also, the validity of using first-order theory may break down for surface states. This has been analysed in one dimension by Noguera.[10] A simple way to picture the problem is that the first-order theory assumes that the electronic population is always in equilibrium. However, electrons added to a surface state cannot in principle reach equilibrium with the bulk, at least within one-electron theory in one dimension and at zero temperature. Therefore, such states cannot contribute to the current, in contradiction to the first-order theory.

In three dimensions, this problem disappears for metals, where there are not true gaps. (There may be gaps in the wavevector-projected bandstructure, but wavevector is not exactly conserved, because of defects and even because of the tip-surface interaction.)

The problem however remains for states which fall in the band gap of a semiconductor. Including electron-electron and electron-phonon interactions might lead to sufficiently rapid equilibration to alter the conclusions of Ref. 10, but this question is beyond the scope of a review.

2.4.5 *Shape of the STM Tip.* The shape of the STM tip may also play a role beyond what has been considered so far here. For example, at the edge of very deep structures, tunneling to the side of the tip may become important, so that the profile of the tip becomes a limitation in one's ability to observe the profile of a large step or hole in the surface.

A distressingly common problem is the occurence of multiple distinct tunneling sites, i.e. multiple tips. This effect has been noted by several authors,[11] and seen (but never discussed) by many others. The result is a nonlinear superposition of two or more images, which preserves the periodicity of the surface (unless the tips are on different sides of a grain boundary or such). Such images can easily lead to confusion. Tip-to-tip reproducibility of images is therefore an essential requirement in STM. Also, the presence of a point defect in the image makes such artifacts much easier to spot.

3. STM Images and Their Interpretation

3.1 Modes of Imaging

In its most general form, STM measures the function $J(\vec{r}_t, V)$. While there is great interest in measuring the dependence on r and V simultaneously,[12,13] it is usually more convenient to consider $J(\vec{r}_t)|_V$, i.e. the tunneling current as a function of tip position, at fixed voltage.

Even this simplification, however, is not enough, since it is not generally practical to map out the full three-dimensional function. The elegant approach of Binnig et al.[1] was to map out the two-dimensional surface $z(\vec{x})$ defined implicitly by the condition $J(\vec{r}_t) = J_f$, where z and $\vec{x}$ are the vertical and lateral components of the tip position $\vec{r}_t$. A feedback circuit is used to constantly adjust the tip height z, as x and y are varied, so as to maintain a constant current J_f.

This constant-current mode of imaging has two crucial advantages over any other approach. One is that this mode avoids the need to measure currents varying over orders of magnitude. (The current changes by roughly one order of magnitude for every angstrom of vertical motion of the tip relative to the surface.)

The second advantage is that the resulting image will, under certain conditions, correspond closely to a "topograph" of the surface.[1] Examples are given in Sections 3.2-3.4 below, both of cases where the image corresponds closely to a topograph, and cases where interpreting the image as a topograph would be grossly misleading.

It is sometimes convenient, with very flat surfaces, to operate in a "constant height" mode.[14] Often this mode is used when tunneling under "dirty" conditions, in air or some other fluid, especially with a graphite surface. While such studies can be useful, a *quantitative* interpretation of STM images under these conditions is not yet possible. One of the problems which may arise under these conditions is discussed in Section 5.

Because of the general superiority of the constant-current imaging mode for well-controlled quantitative investigations, only that mode is specifically considered in the rest of this section. However, most of the results can be directly carried over to the constant-height mode.

3.2 Imaging of Metals

For simple metals, there is typically no strong variation with energy of the local density of states or wavefunctions near the Fermi level. For purposes of STM, the same is presumably true for noble and even transition metals, since the d shell apparently does not contribute significantly to the tunneling current.[15] It is therefore convenient in the case of metals to ignore the voltage dependence, and consider the limit of small voltage, (6). (Effects of finite voltage are discussed in Sections 3.3 and 4.1 below.)

This is particularly convenient, since we then require only the calculation of $\rho(\vec{r}_t, E_F)$, a property of the bare surface. Nevertheless, for reasons discussed briefly in Section 2.4 above, even this calculation is quite demanding numerically. In fact, I know of only one case of the STM image being calculated for a real metal surface, and compared with experiment. That is the case of Au(110) 2x1 and 3x1, discussed in Ref. 9.

The real strength of STM is that, unlike diffraction, it is a local probe, and so can be applied even to disordered surfaces, or to isolated features such as defects. In order to interpret the resulting images quantitatively, it is often necessary to calculate the image for a proposed structure or set of structures, and compare with the actual image. However, while the accurate calculation of $\rho(\vec{r}_t, E_F)$ is difficult even for Au(110) 2x1, it is out of the question for surfaces with large unit cells, and *a fortiori* for disordered surfaces or defects. It is therefore highly desirable to have a method, however approximate, for calculating STM images in these important but intractable cases.

Such a method has been suggested and tested in Ref. 9. It consists of approximating (6) by a superposition of spherical atomic-like densities. This approach is expected to work very well for simple and noble metals, and was tested in detail[9] for Au(110). The success of the method relies on the fact that the model density, by construction, has the same analytical properties as the true density, so that if the model is accurate near the surface, it will automatically describe accurately the decay with distance.

An example of where this approach can be useful has also been presented.[9] The image expected for Au(110) 3x1 was calculated for two plausible models of the structure, differing only in the present or absence of a missing row in the second layer. The similarity of the model images at distances of interest suggested that the structure in the second layer could not be reliably inferred from experimental images. Quantifying the limits of valid interpretation in this way is an essential part of the analysis of STM data.

While this method is intended primarily for metals, Tromp et al.[16] applied it to Si(111) 7x7 with remarkable success. They simulated the images for a number of different proposed models of this surface, and compared them with experimental images. The so-called "Dimer-Adatom-Stacking fault" model gives an image which agrees almost perfectly with experiment, while most models lead to images with little similarity to experiment. Thus the usefulness of such image simulations must not be underestimated, although few such applications have been made to date.

3.3 Imaging of Semiconductors

At very small voltages, the s-wave approximation for the tip led to the very simple result (6). At larger voltages, one might hope that this could be easily generalized to give a simple expression such as

$$J \sim \int_{E_F}^{E_F + V} \rho(\vec{r}_t, E). \tag{7}$$

This is not strictly correct for two reasons. First, the matrix elements and the tip density of states are at least slightly energy dependent, and any such dependence is neglected in (7). Second, the finite voltage changes the potential, and hence the wavefunctions, outside the surface. A more careful discussion, especially of the latter effect, is given in Section 4. Nevertheless, there is considerable evidence that (7) is a reasonable approximation for many purposes,[17] as long as the voltage is much below the workfunction. We shall therefore use (7) in discussing qualitative aspects of STM images of semiconductors at modest voltages.

Unlike metals, semiconductors show a very strong variation of $\rho(\vec{r}_t, E_F + V)$ with voltage. In particular, this quantity changes discontinuously at the band edges. With negative sample voltage, current tunnels out of the valence band, while for positive voltage, current tunnels into the conduction band. The corresponding images, reflecting the spatial distribution of valence and conduction-band wavefunction respectively, may be qualitatively different.

A particularly simple and illustrative example, which has been studied in great detail, is GaAs(110). There, it was proposed[9] that since the valence states are preferentially localized on the As atoms, and the conduction states on the Ga atoms, STM images of GaAs(110) at negative and positive bias should reveal the As and Ga atoms respectively. Such atom-selective imaging was confirmed by direct calculation of (7),[9] and was subsequently observed experimentally.[18]

In a single image of GaAs(110), whether at positive or negative voltage, one simply sees a single "bump" per unit cell. In fact, the images at opposite voltage look quite similar. It is therefore crucial to obtain both images *simultaneously*, so that the dependence of the absolute position of the "bump" on voltage can be determined. Combining the two images, the zig-zag rows of the (110) surface are clearly seen.[18]

Even in this simple case, however, the interpretation of the voltage-dependent images as revealing As or Ga atoms directly is a bit simplistic. A detailed analysis[18] shows that the apparent positions of the atoms in the images deviate significantly from the actual positions. This deviation could be viewed as an undesirable complication, since it makes the image even less like a topograph.

Alternatively, it is possible to take advantage of this deviation. The apparent position of the atom turns out to be rather sensitive to the degree of buckling associated with the (110) surface reconstruction, so that it is possible to infer the surface buckling quantitatively from the apparent atom positions.[18] Thus the images are actually quite rich in information, but the quantitative interpretation requires a more detailed analysis than is often feasible.

Even for semiconductors, there may be cases where the image (at least at some voltage) corresponds fairly closely to a simple topograph of the surface. A striking example of this is Si(111) 7x7, discussed above.[16] Similarly, for Si(111) 2x1, the image at low voltage is rather peculiar,[19] for reasons discussed in Section 3.4; but at higher voltage, the image begins to look a bit more like a simple topograph.[20]

In tunneling to semiconductors, there is an added complication not present in metals: there may be a large voltage drop associated with band-bending in the semiconductor, in addition to the voltage drop across the gap.[21] This means that the tunneling voltage may be substantially less than the applied voltage, complicating the interpretation. Moreover, local band-bending associated with defects or adsorbates on the surface can lead to striking non-topographic effects in the image.[21]

3.4 Imaging Band-edge States

A particularly interesting situation can arise in tunneling to semiconductors at low voltages.[19,22] At the lowest possible voltages, only states at the band edge participate in tunneling. These band edge states typically (though not necessarily) fall at a symmetry point at the edge of the surface Brillouin zone. In this case, the states which are imaged have the character of a standing wave on the surface.

This standing-wave character leads to an image with striking and peculiar properties.[22] The corrugation is anomalously large, and unlike the normal case, it does not decrease rapidly with distance from the surface. This gives the effect of unusually sharp resolution. For example, in the case of graphite (a semimetal which also satisfies these conditions), the unit cell is easily resolved despite the fact that it is only 2 Å across. This effect was also seen on Si(111) 2x1.[19]

In most cases the image can be described by a universal form,[22] consisting of an array of sharp dips with the periodicity of the lattice. (The dips however are broadened by a variety of effects, and in any case may not be well resolved because of instrumental response time[22]). Such an image can be extremely misleading. Specifically, when there is one "bump" per unit cell, it is tempting to infer that there is one topographic feature per unit cell. However, the image may in fact carry no information whatever regarding the distribution of atoms within the unit cell in this case. It is strange indeed that the sharpest image with the best apparent resolution sometimes carries the least structural information.

3.5 Spatial Resolution of STM

In any microscopy or spectroscopy, the image or spectrum is viewed as an "ideal" image containing the desired information, convoluted with an instrumental resolution function or lineshape. The resolution is therefore one of the most crucial properties characterizing the instrument. This Section reviews the analysis of Ref. 23, where the spatial resolution of STM is treated in some detail.

3.5.1 *General Treatment.* Defining the resolution of STM raises tricky issues for two reasons. First, STM is inherently nonlinear, so the usual definition of resolution in terms of convolution with an instrumental function cannot be applied directly. Second, resolution can only be defined relative to what the instrument should ideally measure. For STM, it is not obvious what is meant by the ideal image.

Formally, one defines resolution by assuming that there exists some ideal image or spectrum $I_0(x)$, which would be seen in the case of perfect instrumental resolution. The variable x here may represent one or more dimensions of space, time, energy etc., depending upon the microscopy or spectroscopy in question. The actual measured image $I(x)$ is then

$$I(x) = \int I_0(x - y)F(y)dy . \tag{8}$$

Equation (8) should be viewed as the definition of the resolution function $F(x)$.

It is often convenient to Fourier transform (8) to obtain

$$\tilde{I}(q) = \tilde{I}_0(q)\tilde{F}(q) . \tag{9}$$

Here $\tilde{I}(q)$ and $\tilde{F}(q)$ are the Fourier transforms of $I(x)$ and $F(x)$, e.g.

$$\tilde{I}(q) = \int I(x) \exp(iqx)dx ,$$

neglecting normalization.

Let us rewrite $J(\vec{r}_t, V)$ as $J(\vec{x}, z)$, where we separate lateral and vertical tip position as $\vec{r}_t = (\vec{x}, z)$, suppressing the voltage dependence for notational simplicity. The STM image $z(\vec{x})$ in the constant-current (topographic) imaging mode is implicitly defined by

$$J(\vec{x}, z) = J_f , \tag{10}$$

where J_f is the fixed tunneling current at which the microscope is operated.

Unfortunately, this imaging process is inherently nonlinear, whereas resolution is only well defined for a linear measurement. It is therefore useful to work in the limit of weak corrugation, so that the imaging process can be linearized. One then writes

$$z = z_0 + \zeta(\vec{x}, z_0), \tag{11}$$

where z_0 is an average tip height defined below, which depends on J_f but is not easily accessible experimentally, and ζ is the small corrugation which constitutes the image, and which depends on the tip height z_0 or equivalently on J_f. Expanding $J(\vec{x}, z)$ about $z = z_0$, (10) and (11) give

$$\zeta(\vec{x}, z_0) \simeq [J_f - J(\vec{x}, z_0)] / \frac{d}{dz} J(\vec{x}, z_0). \tag{12}$$

Because of the approximately exponential decay of the wavefunction (4), for weak corrugation one can write

$$dJ(\vec{x}, z_0)/dz \simeq -2\kappa J(\vec{x}, z_0). \tag{13}$$

Also, the characteristic tip height z_0 is defined by the condition that the lateral average of $J(\vec{x}, z_0)$ is

$$A^{-1} \int J(\vec{x}, z_0) d\vec{x} \equiv \tilde{J}_0(z_0) = J_f, \tag{14}$$

where A is the area of integration. Thus z_0 represents the average height of the tip.

To lowest order in the small quantity $[J(\vec{x}, z_0) - J_f]/J_f$ one can then write

$$\zeta(\vec{x}, z_0) \simeq [J(\vec{x}, z_0) - J_f]/2\kappa J_f. \tag{15}$$

Equation (15) states that, in the limit of weak corrugation, the image $\zeta(\vec{x}, z_0)$ is simply proportional to the fractional variation of $J(\vec{x}, z_0)$ about its mean value in the plane $z = z_0$.

In analogy with (15), it was proposed[23] to identify I_0 with the fractional variation of $J(\vec{x}, z_0)$ about its mean value, evaluated in some *reference plane* which is taken here as the origin. Moreover, it seems only reasonable to define the ideal image I_0 as that obtained with an ideal tip, so that $J(\vec{x}, z) \propto \rho(\vec{x}, z)$ [Equation (6)]. Thus

$$I_0(\vec{x}) \equiv [\rho(\vec{x}, 0) - \tilde{\rho}_0(0)]/2\kappa\tilde{\rho}_0(0). \tag{16}$$

Note that this convention does not require that the corrugation be weak in the reference plane. The implicit dependence of κ on z due to the spatial variation of the potential is neglected, since this variation is actually rather weak except very near the surface.

There remains an arbitrariness in the choice of the reference plane $z = 0$. The treatment is most rigorous when the plane is chosen relatively far from the surface. However, this gain in rigor is balanced by a loss of substantive content, since more of the smoothing of the wavefunctions is then included in I_0 instead of in the resolution function, where it intuitively belongs.

It is now convenient to assume that the surface is periodic, and to work with the Fourier transformed quantities. The periodicity can later be taken to be arbitrarily large, to include nonperiodic surfaces. Then

$$\rho(\vec{x}, z) = \sum_G \tilde{\rho}_G(z) \exp(i\vec{G} \cdot \vec{x}), \tag{17}$$

and J is similarly expanded. Here G are the *surface* reciprocal lattice vectors, a tilde indicates a reciprocal-space quantity, and $\tilde{J}_0$ was defined above to be just the $G=0$ term of (17), with $\tilde{\rho}_0$ similarly defined. Now (15) and (16) may be rewritten as

$$\tilde{\zeta}_G(z_0) \simeq \tilde{J}_G(z_0)/2\kappa \tilde{J}_0(z_0), \tag{18}$$

$$I_0(G) = \tilde{\rho}_G(0)/2\kappa\tilde{\rho}_0(0). \tag{19}$$

Combining (18) and (19) with (9) gives the desired expression for the resolution function:

$$F(G) \simeq \tilde{J}_G(z_0)\tilde{\rho}_0(0) / \tilde{J}_0(z_0)\tilde{\rho}_G(0). \tag{20}$$

This formula may of course be Fourier transformed to give an explicit lineshape $F(\vec{x})$.

Equation (20) is an explicit statement of how the resolution of STM depends both on the tip height, and on the sample electronic structure. Any more specific statement requires detailed knowledge of the specific sample, i.e. of $\tilde{\rho}_G(z)$, and of the tip.

While (20) is derived under the assumption of weak corrugation, this restriction represents an unavoidable limitation, reflecting the limited applicability of the very concept of resolution to the highly non-linear STM measurement. For practical purposes, (20) can no doubt be usefully applied well beyond the range of strict validity, as a working definition of resolution.

3.5.2 *Application to Metals.* The resolution (20) evidently depends on the tip, but the role of the tip has not yet been fully treated. The discussion below is therefore restricted to the idealized s-wave tip model, which gave (6). Then (20) reduces to

$$F(G) \simeq \tilde{\rho}_G(z_0)\tilde{\rho}_0(0) / \tilde{\rho}_0(z_0)\tilde{\rho}_G(0). \tag{21}$$

In principle, a precise evaluation of the resolution from (21) requires a detailed knowledge of the surface electronic structure. However, for metals, it has previously been noted[9,23] that one can derive some results concerning the resolution from fairly general considerations. In particular, an *ansatz* based on the superposition of atomic-like densities gives a description which corresponds remarkably well to the exact asymptotic behavior.[9,23]

We imagine that $\rho(\vec{r}, E_F)$ can be mimicked by a sum of spherical densities of the form $C \exp(-2\kappa r)/r$, perhaps with a different coefficient C for each inequivalent atom. Then an arbitrary sum of such "atoms", restricted to the half-space $z<0$, gives a charge density which in the half-space $z>0$ can be Fourier transformed to yield[9]

$$\tilde{\rho}_G(z) = B_G \exp[-(4\kappa^2 + G^2)^{1/2}z], \tag{22}$$

where B is a constant for each G, depending only on κ and on the arrangement of atoms, and the values of their coefficients C.

Combining (20) and (22), one may immediately obtain an explicit model form for the resolution function,

$$\tilde{F}(G) \simeq \exp[2\kappa z_0 - (4\kappa^2 + G^2)^{1/2} z_0] . \tag{23}$$

Since κ is proportional to the square root of the workfunction, which itself doesn't vary too much among metals of interest here, (23) represents a resolution function which is nearly independent of the specific sample.

The "justification" for this *ansatz* is relatively simple, and is discussed in detail elsewhere.[9,23]

In general, for metals the observable G components obey $G \ll 2\kappa$, i.e. typically $G \leq 0.8 \text{Å}^{-1}$, while $2\kappa \sim 2\text{Å}^{-1}$. Then (23) may be expanded as

$$\tilde{F}(G) \simeq \exp(-G^2 z_0 / 4\kappa) . \tag{24}$$

This simplified model resolution function is simply a Gaussian with rms width $(z_0/2\kappa)^{1/2}$.

3.5.3 Application to Semiconductors.

Equation (23) gives an explicit result for the resolution function, i.e. the instrumental lineshape, for the case of STM of metal surfaces. However, it can be shown more generally,[23] that (23) is approximately valid even for semiconducting surfaces, *except* for the lowest Fourier component of the image. However, the lowest component F(g) (where g is the smallest reciprocal lattice vector) may in general have a value which deviates grossly from (23), and which depends sensitively upon the electronic structure of the surface.

This effect distorts the image in a manner which is simple, but which can have profound consequences for the interpretation of STM images. For example, in the extreme case that F(g) is greatly enhanced relative to the higher components, the image will have a simple universal form,[22] with the periodicity of the surface, giving the misleading appearance of a single topographic feature per unit cell. This effect has been observed on Si(111) 2x1, GaAs(110), and 1T-TaS$_2$ surfaces.[18,19,22] Fortunately, in each case a prior knowledge of the electronic structure permitted a correct interpretation.

This extreme case of distortion can easily arise due to tunneling to states at the edge of the surface Brillouin zone, $k_\parallel = g/2$, where g is again the smallest G. Substituting this into (4), and assuming reflection symmetry, one finds that

$$\psi = \sin(k_\parallel x) \exp[-(\kappa^2 + k_\parallel^2)^{1/2} z]$$

plus higher Fourier components.

Substituting into (6) gives $\tilde{\rho}_0 = 2\tilde{\rho}_g = \frac{1}{2} \exp[-2(\kappa^2 + k_\parallel^2)^{1/2} z]$. Thus the g Fourier component decays no more rapidly than the zero Fourier component, and substituting into (21) gives $F(g) = F(0) = 1$, independent of z. For $G > g$, however, $\tilde{\rho}_G$ decays faster with increasing G, as expected, so for $G > g$, $\tilde{F}(G)$ still decreases with increasing z as for a metal.

This is a very peculiar result. It implies that, for large z (i.e. large tunneling distance or tip radius), the ability to resolve structure within the unit cell decreases and is lost, just as for a metal surface, since this structure corresponds to $G > g$. But because $F(g) = 1$, the unit cell itself is well resolved even if it is very small (large g), and even if the tip is relatively far from the surface, as long as the model of Ref. 9 is applicable.

This enhanced resolution of the unit cell is particularly striking for graphite,[22] where the 2 Å unit cell is easily resolved, even though such small structures are not resolved on metal surfaces. In fact, there is evidence[23] that, for most semiconductor surfaces, the resolution is enhanced over that expected for metals by electronic structure effects.

4. Spectroscopy

4.1 Qualitative Theory

Tunneling spectroscopy in planar junctions was studied long before STM.[24] However, the advent of spatially-resolved STM spectroscopy has led to a resurgence of interest in this area. Because of the difficulty of calculating $J(\vec{r}_t, V)$ in general, most detailed analyses have instead focused simply on $J(V)$, without regard to its detailed spatial dependence.[17] Moreover, the important issue of inelastic tunneling spectroscopy is only beginning to be discussed theoretically in the context of STM.[25]

Selloni et al.[26] suggested that the results of Tersoff and Hamann[9] for small voltage could be *qualitatively* generalized as

$$J(V) \propto \int_{E_F}^{E_F+V} \rho(E) T(E,V) dE, \qquad (25)$$

where $\rho(E)$ is the local density of states (6) at or very near the surface, and assuming a constant density of states for the tip. This is similar to (7), except that the qualitative effect of the finite voltage on the surface wavefunctions is included through a barrier transmission coefficient $T(E,V)$.

Unfortunately, despite suggestions in Ref. 26, this simple qualitative model still does not lend itself to a straightforward interpretation of the tunneling spectrum.[17] In particular, the derivative dJ/dV has no simple relationship to the density of states $\rho(E_F + V)$, as might have been hoped. At best, one can say that a sharp feature in the density of states of the tip or sample, at an energy $E_F + V$, will lead to a feature in $J(V)$ or its derivatives at voltage V.

Even this rather weak statement may prove unreliable in practice, where spectral features have considerable widths. The reason for this problem is that $T(E,V)$ is very strongly V-dependent when the voltage becomes an appreciable fraction of the workfunction. Thus the V-dependence of $T(E,V)$ may distort features in the spectrum.[17]

Stroscio, Feenstra and coworkers[19,27] proposed a simple but effective solution to this problem. They normalize[19] dJ/dV by dividing it by J/V. This yields $d\ln J/d\ln V$, and so effectively cancels out the exponential dependence of $T(E,V)$ on V. At semiconductor band edges, where the current goes to zero, a slight smoothing of J/V eliminates the singular behavior at the band edge.[27]

This normalization is however both unnecessary and undesirable at small voltages; in that case, J/V is well behaved, whereas (dJ/dV)/(J/V) is identically equal to unity for ohmic systems, and so carries no information. Thus the appropriate way of displaying spectroscopy results depends on the problem at hand, and a variety of approaches for collecting and displaying data have been considered.[19,21,27]

4.2 Quantitative Theory

A proper treatment of the tunneling spectrum in STM requires calculation of the wavefunctions of surface and tip at finite voltage. This is a difficult problem, which is only now being tackled directly.[35]

A natural approximation[17] is therefore to use (5) with the zero-voltage wavefunctions, but to shift all the surface wavefunctions in energy relative to the tip, by an amount corresponding to the applied voltage V. Unfortunately, the result then depends on the position of the surface of integration for (3).

Lang showed[17] that, by positioning the surface of integration half-way between the two planar electrodes, the resulting error is second order in the voltage, and rather small as long as the voltage is much less than the workfunction. This result assumes that surface and tip have equal workfunctions, and is derived for a one-dimensional model only. Nevertheless, it seems safe to assume that the conclusion is more generally valid.

With this approximation, calculation of the tunneling spectrum becomes relatively straightforward. Such calculations[17] confirm the qualitative applicability of simple models such as (25). Moreover, they confirm that the spectrum (dJ/dV)/(J/V) mimics the density of states reasonably well, as proposed by Stroscio et al.[19]

5. Mechanical Tip-sample Interactions

Ideally, in STM the tip and surface are separated by a vacuum gap, and are mechanically non-interacting. However, sometimes anomalies are observed, which are most easily explained by assuming a mechanical interaction between the tip and surface.

In particular, since the earliest vacuum tunneling experiments of Binnig et al,[28] it has been observed that for dirty surfaces, the current varies less rapidly than expected with vertical displacement of the tip. Coombs and Pethica[29] pointed out that this behavior can be explained by assuming that the dirt mediates a mechanical interaction between surface and tip.

The current is expected to vary with tip height z as $J \propto \exp(-2\kappa z)$, where $\hbar^2\kappa^2/2m = \phi$, ϕ being the workfunction. Thus in principle ϕ can be determined from $d\ln J/dz$. For dirty surfaces, this dependence is weaker than expected, leading to an inferred workfunction which is unphysically small. In liquids, this "effective workfunction" is often 0.1 eV or less.

The explanation proposed by Coombs and Pethica is that some insulating dirt (e.g. oxide) is squeezed between the surface and tip, acting in effect as a spring. Tunneling might take place from a nearby part of the tip which is free of oxide, or from a "mini-tip" poking through the dirt. As the tip is lowered, the dirt becomes compressed, pushing down the surface or compressing the tip if they are sufficiently soft. As a result, the surface-tip separation does not really decrease as much as expected from the nominal lowering of the tip, and so the current variation is correspondingly less.

This issue gained renewed importance with the observation of huge corrugations in STM of graphite.[30,31] While theoretical calculations[26] suggested corrugations of at most 1 Å or so, ridiculously large corrugations were sometimes observed, up to 10 Å or more vertically within the 2 Å wide unit cell of graphite.

Soler et al.[30] at first attributed these corrugations to direct interaction of the tip and surface, but a detailed study by Mamin et al.[31] suggests that in fact the interaction is mediated by dirt, consistent with the earlier proposal of Coombs and Pethica.[29] Clean UHV measurements in fact gave corrugations of under 1 Å.[32]

For the model of direct interaction,[30] there is a complex nonlinear behavior. But for the dirt-mediated interaction model,[29] a linear treatment is appropriate. In this case, assuming the mechanical interaction has negligible corrugation (e.g. because of the large interaction area), the image seen is simply the ideal image, with the vertical axis distorted by a constant scale factor.

It seems surprising that it is possible to obtain any image at all, when there is dirt between surface and tip, since some scraping might be expected as the tip is scanned. For graphite, one could imaging that the dirt (e.g. metal oxide) moves with the tip, sliding nicely along the inert graphite surface. However, any model here is based on indirect inference, and this phenomenon must be considered as not really understood at present.

This lack of a complete understanding is most dramatically pointed out by the recent observation of close-packed metal surfaces, in which individual atoms were resolved.[33,34] It is possible[34] that these cases, like graphite, may represent an enhancement of the corrugation by mechanical interactions between tip and surface. However, no dramatic lowering of the "effective workfunction" was observed,[34] so the model of Coombs and Pethica for the interaction may not be directly applicable. Moreover, it is hard to imaging how mechanical contact could fail to disrupt the metal.

6. Conclusion

STM images at low voltage are relatively well understood at present, although the detailed dependence on tip shape has been considered only qualitatively. Tunneling spectra are also understood in principle, although they bear only a qualitative resemblance to the density of states.

However, this rather satisfactory state of affairs applies only to clean surfaces in ultra-high vacuum. A great deal of STM work is actually done in air or liquids, because of the tremendous reduction in labor and expense relative to vacuum. In this case, the understanding is much less complete.

In particular, mechanical interactions between surface and tip, mediated either by solid dirt or by the surrounding fluid, appear to play a crucial role. Such interactions are not understood at a microscopic level. They may not even be reproducible, depending upon their origin.

Mechanical interactions between surface and tip also provide the mechanism underlying atomic force microscopy (AFM), an important related subject which has not been treated here. Thus mechanical interactions between tip and surface are one of the most important, and least understood, outstanding issues in STM and related microscopies.

REFERENCES

1. G. Binnig, H. Rohrer, Ch. Gerber, and E. Weibel, Phys. Rev. Lett. 49, 57 (1982). For reviews, see G. Binnig and H. Rohrer, Rev. Mod. Phys. 59, 615 (1987); P. K. Hansma and J. Tersoff, J. Appl. Phys. 61, R1 (1987).
2. J. K. Gimzewski and R. Möller, Phys. Rev. B 36, 1284 (1987).
3. N. D. Lang, Phys. Rev. B 36, 8173 (1987).
4. M. D. Stiles and D. R. Hamann, Phys. Rev. B 38, 2021 (1988).
5. N. Garcia, C. Ocal, and F. Flores, Phys. Rev. Lett. 50, 2002 (1983); E. Stoll, A. Baratoff, A. Selloni, and P. Carnevali, J. Phys. C 17, 3073 (1984).
6. J. Bardeen, Phys. Rev. Lett. 6, 57 (1961).
7. For a unique exception, see Y. Kuk, P. J. Silverman, and H. Q. Nguyen, J. Vac. Sci. Technol. A 6, 524 (1988).
8. N. D. Lang, Phys. Rev. Lett. 56, 1164 (1986); and Phys. Rev. Lett. 55, 230 (1985).
9. J. Tersoff and D. R. Hamann, Phys. Rev. B 31, 805 (1985); and Phys. Rev. Lett. 50, 1998 (1983).
10. C. Noguera, (unpublished); see also W. S. Sacks and C. Noguera (unpublished).
11. S. Park, J. Nogami, and C. F. Quate, Phys. Rev. B 36, 2863 (1987); H. A. Mizes, S. Park, and W. A. Harrison, Phys. Rev. B 36, 4491 (1987).
12. R.J. Hamers, R.M. Tromp, and J.E. Demuth, Phys. Rev. Lett. 56, 1972 (1986).
13. R. M. Feenstra, W. A. Thompson, and A. P. Fein, Phys. Rev. Lett. 56, 608 (1986); J. A. Stroscio, R. M. Feenstra, D. M. Newns, and A. P. Fein, J. Vac. Sci. Technol. A 6, 499 (1988).
14. A. Bryant, D. P. E. Smith, and C. F. Quate, Appl. Phys. Lett. 48, 832 (1986).
15. N. D. Lang, Phys. Rev. Lett. 58, 45 (1987).
16. R. M. Tromp, R. J. Hamers, and J. E. Demuth, Phys. Rev. B 34, 1388 (1986).
17. N. D. Lang, Phys. Rev. B 34, 5947 (1986).
18. R. M. Feenstra, J. A. Stroscio, J. Tersoff, and A. P. Fein, Phys. Rev. Lett. 58, 1192 (1987).
19. J. A. Stroscio, R. M. Feenstra, and A. P. Fein, Phys. Rev. Lett. 57, 2579 (1986).
20. R. M. Feenstra and J. A. Stroscio, Phys. Rev. Lett. 59, 2173 (1987).

21. R. M. Feenstra and J. A. Stroscio, J. Vac. Sci. Technol. B 5, 923 (1987); J. A. Stroscio and R. M. Feenstra, J. Vac. Sci. Technol. B 6, 1472 (1988); R. M. Feenstra and P. Mårtensson, Phys. Rev. B 39, 7744 (1989).
22. J. Tersoff, Phys. Rev. Lett. 57, 440 (1986).
23. J. Tersoff, Phys. Rev. B 39, 1052 (1989).
24. C. B. Duke, *Tunneling in Solids*, Suppl. 10 of *Solid State Physics*, edited by F. Seitz and D. Turnbull (Academic, New York, 1969), p. 1.
25. B. N. J. Persson and A. Baratoff, Phys. Rev. Lett. 59, 339 (1987); B. N. J. Persson and J. E. Demuth, Solid State Comm. 57, 769 (1986); G. Binnig, N. Garcia and H. Rohrer, Phys. Rev. B 32, 1336 (1985).
26. A. Selloni, P. Carnevali, E. Tosatti, and C. D. Chen, Phys. Rev. B 31, 2602 (1985).
27. R. M. Feenstra and P. Mårtensson, Phys. Rev. Lett. 61, 447 (1988).
28. G. Binnig, H. Rohrer, Ch. Gerber, and E. Weibel, Appl. Phys. Lett. 40, 178 (1982).
29. J. H. Coombs and J. B. Pethica, IBM J. Res. Develop. 30, 455 (1986).
30. J. M. Soler, A. M. Baro, N. Garcia, and H. Rohrer, Phys. Rev. Lett. 57, 444 (1986).
31. H. J. Mamin, E. Ganz, D. W. Abraham, R. E. Thomson, and J. Clarke, Phys. Rev. B 34, 9015 (1986).
32. R. J. Hamers, (unpublished).
33. V. M. Hallmark, S. Chiang, J. F. Rabolt, J. D. Swalen, and R. J. Wilson, Phys. Rev. Lett. 59, 2879 (1988).
34. J. Wintterlin, J. Wiechers, H. Brune, T. Gritsch, H. Höfer, and R. J. Behm, Phys. Rev. Lett. 62, 59 (1989).
35. N. D. Lang, (unpublished).

Theory of tunneling from transition metal tips

G. DOYEN, E. KOETTER, J.BARTH [1] AND D. DRAKOVA [2]
Fritz-Haber-Institut der Max-Planck-Gesellschaft
Faradayweg 4-6
D-1000 Berlin 33, FRG

ABSTRACT. A transition metal tip consisting of a single tungsten atom adsorbed on a W(110) surface is modelled. Chemisorpion theory is applied to treat the interaction of the tip atom with the flat W(110)-surface. The basis set on the tip atom includes 6s-, 6p- and 5d-orbitals. The importance of the d-electrons for the tunneling to an Al(111)-surface is investigated. Inclusion of the d- orbitals affects strongly the local electronic structure and modifies the current. For short distances a significant part of the total tunnel current flows directly via the tip d-orbitals. The theory is used to investigate the 'barrier height' of the clean surface. A corrugation of the Al(111)-surface is obtained in STM-theory, if a elastic deformation of the tip is taken into account.

1 Introduction

This article suggests a method to calculate STM images of metal surfaces within a model which is truly 3-dimensional and allows inclusion of d-electrons on the tip. Hence this theory moves closer to experiment. Three-dimensional methods in the literature rely mostly on the transfer Hamiltonian approach [1,2] suggested by Bardeen [3]. We apply a way of calculating the tunnel current which avoids this kind of approximation. The theory has already been applied to treat tunneling between a transition metal surface with an adsorbate [4,5] and with a line (step) defect [6] and a tungsten tip with a single protruding host atom on a W-surface. This contribution will explain the formalism and illustrate it with some other interesting examples. Approximations are inevitable in attacking such a complicated problem. These will be explained in the following. Approximations which are made but not discussed further involve:

1. Neglect of the distortion by the electric field resulting from the different work functions or the applied voltage.

2. Assumption that the non-equilibrium situation established by applying a voltage is maintained through processes which are fast compared to the tunnel current.

3. Tunneling from the base of the tip on which the single tungsten atom sits is neglected.

---

[1] Institut für Physikalische Chemie der Universität München, D-8000 München 2, Theresienstr. 37
[2] Department of Inorganic Chemistry, University of Sofia, Bulgaria

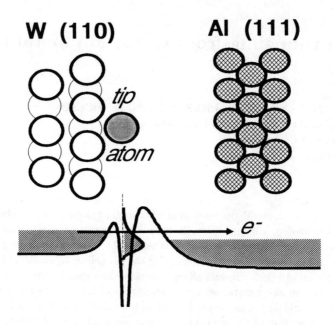

Figure 1: Geometrical arrangement of atoms and schematic effective potential for a tunneling process

As examples of applying this method we shall discuss the influence of the tip d-wave functions on the tunnel current and on the so called 'barrier height' which is obtained by forming the logarithmic derivative of the tunnel current. An elastic deformation of the tip is investigated in order to look for a possible explanation of the experimentally found corrugation on Al(111) [7,8]. For this latter theoretical study of imaging of the Al(111) - surface d-functions on the atom forming the apex of the tip have not been included.

2 The adsorption model: electronic structure of the tip

An ideal STM tip will be one atom adsorbed on a group of other meal atoms. For a first theoretical attempt we will model this as a single W-atom adsorbed on a flat W(110) - surface.

Figure 1 illustrates schematically the geometrical arrangement of atoms as well as the effective one-electron potential at the tunnel junction. For the indicated bias voltage tunneling takes place from filled states on the left to unfilled states on the right. The current flows via tip orbitals which are resonances in the continuous spectrum of W-metal states. In calculating the STM-current we will have to determine the electronic structure of the tip atom. The wave functions for the W-surface together with its adsorbed tip atom are just those of the one-atom chemisorption problem.

The philosophy is quite similar to Lang's approach [2]. The difference is that we take

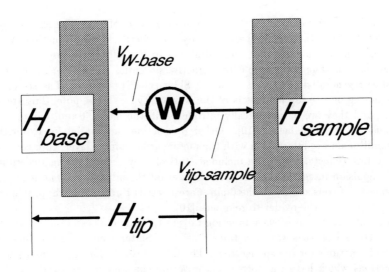

Figure 2: Structure of the Hamiltonian

into account the atomic structure of the tip and the sample and that we do not use the transfer Hamiltonian approach to evaluate the tunnel current. We cannot use first principle methods to solve this complicated problem. Rather we use a model Hamiltonian where the one- and two-electron integrals are chosen in a semiempirical way such as to reproduce experimental and theoretical data for the non-interacting system.

Figure 2 illustrates the structure of the Hamiltonian which is used to solve the chemisorption problem. H_{base} stands for the flat W(110)-surface, H_{sample} represents the sample surface without the tip present. $V_{tip-sample}$ is the interaction between the tip and the sample, V_{W-base} indicates the coupling of the tip atom to its base.

The model Hamiltonian has been described in detail [9,10]. Here we mention only briefly its major characteristics, The metal surface is semi-infinite. The sp-electrons are described in the effective one-electron model of Sommerfeld. A discretization of the metal sp-band is performed, permitting a reduction of the dimensions of the Hartree-Fock matrices involved. The discretization procedure [11] defines a set of more or less localized metal wave functions, which are obtained by projecting onto adorbitals and which are the only metal states coupling to the adparticle. The d-electrons are explicitly introduced by means of localized wave functions centered at the lattice sites. The description of the metal surface requires input values which are taken from experiment and band structure calculations available in the literature: the work function [12], the inner potential depth, the energetic position and the width of the sp-d hybridization gap, the lattice constants, and the energetic position of the d-band. The electronic structure of the adparticle is described by a basis set of ab initio wave functions found in the literature. Coulomb repulsion integrals, exchange integrals and core energies on the adparticle are parameterized in a way to reproduce several experimentally determined ionization and affinity energies. The interaction between the adparticle and the metal surface involves short-range and long-range effects. The short-range effects are calculated within the region overlapped by the adparticle wave functions and include electron-electron interactions and a potential due to the adparticle core and its image in the metal surface. The adparticle core attractive potential felt by the metal electrons is described as a generalized Göppert - Mayer - Sklar (GMS) potential. The screened adcore potential acting on overlap charges is of the Yukawa-type. These potentials are adorbital dependent and they are parametrically introduced in the model Hamiltonian. They have been extensively studied before [10]. Other interactions accounted for explicitly in the model Hamiltonian are interaction with the metal surface dipole layer, core-core repulsion, and image force effects, which are included statically by renormalizing the values of the Coulomb repulsion integrals and the core energies on the adparticle. The model Hamiltonian is solved self-consistently in the (spin- unrestricted) Hartree-Fock approximation. For more details we refer the reader to reference [10].

An accurate treatment of wave function tails in the barrier is essential for any theory of tunneling. Our set of basis wave functions consists of Slater orbitals for the d-electrons and Sommerfeld functions for the sp-electrons. The latter provide *for every energy in the band* basis functions which have the correct asymptotic behaviour in the barrier and artifacts which might occur with calculations based on gaussian or plane wave basis sets are more unlikely.

The electronic structure of the adsorbed W-atom representing the apex of the tip has a decisive influence on the tunnel current. The important quantity to look at is the modification of the local density of states due to the presence of the tip atom, or, more precisely, the spectral resolution of the tip orbitals in the eigenstates of H_{tip}. An isolated W-atom has 4 electrons in the 5d-shell and 2 electrons in 6s-orbitals, all deeper shells being completely filled.

In figure 3a we show the spectral resolution of the 6s-tip orbital for a calculation where the d-functions on the tip atom have been omitted from the basis. One recognizes two peaks, one below the Fermi level and one above, simulating a bonding and an anti-bonding state. The situation is quite different from an adsorbed Na-atom which has been used as tip-atom by Lang [2] and which shows a resonance corresponding to the 3s valence level mostly above the Fermi level indicating that the 3s electron of the Na has been largely lost to the metal.

The tunnel current will depend crucially on the spectral weight of this spectral function at the Fermi level. Due to the bonding - anti-bonding character of the spectral function the weight near the Fermi level is quite small and will vary strongly with distance of the tip atom from the base and with any modification of the surrounding electron density. A large effect is therefore found, if we include the d-orbitals on the tip atom in the basis (cf. figure 3b). Both peaks broaden and shift to lower energy increasing the spectral weight near the Fermi level.

If the tip is very sharp and consists of a single atom, our results suggest that the spectral function might exhibit complex structure which could change, if for example the tip is only slightly deformed in an elastic way during the tunneling experiment. In this case the spectral density could have an important bearing on the tunneling results and would have to be calculated reliably in any theory.

The potential energy curves of the tip atom depend drastically on the tip basis (cf.

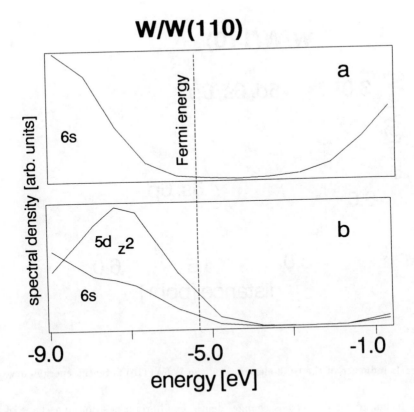

Figure 3: Spectral resolution of the tip orbitals. a) d-electrons on the tip atom neglected; b) including d-electrons on the tip atom. The d-electrons on the W-substrate atoms are included in both cases.

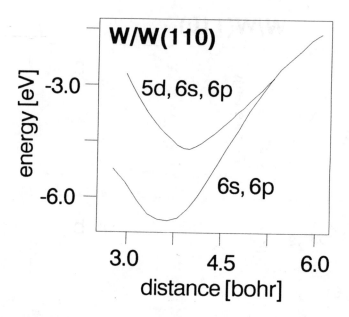

Figure 4: Influence of the tip-d-electrons on the W/W(110) potential energy curve

figure 4). Including the d-electrons changes depth, equilibrium position and shape of the curve. Without the d-electrons present, the curve becomes quite soft for displacements of the tip atom from the equilibrium position towards the surface. This is due to two nearly compensating effects. As a consequence of the image interaction the $6p_z$-orbital shifts below the Fermi level near the equilibrium position. This increases the repulsive electron - electron interaction. The counterbalancing effect is electron - core attraction which lowers for smaller separations. Including the d-electrons increases the electron-electron repulsion much more strongly for smaller separations leading to a steeper increase of the potential energy curve towards the base.

3 The tunnel current

The aim of a theory of tunneling microscopy is to calculate the current flowing from the tip to the sample surface or vice versa depending on the applied voltage. In a one particle picture this can be viewed as a scattering process where e.g. an electron incident from the interior of the tip metal scatters from the tunnel junction and has a certain probability P of penetrating into the sample surface. This probability is the modulus squared of the projection of the scattering function $\mid \vec{i}+ >$ on a current carrying state $\mid \vec{f} >$ in the sample metal (the states are labelled by the incident momentum $\vec{i}$ and the final momentum $\vec{f}$, respectively):

$$P = |<\vec{f}|\vec{i+}>|^2 \tag{1}$$

$|\vec{i+}>$ is here an eigenfunction of the total Hamiltonian including the interaction $V_{tip-sample}$ between the tip and the sample surface. $|\vec{f}>$, however, should be considered as the state in which the electron is prepared when a measurement of the current is performed. Measuring the current is equivalent to determining the velocity of the electron. As the current is measured away from the tunnel junction, $|\vec{f}>$ should not include the tip-sample interaction $V_{tip-sample}$. This means that $|\vec{f}>$ has to be an eigenstate of the momentum operator deep inside the sample metal. For a free electron metal $|\vec{f}>$ would then at the same time be an eigen state of the metal Hamiltonian H_{sample} in the bulk.

V_{sample} is constant for a free electron metal. For a more realistic description of the sample metal V_{sample} has to be treated as a perturbing potential for the scattering process. However, according to the theorem by Gell-Mann and Goldberger [15,16,17] V_{sample} can be removed from the scattering potential, if it is included in determining the final state function $|\vec{f}>$, i.e., if $|\vec{f}>$ is chosen to be an eigenstate of H_{sample}. In the following $|\vec{f}>$ is interpreted in this way.

The current itself is the number of electrons per unit time being detected in the state $|\vec{f}>$ times the electron charge e:

$$J = e\frac{d}{dt}|<\vec{f}|\vec{i+}>|^2 \tag{2}$$

Using Lippmann's generalization [13] of Ehrenfest's theorem [14] we obtain the current in the form:

$$J = 2\frac{\pi e}{\hbar}|<\vec{f}|V_{tip-sample}|\vec{i+}>|^2 \delta(E_f - E_i) \tag{3}$$

This expression is exact. It reduces to Fermi's golden rule, if the exact scattering state $|\vec{i+}>$ is replaced by an eigenstate $|\vec{i}>$ of H_o (cf. figure 2):

$$H_0 = H_{tip} + H_{sample} \tag{4}$$

In H_o the interaction between the tip and the sample surface is neglected.

Theoretical studies of STM in the literature are mostly based on Bardeen's work [3], which appeared before the fundamental studies by Lippmann about scattering theory have been published. Bardeen assumes that $|\vec{i+}>$ is an eigenvector of H_o and approximates $|\vec{f}>$ in a special way. With the additional assumption introduced by Tersoff and Hamann [1] that the initial state is a spherical symmetrical tip-orbital one obtains:

$$J \propto \sum_{\vec{f}} |\psi_f(\vec{r}_o)|^2 \delta(E_f - E_i) \tag{5}$$

According to this result the tunnel current should be proportional to the charge density at the center of the tip atom. It appears that this approximation is somewhat too drastic to be of general value for a further development of STM-theory. Obviously the observed corrugation on Al(111) cannot be understood within this framework as the charge density on this surface is known to be flat [18,19].

An exact expression for the transition matrix element is:

$$< \vec{f} \mid V_{tip} \mid \vec{i}+ > = \sum_{A,B} V_{fA} G_{AB} V_{Bi} \qquad (6)$$

The label A and B indicate here basis functions which form a conplete set in the barrier region. V_{fA} and V_{Bi} are matrix elements of the interaction potentials V_{W-base} and $V_{tip-sample}$, respectively (cf. figure 2). G_{AB} indicates a matrix element of the Green operator. The important theoretical task is to evaluate the exact Green functions G_{AB} in the tip region.

Using equation (6) with a set of basis orbitals $\{\mid A >\}$ centered on the tip atom and performing the summation over all states between the two Fermi levels, we end up with the following formula for the current:

$$J = \frac{2\pi e}{\hbar} \frac{1}{\Delta E} \sum_{A,B} \tilde{S}_{AB}^{tip} S_{AB}^{sample} W(A) W(B) \qquad (7)$$

with

$$\tilde{S}_{AB}^{tip} = \sum_{k}^{tip} < \tilde{A} \mid \vec{k}+ > < \vec{k}+ \mid \tilde{B} >$$

$$S_{AB}^{sample} = \sum_{k}^{sample} < A \mid \vec{k} > < \vec{k} \mid B >$$

ΔE is the energy between the two Fermi levels. $W(A)$ is the averaged potential experienced by a tunneling electron in the tip orbital $\mid A >$. $\tilde{S}_{AB}^{tip}$ and S_{AB}^{sample} are matrix elements of the projection operators on the tip and the sample wave functions, respectively. The diagonal matrix elements might be interpreted as integrated local densities of states. $\tilde{S}_{AA}^{tip}$ is the projection of the local state $\mid A >$ in the barrier on the eigenstates of the tip and S_{AA}^{sample} is the projection of this local state on the eigenfunctions of the sample surface. For example, $\tilde{S}_{6s,6s}^{tip}$ is essentially the tip induced density displayed in figure 3, integrated over the energy region bewteen the two Fermi levels.

We use equation (7) to evaluate the tunnel current from the W-tip to the Al(111)-surface. The effect of the d-electrons on the tunnel current is demonstrated in figure 5. For the upper curve the d-electrons on the tip have been included. The d-electrons on the substrate atoms of the base are included for both calculations. As the current flows dominantly via the 6s - tip orbital, the increase in the tunnel current arises essentially from the increase of $\tilde{S}_{6s,6s}^{tip}$. The tip d-electrons enhance the electron - electron repulsion in the barrier and shift the spectral distribution of the 6s - tip orbital to lower energy thus increasing the spectral weight of this orbital at the tip Fermi level (cf. figure 3). Only at very close distances does the current flow directly via the tip - d - orbitals.

$\tilde{S}_{AB}^{tip}$ is constant for a rigid tip but changes, of course, if there is an elastic deformation of the tip. This effect will be investigated in section 5 which considers a deformable tip. In the next section a static tip is used to investigate the logarithmic derivative of the tunnel current.

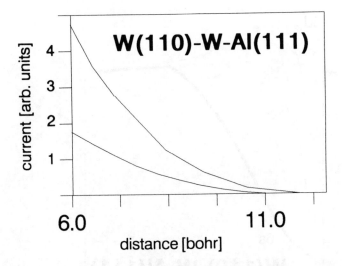

Figure 5: Influence of d-electrons on the tunnel current: the upper curve includes tip - d -electrons, the lower curve does not. The bias voltage is 0.1 eV. The distance is measured from the center of the tip atom to the first layer of Al - atoms.

4 'Barrier height' for clean Al(111)

For a one-dimensional square barrier of width a and height ϕ above the Fermi level, the tunnel current J would depend exponentially on these two quantities:

$$J = C \cdot e^{-2\sqrt{2\phi} \cdot a} \tag{8}$$

Solving this equation for ϕ one obtains in the case of a constant pre-exponential factor C:

$$\phi = \frac{1}{8}[dlnJ(a)/da]^2 \tag{9}$$

The expression on the r.h.s. of this equation will be termed the 'barrier height' also for a general 3-dimensional shape of the barrier when an interpretation as suggested by this simple example is not allowed. This expression can be determined experimentally and theoretically.

Figure 6 shows the result of the calculation corresponding to a bias voltage of 0.1 eV with the current flowing from the tip to the sample. The thick curve represents the 'barrier height' as a function of the distance of the tip atom from the Al(111)-surface. The other curves demonstrate how the 'barrier height' would behave, if the tunnel current flows only via a single tip-orbital. For these calculations the sum in equation (7) has been restricted to a single tip-orbital, but the electronic structure was the correct one including all basis functions. There are three points which have to be explained in the behaviour of the 'barrier height':

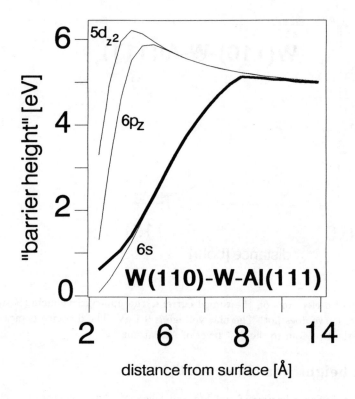

Figure 6: 'Barrier height' for Al(111). The thick curve gives the complete result. The other curves indicate the results obtained, if the current would flow exclusively over the indicated tip orbital.

- The 'barrier height' is above the work function for large separations.
- The 'barrier height' is below the work function for small separations.
- The dependence of the 'barrier height' on the angular momentum of the tip wave function.

The first point arises from the pre-exponential in equation (10) which depends inversely on the barrier width a [20]. This is already true for a one-dimensional square barrier where the 'barrier height' would then assume the form:

$$\text{'barrier height'} = \phi + \sqrt{\frac{\phi}{2}\frac{1}{a} + \frac{1}{8a^2}} \qquad (10)$$

The second point results from the behaviour of the sample projected tip-orbital density S_{AA}^{sample}. When the tip approaches the sample surface, at some distance the tip-orbital starts probing inside the metal where the sample wave functions oscillate. From this distance on, S_{AA}^{sample} does no longer increase exponentially but saturates. The tunnel currents saturates as well and the logarithmic derivative of the current tends to zero.

The third point depends on the behaviour of the wave functions parallel to the surface. Due to its angular dependence the $5d_{z^2}$-orbital e.g. is more contracted parallel to the surface than the 6s-orbital. This means that it samples larger $k_\parallel$ more efficiently than the 6s. A sample wave function for non-zero $k_\parallel$ has the form:

$$\Psi(E_F, \vec{k}_\parallel) \propto \sqrt{2E_F - k_\parallel^2} \cdot e^{-\sqrt{2\phi + k_\parallel^2}\, z} \cdot e^{i\vec{k}_\parallel \cdot \vec{r}_\parallel}$$

The exponent in the exponential for perpendicular decay contains $k_\parallel$ and therefore the 'barrier height' increases, if these wave functions are probed by the tip. Of course, the $5d_{z^2}$-orbital 'sees' sample wave functions of any $k_\parallel$, but the weight of larger $k_\parallel$ is greater than for 6s and therefore the 'barrier height' derived for the $5d_{z^2}$-orbital alone is larger.

The results for the 'barrier height' obtained by Lang [21] also show the characteristics discussed as the first two points above.

5 Elastic deformation of the tip: A clue to the corrugation on Al(111)?

Being a simple metal, the close packed Al(111) surface has a practically flat charge density without any surface state near the Fermi energy [18]. He scattering shows a very small corrugation ($\langle 0.02 \text{Å}$) [19]. We assume that the tip atom can move, if a force is exerted on it (cf. figure 7). In order to model a soft vibration we use a tip-tungsten interaction potential derived by neglecting the d-electrons. This is quite flat near the equilibrium minimum.

In our model the Al-sp-band charge density does not exhibit any corrugation at all. If the W-tip atom approaches close to the surface, it will interact with the core electrons of the Al-atoms. this leads to a corrugation of the interaction potential in the repulsive region as shown in figure 8. As is obvious by physical intuition, the potential is more repulsive for the top position than for the center position inbetween three Al-atoms. Depending on the lateral position the tip atom recedes to a different extend towards the W-surface.

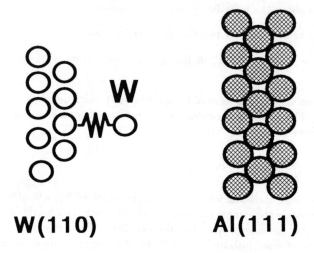

Figure 7: Simplest possible model for an elastically deformable tip

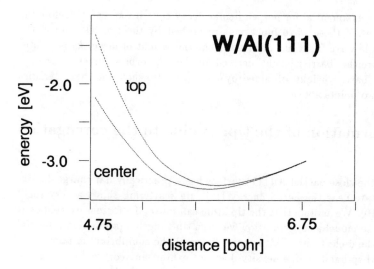

Figure 8: Corrugation of the W/Al(111)-potential

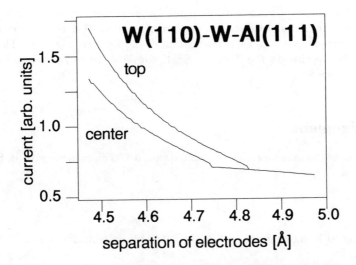

Figure 9: Tunnel current as function of elctrode separation for Al(111) with an elastically deformable tip

The total potential is obtained by superposing the W-Al and the W-W-interaction potential. For a given electrode separation the W-tip atom will find its equilibrium position between the two electrodes such that the net force is zero.

How does this influence the tunnel current? The tunnel current flows dominantly via the 6s-tip orbital so that the current is approximately given by the formula:

$$J = \frac{2\pi e}{\hbar} \frac{1}{\Delta E} \tilde{S}_{6s,6s}^{tip} S_{6s,6s}^{sample} W^2(6s) \tag{11}$$

It is the product of the 6s-density in the tip states times the 6s-density in the sample states. Both quantities vary with distance. The 6s-tip densitiy varies much more strongly than the 6s-sample density. if the W-tip atom recedes towards the surface $S_{6s,6s}^{tip}$ will increase strongly, whereas $S_{6s,6s}^{sample}$ will decrease slightly. The net effect is hence an increase of the tunnel current which is larger in the top position than in the center position. The result is a corrugation in the tunnel current as shown in fig. 9. This means that in the constant current mode the separation of the electrodes has to be increased in the top position as compared to the center position. There is only a small increase of the corrugation with tunnel current . We do not obtain an exponential variation of the corrugation amplitude with electrode separation.

6 Conclusions

A formalism has been presented which permits the inclusion of d-electrons on the tip. The evaluation of the tunnel current avoids the approximations involved in the transfer

Hamiltonian approach. The d-wave functions influence the tunnel current by modifying the electronic structure and, at short distances, by directly contributing to the current. The 'barrier height' depends on the angular momentum of the tip wave function. The elastic deformation of the tip changes the 'barrier height' and leads to corrugation on sp-band metals at short distsnces.

Acknowledgement

We thank J. Behm for assistance. G. D. profited from useful discussions with M. Scheffler and J.P. Vigneron.

References

[1] J. Tersoff and D.R. Hamann, Phys. Rev. Lett. **50**, 1998 (1983); Phys. Rev. **B 31**, 805 (1985).

[2] N.D. Lang, Phys. Rev. Lett. **55**, 230 (1985); N. D. Lang, Phys. Rev. Lett., **56** (1986) 164; Phys. Rev. **B34**, 5547 (1986).

[3] J. Bardeen, Phys. Rev. Lett. **6**, 57 (1961).

[4] G. Doyen, D. Drakova, E. Kopatzki and R.J. Behm, J. Vac. Sci. Technol. **A6**, 327 (1988).

[5] E. Kopatzki, G. Doyen, D. Drakova, and R.J. Behm, Journal of Microscopy, **151**, 687 (1988).

[6] G. Doyen and D. Drakova, Surface Sci., **178**, 375 (1986).

[7] J. Wintterlin, J. Wiechers, H. Brune, T. Gritsch, H. Höfer, and R. J. Behm, Phys. Rev. Lett. **62**, 59 (1989).

[8] J. Wintterlin, PhD-Thesis, Freie Universität Berlin, 1988.

[9] D. Drakova, G. Doyen and F. von Trentini, Phys. Rev. **B 32**, 6399 (1985)

[10] D. Drakova, G. Doyen and R. Hübner, J. Chem. Phys., **89**, 1725 (1988).

[11] G. Doyen and G. Ertl, Surface Sci., **65**, 641, (1977); G. Doyen and G. Ertl, J. Chem. Phys., **68**, 5417 (1978).

[12] J.Hölzl and F.K. Schulte, in: 'Solid Surface Physics', ed. G. Höhler, Springer Tracts in Modern Physics, Vol. 85,p. 1, 1979, Springer Verlag, Berlin, Heidelberg, New York.

[13] B. A. Lippmann, Phys. Rev. Lett. **15**, 11 (1965); B. A. Lippmann, Phys. Rev. Lett. **16**, 135 (1966).

[14] P. Ehrenfest, Z. Physik, **45**, 455 (1927).

[15] M. Gell-Mann and M.L. Goldberger, Phys. Rev. **91**, 398 (1953).

[16] G. Doyen and D. Drakova, in preparation.

[17] R. D. Levine, 'Quantum Mechanics of Molecular Rate Processes', Oxford University Press, Oxford (1969).

[18] K. Mednick and L. Kleinman, Phys. Rev., **B 22**, 5768 (1980).

[19] J. P. Toennies and A. Lock, cited by J. Wintterlin, ref. [8].

[20] J. G. Simmons, J. Appl. Phys., **34**, 1793 (1963); J. Appl. Phys., **34**, 2581 (1963).

[21] N. D. Lang, Phys. Rev., **B 37**, 395 (1988).

TIP- SURFACE INTERACTIONS

S. Ciraci
Department of Physics,
Bilkent University,
Bilkent 06533 Ankara, Turkey.

ABSTRACT In the earlier studies of Scanning Tunneling Microscopy the tip-sample distance was generally assumed to be large enough to disregard the tip-surface interaction effects, and the electrodes were considered to be independent. However, modifications of the electronic structure depending upon the tip-surface separation have led to the identification of different regimes (i.e. independent electrode, electronic contact and point contact regimes) in the operation of STM. As the tip approaches the sample surface, the tip-sample interaction gradually increases and the potential barrier is lowered. The charge density is rearranged and the ions of the tip and sample are displaced to attain the minimum of the total energy at the preset tip-sample distance. Owing to the overlap of the tip and sample states site-specific localized states appear and provide a net binding interaction, whereby the tip and sample are connected electronically. Upon further approach of the tip to the sample surface the tunneling barrier is perforated even before the tip enters in the strongly attractive force region. Eventually, the point contact regime is initiated, and new channels of conduction are opened through the electronic states localized in the gap. The conductance undergoes a qualitative change and its commonly accepted proportionality to the local density of states of the unperturbed sample surface is invalidated. In this lecture note, a microscopic analysis of the tip-surface interaction is presented and recent STM studies are examined within the framework of this analysis.

1. Introduction

The invention of Scanning Tunneling Microscopy (STM) by Binnig and Rohrer[1] provided a powerful technique with the real space imaging capability and atomic scale resolution. It has created a great deal of interest in revealing atomic and electronic structure of solid surfaces, and as well as molecules and biological substances without invoking the atomic periodicity. The atomic resolution has been reported not only in ultra high vacuum (UHV), but under atmospheric as well as aqueous conditions[2-4]. Although only a few years have passed since the first operation of STM, our understanding about the physics underlying this new technique has greatly advanced, and the scanning tunneling microscope has developed as a real-space imaging tool with important applications in surface science and technology. In the meantime several new applications have already been introduced. For example, scanning tunneling spectroscopy proposed by Baratoff[5] and Selloni et al.[6] has been already operational with useful results. It appears that in some cases STM combined with spectroscopy can be superior to the other experimental techniques in the investigation of the electronic and atomic structure of surfaces. Either by changing the polarity of the applied bias voltage V, or by changing its value for a given polarity one is able to probe surface–localized states in the electronic energy spectrum[7]. There have been also efforts to identify adatoms by using STM. The Atomic Force Microscopy (AFM) introduced by Binnig, Quate and Gerber[8] is another important invention emerged after STM. Instead of the tunneling current, AFM measures the force generated between the tip and sample surface and its operation is usually combined with a scanning tunneling microscope. To achieve a meaningful force measurement AFM has to operate with a small tip-sample distance to yield significant tip-sample interaction. As far as the tip-surface interaction effects are concerned STM and AFM have many common features, and hence both has to be treated in the same context.

In the early years of STM, the focus of attention has been the nature and resolution of the observed images. Starting from the transfer Hamiltonian formalism proposed by Bardeen[9] and representing the tip by a single s-wave Tersoff and Hamann[10] showed that the tunneling current and the STM images can be related to the local density of the electronic states of the free sample evaluated at the center of the tip and at the Fermi level, $\rho_s(\mathbf{r}_o, E_F)$. This quantity may differ from the total surface charge density $\rho_s(\mathbf{r})$, especially in the case of semiconductors and semimetals. In fact, intriguing evidence of their theory has emerged from the STM images of graphite surface. The symmetry lowering due to the shift of atomic positions in the consecutive layers and thus weak interlayer interaction between certain type of carbon atoms has been resolved by STM yielding only three protrusions rather than six in a surface hexagon[11]. Based on the Tersoff-Hamann theory, Batra et al.[12] suggested that three atomic sites which do not have atoms directly below and for which $\rho_s(\mathbf{r}_o, E_F)$ is significant are more likely to be probed. The crucial effect of the shape of the tip on the STM images has been demonstrated recently by Mizes and Harrison[13].

The first-order perturbation approach indigenous to the above theories assumes sufficiently spaced, nearly independent electrodes. In fact, even if the wave function overlap or, equivalently the tunneling probability is small, there exist the attractive Van der Waals (or electron correlation) energy and the repulsive Pauli exclusion energy yielding a net weak binding energy between the tip and surface. The observation of force variation of order 10^{-9} N while the tip is scanned under typical STM operating conditions demonstrates the existence of a tip-sample interaction with varying degree of importance depending upon the tip-sample separation d. This points to the question that in what range of d a simple relation between STM images and the electronic structure of the unperturbed sample is valid.

The importance of the tip-surface interaction was brought about by the seminal work of Soler et al.[14]. They showed that STM operating at very small bias voltage can yield corrugation which is much larger than that can be deduced from $\rho_s(\mathbf{r}_o, E_F)$ of the free graphite surface. The observed giant corrugation has been explained by the elastic deformation indicating a significant tip-surface interaction due to the close proximity of the tunneling tip to the graphite surface. Later, self-consistent force and charge density calculations by Ciraci and Batra[15] have justified the strong interaction induced by the tip. They demonstrated that for $d \simeq 3$ a.u (1.5Å) the tunneling barrier collapses and a chemical bond between the tip and surface atom is formed. Quite recently, increasingly stronger attraction up to the adhesion minimum was shown to occur over essentially the same d-range in a combined STM-AFM experiment by Dürig et al.[16]. Atomic resolution observed for the nominally flat (111) surface of the close-packed noble and simple metals with a corrugation much larger than one would deduce from $\rho_s(\mathbf{r}_o, E_F)$, provide also evidence for unusually strong tip-surface interaction[17,18]. Whether force variations and induced deformations along STM scans and/or changes in electronic structure must be invoked to explain such observations is understandably a matter of increasing concern.

As the tip approaches the surface, the tip-sample interaction gradually increases and the potential barrier is lowered. The charge density is rearranged and the ions of the tip and sample are displaced to attain the minimum of the total energy at the preset d. Owing to their significant overlap at small d, the tip and sample states are combined to form site-specific localized states and provides a net binding interaction. The ab-initio force and electronic structure calculations performed recently by Ciraci, Baratoff and Batra[19] indicate that the corrugation amplitude observed in the STM images is slightly reduced by the elastic deformation of the tip in the weak attractive force region. In contrast, the tip-induced local modifications of the electronic structure have a much stronger effect on the STM images, whereby the tip and sample is electronically connected. Tekman and Ciraci[20] studied effects of the tip sample interaction for the graphite surface within the empirical tight binding approximation, and provided additional insight into the formation and evolution of tip-induced states as a function of d. They showed that STM contrast can be significantly enhanced in the presence of the tip induced local perturbation in the electronic structure. This was demonstrated in calculations based on the generalization of the Tersoff-Hamann approach, which properly includes contributions from the tip induced localized states concentrated near the tip or the sample. Dramatic effects of the disturbances in the electronic states due to the close proximity of the tip to the sample surface are clarified by the recent study of Ciraci, Baratoff and Batra[21]. They showed that for the graphite surface the tunneling conductance, G_t at small voltage, which is determined by states at the corners of the Brillouin zone (BZ) if tip-sample interactions are neglected, becomes dominated by the admixture of other states near E_F. In this fashion a significant tunneling current can arise even if the tip is at the hollow site, which is normally has very low charge density.

Upon further approach of the tip to the sample the perforation of the tunneling barrier will set on even before it enters in the strongly attractive force region[15,19,21,22]. This will lead to the point contact[23]. Since new channels of conduction[24] are opened through the electronic states localized in the gap, the character of the conductance, G undergoes a qualitative change and its commonly accepted proportionality to the local density of the free sample surface is invalidated. The gradual collapse of the barrier as d is decreased and its effect on the current have been illustrated experimentally by Gimzewski and Möller[25] in studies of the transition from tunneling to essentially point contact between metallic tip and sample, and theoretically by Lang[26] in calculations for adatoms on jellium substrate.

The interpretation of the corrugation measured in AFM and thus the order of the force exerting on the tip - which is a direct measure of the tip-surface interaction - are also subject of interest. Atomic resolutions have so far been achieved with repulsive forces of 10^{-6} to 10^{-8} N in the case of graphite[27]. Theoretical results by Abraham, Batra and Ciraci[28] indicate, however, that such strong forces are sufficient to induce plastic deformation[15,28]. It is commonly assumed that force variations are primarily determined by the repulsive interaction which is in turn proportional to the total electronic charge density of the free surface of the sample, $\rho_s(\mathbf{r})$. Recent theoretical results have been at variance with this simple picture. Atomic force calculations indicated that the ion-ion repulsion dominates the force acting on the outermost tip atom in the repulsive or strongly attractive range. Beyond a certain separation the magnitude of the force on the tip at the hollow site (i.e. center of the surface hexagon) of graphite can exceed that at the atomic site.

In this lecture note a microscopic analysis of the tip-surface interaction, which comprises the disturbances of the electronic structure of the tip and sample, as well as the variation of the force field occurred due to the close proximity of the tunneling tip is presented. Depending upon the value of d the strength of the tip-surface interactions varies. As a result one identifies different regimes (i.e. independent electrode, electronic and point contact regimes) in the operation of STM. The dividing line between these regimes are not sharp and in each regime the origin of the conductance is different. Beyond the independent electrode regime the effects of this interaction appear as either a weak attractive force field and states localized in the vicinity of the gap modifying the tunneling current, or as strongly attractive (or even repulsive) force field and chemical bond between the tip and surface atoms. The strong tip-sample interaction may lead to elastic (even plastic) deformation imposing dramatic changes in the corrugations observed by STM. Accordingly, present analysis covers a wide range of d starting from weak attractive up to strongly repulsive force region.

2. Experimental Studies

The graphite surface has been the center of attention in STM for the following reasons. First the flatness and inertness make the (0001) cleavage plane of graphite an ideal sample in STM. Secondly, STM images of the graphite surface reveal anomalous features, namely the asymmetry and the giant corrugation at small bias voltage. The bulk graphite has a layered structure with a weak interlayer coupling. Since the atomic positions are shifted in the consecutive layers, two inequivalent atomic sites occur in a given layer. These sites are usually denoted as the A- and the B-sites. The A-site has carbon atoms directly below and above it in the adjacent layers, whereas the B-site does not. In each layer, the A-, and the B-site atoms form hexagons, the center of which are denoted as the H-site. While an individual graphite layer (unsupported monolayer) has sixfold rotation symmetry with zero charge density at the H-site and with the Fermi surface collapsed to a single point at the K-point of BZ, for a graphite slab the symmetry is lowered to the threefold rotation, and the Fermi surface changes into the small cigar-like pockets around the H-K line of the bulk BZ. Four bands near E_F contribute to the Fermi surface[29]. The $\pi-$ and π^*- bands lying below and above the Fermi level, respectively display small dispersion, and originate from the bonding and antibonding combination of the p_z-orbitals located at the A-sites of the consecutive layers. In contrast, the p_z-orbitals at the B-site can not form bonding or antibonding combination[12], and hence yield flat bands[30], which practically coincide with the Fermi level along the H-K direction. Consequently, $\rho_s(\mathbf{r}_o, E_F)$ is dominated by the states of the flat bands with electronic charge concentrated on the B-site atoms, which, in turn, sampled by STM operating either in the current-imaging or in the topographic imaging mode with large bias voltage. The observed corrugation is found to be in reasonable agreement with that deduced from the tunneling-current formalism of Tersoff and Hamann[10] using the calculated $\rho_s(\mathbf{r}_o, E_F)$[6,15]. On the other hand the calculated corrugation of the total surface charge density, $\rho_s(\mathbf{r})$ is quite small and is only one forth of that calculated for $\rho_s(\mathbf{r}_o, E_F)$. This dramatic difference between $\rho_s(\mathbf{r}_o, E_F)$ and $\rho_s(\mathbf{r})$ has made graphite an interesting sample for STM.

Soler et al.[14] observed unusually large corrugation for the graphite surface by decreasing both the bias voltage and the tunneling current (or equivalently by decreasing the tip-sample distance) in STM operating in the topographic imaging mode. Moreover, they showed that at constant bias the contrast of the STM images increases with the increasing tunnel-current. These results have been the first indication of the fact that the commonly accepted proportionality of the tunnel-current to the electronic structure of the bare sample can break down in the presence of the tip-surface interaction. The observed huge corrugation was attributed to the enhancement of the corrugation - which is normally derived from $\rho_s(\mathbf{r}_o, E_F)$ - by the local elastic deformation of the surface. Using the elastic constants of graphite and approximating the tip-surface interaction by an empirical Morse-potential, they obtained the variation of the force exerted on the sample surface as a function of d. Then the perpendicular displacement of the surface layer was estimated within the theory of elasticity. With this oversimplified model they showed that the tip following the contours of $\rho_s(\mathbf{r}_o, E_F)$ lies in the repulsive region when it is at the H-site, but enters in the attractive region when it is retracted by going to the B-site. At the same time, because of the interaction set in between the sample and tip, the surface layers are elastically deformed as if a membrane attached to the tip. They argued that the elastic deformation of the surface layers are included in the apparent vertical displacement of the tip, which follows the contours of $\rho_s(\mathbf{r}_o, E_F)$ to keep the tunnel current constant at the value set by the apparatus. Like the springs connected in series the displacement of the surface is further increased resulting in an enormously enhanced apparent displacement of the tip. As it is noted neither disturbances of the electronic states induced by the close proximity of the tip, nor the lateral rigidity of the graphite layers imposing significant resistance for the perpendicular displacement are taken into account in the work of Soler et al.[14].

This explains why large displacements are induced by small tip forces of the order of 10^{-10} Newtons estimated in this study. While the interpretation based on the elastic deformation is revised by including the effect of the dirts, or a particle of dust left between the tip and sample, or other explanations are proposed[31], the observed huge corrugation have already created a strong motivation to investigate the tip-surface interaction.

Most recently, Wintterlin et al.[18] reported the atomic resolution with a corrugation amplitude of 0.6-1.5 a.u. (0.3-0.8 Å) for the Al(111) surface obtained by STM operating in the topographic mode. Being a simple metal, the close packed Al(111) surface is practically flat with negligible corrugation of charge density. Moreover no surface state was found near the Fermi level. The contours of $\rho_s(\mathbf{r})$ do not deviate significantly from those of $\rho_s(\mathbf{r}_o, E_F)$. The He-scattering potential, which is directly associated with $\rho_s(\mathbf{r}_o, E_F)$ of Al(111) surface at the turning point of the scattered He, exhibits[18] a corrugation of 0.04 a.u. The large corrugation observed over a range of gap-width of ∼5 a.u. which is far outside the turning point for He scattering, and which is detected only for certain treatments of the tip, is rather unexpected for Al(111) surface. Wintterlin et al.[18] sought the origin of the large corrugation in the tip-surface interaction, and argued that the forces exerting on the tip atoms give rise to the elastic distortions. This distortion becomes significant when the constant force contours deviate from the line-scans of the tip to be obtained in the absence of the tip-sample interaction. The apparent vertical displacement of the tip is amplified by the site-dependent elastic deformation of the tip atoms[18]. Clearly this observation, as well as the earlier STM of Au(111) surface[17], both revealing large corrugations at certain conditions provide evidence for a strong tip-surface interaction. However, later it will be shown that the STM contrast is reduced by the deformation of the tip atoms lying in the attractive force region. Furthermore, at the given tip-sample distance, the potential barrier collapses leading to an electronic contact, the strength of which varies with the lateral position of the tip.

A direct way to investigate the tip-sample interaction has been taken by Dürig et al.[16]. In a combined STM-AFM experiment they measured the gradient of the force acting between the Ir tip and Ir sample in the form of a cantilever as a function of d. Using the data for the force gradient they were able to estimate the force curve. Their observation that the gap width is considerably smaller than one generally assumes (of the order of 2-8 a.u. or 1-4 Å) and the tip-sample interaction forces are substantial (of the order of 10^{-9} N) demonstrate the important role of the tip-sample interaction effects.

3. Theoretical Analysis

With the STM images showing dramatic deviation from the theory based on the electronic structure of the bare sample surface and also with measurements of AFM, attention has turned to the nature of the tip-surface interaction[15]. Some theoretical studies attempted to include the tip-induced effects either by reformulating the transfer Hamiltonian approach[20] or by developing a new formalism beyond the perturbation theory.[32] By going beyond the perturbation theory, Lang[33] computed the tunneling current directly from the electronic structure of a combined tip-sample system. Since the system he used consists of a jellium metal for the sample, and a tip modeled by a single atom attached to the jellium metal, the connection between the electronic and atomic structure is omitted.

3.1 AB-INITIO MICROSCOPIC THEORY

The important result emerging from several experimental STM and AFM works is, however, that the tip-surface interaction is sample-specific and strongly dependent on the detailed atomic structure of the surface. It becomes also clear that rather than rigid electrodes, the tip-induced disturbances in the electronic and atomic structures have to be considered concomitantly. Therefore, a theoretical analysis on the nature of the said interaction and its effects has to be in the atomic scale. Although a universal formalism can not be developed due to the sample-specific nature of the interactions, we nevertheless explore the tip-surface interactions on certain prototype systems. This way, a more realistic and deeper understanding of interatomic forces on the tip atoms of the surface can be provided. Also a more systematic account of changes in the potential energy, charge density and electronic structure as a function of the tip position can be given in a self-consistent framework.

3.2 LOCAL DENSITY APPROACH

A solution of the charge distribution and atomic structure of a combined tip-sample system at a preset d and under a small bias voltage, V can be obtained from the ground state total energy E_T. Hohenberg and Kohn[34] derived a powerful formulation of this many-electron problem. They showed that the ground state total energy of an interacting inhomogeneous electron gas confined in the field of the constituent ions can be written as a functional of the electron density $\rho(\mathbf{r})$. Further, they showed that $E_T[\rho]$ assumes a minimum value for the correct $\rho(\mathbf{r})$, if admissible density function conserves the total number of electrons. Thus, $\rho(\mathbf{r})$ can be determined from $(\delta/\delta\rho)E_T[\rho] - \mu N = 0$ with $\mu = \delta E_T/\delta\rho$ and $N = \int \rho(\mathbf{r})d\mathbf{r}$.

3.2.1 SCF-Pseudopotential Method

Following the density functional formalism Ihm, Zunger, Yin and Cohen[35] obtained an expression of the total energy of the crystal (defined as the total energy difference between the solid state and isolated atoms) in the pseudopotential theory.

$$E_T = T + V + \int E_{xc}(\mathbf{r})d\mathbf{r} \qquad (3.1)$$

The total kinetic energy, T is expressed in atomic units

$$T = \sum_i^{occ} \psi_i^*(\mathbf{r})(-\nabla^2)\psi_i(\mathbf{r})d\mathbf{r} \qquad (3.2)$$

and the potential energy, V is

$$V = \sum_{i,\mu,l} \int \psi_i^*(\mathbf{r})U_l(\mathbf{r}-\mathbf{R}_\mu)\mathcal{P}_l\psi_i(\mathbf{r})d\mathbf{r} + \int\int \frac{\rho(\mathbf{r})\rho(\mathbf{r'})}{|\mathbf{r}-\mathbf{r'}|}d\mathbf{r}d\mathbf{r'} + \sum_{\mu\neq\nu} \frac{Z^2}{|\mathbf{R}_\mu - \mathbf{R}_\nu|} \qquad (3.3)$$

In these equations, $U_l(\mathbf{r}-\mathbf{R}_\mu)\mathcal{P}_l$ is the non-local pseudopotential and angular momentum projection operator at the ion-core position $\mathbf{R}_\mu$, and Z is the valency. The total charge density is given by $\rho(\mathbf{r}) = \sum^{occ} \psi_i^*(\mathbf{r})\psi_i(\mathbf{r})$. The wave functions are the eigenstates of the one electron Schrödinger equation, which is obtained variationally from $E_T[\rho]$.

$$[-\nabla^2 + \sum_{\mu,l} U_l(\mathbf{r}-\mathbf{R}_\mu)\mathcal{P}_l + \int \frac{2\rho(\mathbf{r'})}{|\mathbf{r}-\mathbf{r'}|}d\mathbf{r'} + \mu_{xc}(\mathbf{r})]\psi_i(\mathbf{r}) = \epsilon_i\psi(\mathbf{r}) \qquad (3.4)$$

The local exchange-correlation potential, $\mu_{xc}(\mathbf{r})$ is expressed in several approximate forms. By using the periodic boundary conditions and the translational symmetry of the crystal the solution of the time independent, one-electron Schrödinger equation is greatly simplified, and the wave functions are obtained in the form of Bloch states, $\psi_{n,k}(\mathbf{r})$, with n and k being the band index and the wave vector in the reduced BZ. Owing to the weak pseudopotential $U_l(\mathbf{r})$, the Bloch wave functions can be expressed by the linear combination of a limited number of plane waves,

$$\psi_{n,k}(\mathbf{r}) = \sum_\mathbf{G} \psi_n(\mathbf{k}+\mathbf{G})e^{i(\mathbf{k}+\mathbf{G})\cdot\mathbf{r}} \qquad (3.5)$$

where $\mathbf{G}$ is the reciprocal lattice vector and $\psi_n(\mathbf{k}+\mathbf{G})$ represents the Fourier component corresponding to the wave vector $(\mathbf{k}+\mathbf{G})$. With this expansion, the total energy of the system is expressed in momentum space, in which the positions of the individual ions are the parameters corresponding to a point in the Born-Oppenheimer surface. However, if the system is under an external force this point may not coincide with a local minimum.

3.3 PERIODIC MODEL

The application of the SCF-pseudopotential method on the combined tip-sample system in STM or in AFM is achieved by using repeated slab model. Each slab, which is separated by a large vacuum region from the adjacent ones, is made of periodically repeated tip-sample system in a supercell. The periodic boundary conditions imposed are artificial, but allow one to use a plane-wave basis set. The major disadvantage of this model is the artificial intertip interaction which, in turn, can be reduced significantly by using a supercell with large lateral lattice constants. In the calculations, the ion cores are represented by nonlocal, normconserving ionic pseudopotentials given by Hamann et al.[36], and the exchange-correlation potential given by Ceperley and Alder[37]. Other details of the calculations can be found in Ref's 15,38.

3.4 INTERACTION ENERGY AND ATOMIC FORCES

The interaction energy and the atomic force exerting on the tip and surface atoms are quantities which measure the tip-sample interaction. The interaction energy is defined as

$$\mathcal{E}(d) = E_T[t+s] - E_T^o[s] - E_T^o[t] \quad (3.6)$$

and requires separate computations of the total energies of the bare sample $E_T^o[s]$, the bare tip $E_T^o[t]$, and also that of the combined tip and sample system, $E_T[t+s]$. To reduce errors due to the finite kinetic energy cutoff determining the number of plane-waves in the Bloch wave functions, all calculations are carried out in the same supercell. The interaction energy is repulsive for small tip-sample distance d, but becomes attractive for large d. The minimum of the interaction energy with respect to d, i.e. $\partial \mathcal{E}(d)/\partial d = 0$, yields the binding energy E_b. It should be noted that the calculation of the interaction energy $\mathcal{E}$ within the local density approximation (LDA) cannot be carried out for large d. First lateral interaction between the artificially repeated tips may exceed the tip-surface interaction. Secondly, in the resulting thick vacuum region between the surface and the outermost tip atom the tails and the residual overlap of wave functions are poorly represented. Because the charge density becomes very low ($\rho \simeq 10^{-5} - 10^{-6}$ a.u.) the local density approximation to the exchange-correlation potential becomes inappropriate. For large d the interaction between the sample surface and the tip is better represented by the sum of the attractive Van der Waals energy and the repulsive energy arising from the orthogonality of weakly overlapping tip and substrate states. The sum of these two contribution always yields a bound state, which has to join to that obtained by LDA as d decreases[39].

Following the Hellman-Feynman theorem, the ith component of the forces exerting on the atom j at τ_j in the supercell is obtained from $\partial E_T[t+s]/\partial \tau_{j,i}$. In the pseudopotential method implemented in momentum space within the LDA the atomic forces have two components. The ionic one, F_{io} originates from the ion-ion repulsion and is calculated by using the Ewald procedure.[35]. The electronic one, F_{el} is due to the interaction of electrons with the ionic pseudopotential, and is obtained from the negative gradient of the expectation value of the valence ionic pseudopotential

$$<U> = \sum_{n,k}\sum_{\mu,s}\sum_{l} <\psi_{n,k}(\mathbf{r}) \mid U_l(\mathbf{r} - \mathbf{R}_\mu - \tau_j)\mathcal{P}_l \mid \psi_{n,k}(\mathbf{r})> \qquad (3.7)$$

with respect to τ_j[38].

The force gradient, which is preferably measured in the combined AFM-STM experiments, is obtained from the gradient of F_{io} and F_{el} terms. It should be noted that an accurate computation of the atomic force is sensitive to the self consistency of the electronic charge and thus is achieved with very stringent SCF convergence criterion of 10^{-7} Ry (rms deviation in potential).

4. Results

In this Section the results of earlier ab-initio calculations[15,19,21] is outlined and the effects of the tip- surface interaction revealed thereof are analyzed.

4.1 CARBON-TIP AND GRAPHITE-SAMPLE

Representing the graphite sample by a three-layer slab and the tip by a single carbon atom, Ciraci and Batra[15] carried out SCF-pseudopotential calculations within a (1x1) supercell structure. They calculated $E_T[t+s]$, the perpendicular component of the force exerting on the tip, charge density and the potential for d ranging from 2.7 a.u. to 7.6 a.u. The potential barrier between the tip and surface were obtained from the planarly averaged potential, $\bar{V}(z)$. As seen in Fig.1 for the B-site the potential barrier is lowered as d decreases, but totally collapses for d=2.7 a.u. allowing the formation of a bond between the tip and nearest surface atom. Moreover, a strong repulsive force of 1.5×10^{-8} N is exerted on the tip atom at this tip-sample distance. This force is balanced by resultant reaction force distributed over the layers of the graphite. It is large at the surface, but decays rapidly as one goes deeper into the bulk. The calculated values indicate that the small d is a non-equilibrium state, and $\mathcal{E}_T$ and as well as the repulsive F_t decrease with increasing d. For example, the force is still repulsive for d=4.7 a.u., but its magnitude

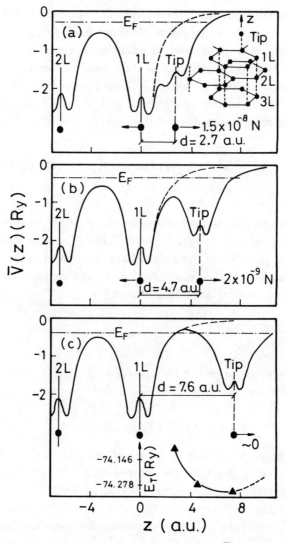

Figure 1 Planarly averaged 1D pseudopotential $\bar{V}(z)$, interatomic forces for $d = 2.7$, 4.7 and 7.6 a.u. The inset in the top panel describes three layer graphite slab and single carbon atom representing the tip. The variation of the total energy with z (or d is shown in the bottom panel).

is dramatically reduced. Further retracting of the tip (d=7.6 a.u.) causes the magnitude of the force to diminish. This is where the minimum value of $E_T[t+s]$ occurs for the B-site. For the H-site, the zero-force distance occurs at d=6.7 a.u.

Note that these zero-force distances were defined only for a single-atom tip, and hence are expected to be different for the macroscopic one. The work of Ciraci and Batra[15] was the first ab-initio calculation of the tip force and the potential barrier obtained from a simple model, and has nevertheless provided evidence for the strong tip-sample interaction leading to the point contact regime.

4.2 ALUMINUM-TIP AND GRAPHITE SAMPLE

The interaction between a metal-tip and the graphite sample is investigated in the supercell of the periodically repeated tip-sample system consisting of four unit cells of the graphite surface, and a single Al-atom representing the tip. Since the previous calculations showed that tip-surface interaction forces predominantly act on the top graphite layer and the interlayer interactions are immaterial at small d, the graphite sample is represented by a monolayer. In this case the A-, and the B-site become equivalent, and are denoted as the T-site. In this (2x2) model the lateral repeat period is 9.3 a.u. (4.92 Å). The SCF-total energy, force and electronic structure calculations are carried out for the tip atom positioned above the T-, and the H-site with d varying in the range from 2.7 to 8 a.u. (1.43 to 4.23 Å). For purposes of comparison, a few calculations are also made with a (3x3)-supercell consisting of 9 graphite monolayer unit cell and either one or four Al atoms representing the metal tip with lateral repeat period of ∼14 a.u.

The calculated interaction energies are illustrated in Fig.2. The binding energies are rather small: $E_b \simeq$ -0.33 eV for the T-site, and $E_b \simeq$ -0.61 eV for the H-site. As expected $\mathcal{E}^H \to \mathcal{E}^T$ as $d \to \infty$. The small value of the binding energy at either site is the manifestation of the fact that the graphite surface is inert, and a small number of graphite states near E_F can effectively interact with Al states. As will be seen later in this section, in the region of d where the interaction energy is significant (i.e. where the minimum of $\mathcal{E}$ occurs) a chemical bond between the outermost tip atom and the atoms of the sample surface may form. At small d, ionic repulsion between the carbon and Al atoms is larger at the T-site than at the H-site. Consequently, $\mathcal{E}^T > \mathcal{E}^H$ for $d \leq 3$ a.u. In the (3x3) supercell with a tip containing four Al atoms $|\mathcal{E}^H - \mathcal{E}^T|$ is found to be 40% smaller than for the (2x2) supercell having a single Al atom for the metal tip. It appears that the metallic binding among four Al atoms representing the tip redistribute the electronic charge causing the interaction between the tip and the sample to weaken, and the Al-graphite bond to be less localized. As a result, the monolayer feels the interaction of a metal cluster, and the difference between the T-site and H-site becomes less pronounced. At large d ($d \geq 7$ a.u.) the two curves corresponding to the T-, and the H- sites essentially merge leaving a site-independent attraction.

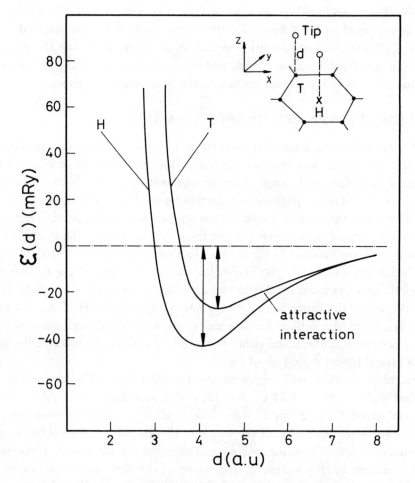

Figure 2   Interaction energy $\mathcal{E}(d)$ calculated for the single Al atom and the graphite monolayer. The inset describes the T-, and the H-site positions. Binding energies are indicated by arrows.

The calculated force which acts on the tip atom is directly related to AFM. There are several important issues related with the tip force, which are worth exploring. The fundamental issue is what physical property of the surface determines the force field. Is it the total surface charge density, or $\rho_s(\mathbf{r_o}, E_F)$? The second one, which is of course related to the former, is the force corrugation and the atomic resolution. Other questions to be answered are why the calculated forces are so much different from the measured ones, and how the deformation induced by the tip force affects the corrugation obtained in STM.

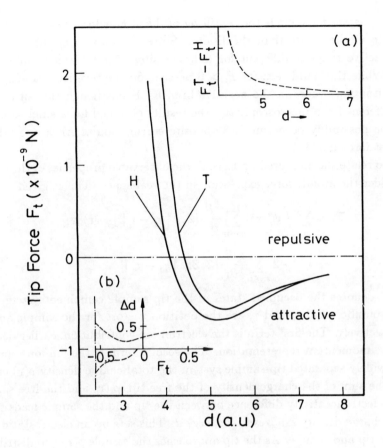

Figure 3 Perpendicular component of force, F_t exerted on the tip represented by a single Al atom versus the tip-sample distance d.

The vertical component of the tip force F_t, calculated for the T-, and the H-site illustrated in Fig.3. It is seen that for a given d in the repulsive range the vertical component of the tip force at the T-site, F_t^T, is larger than that at the H-site, F_t^H. By contrast, in the attractive force region the magnitude of the tip force at the H-site is larger. Moreover calculations imply that at large d (i.e $d \geq 8a.u$) the force curves $F_t^T(d)$ and $F_t^H(d)$ cross and $\mid F_t^T \mid$ becomes larger than $\mid F_t^H \mid$. However, the crossing of the force curve is sample specific and occurs in the region where LDA breaks down. In the inset (a) of Fig.3 $F_t^T - F_t^H$ is plotted as a function of d. The inset(b) shows how the corrugation Δd (i.e. the displacement of the tip by going from the T-site to the H-site under the constant force F_t) varies with F_t. Apparently, $\Delta d(F_t)$ is a single valued and positive function for $F_t > 0$, and indicates that in the constant force mode the cantilever of the atomic force microscope would deflect more at the T-site than the H-site. Since the maximum value of $|F_t|$ in the attractive range is different for different sites, $\Delta d(F_t)$ is a double valued function. While the trend seen for $F_t > 0$ persists on the positive branch, nearly a reverse situation occurs in the negative branch. Even though this situation is perhaps different for a macroscopic tip, the results obtained for a single-atom tip suggests the possibility observing two opposite corrugation of AFM operating in the constant force mode.

In order to relate the measured tip force to the observable properties of the sample let us consider the atomic force expression in the real space. This is given by

$$\mathbf{F_t} = \sum_i^{occ} \int \psi_i^*(\mathbf{r}) \{ \sum_{j \in tip} \frac{\partial}{\partial \tau_j} [Z_j/|\tau_j - \mathbf{r}|] \} \psi_i(\mathbf{r}) d\mathbf{r}$$

$$- 2 \sum_{s,j \in tip} \frac{\partial}{\partial \tau_j} [Z_j Z_s/|\tau_j - \tau_s|] \quad (4.8)$$

where $\psi_i(\mathbf{r})$ denotes the occupied states of the tip-sample combined system. Z is the atomic number, and τ_s and τ_j are the position vectors for the sample and tip atoms, respectively. The first term is the electron-ion contribution, earlier denoted as F_{el} in the momentum representation. The second term is the ion-ion repulsion F_{io}. For a widely separated tip-sample system the total charge density $\rho(\mathbf{r})$ can be viewed as the sum of the charge density of the free tip, $\rho_t(\mathbf{r})$ and the free sample, $\rho_s(\mathbf{r})$. The electronegativity difference between the tip and the sample leads to the transfer of charge density $\Delta \rho_\infty$ even for large d. This sets up an electrostatic force between the tip and sample. As the tip approaches the sample $\rho(\mathbf{r})$ is redistributed to minimize the total energy for a given ionic position. Therefore the total charge density is expressed as $\rho(\mathbf{r}) = \rho_s(\mathbf{r}) + \rho_t(\mathbf{r}) - \Delta \rho(\mathbf{r})$. Clearly, $\Delta \rho(\mathbf{r})$ is the difference charge density arising from the tip-sample interaction, and its integral over the space is equal to zero. Assuming that $\Delta \rho_\infty$ is split between ρ_s and ρ_t and taking certain cancellations into account one finally obtains the perpendicular component of the force acting on the tip:

$$F_t \simeq 2 \int [\rho(\mathbf{r}) - \Delta \rho(\mathbf{r})] \{ \sum_{j \in tip} \frac{\partial}{\partial \tau_{j,z}} [Z_j/|\tau_j - \mathbf{r}|] \} d\mathbf{r}$$

$$- 2 \sum_{s,j \in tip} \frac{\partial}{\partial \tau_{j,z}} \{ Z_j Z_s/|\tau_j - \tau_s| \} \quad (4.9)$$

We first drop the summations over all the tip atoms and consider only the outermost tip atom. The first limiting case we consider is very small d, where $F_{io} > |F_{el}|$, and thus F_t is positive and repulsive. In the strongly repulsive regime F_{io}^T considerably exceeds $\mid F_{el}^T \mid$ at the top site in spite of the substantial value of $\Delta\rho(\mathbf{r})$. Because of the relatively larger $\mid \tau_s - \tau_t \mid$ it also larger than F_{io}^H at the H-site. As a result $F_t^T > F_t^H$ and the tip mainly feels the ionic repulsion imaging the atomic sites. As d increases $\mid F_{el} \mid$ decays more slowly than F_{io}, and F_t (which is $F_{el} + F_{io}$) changes sign where $\mathcal{E}(d)$ attains its minimum value. In this intermediate range of d $\Delta\rho(\mathbf{r})$ provides an attractive interaction by the charge accumulated between the tip and the surface and plays a crucial role in determining the force corrugation. In general, $F_{io}^T - F_{io}^H$ is large enough to compensate the appreciable difference arising from $F_{el}^T - F_{el}^H$ to yield less attraction at the top-site, however. Consequently, in the repulsive and strongly attractive force regions the force exerting on the outermost tip atom is dominated by the ion-ion repulsion. Since $\Delta\rho(\mathbf{r}) \simeq 0$ and $F_t < 0$ in the limit of very large d ($d > 10 - 15$ a.u.), F_t images the total charge density of the surface. This is also true for a multiatom tip provided that d is large.

Having discussed the tip force let us now pass to the effects of the tip on the electronic structure. In the independent-electrode regime the tip and the sample are separated by a substantial potential barrier ϕ_o. Because of the conservation of the parallel momentum the effective barrier, ϕ_{eff} may even be larger than ϕ_o. However, as the tip approaches the sample, the potential barrier is lowered and the overlap of the tip and sample wave functions increases. The contours of the SCF-potential, $V(x,y,z = d/2)$ are illustrated in Fig.4. A dramatic lowering of the barrier is seen to occur even for $d=6$ a.u. The effect of the barrier lowering on the electronic structure can be described within the tight-binding scheme. To this end we consider unperturbed sample and tip wave functions φ_s and φ_t with energies $\epsilon_s = < \varphi_s \mid \mathcal{H}_s \mid \varphi_s >$ and $\epsilon_t = < \varphi_t \mid \mathcal{H}_t \mid \varphi_t >$, respectively. $\mathcal{H}_s$ and $\mathcal{H}_t$ denote the Hamiltonians of the free sample and the free tip. For the interacting tip-sample system, the total Hamiltonian $\mathcal{H}$ is different from the sum of $\mathcal{H}_s$ and $\mathcal{H}_t$, and the interaction energy at a given d is $\mathcal{U}(d) = - < \varphi_s \mid \mathcal{H} \mid \varphi_t >$. The energies of the unperturbed tip and sample shift slightly when the interaction is small. On the other hand, if $\mathcal{U}$ is significant the interacting states become bonding and antibonding combinations of the tip and sample wave functions, $\Phi = c_s\varphi_s + c_t\varphi_t$. Depending upon the degree of the barrier lowering Φ is localized either in the gap or in one of the electrodes. While the energy of the bonding state is lowered, and in some cases dips below the Fermi level, the energy of the antibonding state rises. Nevertheless, the charge density in the gap, as well as the density of states at the Fermi level undergo a change in both case. This means that $\Delta\rho(\mathbf{r})$ is substantial.

The states of the combined tip-sample system, which are calculated self-consistently for various d, are analyzed. The evolution of the states at the Γ- and M-point, which are folded to the Γ-point of the (2x2) BZ, are illustrated in

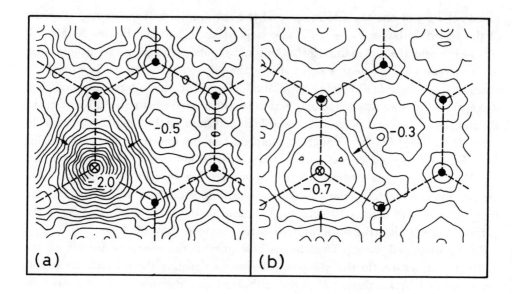

Figure 4   Contour plots of the potential $V(x, y, z = d/2)$ in a horizontal dividing plane. (a) An Al atom facing the T-site and d=3.8 a.u. with the contourspacing of -0.08 Ry. (b) Same for d=6 a.u. with the contourspacing of -0.07 Ry. Potential is lowered in the direction indicated by small arrows. Al and carbon atoms are represented by the cross and the filled circle, respectively.

Fig.5. In comply with the above arguments the tip and the graphite states are split apart in energy, and some states which are normally unoccupied for the free sample dip below the Fermi level. This gives rise to a change in the density of states. While $\rho_s(\mathbf{r}_o, E_F)$ of free graphite is determined by the states near the corner of BZ, the states derived from the M-point contribute to the density of states at E_F. For example, the unoccupied M-derived state forms a bonding combination by acquiring appreciable admixture of Al$p_{x,y}$-orbitals, and dip below E_F. The tip and substrate states near the Fermi level have a good deal of mixing in the range $4 < d < 6$ a.u. Earlier, these states were identified as the tip-induced localized states (TILS)[20], and also were shown to play a crucial role in STM. The tip induced states can also be viewed as the impurity states, which move with the tip. Figure 6 illustrates the contour plots of the charge density calculated for the tip-induced states.

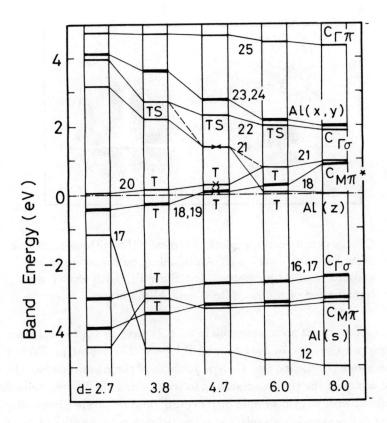

Figure 5 Evolution of the states of the Al-graphite system as a function of d. The Al-atom represents the tip and faces the graphite surface at the T-site. Numerals denote the band index and TILS are marked by T. C_M and C_Γ denote states originating from the M- and Γ-points of the graphite BZ, respectively. States, which are degenerate or very close in energy are shown by a thick bars.

4.3 ALUMINUM-TIP AND ALUMINUM-SAMPLE

We next consider a prototype system consisting of the Al(111) sample and the Al-tip, which are treated in a (2x2) supercell periodicity. As described by inset in Fig.7 the Al-sample is represented by the three Al(111) layer with the standard A-, B-, and C-type layer sequence. The tip consists of four Al-layers (C,A,B,C), except that from every four atoms in both C-layers three Al atoms are removed. Hence the

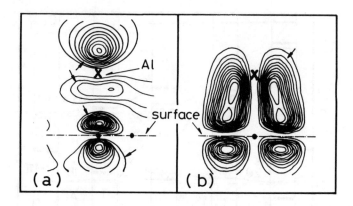

Figure 6 Charge density contour plots of typical TILS with energies near E_F. (a) State-Γ_{21}^T with d=4.7 a.u. Antibonding combination of the M-derived graphite and Al-p_z states. (b) State-Γ_{18}^T with d=4.7 a.u. Bonding combination of the graphite π- and Al-$p_{x,y}$ states.

tip has single outermost atom arranged in a (2x2) array in the frontal layers, which are anchored on the back by two A-, and B-layers. The repeat period is 10.8 a.u. In the slab geometry the A-, and C- type surfaces of the Al-sample face the C-type outermost atoms of the tip slab resulting in the T-, and the H-site positions in the same supercell with the tip-sample distances, d^T and d^H, respectively. By varying the spacing of the sample-slab relative to the tip-slab we were able to calculate the tip-sample interaction for d ranging from 4 to 8 a.u. without changing the size of the supercell. The experimental data by Wintterlin et al.[18] provides evidence that Al atoms of the (111) surface of the sample are transferred to cover the probing tip upon certain treatments. Therefore, the present model is realistic and relevant to the experiment.

The perpendicular components of the forces acting on the outermost tip atom at the T-, and H-site positions are presented in Fig.7(a). In the attractive region the zero-force gradient of the H-site occurs again at smaller distance, but at larger force relative to the T-site. The elastic deformation of the tip under the force field generated by the tip-sample interaction is calculated through the variation of the total energy and the force as a function of the perpendicular displacement of the outermost tip atom. The force-displacement curve is presented in Fig.7(b). In order to reveal the effect of the displacement of the tip atom we first assume that the

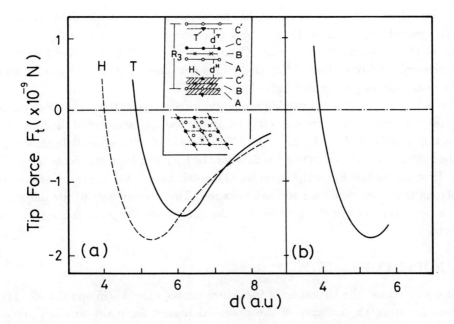

Figure 7 (a) Perpendicular component of the force exerted on the outermost tip atom versus d. (b) Force-displacement curve of the outermost tip atom. The model representing the tip-sample system is described by the inset.

tunneling current is set at the H-site at a given d^H within the topographic mode of STM. We also assume that d^H lies in the attractive force region ($h \geq 8$ a.u.), where the Tersoff-Hamann theory is still valid. The tip scanning from the H-site to the T-site has to be retracted by the corrugation amplitude $\Delta d \propto \rho_s(\mathbf{r}_{o,T}, E_F) - \rho_s(\mathbf{r}_{o,H}, E_F)$. On the other hand, the perpendicular and upward motion of the whole tip causes the attractive forces on the tip (and also on the sample atoms) to decrease. Consequently, the tip atoms mainly the outermost one experience upward displacement relative to their equilibrium positions at the H-site, yielding an effective gap $d^{eff} > d^H + \Delta d$. The gradient of $F_t(d)$ is negligibly small in the range of $d > 10$ a.u., so the displacement of the tip atoms further away from the vertex is not significant. The total relative displacement of the atoms is calculated from Fig.7(b) to be ~0.1 a.u. Since the the tunneling current will decrease owing to the upward displacement, the feedback mechanism lowers the tip to recover the value of the preset constant-current. Clearly, because of the attractive force field the recorded upward motion of the whole tip is not larger, but slightly smaller than the corrugation amplitude Δd derived from $\rho_s(\mathbf{r}_o, E_F)$.

The deformation of the tip would have pronounced effect in the region of d, where $\partial F_t/\partial d < 0$, if the current were directly related to the displacement of the tip atom. In fact, at $d \sim 5$ a.u. $F_t^H - F_t^T \simeq$ -1.5×10^{-9} N. This implies an upward displacement of the outermost tip atom as large as $6\Delta d$ for the tip going from the H-site to the T-site. However, in this range the tip induced disturbances in the electronic states are so strong that they dominate the conductance. This corresponds to the initiation of the point contact. In the next section the analysis of the point contact will be presented.

4.4 POINT CONTACT

Figure 8 illustrates the formation of the orifice between the Al-tip and the Al(111)-surface following the collapse of the potential barrier for a tip-sample distance as large as 8 a.u. Self-consistent calculations indicate high charge density in the dividing plane in the region where $V(x,y,z = d/2) < E_F$. This situation occurs in the range where the transition from the tunneling to the ballistic transport takes place. Upon further approach of the tip to the sample surface the orifice expands by the induced plastic deformation or by mechanical instability, and thus the point contact is initiated by the occupation of the states quantized in this orifice. This way new channels of conduction are opened, and the character of the conductance undergoes a qualitative change. This way, the theory of STM developed by Tersoff and Hamann[10] is no longer valid, and thus the commonly accepted proportionality to $\rho_s(\mathbf{r}_o, E_F)$ of the bare sample is invalidated. The gradual collapse of the barrier and the initiation of the point contact have been studied experimentally by Gimzewski and Möller[25] for the Ir-tip and the Ag-sample. Their log$I$ versus d plot clearly shows that the current, I increases first exponentially with decreasing d. This implies the tunneling behavior. The discontinuity observed at small d was attributed to the plastic deformation[40]. The recorded values of the conductance just after the discontinuity was only 80 % of the unit of conductance[41] ($2e^2/h$). Upon further decrease of d the current continues to increase and exhibits an oscillatory behavior.

Lang[26] simulated the point contact realized in the above experiment by two jellium electrodes, one of them having a Na atom attached to the jellium edge and thus representing a single atom tip. The conductance he calculated saturates at the

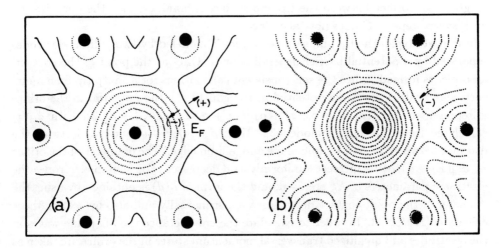

Figure 8 Contour plots of the potential $V(x,y,z=d/2) - E_F$ in the horizontal dividing plane between the tip and sample. (a) $d=8$ a.u. (b) $d=7$ a.u. Negative contours are shown by dotted lines. Contourspacings are 20 mRy.

value $\xi 2e^2/h$ when d is in the range of the distance from the Na core to the positive background core edge of the tip electrode. The value of ξ is only 0.4 for Na, and is found to depend on the identity of the tip. Within the tight-binding approximation and using the non-equilibrium Green's function method Ferrer et al.[42] also found that G_t saturates at $\leq 2e^2/h$. In both calculations, which predict G_t saturating at a value $\leq 2e^2/h$, the mechanism related to the experimentally observed increase of G_t following a discontinuity is not included. However, predicting G_t which is less than $2e^2/h$ indicates the tunneling regime, and thus cannot be reconciled with the point contact in the sense described by Sharvin[23]. Whereas, Gimzewski and Möller[25] give an estimate for the dimension of the contact area, which lies in the range of λ_F. If so the observed transport beyond the discontinuity has to be associated with the quantum ballistic transport[22].

Ciraci and Tekman using the results of the ab-initio calculations in a model based on the jellium approach studied the point contact[24]. They showed that after the point of discontinuity the conductance normally increases as the tip continues to be pushed towards the surface. However, the decrease of G is also a possibility depending upon the shape of the tip and the character of the plastic deformation induced by the tip. More importantly, the observed behavior beyond the point of discontinuity was found to be compatible with the ballistic transport. Since the length of the orifice between the tip and surface is small ($< \lambda_F$), the quantization with sharp steps in G does not take place.

In that study[24] the tip and sample system is modeled by two jellium electrodes separated by a potential barrier depending on d. Only at the point of the contact, an orifice is formed. Since the self-consistent calculations for a sharp tip on a metal surface indicate a parabolic potential in the gap[19,21], the potential in the orifice is represented by $V(\rho,z) = \phi_m(z) + \alpha\rho^2\theta(z+d_o/2)\theta(z-d+d_o/2)$, with $\phi_m(z)$ calculated from the jellium model[43]. Here $\rho = \sqrt{x^2+y^2}$, and d_o is twice the distance between the first atomic plane and jellium edge. The diameter of the orifice $\rho_m \simeq [(E_F - \phi)/\alpha]^{1/2}$, increases uniformly as d decreases. For jellium electrodes the electronic parameters of Ag are used, and the form of $\alpha(d)$ is obtained by using the diameter of the point contact determined experimentally and also by scaling those calculated[19] for Al. In order to calculate the conductance the 3D plane waves in the electrodes and quantized transversal momentum states in the orifice are taken as the current carrying states. The wave functions of the orifice states are the products of the 2D harmonic oscillator solutions and the 1D solutions of the potential $\phi_m(z)$. The coefficients of these states were obtained from the boundary conditions at the jellium edges. Finally the conductance expression derived earlier[44] are used.

$$G = \frac{e^2}{\pi h} \int_{F.S} \frac{d\mathbf{k}_\|}{k_z(\mathbf{k}_\|)} \{[\Theta^\dagger(\mathbf{k})\Lambda_R\Theta(\mathbf{k}) - \Delta^\dagger(\mathbf{k})\Lambda_R\Delta(\mathbf{k})] + 2\mathrm{Im}[\Theta^\dagger(\mathbf{k})\Lambda_I\Delta(\mathbf{k})]\} \quad (4.10)$$

where Θ and Δ are the matrices of the coefficients for the right- and left-going orifice states, corresponding to an incident plane wave vector $\mathbf{k} = \mathbf{k}_\| + k_o\hat{\mathbf{z}}$, respectively. The ballistic transport of electrons through a quasi 1D constriction was studied by using this expression but with different basis set[44]. It was found that the conductance is quantized with the sharp step structure if the length of the constriction is long enough ($l \geq \lambda_F$) and if its width is uniform. In the present case the orifice between two electrodes is considered as constriction through which the electrons pass from one electrode to the other.

Figure 9 presents the results of calculations. In agreement with the previous calculations[26,42] the conductance associated with a uniform orifice set up by a single atom at the vertex of the tip has a value less than $2e^2/h$. Since the length of the orifice is finite and in the range of the internuclear distance a_o, this result implies that the energy of the first subband ϵ_1 is above E_F, and the conductance is dominated by tunneling. According to experiment[25] it appears that the ballistic regime starts subsequent to the structural instability occurring at $d > a_o$. If one continues to push the tip further, a_o is approximately conserved, but the contact aperture expands with an enhanced plastic deformation followed by the adhesion

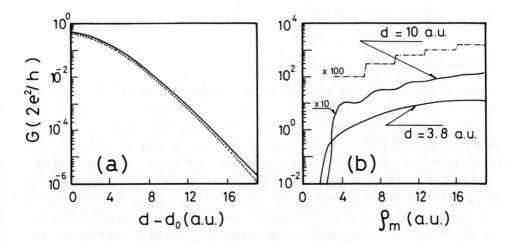

Figure 9 (a) G_t versus d curve calculated for two Ag electrodes. The dotted curve is the contribution of the first subband. (b) G versus ρ_m for the Ag electrodes calculated for $d_o \simeq 4$ a.u., $d_o \simeq 10$ a.u. ($\simeq \lambda_F$) and for ideal quantization.

of several atoms. In this case, several subbands of states, which are quantized in the orifice, can be occupied. Each subband occupied contributes to G by $2e^2/h$ (n is the degeneracy) and yields a step structure. The perfect quantization with sharp steps occurs if $d \gg \lambda_F$ and $\phi_m(z) = 0$. Earlier the observed oscillations of the $\log I(d)$ curve was attributed to the quantization of the ballistic transport[22].

In order to cover the plastic deformation region, where a significant hysteresis is observed in the excursion of the tip[25], the model is extended and G is calculated as function of ρ_m (i.e. $\rho_m = \sqrt{E_F/\alpha}$) with fixed length of the orifice. In Fig. 9(b), the smeared out step structure (or quantum oscillations) of $G(\rho_m)$ curve is apparent for $d = \lambda_F$ (~10 a.u.) and are compared with the sharp steps corresponding to the exact quantization. However, for $d \simeq a_o$ (≃3.8 a.u.) as in the experiment $G(\rho_m)$ is featureless. It appears that the point contact through a short constriction prevents G from quantization (or from quantum oscillations), and the ballistic transport occurs in the Sharvin regime. Therefore, these results suggest that the observed oscillations in the $\log I$ versus d curve cannot be associated by the quantum oscillations. Possibly they originate from the atomic motion introduced by the plastic deformation. Each tip or sample atom entering in the strongly repulsive force region are displaced irreversibly, leading to discrete changes in α. As a result, α undergoes discrete variations, and in general ρ_m increases discontinuously as the tip is pushed towards the surface.

5. Conclusions

In view of the above results, two regimes are identified for STM. If the outermost atom is in the weak attraction, which corresponds to the electronic contact regime, electronic states are disturbed by the tip-induced states. These states can be viewed as if the states of a point defect in the surface, which move with the tip. The effect of the local perturbation of electronic states can be sought in the expression of tunneling current, $I \propto \int_{E_F}^{E_F+eV} \mathcal{D}_t(E)\mathcal{D}_s(E+eV)T(E,V)dE$. Apparently, the density of states of the tip and sample ($\mathcal{D}_t$ and $\mathcal{D}_s$, respectively), as well as the transmission function $T(E,V)$ are modified locally by the tip-sample interaction and thus deviate from those corresponding to the bare tip and the bare sample. Due to the strong tip-sample interaction the parallel wave vector $\mathbf{k}_{\parallel}$, which is normally conserved in the independent electrode regime, is no longer a good quantum number. As a result, the difference between the effective barrier height ϕ_{eff} and the potential barrier ϕ is substantially reduced. The transport occurs via tunneling through a barrier which is already lowered due to the tip-sample interaction. The STM images are affected by the local disturbance of the electronic states, and hence the observed corrugation may not be directly proportional to $\rho_s(\mathbf{r}_o, E_F)$.

Based on the above conclusions it can be argued that the atomic resolution obtained from the close-packed metal surfaces (Al and Au) is associated with the local and strong disturbance of the density of states. The effect of the elastic deformation induced by the tip is found to be insignificant and to reduce the corrugation derived from $\rho_s(\mathbf{r}_o, E_F)$ [19].

Following the collapse of the potential barrier the point contact is initiated if the plastic deformation is induced by the tip in the repulsive or strongly attractive region. The states, which are quantized in the orifice between the tip and sample become occupied and allow the ballistic transport if the aperture is large enough. However, since the length of this orifice is less than λ_F the conductance is not quantized, and thus it is featureless as in the Sharvin's conductance.

Ab-initio force calculations indicate that the ion-ion repulsion plays a dominant role in determining the force corrugation in the repulsive and strong attractive region. This leads to two types of force corrugation between the top and hollow sites in the constant force mode of AFM operating in the attractive region.

The forces on the tip atoms far away from the outermost one make important contributions to the total tip force. In some cases, their contribution might become so dramatic that while the total tip force is attractive, the outermost tip atom is in the strong repulsive force region causing plastic deformation in the sample surface. Therefore, the repulsive force on the outermost tip atom is always underestimated in AFM measurements.

The calculated values of force are found to be two orders of magnitude smaller than the loading force in typical AFM measurements. On the other hand measured tip forces of $10^{-6} - 10^{-8}$ N is enough to destroy the sample surface. The source of this discrepancy between the theory and experiment can be sought in the shape of the tip. A flat tip having several atoms facing the sample surface is expected to experience a large repulsive force. While the sharp tip with a single atom at the vertex can yield atomic resolution, the images obtained by a flat tip can be quite complicated and requires a detailed Fourrier analysis. For example a tip with two outermost atom at the same height fails to yield atomic resolution and may yield protrusions located between two adjacent surface atoms. Moreover, images are strongly dependent on the orientation of the blunt tip. Nevertheless, either the flat tip or the blunt tip is capable of imaging the unit cell of the sample surface[28].

ACKNOWLEDGEMENT. The author thanks Dr. I. P. Batra, E. Tekman and Prof. A. Baratoff for their contributions in the works, which are discussed in this lecture note; and Dr. H. Rohrer for his interest on this work; and O. Gülseren for his valuable help in the preparation of the manuscript. This work is partially supported by the Joint Study Agreement between Bilkent University and IBM Zurich Research Laboratory.

References

1. G. Binnig, H. Rohrer, Ch. Gerber, and E. Weibel, Phys. Rev. Lett. **49**, 57 (1982).

2. C. F. Quate, Phys. Today **39**, 26 (1986).

3. J. Schneider, R. Sonnenfeld, P. K. Hansma, and J. Tersoff, Phys. Rev. **B39**, 4979 (1986).

4. R. Sonnenfeld and P. K. Hansma, Science **232**, 211 (1986); P. K. Hansma and J. Tersoff, J. Appl. Phys. **R1**, 6148 (1987).

5. A. Baratoff, Physica **127B**, 143 (1984).

6. A. Selloni, P. Carnevali, E. Tosatti and C. D. Chen, Phys. Rev. **B31**, 2602 (1985).

7. R. M. Feenstra, D. M. Martensson, Phys. Rev. Lett. **61**, 447 (1988).

8. G. Binnig, C. F. Quate, and Ch. Gerber, Phys. Rev. Lett. **49**, 57 (1986).

9. J. Bardeen, Phys. Rev. Lett. **6**, 57 (1961).

10. J. Tersoff and D. R. Hamann, Phys. Rev. Lett. **50**, 1998 (1983).

11. G. Binnig, H. Fuchs, Ch. Gerber, H. Rohrer, E. Stoll, and E. Tosatti, Europhys. Lett. **1**, 31 (1985).

12. I. P. Batra, N. Garcia, H. Rohrer, H. Salemink, E. Stoll, and S. Ciraci, Surf. Sci. **181**, 126 (1987).

13. H. Mizes, Sang-Il Park and W. A. Harrison, Phys. Rev. **B36**, 4491 (1987).

14. J. M. Soler, A. M. Baro, N. Garcia, and H. Rohrer, Phys. Rev. Lett. **57**, 444 (1986).

15. S. Ciraci and I. P. Batra, Phys. Rev. **B36**, 6194 (1987); I. P. Batra and S. Ciraci, J. Vac. Sci. Technol. **A6**, 313 (1988).

16. U. Dürig, O. Züger and D. W. Pohl, J. Microscopy (to be published).

17. V. M. Hallmark, S. Chiang, J. F. Rabolt, J. D. Swalen, and R. J. Wilson, Phys. Rev. Lett. **59**, 2879 (1987).

18. J. Wintterlin, J. Wiechers, H. Brune, T. Gritsch, H. Höfer and R. J. Behm, Phys. Rev. Lett. **62**, 59 (1989).

19. S. Ciraci, A. Baratoff and I. P. Batra, Bull. Am. Phys. Soc. **34**, 538 (1989).

20. E. Tekman and S. Ciraci, Physica Scripta, **38**, 486 (1988).

21. S. Ciraci, A. Baratoff and I. P. Batra, Phys. Rev. **Bxx**, xxxx (1989).

22. N. Garcia, in *STM' 87 Conference* held at Oxnard, California (1987).

23. Yu. V. Sharvin, Zh. Eksp. Teor. Fiz. **48**, 984 (1965).

24. S. Ciraci and E. Tekman, (to be published).

25. J. K. Gimzewski and R. Möller, Phys. Rev. **B36**, 1284 (1987).

26. N. D. Lang, Phys. Rev. **B36**, 8173 (1987).

27. G. Binnig, Ch. Gerber, E. Stoll, T. R. Albrecht, C. F. Quate, Europhys. Lett. **3**, 1281 (1987); T. R. Albrecht and C. F. Quate, J. Vac. Sci. Technol. **A6**, 271 (1988).

28. F. F. Abraham, I. P. Batra and S. Ciraci, Phys. Rev. Lett. **60**, 1314 (1988).

29. R. C. Tartar and S. Rabii, Phys. Rev. **B25**, 4126 (1982).

30. D. Tomanek, S. G. Louie, H. J. Mamin, D. W. Abraham, R. E. Thomson, E. Ganz, and J. Clarke, Phys. Rev. **B35**, 7790 (1987).

31. J. P. Pethica, Phys. Rev. Lett. **57**, 3235 (1986); H. J. Mamin, E. Ganz, R. E. Thomson, and J. Clarke, Phys. Rev. **B34**, 9015 (1986).

32. T. Feuchtwang, Phys. Rev. **B10**, 4121 (1974).

33. N. D. Lang, Phys. Rev. Lett. **56**, 1164 (1986); Phys. Rev. Lett. **58**, 45 (1987).

34. P. Hohenberg and W. Kohn, Phys. Rev. **136**, B864 (1964).

35. J. Ihm, A. Zunger and M. L. Cohen, J. Phys C. **12**, 4409 (1979); M. T. Yin and M. L. Cohen, Phys. Rev. **B26**, 3259 (1982).

36. D. R. Hamann, M. Schlüter and C. Chiang, Phys. Rev. Lett. **43**, 1494 (1979).

37. D. M. Ceperley and B. J. Alder, Phys. Rev. Lett. **45**, 566 (1980).

38. I. P. Batra, S. Ciraci, G. P. Srivastava, J. S. Nelson and C. Y. Fong, Phys. Rev. **B34**, 8246 (1986).

39. J. Harris and A. Liebsch, J. Phys. C. **15**, 2275 (1982).

40. J. B. Pethica and W. C. Oliver, Phys. Scr. **19A**, 61 (1987).

41. R. Landauer, IBM J. Res. Develop. **1**, 233 (1957).

42. J. Ferrer, A. Martin-Rodero and F. Flores, Phys. Rev. **B38**, 10113 (1988).

43. J. Ferrante and J. R. Smith, Surf. Sci. **38**, 77 (1973).

44. E. Tekman and S. Ciraci, Phys. Rev. **B39**, 8772 (1989) and references therein.

ON THE QUANTIZED CONDUCTANCE OF SMALL CONTACTS

L. ESCAPA
Departemento di Fisica de la
Materia Condensada
Universidad Autónoma de Madrid
Madrid-28049
Spain

N. GARCÍA*
IBM Research Division
Zurich Research Laboratory
8803 Rüschlikon
Switzerland

ABSTRACT. The electron elastic conductance of small contacts on constrictions between 2-D reservoirs is investigated. Experimental results in two-dimensional electron gas (2 DEG) GaAs/GaAlAs structures show that this conductance, to a good approximation, is quantized. We present calculations for general geometries of the contacts that describe the experimental data reasonably well. Our calculations also show how the resonant scattering structure superimposed on the quantized conductance for some geometries tend to be washed out when the contacts are made smooth. Finally we discuss the interpretation of those experiments in terms of ballistic transport. Calculations using a first iterative selfconsistent procedure show that it may not be correct to use ballistic transport for all the modes in the constriction. The importance of the electronic mean free path concept in the constriction when compared with that of the reservoirs is also discussed. We conclude that for the higher mode in the constriction the transport may be sequential.

Introduction

The elastic resistance of small contacts was first calculated by Sharvin [1] and has been the subject of much theoretical work [2-5]. The development of nanometer technology has made measurements possible, and quantized conductance in 2 DEG GaAs/GaAlAs heterostructures has been observed [6, 7].

In Section I we will be discussing the behavior of a contact elastic resistance, always present due to the contact bridging two reservoirs. In Section II we present quantum mechanical calculations that explain the experimental results. In Section III, using a first iterative selfconsistent procedure we show that the interpretation of those experiments in terms of ballistic transport may not be correct. We also study the electronic mean free path in the constriction. As it turns out to be much smaller than the one in the reservoirs we conclude that for the higher mode in the constriction the transport could be sequential.

---

*Permanent address: Departemento di Fisica de la Materia Condensada, Universidad Autónoma de Madrid, Madrid-28049, Spain

Section I

Suppose we have a small contact bridging two 2-D reservoirs (see Fig. 1). If we want to establish a current between them this contact will show a resistance that can be approximated by

$$R = R_i + R_d + R_e \ . \tag{1}$$

R_i is the inelastic resistance, given by

$$R_i = (\rho_i \ell_i / W^2) \, \theta(\ell - \ell_i) \, \theta(W - \ell_i) \ , \tag{2}$$

where ρ_i is the resistivity due to electron-electron and electron-phonon collisions, ℓ_i is the electron mean free path at the Fermi level and θ is the Heaviside function.

R_d is the residual or impurity resistance. R_e is an elastic coherent resistance that shows up because the electrons are injected into the constriction and ejected from it. Sharvin [1] has shown that in the classical limit, when the electrons are treated like bullets going through a hole, $(W, \ell \ll \ell_i)$,

$$G_e = \frac{1}{R_e} = 2 G_0 W \ , \tag{3}$$

where $G_0 = 1/R_0$, $R_0 = h/2e^2 \sim 12{,}900 \, \Omega$ being the quantum of resistance, and W is measured in units of the Fermi wavelength, λ. However those calculations did not account for the quantum interferences that should be present when the electrons diffract with small contacts, i.e. $W, \ell \sim \lambda$.

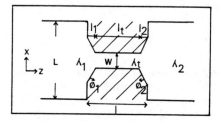

Figure 1. Geometry of the contact and the reservoirs. Note the two trumpets connecting the reservoirs and the constriction of width W. These trumpets are defined by ℓ_1, ℓ_2, ϕ_1 and ϕ_2 and are part of the reservoirs 1 and 2, respectively. $\ell = \ell_1 + \ell_t + \ell_2$ is the total length of the contact.

We have performed exact quantum mechanical calculations for this elastic conductance of small contacts. The general geometry is depicted in Fig. 1. Two 2-D reservoirs, 1 and 2, with carrier concentrations defined by their Fermi wavelengths λ_1, λ_2, are connected by a constriction, defined by its width W, length ℓ_t and Fermi wavelength λ_t. The junction between the reservoirs and the constriction can be made smoother with what we call trumpets, defined by their lengths ℓ_1, ℓ_2 and angles ϕ_1, ϕ_2. The trumpets are part of the reservoirs that are supposed to have a finite size L, $L \gg W$. In what follows, all the lengths are expressed in terms of the Fermi wavelength, $\lambda = 1$. The results are obtained by solving Schrödinger's equation with the boundaries defined by the contact constriction and the reservoir walls with a similar method to the one used in Ref. 4 to calculate quantum elastic resistances of interfaces. In this case, because the reservoirs have dimension L, we have expanded the solution in the reservoirs into a series of sines and cosines vanishing at the walls thereof. The solution in the constriction (ℓ_t, W) is also expanded in sine and cosine owing to the characteristic modes of the width W. At the trumpets, i.e., the opening regions of the contacts, defined by $\ell_1, \ell_2, \varphi_1$ and φ_2, the solution is expanded into sine and cosine linear combinations of the reservoir 1 and 2 respectively. This solution in the trumpets is within the Rayleigh assumption [8] and the results should be checked against the unitarity rules in the reflected and transmitted intensity through the contact. The solution in all space is then obtained by matching the wave function reservoir 1, the constriction and reservoir 2, and at the same time this wave function has to vanish at the boundaries defined by the contact.

The conductance is obtained by

$$G = G_0 \sum_i T_i \ , \qquad (4)$$

where T_i is the total transmittivity for the electron i impinging the contact and going through it. Also

$$T_i = \sum_j T_{ij} \ , \qquad (5)$$

where T_{ij} is the transmittivity carried out by the j-component of the propagating wave of the transmitted electron (see Ref. 4). The expansion of the wave function has a finite number of propagating terms N_p and an infinite number of evanescent waves that has to be truncated to N_e terms, and the convergence has to be studied as a function of these evanescent terms that play a very important role in the convergent solutions. The index i in Eqs. (4) and (5) runs over all states at the Fermi level.

When there is no constriction ($\ell_t = 0, \ell_2 = 0$) we observe just a contact between the two reservoirs (see inset in Fig. 2a). Figure 2a presents typical results for the elastic resistance and conductance. $G_e = 1/R_e$, vs. W for $\ell_1 = \lambda_1/2$, $\phi_1 = 90°$. The oscillations or plateaus observed in the values of R_e and G_e are due to quantum effects because of the small size of the contact and can be correlated with the different quantization modes appearing in the transversal direction of the contact [3]. Each time W increases in $\lambda/2$ a new mode appears in the contact and the conductivity increases in a quantum value G_0. We find that for $\phi_1 = 90°$ and $\ell_1 = \lambda_1/2$ the conductivity has well-defined steps of height G_0, but when $\phi_1 < 45°$ these jumps in conductivity do not exist and it exhibits a more oscillatory behavior (see Fig. 2b). This was observed much earlier for scanning tunneling microscopy (STM) calculations [5]. From many calculations we find that the average formula for the elastic conductance can be written approximately

$$G_0 = 2 G_0(W/\lambda - p) \ , \qquad (6)$$

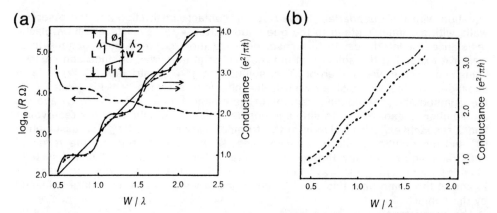

Figure 2. a) Elastic conductance and resistance as a function of the contact size W between reservoirs 1 and 2. $L = 10\,\lambda_1$, $\ell_1 = \lambda_1/2$, $\phi_1 = 90°$. The inset shows the geometry for this "point contact." Note that there is no constriction. Solid line $\lambda_1 = \lambda_2$; dashed line $\lambda_2 = 0.8\,\lambda_1$. The straight line is the Sharvin resistance Eq. (3) for $p = 1/4$. **b)** Values of the conductance for the same geometry as in Fig. 2a with $\phi_1 = 30°$ (solid line) and $\phi_1 = 45°$ (dashed line), $\lambda_1 = \lambda_2$. This may be the appropriate geometry for STM. Note that the steps in Fig. 2a are smoothed to oscillations around the Sharvin resistance.

i.e., the same value Sharvin obtained but with $(W/\lambda - p)$ changed into (W/λ). From our results we notice that $p = 1/4$ for $\phi_1 = 90°$, $\ell_1 = \lambda/2$ and $p \to 0$ for $\ell_1 \to 0$ or $\phi_1 < 45°$ and $\ell_1 < \lambda/2$ (see Fig. 2b). We obtain, then, when $L \to 0$ (infinitesimally small contact) oscillations around the Sharvin conductance (Eq. 3). We have performed calculations for the resistance for random sets of small contacts through an interface as a function of the Fermi wavelength λ_2 and find that for finite L ($10\,\lambda_1$ and $15\,\lambda_1$) the resistance also oscillates as a function of λ_2. Therefore, R_e defined by Eq. (6) always exists, independent of the material, impurities and temperature. We can imagine that by miniaturization and lowering the temperature we would obtain $\ell_j > \ell$ and $\ell_j > W$ (see Eq. 2) and, therefore, $R_j = 0$. Also for good purity and perfect crystals, where the average distance between impurities is larger than ℓ and W, the term R_d can be neglected as well.

We have a peculiar situation: while R_i and R_d are reduced and can be neglected by reducing the contact and increasing the purity of the material, the value R_e has an opposite trend and increases inversely with the value of W. For example, if we assume semiconducting reservoirs with typical values of $\lambda \sim 10$ nm when $W \sim 100$ nm, then $R_e \sim 600\,\Omega$, which is certainly not negligible. Equation (6), as well as Eq. (4), are obtained assuming that $1/R_e \sim T$ (T is the electron transmittivity through the contact), but Landauer [2] has proposed for very thin wires that $1/R_e \sim T/(1-T)$ and, therefore, $R \to 0$ when $T \to 1$. This corrective formula will produce a reduction in the resistance by factors on the order of 5 for thick wires as estimated with a generalized formula by Büttiker et al. [3] and García and Stoll [4]. However, this is balanced again if the reservoirs have Fermi wavelengths λ_1 and λ_2 that differ by 20% [4]. Dashed lines in Fig. 2a show the calculated conductance for $\lambda_2 = 0.8\,\lambda_1$.

The STM is very well suited to study this conductance of small contacts. At very short distances between sample and tip, the tunnel barrier disappears, forming a point contact, and the current between sample and tip can be measured at different applied voltages. The dimension of the contact can be changed by pressing the tip against the sample, thus increasing the size of the point contact. Experiments of this kind were performed by Gimzewski and Möller [9] where they study the transition from tunneling regime to point contact. In Fig. 2 of Ref. 9 one can see the oscillatory behavior of the conductance for the larger values of the current. These oscillations were in agreement with earlier calculations by García [5]. In the STM case, because of the geometry of the tip, the angle ϕ_1 would be smaller than 90°. Figure 2b shows conductance calculations for $\ell_1 = \lambda/2$ and $\phi_1 = 30°$ and 45°. As mentioned above, the steps in this case are not pronounced and have been smoothed out to oscillations around the Sharvin conductance. We contend [5] that this oscillatory behavior could be the explanation for the experimental data in Fig. 2 of Ref. 9.

Section II

Recent work on high mobility 2 DEG in GaAs/GaAlAs heterostructures [6, 7] has shown that the conductance G of point contacts exhibits a quantized behavior as a function of the width of the contact W (see Fig. 1). The contacts are defined by applying a negative voltage to the gate described by the shaded region that creates an infinitely large repulsive potential for the electrons in the 2 DEG. Using the method described above, Fig. 3a shows results for three characteristic cases. In all

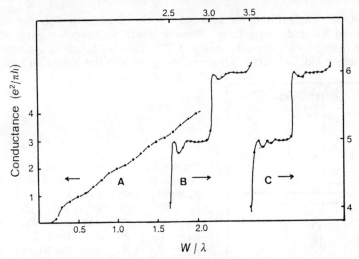

Figure 3a. A: oscillatory behavior of the conductance vs. W for $\ell_t = 0.001$ and no trumpets. B and C: same as in A, but $\ell_t = 2$ and 5, respectively. In these cases well-defined steps and plateaus develop as well as a superimposed resonant structure. In all cases $L = 10$ and no appreciable changes were observed for larger values of L. The lower and left-hand axes apply to curve A and the upper and right-hand label curves to B and C.

of them we have assumed that the contact is defined by a perfect constriction or tube of constant width ($\ell_1 = \ell_2 = 0$). Figure 3a, curve A, shows the case of $\ell_t = 0.001\,\lambda$, for which only very weak and small oscillations are observed of period 0.5 λ and the averaged slope of the curve is 2, in agreement with Eq. (3) and results in Section I: i.e., for a pure flat contact the quantization is difficult to define. In Fig. 3a, curve B, we present calculations for $\ell_t = 2\,\lambda$ and we can now observe plateaus in the conductance at integral values of G with periods of 0.5 λ; however, we also notice the appearance of a resonant structure superimposed on the plateaus. This resonant structure appears only when the value of ℓ_t is larger than a critical value of about 0.7, in agreement with the results in Section I, where we did not find resonances when $\ell_t < 0.5\,\lambda$. These resonances are well understood in terms of the resonant character of the transmittivity of a step of length ℓ_t and height V [11]; here the height of the step is the energy of the different propagating modes n in the constriction of energy $(\hbar^2/2m^*)\,(n\pi/W)^2$, m^* being the effective mass of the electron. Figure 3a, curve C, shows calculations similar to those of Fig. 3a, curve B, but with $\ell_t = 5\,\lambda$, and certainly we observe that the number of resonances increases. The jumps or quantized steps in the conductance can, as mentioned before, be understood in terms of the number of propagating or conducting modes owing to the contacts that are defined by int $(2\,W/\lambda_t)$. The conductance of wires and tubes has been clearly discussed in (3).

However, the above-calculated cases show resonances that are quite fictitious, because for real-life devices the width of the contact should change smoothly in the reservoir region because the depletion of electrons by applying the gate voltage should vary slowly. In other words, the real contacts should show something like the <u>trumpets</u> we define in Fig. 1 by the angles ϕ_1 and ϕ_2 and the lengths ℓ_1 and ℓ_2.

In Fig. 3b we analyze the behavior of the resonances in Fig. 3a, curve B, for $\ell_t = 2\,\lambda$ when we add trumpets of different length; to keep the parameters under control we take in all cases $\phi_1 = \phi_2 = 30°$. Figure 3b, curve A, represents an augmentation of Fig. 3a, curve B, to be compared with the following cases: Fig. 3b,

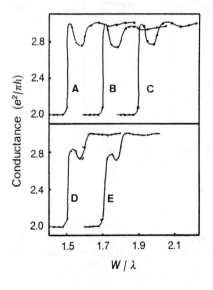

Figure 3b. A: same curve as B in Fig. 3a; curves B, C, D and E: same as in A, but with trumpets. For curve B, $\ell_1 = 0.5$, $\phi = 30°$, $\ell_2 = 0$; for curve C, $\ell_1 = \ell_2 = 0.5$, $\phi_1 = \phi_2 = 30°$, curve D, $\ell_1 = \ell_2 = 0.75$, $\phi_1 = \phi_2 = 30°$ and curve E, $\ell_1 = \ell_2 = 1$, $\phi_1 = \phi_2 = 30°$. Note the trumpet effect: the resonant structure is washed out; compare curve A with curves D and E.

curve B, shows the resonant structure for a trumpet, connecting reservoir 1 and constriction, with length $\ell_1 = 0.5\,\lambda$; Fig. 3b, curve C, presents the resonant structure with two trumpets of length $\ell_1 = \ell_2 = 0.5\,\lambda$. It can be observed that the resonant structure remains practically unchanged for trumpet lengths smaller than $0.5\,\lambda$. However this is not the case when the trumpets are longer. For $\ell_1 = \ell_2 = 0.75\,\lambda$ and $\ell_1 = \ell_2 = 1$, corresponding to Fig. 3b, curves D and E, respectively, the strong resonance together with the weak one are clearly washed out and tend to disappear in such a way that the trend is that the jump in the conductance step is smoothed out. We have tried to increase the length of the trumpets for ℓ_1 and ℓ_2 larger than $1\,\lambda$, but our results do not converge. Physically it is very easy to understand this disappearing behavior of the resonances for long trumpets (trumpet effect) and it is simply related to the behavior of the transmittivity of a particle through a smooth step. Now the step in the potential V mentioned above takes place in a wavelength distance and the resonant structure disappears [10].

It has been mentioned (lecture by Jacobi at this school) that the trumpet connecting the constriction and the reservoir would produce a focusing of the electron beam due to the adiabatic expansion. We have calculated the focusing obtained in these devices with and without trumpets. For trumpets of length λ and angles 30° and 45° we have found no effect: in both cases the angular width of the beam is $\simeq$ 120° [11]. Furthermore, recent calculations show that to have some appreciable focusing the length of the trumpets needs to be more than $5\,\lambda$.

In conclusion, we have calculated in Section I and II the elastic ballistic resistance of electrons through small contacts. Our results show that for short contacts the conductance shows a periodic oscillatory behavior that transforms into plateaus and steps of height $2e^2/h$ when the length of the contact is on the order of λ. The plateaus also show a superimposed resonant structure for the cases in which the interfaces between the contact and the reservoirs have sharp edges. However, these resonances disappear for general geometries by introducing what we have called the trumpet effect. In our opinion, experimental results should not show resonances, because the electron depletion between the constriction defining the contact and the reservoir should vary slowly and be in a Fermi wavelength.

Section III

In this section we discuss the consequences of the approach used in Sections I and II and its relation to the electronic mean free path ℓ_j. Experiments have shown that at very low temperatures and high purity samples, the GaAs/GaAlAs heterostructures of very large samples show large mobility and therefore large values (up to 10 μm) of ℓ_j. This means that in Eq. (1) we can neglect R_j and R_e which, together with the possibility of producing well-defined contacts (W, ℓ), has led to the interpretation of the observed quantized conductance in terms of ballistic conductance [6, 7]. In other words the electron travels ballistically, with phase coherence, between reservoirs 1 and 2 throughout the constriction because $\ell_j \gg W$ and $\ell_j \gg \ell$. This is the approach we took in Section II, as well as in a large amount of calculations [12], to explain the experimental data [6, 7]. All the calculations produce practically the same results, although they use different sets of functions to expand the wave function to obtain the transmitivities that define G through the constriction. However, it should be said that, indeed, the mean free path in the constriction has not been measured.

It is also very important to stress that all the calculations in Ref. 12, like those in Sections I and II, are not selfconsistent at all, the wave functions being obtained for the same potential in reservoirs 1 and 2 and in the constriction. In our opinion this could be a limiting hypothesis because our findings show that at the edges of the constriction there is strong scattering with evanescent waves that are needed to perform the expansion of the wave functions to obtain convergence. For example, we notice that 50 evanescent waves are needed for $\ell_t = 1$, $W = 2.4$, $\ell_1 = \ell_2 = 0$. It is evident to us that this will change the initial Fermi energy of the constriction in a selfconsistent procedure. Also strong long-range Friedel oscillations and pileup of charge should exist at the edges of the constriction. This was discussed earlier by Büttiker [13] as an especially important effect to be considered in the many channel formula for the conductance by Büttiker, Imry, Landauer and Pinhas [3], who do not properly treat the pileup of charge and the Friedel oscillations. In this section we analyze the first iterative solution and discuss its consequences on G. We also mention some ideas about the value of ℓ_i in the constriction.

We proceed by calculating the wave functions in zero order, as in Section II, for all occupied states $\psi(R, N, E)$, where $R(x, z)$ is a two-dimensional vector with x and z defining the direction of the constriction (see Fig. 1); N and E describe the different bands and energies of the electron states. The total electronic charge in any point $\rho(R)$ reads

$$\rho(R) \propto \sum_{N,E}^{E < E_F} |\psi(R, N, E)|^2 , \qquad (7)$$

where the summation is done for all values of N bands and energies E such that $E < E_F$, and adding enough values (energy interval small enough) that $\rho(R)$ converges.

Figure 4a shows the results for the averaged value of $\rho(R)$ in the transversal direction x, $\rho(<x>, z)$ as a function of z for $L = 10.4$, $W = 2.4$ and three different values of $\ell = 1$, 2.5 and 5. Large Friedel oscillations are observed at the edges of

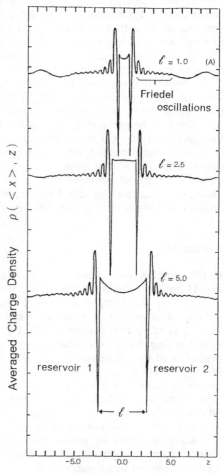

Figure 4a. Values of the averaged charge densities in the x-direction for $W = 2.4$, $L = 10.4$, no trumpets and $\ell = 1$, 2.5 and 5 (see text). The axis $z = 0$ is at the center of the constriction.

the constriction as well as built-up charge owing to the localized evanescent states. The little bumps labeled A for $\ell = 1$ are due to the reflections of the waves at the edges of the reservoir of width $L = 10.4$. The same structure appears for $\ell = 2.5$ and $\ell = 5$, but are outside the view of the figure. A more detailed plot of the charge density distribution is presented in Fig. 4b for cuts on different lines of constant x_j as a function of z, $\rho(x_j, z)$ for $\ell = 1$. The top line is at the axis of the constriction, $x_0 = 0$ and each line is separated by $\Delta x_j = 0.2$. The seventh line is exactly on the constriction wall. We find that the total charge in the region (W, ℓ) is 19% larger than in the reservoirs and therefore the Fermi energy increases by 19% in the constriction with respect to that of the reservoirs. Analogous results are obtained for $\ell = 2.5$ and $\ell = 5$. To see how these new densities affect the values

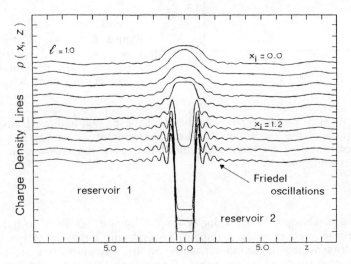

Figure 4b. Charge density lines for $\ell = 1$. Note the pileup of charge at the interfaces. The separation between lines is 0.2 and the upper lines is at the axis of the constriction.

of G we have recalculated it assuming that the Fermi energy changes according to the density, i.e., we have the same Fermi level but different band bottoms. This implies that the constriction has a potential energy 19% deeper than reservoirs 1 and 2. Figure 5 shows values of G vs. W for $\ell = 1$ and the new densities. The quantization is clearly lost and the resonance is deeper than before. Also the jumps in the conductance do not occur at $W = J/2$ where J is an integer. It seems clear to us that the first iterative term toward consistency departs from the previous agreement with experimental results. We do not know what would happen when full selfconsistency is reached, but we have some ground on which to say that the signature of the experiments [6, 7] is not due to ballistic transport. By iterating the selfconsistent series, the electron screening will tend to smooth the pileup of charges and the Friedel oscillations, but they will still be present even if reduced. In our opinion the phase of the electrons in the higher mode leaving the constriction has no memory of those entering it.

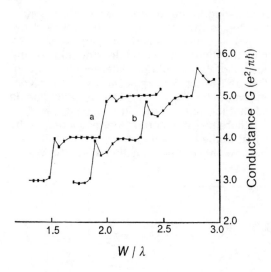

Figure 5. G vs. W for $W = 2.4$, $L = 10.4$, $\ell = 1$, no trumpets for zero and first order iterative procedures, curves a and b, respectively.

The real physical parameter involved in ballistic or non-ballistic transport is the mean free path of the electrons, ℓ_j. The experimental values for it are obtained from collision time (τ) by measuring the mobility of the carriers in very large samples. The value of ℓ_j is

$$\ell_j = \tau \, V_f \, , \tag{8}$$

where V_f is the isotropic Fermi velocity for the infinite 2-D electron gas. This is a well-defined averaged quantity owing to the isotropic behavior of the gas, but it should be stressed that in taking averages, all the wave functions are propagating states without evanescent components. Remember that the evanescent components are extremely important in the case of the constriction bridging two reservoirs. The question to be raised is: what is the mean free path in the constriction? There are as many mean free paths as occupied bands or modes in the constriction. Consider that the constriction has no impurities (or that its purity is as good as the one in the reservoirs) and assume that τ is the same everywhere (which is not true, because, owing to the Coulomb scattering with the built-up charge at the constriction edges, it should be smaller in the constriction). Then the only mean free path relevant in the narrowing is the one in the direction along the axis of the constriction, z. In the other directions, the modes are bounded. The mean free path can be defined for each mode at Fermi energy $\ell_j(N)$:

$$\ell_j(N) = \tau * V_f(N) \, , \tag{9}$$

where $V_f(N)$ is now the Fermi velocity of the N mode in the constriction along the z-direction. τ has different contributions, arising from impurities-electron scattering, phonon-electron interactions and inter- and intra-band electron-electron interaction. Hence τ can be expressed as

$$\frac{1}{\tau} = \frac{1}{\tau_{imp}} + \frac{1}{\tau_{ph}} + \frac{1}{\tau_{intrab}} + \frac{1}{\tau_{interb}} \ . \tag{10}$$

Note that to have a well-defined step in G the relevant $\ell_i(N)$ is that of the larger allowed N. In fact it has its most relevance at the threshold condition, when a mode passes from evanescent to propagating by increasing W slightly (we call this mode N_t), but then $V_f(N_t)$ tends to zero, resulting in a zero mean free path value for these electrons. It could be argued that if $V_f(N_t) \to 0$, $\tau \to 1/[V_f(N_t)]$ so that $\ell_i(N_t)$ is different from 0 and finite. This would be true for $\tau = \tau_{imp}$ or $\tau = \tau_{intrab}$, but it may not be the same for the contributions due to phonon-electron scattering and electron-electron interband interactions: even if the electron in level N_t cannot lose energy, it can gain it from phonons and electrons in other bands. Also, these electrons are the ones giving rise to the resonances and they spend a great deal of time moving backwards and forwards in the constriction which in turn will scatter them with the pileup charge and equilibrate them through inelastic processes in the constriction.

The other electrons have $\ell_i(N) \gg \ell$, but that does not affect the observability good steps in G because they contribute with unity transmittivity. In other words, the step is defined by the transmittivity of N_t, and it is only necessary to equilibrate the phase of a small number of electrons (on the order of 5%) to destroy the steps in G.

Because of the calculations and arguments given above, we think that the idea of extrapolating the value of ℓ_i for the infinite gas into the constriction is not correct and the mean free paths may be controlled by the processes going on at the interfaces. The chemical potential oscillations near the constriction have to be considered for the appropriate conductance value [14].

To check the consequences of sequential transport through the constriction we have performed calculations for G vs. W in zero order and also in the first iterative correction toward selfconsistency in analogy to the above-mentioned ballistic case. We consider now that electrons travel from reservoir 1 and equilibrate in the constriction, then the electrons have to be taken out and put into reservoir 2 without memory. The transport not being ballistic, the calculation consists then, ultimately, of taking electrons out of the tube defining the constriction and putting them into reservoir 2. The initial states of the problem are not states of the region 1, but modes from the constriction, i.e., the problem is the case of a contact between a constriction and reservoir 2. In Fig. 6a we show the averaged charge density in the x-direction as a function of

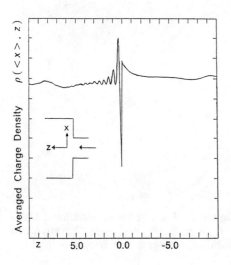

Figure 6a. Averaged charge densities in the x-direction vs. z for non-ballistic transport. $z = 0$ is at the edge constriction reservoir.

z, for $W = 2.4$ and $L = 10.4$ (the same parameters as that for the ballistic case). The constriction is found where $z < 0$ and the interface is at $z = 0$. Once again we notice build-up of the charge and strong Friedel oscillations at the interface. The charge build-up is now 15% larger in the constriction than in the reservoir (remember it was 19% in the ballistic case). Figure 6b shows charge density in the same way as in the ballistic case in Fig. 4b. Finally Fig. 7 shows the values of G

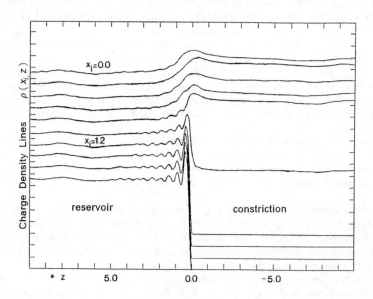

Figure 6b. Same as in Fig. 4b for non-ballistic transport.

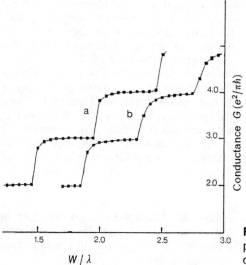

Figure 7. G vs. W for non-ballistic transport in zero and first order iterative procedures, curves a and b, respectively.

vs. W in zero order and in first iterative order, when the Fermi energy in the constriction is 15% larger than in reservoir 2. These results do not show resonances because of non-ballistic transport and the resonant electrons do not travel backwards and forwards reducing the loss of phase coherence with the extra localized charge at the interface. The values of G now bear a closer resemblance to the data in [6, 7] although the edges of the conductance steps seem to be smoother in the calculations. We notice also that the steps appear displaced by a fixed amount from $W = J/2$ but this does not contradict experiments because they do not measure W directly. The quantization is now better although probably the transport is such that it is sequential for the higher modes in the constriction and ballistic for the lower ones.

Our calculations appear to indicate that to have a good quantized conductance it is necessary to have a device geometry with smoothed-out contact edges so that the reservoir regions are not well-defined and the scattering is weaker owing to evanescent waves. But one thing seems clear to us: the observed step behavior of G vs. W is not due to ballistic transport through the constriction for the higher constriction modes that have a very small mean free path.

In conclusion we have discussed ballistic and non-ballistic transport and their consequences, calculating zero and first order in the iteration series toward self-consistency. The results favored non-ballistic transport to explain the experiments. The mean free paths of the electrons at the constriction are also discussed and we claim that these do not necessarily resemble those of the reservoirs. Key experiments in these devices consist of measuring the mean free paths in the constriction and it will not be surprising if the mean free path decreases by orders of magnitude for the higher mode. Also voltametric curves as those performed for grain boundaries [15] are needed to see where the potential drops.

References

[1] Sharvin, Yu. V. (1965), Zh. Eksp. Teor. Fiz., 48, 984; (1965) Sov. Phys. JETP, 21, 655.
[2] Landauer, R. (1957) IBM J. Res. Dev., 1, 223; (1970) Philos. Mag. 21, 863.
[3] Büttiker, M., Imry, Y., Landauer, R., Pinhas, S. (1985) Phys. Rev. B, 31, 6207.
[4] García, N. and Stoll, E. (1988) Phys. Rev. B, 37, 445.
[5] García, N. (1987) "Oscillatory Behavior in the Quantum Elastic Resistance of Small Contacts," STM '87 conference, Oxnard, California, July 1987; STM Workshop, Trieste 1987; also presented by H. Rohrer at the General Conference of the European Physical Society, Helsinki, August, 1987.
[6] Van Wees, B.J., Van Houten, H., Beenakker, J., Williamson, J.G. Kouwenhoven, L.P., van der Maret, D., and Foxon, C.T. (1988), Phys. Rev. Lett., 60, 848.
[7] Wharam, D.A., Thorton, T.J., Newbury, R., Pepper, M., Ahmed, H., Frost, J.E.F., Hasko, D.G. , Peacock, D.C., Ritchie, D.A., and Jones, G.A.G. (1988), J. Phys. C, 21, L209.
[8] García, N. and Cabrera, N. (1978) Phys. Rev. B, 18, 576.
[9] Gimzewski, J.K. and Möller, R. (1987) Phys. Rev. B, 36, 1284.
[10] Landau, L.D. and Lifshitz, E.M. (1969) Quantum Mechanics (New York, Pergamon).
[11] Escapa, L., De Raedt, H., García, N., and Sáenz, J.J., in preparation.

[12] This reference contains all recent calculations on ballistic resistance of which the authors are aware, in alphabetical order: Escapa, L. and García N. (1989) J. Phys. C, 1, 2125; García N. and Escapa, L. (1989) Appl. Phys. Lett., 54, 1418; Haanappel, E.G. and van der Marel, D. (1989) Phys. Rev. B, 39, 5484; Kirkeenow, G. (1989) Solid State Commun., 68, 715; Szafer, A. and Stone, D. (1989) Phys. Rev. Lett., 62, 300; Tekman, E. and Ciraci, S. (1989) Phys. Rev. B, 39, 8772.

[13] Büttiker, M. (1986) Phys. Rev. B, 33, 3020; Büttiker, M. (1986) IBM J. Res. Dev., 32, 317.

[14] Büttiker, M. (1986) "Chemical Potential Oscillations Near a Barrier in the Presence of Transport" (preprint).

[15] Kirtley, J.R., Washburn, S., and Brady, M.J. (1988) Phys. Rev. Lett., 60, 1546.

ADIABATIC EVOLUTION AND RESONANT TUNNELING THROUGH A ONE DIMENSIONAL CONSTRICTION

E. Tekman and S. Ciraci
Department of Physics,
Bilkent University,
Bilkent 06533 Ankara, Turkey.

ABSTRACT. The effect of geometry on the quantized steps and resonance structure is investigated for a quasi 1D constriction within the assumption of ballistic transport and by using the "mixed basis" boundary matching technique.

1.Introduction

The effects of the quantization of the transversal momentum in the infinite 1D electron waveguide were investigated theoretically[1] and the conductance of this constriction, G_c, was predicted to be quantized as the multiples of $2e^2/h$. Two recent experiments by van Wees et al.[2] and Wharam et al.[3] demonstrated that the conductance through a constriction between two 2D EG reservoirs increases with the gate voltage V_g (or equivalently with the width w of the constriction) approximately in steps of $2e^2/h$. Subsequently several theoretical explanations have been proposed[4,5,6,7,8] based on the calculations of conductance by using direct boundary matching[6], tight-binding[7] and "mixed basis" boundary matching[4,5,8] techniques. In the "mixed basis" boundary matching method, which was shown to be the most effective and versatile one[8,9], the eigenstates are represented by plane waves in the 2D EG reservoirs and by the quantized transverse momentum states in the constriction. Then the conductance is calculated within the linear response theory. The focus of attention of theoretical studies has been the resonance structure due to the interference of the multiple reflected waves from the boundaries[4,5,6,7,8]. While the lack of the resonance structure in the experiments were attributed to the finite temperature effects[5,8] or to scattering from the nonuniform constriction potential[6,9], García objected to the ballisticity of the transport[10]. On the other hand more refined experiments presented evidence corroborating that the transport is ballistic[11,12].

In this paper, we will investigate the effects of the geometry of the constriction on the observed conductance. We will place emphasis on the adiabatic evolution (which washes out the resonance structure without destroying the quantized steps) and the resonant tunneling (which is due to confined states occurring when the constriction gets wider in the middle).

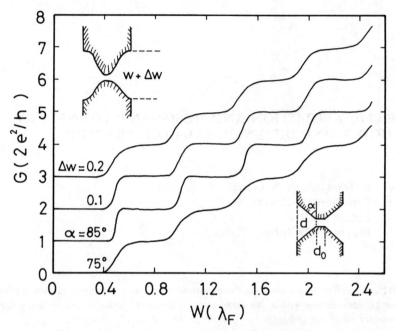

Figure 1 Calculated $G(w)$ curves demonstrating the adiabatic evolution of states for several geometries. The uppermost two curves are for cosine-modulation (length of the constriction is $d = \lambda_F$) and the lowest two curves are for tapered constrictions ($d = 0.5\lambda_F$, $d_o = 0.25\lambda_F$). All lengths are in units of λ_F.

2. Formalism

Here we use the formalism[8,9] of the quantum conductance developed earlier, and provided exact calculation of G. The spatial form of the potential, $V(y,z)$, of the constriction depends on the geometry of the split-gate and gate voltage, as such that it is parabolic for small w, but becomes square well like at large w[13]. For computational convenience, we consider an infinite well potential. Nevertheless, infinite well and square well potentials yield similar eigenstates in the energy range relevant to the experiments. The expression we obtained for the conductance was given in the matrix form[8,9]:

$$G = \frac{e^2}{\pi h} \int_{-k_F}^{k_F} \frac{d\kappa}{k_z(\kappa)} \{[\tilde{\Theta}^\dagger(k)\tilde{\Gamma}_R\tilde{\Theta}(k) - \tilde{\Delta}^\dagger(k)\tilde{\Gamma}_R\tilde{\Delta}(k)] + 2\text{Im}[\tilde{\Theta}^\dagger(k)\tilde{\Gamma}_I\tilde{\Delta}(k)]\}. \quad (2.1)$$

$\tilde{\Theta}(k) = [\tilde{I} - (\tilde{r}e^{i\tilde{\Gamma}d})^2]^{-1}\tilde{t}_k$ and $\tilde{\Delta}(k) = e^{i\tilde{\Gamma}d}\tilde{r}e^{i\tilde{\Gamma}d}\Theta(k)$ are expressed in terms of the reflection matrix $\tilde{r}$ and the transmission vector $\tilde{t}_k$, which are the analogues of the reflection and transmission coefficients for 1D step potential. The diagonal matrix $\tilde{\Gamma}$ has elements $(\Gamma_{ij})^2 = \delta_{ij}2m^*(E - \epsilon_n)/\hbar^2$ for energy E, with the effective mass m^* and constriction eigenenergy ϵ_n, and are expressed as $\tilde{\Gamma} = \tilde{\Gamma}_R + i\tilde{\Gamma}_I$.

In this expression, the interference effects of the reflected waves are built in by the separation of the right going (first term) and left going (second term) states. The evanescent states in the third term contribute as tunneling, and smooths out the sharp rises in G

corresponding to the opening of a new channel. This effect becomes significant at small constriction length, d. As pointed out earlier[8], for the simplest system, which is a uniform quasi 1D constriction between two 2D EG, the quantization of G starts at $d \simeq \lambda_F$ and steps occur exactly at the integer multiples of $2e^2/h$ only for $d > 5\lambda_F$. The case of $d = 0$ corresponds to Sharvin conductance[14], $G_s = (2e^2/h)2w/\lambda_F$, but whole curve is displaced and weak oscillations are superimposed owing to the quantum interference effects. However, the larger is d, the sharper are the quantum jumps and the flatter are the plateaus. In addition, for large d the resonance structure is superimposed on the flat plateaus.

3. Results

3.1 ADIABATIC EVOLUTION OF STATES

The lack of resonance structure in the experimental data may also be sought in the adiabatic evolution of the states through the constriction. Adiabatic matching of a narrow constriction to a wide reservoir was first suggested by Landauer[15]. Glazman et al.[16] analytically showed the validity of such a matching. In their model the narrowing occurs in an infinitely long region. However, in reality the constriction has a finite length. In order to be able to associate the experimental results devoid of resonances with the adiabatic evolution, one has to demonstrate that it can also occur for finite length constrictions with 2D EG boundaries.

In Fig. 1, we present the $G(w)$ curves calculated for various constriction geometries which mimic the adiabatic matching in finite length. The conditions, in which the adiabatic evolution of the states occurs depend on the geometry of the constriction. It appears that a tapering with $\alpha \simeq 85°$ satisfies this condition. Because of the adiabatic evolution, the quantization of conductance is not affected in any essential manner[16], but owing to the mixing of the phases the resonance structure disappeared. For a smoothly varying width (such as sine- or cosine-modulation) adiabatic evolution is almost complete, yielding sharp steps and almost flat plateaus for small Δw, whereas the quantization is not reached and the steps are not sharp for large Δw. Our results are in agreement with those of Yacobi and Imry[17] which are stating that the effects of the abrupt boundaries with the 2D EG are exponentially small for adiabatic constrictions.

3.2 RESONANT TUNNELING THROUGH THE CONSTRICTION

On the contrary to the above geometry, a finite constriction which is relatively narrower at both ends gives rise to the spatially varying subband energies, $\epsilon_n(z)$, which are lowered near the center of the constriction. These subbands can be viewed as the potential wells, in which quasi 0D (or confined) states are formed. A similar confinement can occur even if the widening is abrupt inside the constriction. This situation is reminiscent of a double barrier resonant tunneling (DBRT) structure, and hence may lead to the resonant tunneling. Before the nth channel ($n \geq 1$) is opened, the resonance may occur on the $(n-1)$th plateau whenever a confined state in the well of $\epsilon_n(z)$ matches to the states of the 2D EG. As a result, resonance peaks illustrated in Fig. 2 appear near the edges of the quantum steps of $G(w)$.

We used sine- and cosine-modulation along the constriction with the amplitude Δw. We found that the sine-modulation yields relatively smaller widths w_r (at which resonant

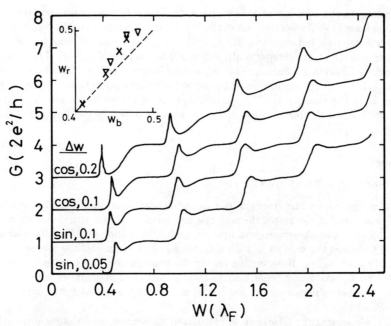

Figure 2   Calculated $G(w)$ curves for the resonant tunneling. Sine- and cosine-modulation is used along the constriction (the length is $d = \lambda_F$). Inset shows the comparison of the infinite constriction (w_b) and finite constriction (w_r) resonances for 8 different geometries. All lengths are in units of λ_F.

tunneling occurs), and broader peaks as compared to those of the cosine-modulation. This can be explained by using the DBRT analogy. Since the sine-modulation is represented by a DBRT structure with a relatively wider well, but narrower barriers as compared to those of the cosine-modulation, the resonance energies are relatively farther from the top of the barrier (or equivalently from the quantized steps in G) and resonance peaks are relatively broader for the sine-modulation. That the DBRT analogy is valid and thus the peaks near the steps are related to the resonant tunneling are shown by calculating the 0D states for an infinite constriction with a widening of the same form as the finite constriction above, and by comparing the positions with the resonance positions w_r in the inset of Fig 2.

4.Conclusions

We investigated the effects of the geometry of the constriction, which plays a crucial role in the quasi 1D ballistic transport, by performing calculations of the conductance. We showed that, if the constriction becomes smoothly narrower inside, the current carrying states evolve adiabatically leading to the quantized conductance without resonance structure. In contrast, quasi 0D (confined) states can form in a local widening inside the constriction, and give rise to resonant tunneling.

ACKNOWLEDGEMENTS. We acknowledge valuable discussions with Professor N. García and A. Yacobi. This work is partially supported by the Joint Project Agreement between Bilkent University and IBM Zurich Research Laboratory.

References

1. R. Landauer, IBM J. Res. and Develop. **1**, 233 (1957); Y. Imry, in *Directions in Condensed Matter Physics*, ed. G. Grinstein and G. Mazenko (World Scientific, Singapore, 1986), vol. 1, p. 102.

2. B. J. van Wees, H. van Houten, C. W. J. Beenakker, J. G. Williamson, L. P. Kouwenhoven, D. van der Marel, C. T. Foxon, Phys. Rev. Lett. **60**, 848 (1988).

3. D. A. Wharam, T. J. Thornton, R. Newbury, M. Pepper, H. Rithcie, G. A. C. Jones, J. Phys **C21**, L209 (1988).

4. G. Kirczenow, Solid State Commun. **68**, 715 (1988).

5. A. D. Stone, A. Szafer, Phys. Rev. Lett. **62**, 300 (1989).

6. N. García, L. Escapa, Appl. Phys. Lett. (in press); L. Escapa, N. García, J. Phys.: Condens. Matt. **1**, 2125 (1989).

7. E. G. Haanapel, D. van der Marel, Phys. Rev. **B39**, 5435 (1989); D. van der Marel, E. G. Haanapel, *ibid*, 7811 (1989).

8. E. Tekman, S. Ciraci, Phys. Rev. **B39**, 8772 (1989).

9. E. Tekman, S. Ciraci, in *Science and Engineering of 1- and 0-D Semiconductors*, ed. S. P. Beaumont and C. M. Sotomayor-Torres (Plenum Press, New York, 1989).

10. N. García, (private communication).

11. L. Spendeler, (private communication).

12. D. A. Wharam, M. Pepper, H. Ahmed, J. E. F. Frost, D. G. Hasko, D. C. Peacock, D. A. Ritchie, G. A. C. Jones, J. Phys. **C21**, L887 (1988).

13. S. E. Laux, D. J. Frank, F. Stern, Surf. Sci. **196**, 101 (1988).

14. Yu. V. Sharvin, Zh. Eksp. Teor. Fiz. **48**, 984 (1965) (Sov. Phys.–JETP **21**, 655 (1965)).

15. R. Landauer, Z. Phys. **B68**, 27 (1988).

16. L. I. Glazman, G. B. Lesorik, D. E. Khmelnitskii, R. I. Shekhter, Pisma Zh. Eksp. Teor. Fiz. **48**, 218 (1988) (Sov. Phys.–JETP Lett. **48**, 238 (1988)).

17. A. Yacobi, Y. Imry, (unpublished).

WHAT DO WE MEAN BY "WORK FUNCTION" ?

RICHARD G FORBES
University of Surrey
Department of Electronic and Electrical Engineering
Guildford, Surrey GU2 5XH, UK

ABSTRACT. This paper is intended as a brief tutorial on work functions. It summarises some basic facts about the concept of "work function" as conventionally used in surface science. It then discusses various other parameters that get called "work function" in the contexts of field emission and scanning tunnelling microscopy.

1. Introduction

The author has for a number of years worked in the general area of field emission. In this area, as a result of discussion over the years, ideas about "work-function" are well developed, and some of the associated difficulties are well understood.

It became clear at the Erice study institute that many members of the STM community are, not surprisingly, less familiar with these things. Consequently, in the context of STM there seems to be a degree of looseness about the deployment of the term "work-function", - and a corresponding possibility of confusion.

This paper is a slightly extended version of part of the seminar the author gave at the end of the study institute. The aim is to provide a brief tutorial on work-functions and related parameters.

2. Basic work-function concepts

Historically, the name "work function" was given to a class of parameters that quantify the work done in removing an electron from a solid, in the absence of any applied electric field. Work-functions are traditionally denoted by ϕ or Φ, and the modern convention is to state them in eV. (Some older literature gives them in V.) In definitions, the electron is always taken from the solid's fermi level, but it is then necessary to ask what is the physical situation outside the solid's surface, and precisely what removal process is being considered.

In particular, in thermionic emission, where the electrons are ejected from a heated surface, a space-charge may build up outside the surface, and this may alter the energy barrier that an electron would need to surmount. In consequence, the old name "thermionic work-function" is usually avoided in surface science, and modern work-function definitions relate to situations in which space charge effects are assumed to be absent.

Various complications are avoided if we initially consider a solid surface that is atomically flat, and (on the atomic scale) is very large in extent. The so-called **local work-function** can be defined as "the work needed to remove an electron from the solid's fermi level to a position somewhat outside the surface". It is implicitly understood that some sort of slow thermodynamic process is being discussed.

The word "somewhat" is crucial in this definition. It implies, firstly, that the final electron position is effectively outside the range of image forces (- which in turn implies that the final position is outside the range of the short-range forces due to the atomic nature of the surface charge distribution). It implies, secondly, that the distance of the final electron position from the surface should be small in comparison with the lateral extent of the flat surface.

It should be noted that real solids, even when neutral, are surrounded by small medium-range electric fields. (I discuss the origin of these below.) Consequently the local work-function cannot in general be identified with the work done to remove an electron from the fermi level to infinity in a slow thermodynamic process. The latter work enters into some thermodynamic arguments and is sometimes refered to as the **total work-function**[1].

For the flat surface of infinite extent, beloved of theoretical physicists, the total work - function does equal the local work-function of the surface.

Work-function is sometimes defined as "the difference in energy between the vacuum level and the fermi level". But this approach hides the question of "how do you define the vacuum level ?" In the vicinity of a real (neutral) solid the vacuum level is a function of position, due to the electric-field system surrounding the solid. The name "vacuum level" suggests to the unwary an absolute level, and this can be misleading.

3. The physical origin of local work-function

Values of local work-functions are determined by two kinds of physical interaction between the solid and the electron being removed:

       (a) "Chemical" (correlation and exchange) forces;
       (b) Electric dipole-layer effects.

Consequently the work-function can be split into **"chemical"** (χ) and **"electrical"** (ψ) **components**:

$$\phi = \chi + \psi \qquad (1)$$

---

[1]Also called the "absolute work-function", but this is a misleading name for a quantity that depends in principle on emitter shape.

Outside the solid surface the correlation and exchange forces are often (loosely) described as "image forces". In origin, the classical image interaction represents only the correlation part of the interaction. However, it is always assumed that, at sufficient distance outside the surface, the image interaction is an adequate approximation for the combined correlation and exchange interaction. Obviously, close to the surface the image interaction is a poor representation, and - as inside the solid - quantum-mechanical treatments are needed.

The "chemical" component of the work-function is thus given by the correlation-and-exchange energy of a fermi-level electron inside the solid. This is determined by the chemical nature and structure of the solid, and is a "bulk" property. It applies to all faces of the solid.

The "electrical" component of the work-function, however, depends both on the chemical nature of the solid and on the surface crystallography of the solid face in question. At the surface the electron charge distribution redistributes itself, and this gives rise to an electric dipole layer, which in turn leads to an electric potential difference between the exterior and interior of the solid.

Following Smoluchowski (1941), it is convenient to think of the electron behaviour in terms of two effects: **"spreading"** and **"smoothing"**. At the surface the electrons "spread" out into the vacuum, and this gives rise to a dipole with the negative end outwards. The electrons also "smooth" sideways into the gaps between the atoms, and this produces a dipole with positive end outwards. The sign of the resultant electric dipole depends on the balance between spreading and smoothing, and both this and the sizes of the effects depend on the details of the surface crystallography.

For a given solid, the extent of spreading and of smoothing will vary as between different faces, due to the different surface atomic arrangements. This produces a corresponding variation in the strength of the dipole layer and hence in the electrical component of work-function. So there is a variation with crystallographic orientation in the value of the local work function. Similar effects can occur if surface reconstruction takes place.

In broad terms, chemical effects are the major determinant of work-function size, and typically contribute several eV. The rest is contributed by electric effects, and these can vary as between faces by typically 0.1 to 1 eV.

4. Patch fields

The question arises: "what happens if a solid surface has a grain structure, and is comprised of various small areas or "patches", each of which has a different local work-function ?" A similar question occurs if the solid has three-dimensional structure, exposing facets of different crystallographic orientations, as with a field emitter tip.

In both cases, the fermi-level is uniform within the solid, because this is the definition of thermodynamic equilibrium for the electrons within the solid. Variations in local work-function mean that the electrostatic potentials outside each of the different faces are different. Thus, a neutral solid is surrounded by a system of electric fields, known as **"patch fields"**. These extend out into the space surrounding the solid, and have an effective range comparable with the dimensions of the areas or facets in question.

5. The effect of adsorbates

Charge transfer to and from adsorbate atoms and molecules also produces an electric dipole layer. Consequently adsorption usually changes the local work-function value. If adsorption does not take place uniformly across the surface, then patch-field effects can occur in this case too.

6. Field electron emission theory and the Fowler-Nordheim plot

We now need to look at how these defined work-functions relate to the the measured quantities called work-function. It is convenient here to look at some issues that arise in connection with field emission measurements of work function. Related issues arise in scanning tunnelling microscopy. We consider first the theory for the case where electrons are field emitted from a large flat <u>metal</u> surface of uniform local work-function.

Field electron emission is a tunnelling process, and the simple Fowler-Nordheim (FN) theory assumes that tunnelling is taking place through a triangular barrier of height h (relative to the fermi level), as shown in Figure 1. Conventional FN theory identifies h with the local work-function φ, but we make a distinction between the two concepts here.

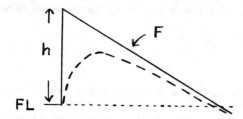

Figure 1.

Triangular barrier assumed in Fowler-Nordheim theory, with reduction by image potential shown as dashed line.

The Fowler-Nordheim result is that the flux of emitted electrons per unit area (j) is given by an expression of the form:

$$j = AF^2 \exp(-bh^{3/2}/F) \qquad (2)$$

where F is the applied field, b is a universal constant equal to 6.8309 eV$^{-3/2}$ V nm^{-1}, and A is a constant of value depending on the nature of the emitter.

The value of the applied field may not be well known, so we write F = βV , where V is the applied voltage and β is a proportionality factor (that depends on the geometry of the tip and its surroundings), take logarithms, and hence obtain:

$$\partial \ln\{j/V^2\}/\partial(1/V) = -bh^{3/2}/\beta \qquad (3)$$

This is the basis of the "Fowler-Nordheim plot" (a plot of $\ln\{j/V^2\}$ vs 1/V); the slope is equal to $-bh^{3/2}/\beta$, so a value of h can be derived if the value of β can be determined.

In reality, the value of β for a given emitter is usually determined by doing a FN plot on a well-defined clean face of assumed known work-function; this β can then be used to determine changes in work-function, for example those resulting from adsorption on the face, using FN techniques.

However, let us suppose that an independent value of β as defined above has been been obtained in some other manner (for example, from field ion emission experiments). The question now is: " If we use the slope of a FN plot and this value of β , do we actually get out the correct physical value of local barrier height for the face in question ?"

7. A basic fault in FN theory - the Burgess et al. correction

The answer to the question just posed is (as you might expect): "Not quite - because the theory is too simple". Let us suppose for the moment that the experiments are well-defined: for example, currents have been measured using a probe hole, centred over a well-defined facet, and that the probe-hole encompasses many atoms but only a small fraction of the facet in question.

A first issue is whether we have correctly described the barrier seen by the tunnelling electron. In reality, the inside edge of the barrier is not the sharp step assumed by elementary FN theory. Rather, the barrier is reduced as a result of the correlation and exchange interaction between the tunnelling electron and the surface. This lowering can be represented (approximately) by an image potential, giving the reduced barrier as schematically shown in Figure 1.

Burgess et al. (1953) incorporated image-potential lowering into field emission theory, and showed that the flux density is given by:

$$j = A'F^2 \exp(-\alpha_B bh^{3/2}/F) \qquad (4)$$

where A' is a modified constant, and α_B is a slowly varying function of h and F . Algebraic manipulation gives the slope of the FN-plot as:

$$\partial \ln[/V^2]/\partial(1/V) = -s_B bh^{3/2}/\beta \qquad (5a)$$

where the factor s_B is related to α_B by:

$$s_B = \alpha_B - F\, \partial\alpha_B/\partial F \qquad (5b)$$

and was tabulated by Burgess et al.

In a field electron experiment, α_B might have the value 0.65 and s_B the value 0.95. Clearly there would be an error by a few percent in the predicted value of h if we used an independent value for β, a Fowler-Nordheim plot, and eq.(3) rather than eq.(5).
In reality, in field emission work, this error can usually be neglected; it is the fact of its existence that is important here.

8. Generalisation - other slope errors

The characteristic feature of the above error is that the slope of the FN plot predicted from a corrected theory is not what simple theory predicts it to be. So use of the simple result (3) to derive barrier height from an observed Fowler-Nordheim plot gives an incorrect value of h. The quantity derived from the simple theory is not in fact a barrier height; rather, it is a **"characterisation parameter"**. For convenience we refer to a mistake of this type (treating a characterisation parameter as a real or model barrier height) as a "slope error".

Fowler-Nordheim plots often are linear (or nearly linear) over the range of current values measured, so the possibility of a slope error may not be apparent. Obviously, if the plot is not linear, then the need to use something better than the simple theory is rather more obvious. A useful question to ask is: "In the context of field electron experiments, what other things might cause slope errors, either over the whole range of current measurements or over part of it ?"

In the case of metals, slope errors can be associated with the following situations:
(1) Space-charge build up near the emitter.
(2) Experiments in which the measured current comes from a combination of areas of different work-functions.
(3) Situations in which the band structure near the fermi-level differs markedly from the simple particle-in-a-box band structure assumed by FN theory.
(4) Situations in which travelling plane waves normal to the emitting surface are not good solutions to the Schrodinger equation near the fermi level, and/or where "surface resonances" occur.
(5) Situations in adsorption where the presence of an adsorbate atomic level near the fermi level gives rise to a distorted barrier, and/or a "transmission resonance".

This is not to say that these situations will necessarily give rise to significant slope errors, merely that if any of these situations exists, then the possibility of slope errors in FN analysis ought to thought about.

In the case of semiconductor field emission, simple FN theory does not apply, but the simplest theory is something resembling it in which the barrier height is measured from a band edge rather than the fermi level. Fowler-Nordheim plots can still be nearly linear. With semiconductors there are some additional causes of discrepancy from simple theory, including the following:
(6) Field penetration and band bending.
(7) Screening by surface states, which can result in band bending in the opposite sense to that expected in their absence.
(8) Thermodynamic de-coupling of the surface states from the bulk states, which can result in the surface-state fermi-level being different from the bulk fermi level by an amount that depends on field (Modinos, 1974).

9. The distinction between barrier height and work function

The preceding section deals with the issue of whether the theory used is good enough to predict the barrier height correctly. There is a separate question of whether it is correct to regard the parameters barrier height and local work-function as equivalent.

The heart of the matter is as follows. Local work-function is in principle defined by a thermodynamic-type argument in which the final position of the electron is important, but the path by which it reaches that final point is not (because the work involved is independent of path). Providing that the final point "somewhat outside the surface" is in a region where potential is laterally uniform, the details of how the potential varies laterally at positions nearer the surface is quite irrelevant.

However, in a tunnelling technique, the details of the lateral potential structure close to the surface obviously may be important.

The essential question seems to be whether or not the technique in question "laterally averages" over the potential structure. With normal field electron emission techniques, (a) the barrier is relatively thick, so lateral potential structure on its inner side has small influence, and (b) probe-holes usually collect electrons from an area that corresponds to many atoms. With such techniques, it seems valid to identify barrier height with local work-function.[1]

However, the whole purpose of atomic imaging techniques, in particular field-ion microscopy (FIM) and scanning tunnelling microscopy, is to detect fine lateral structure. Almost by definition, if the theories of these techniques are developed as tunnelling theories in which a barrier height (or something resembling it) appears, then this parameter must not to be identified a-priori with the local work-function. The barrier height must have the option of varying (on an atomic scale) with position across the surface; the work-function is a property of the surface as a whole.

---
[1]Field electron techniques involving emission from only a few projecting atoms are a different matter. But for such a case it is difficult to formulate a unique definition of local work-function, due to problems with patch fields.

10. Barrier height as a location-dependent parameter

At this point the scale of the argument shifts downwards, since we need to discuss the lateral variation of parameters on a subatomic scale. Various new features appear.

Obviously, there is a now a question of how do we define barrier height at a particular location in the surface lattice cell. And, - if this can be done satisfactorily - there is a second question as to whether this defined barrier height varies with location in the surface lattice cell.

More generally, one can think of one-dimensional sections, normal to the surface plane, taken through the potential structure at various locations in the surface lattice cell, and ask "does the strength of the barrier vary with location?"

Here, it is of interest to note that there is a marked difference between the theories of field-ion imaging and scanning tunnelling imaging. Current FIM theory (e.g. Forbes 1985) explains the imaging process in terms of lateral variations in barrier strength; whereas the Tersoff and Hamann (1985) STM theory takes the barrier height as laterally uniform and explains across-surface tunnelling current variations in terms of variations in local densities of states (LDOS).

Although the presence of the higher field in the FIM situation might justify this difference in theoretical approach (or it might be something to do with the fact that most FIM is done on metals, whereas much STM is done on semiconductors), it is also arguable that FIM theory should take more notice of LDOS effects and STM theory more notice of lateral barrier strength variations.

11. The reliability of "measured" barrier heights.

A final question arises. Suppose that some physical method could be found of associating a barrier height h with a specific location in a surface lattice cell, and that there is some constant barrier height h_p that characterises the scanning probe. Suppose, further, that some simple one-dimensional model barrier (e.g. a trapezoidal barrier) is constructed and theoretically analysed to give an expression of the form (6) for emission flux per unit area:

$$j = B \exp [- f(s,h,h_p)] \qquad (6)$$

where s is barrier width, B is a factor independent of s, and f is some analytical function. Further suppose that we can deduce an explicit expression for a parameter g defined by:

$$g = \partial \ln\{j\}/\partial s = - \partial f(s,h,h_p)/\partial s \qquad (7)$$

and that this parameter can also be measured. (STM does, of course, have a have a procedure of this kind.) Finally suppose that from a knowledge of h_p and g, a value of h can be found via eq.(7).

Given all these things, the question is: "Can the resulting value of h be taken as reliable ?"

The answer, as in the case of Fowler-Nordheim analysis in field emission, must be "not necessarily". As with FN analysis, if the barrier modelling and/or the theoretical analysis is overly simple, then the parameter h_{exp} extracted from a measured value of g may not correspond well with the barrier height h_{mod} inserted into the simple model. As in the case of field emission, the extracted parameter h_{exp} must be regarded simply as a "characterisation parameter".

And, as with the "slope errors" discussed in connection with FN analysis, one might expect that there will be a number of physical situations particularly likely to cause discrepancies between the inserted model barrier height h_{mod} and the extracted characterisation parameter h_{exp}. Image-potential lowering of the basic model barrier is a factor well discussed in STM theory, but some of the situations listed in section 8 may be good candidates for producing discrepancies in STM theory as well as in field emission theory.

A further problem with the STM situation is that, in reality, one-dimensional theories cannot be adequate. The three-dimensional nature of the barrier and the three-dimensional nature of the wave-function in the barrier have to be taken into account. It will presumably be possible to define a central axis for the barrier, but at the very least the full theory must be equivalent to some sort of weighted local averaging of barrier heights, over locations in the surface lattice cell adjacent to the location of the central axis. This will produce further blurring of the interpretation of any characterisation parameter extracted from experimental measurements using an simplified model.

12. Summary

The main points to emerge from this discussion are:
(1) Local work-function as defined in surface science is a property associated with a (well-defined) surface as a whole.
(2) The quantities ϕ used in STM theory should not be closely identified with the surface-science local work-function. It is better always to think to of them as barrier heights or effective barrier heights.
(3) Barrier-height type parameters extracted from tunnelling current measurements should not be assumed equal to the barrier heights inserted into simple barrier models, - at least, not without careful theoretical analysis.

More generally, it is suggested that it might be useful for STM theory to pay more attention to the detailed nature of the tunnelling barrier, and that some of the knowledge gained in connection with the field emission techniques may be of help as a stimulus to thought.

Specific References

Burgess, R.E., Kroemer, H. and Houston, J.M. (1953) "Corrrected Values of Fowler-Nordheim Field Emission Functions v(y) and s(y), Phys. Rev. 90, 515.

Forbes, R.G. (1985) "Seeing Atoms: The Origins of Local Contrast in Field Ion Images", J. Phys. D: Appl. Phys. 18, 973-1018.

Modinos, A. (1974) "Field Emission from Surface States in Semiconductors", Surface Sci. 42, 205-227.

Smoluchokski, R. (1941) "Anisotropy of the Electronic Work Function of Metals", Phys. Rev. 60, 661-674.

Tersoff, J. and Hamann, D.R. (1985) "Theory of the Scanning Tunneling Microscope", Phys. Rev. B31, 805-813.

General References

Some general references on work-functions and the correlation and exchange interaction are:

Jennings, P.J. and Jones, R.O. (1988) "Beyond the Method of Images - the Interaction of Charged Particles with Real Surfaces", Adv. Phys. 47, 341-358.

Knor, Z. (1977) "The Interplay of Theory and Experiment in the Field of Surface Phenomena on Metals", in: M.W. Roberts and J.M. Thomas (eds.), Surface and Defect Properties of Solids, Vol.6, pp.139-178.

Riviere, J.C. (1969) "Work Function: Measurements and Results", in: M. Green (ed), Solid State Surface Science, Vol. 1, Dekker, New York, pp. 179-287.

Some general references to past work on field electron emission are:

Gadzuk, J.W. and Plummer, E.W. (1973) "Field Emission Energy Distribution (FEED)", Rev. Mod. Phys. 45, 487-548.

Gomer, R. (1961) Field Emission and Field Ionization, Harvard Univ. Press, Cambridge, Mass.

Swanson, J.E. and Bell, A.E. (1973) "Recent Advances in Field Electron Microscopy of Metals", Adv. Electr. Electron Phys. 32, 193-309.

SCANNING TUNNELING MICROSCOPY:
METAL SURFACES, ADSORPTION AND SURFACE REACTIONS

R. J. Behm
Institut für Kristallographie und Mineralogie
Universität München
Theresienstr. 41, D-8000 München 2, Fed. Rep. Germany

ABSTRACT: The role and application of scanning tunneling microscopy (STM) for the investigation of clean and adsorbate covered metal surfaces are discussed. STM studies on the periodic structure and structural defects on these surfaces are reviewed. Experimental results related to atomic resolution imaging of close-packed metal surfaces are presented. The contribution of electronic effects to STM imaging of adsorbates is discussed, electronic modifications of the surrounding metal substrate atoms illustrate the effect of the chemical bond between adsorbate and substrate atoms. Time resolved observations of local adsorbate and substrate structures are shown to gain detailed information on various surface processes, in particular they allow direct access to mechanistic details of surface reactions.

1. Introduction

Investigations of structural and electronic properties of well defined surfaces and of processes on these surfaces – the area of surface physics and surface chemistry – belong to the most obvious and straightforward applications of scanning tunneling microscopy (STM). Correspondingly the first STM studies concentrated on the structure of (reconstructed) clean surfaces [1-3]. Relatively early and from various reasons a separation between STM work on metal surfaces and semiconductor surfaces developed, which is reflected by the division into two separate contributions in this volume.

This chapter describes recent results as well as general trends and problems of STM investigations on different aspects of well-defined metal surfaces. From methodic reasons it will be limited to work performed under ultra high vacuum conditions. In-situ STM studies on electrolyte-covered electrode surfaces, which can be carried out under comparatively "clean" conditions and where atomic resolution was equally achieved recently [4], are subject of another chapter in this volume [5]. On the other hand, the mostly very high reactivity of metal surfaces excludes investigations on well-defined surfaces under ambient conditions with very few exceptions, such as the rather inert Au(111) surface [6] or surfaces passivated by an adsorbate adlayer [7]. Despite of the more general review character of this chapter it is closely connected to our own work, and examples from that work are used for discussion where appropriate.

Following a brief account on the experimental setup and procedures microscopy results on clean surfaces are discussed first. They provide insight into a variety of structural properties of these surfaces. Experimental details on atomic resolution imaging of close-packed metal surfaces will be presented, too. In the next section STM measurements on adsorbates and adsorbate covered surfaces are reviewed. The resolution and representation of individual adsorbate species in the STM image is discussed. Here the local interaction between substrate and adsorbate leads to (local) electronic modifications of the surface, and electronic effects must be considered as a significant contribution to the STM image. These measurements also yield information on adlayer mobility and ordering, on the formation of local island structures in such adlayers and on the role of substrate defects on the adsorption process. The last part finally deals with STM observations on surface reactions such as decomposition of adsorbates, structural transformations of the substrate induced by the presence of an adsorbate or reactions between adsorbate and substrate, e.g. formation of a surface oxide. The local observation of these processes often allows conclusions on their microscopic mechanism. This complements integrated spectroscopic and kinetic data on the respective surface reactions.

2. Experiments

STM investigations on well-defined (metal) surfaces generally require facilities for in-situ sample preparation and characterization. In most cases this also necessitates heating, and temperature control via a thermocouple or a pyrometer, is desirable. For most of our own work a 'pocket-size' type STM was used where piezo-ceramics and mechanical tip-sample approach are incorporated into a massive stainless steel block of 5x4x3 cm3 [8,9]. In conjunction with internal vibration damping via a stack of metal plates and viton spacers and an outside hydropneumatic damping stage this very rigid setup attained vertical stabilities of ~ 0.01 Å. For sample preparation we usually followed standard procedures. Repeated brief cycles of Ar^+ ion bombardment ($\sim 1\mu A$, 0.5 -1 kV) and annealing gave good results. Gentle oxidation cycles in the later stages of sample preparation, performed by adsorption of small amounts of oxygen at 300 K and subsequent heating to CO formation, were frequently used to remove last traces of carbon contamination. Analysis of the surface chemical composition was performed by Auger spectroscopy (AES), and the cristallographic quality of the surface was monitored by low energy electron diffraction (LEED).

The tunneling tip is still the least defined part in the STM measurement. In a few cases STM tips were characterized by field ion microscopy (FIM) measurements [10]. Using W tips of a known apex structure Kuk and Silverman found an exponential decay of the corrugation measured on the (1x2) reconstructed Au(110) surface with increasing tip radius [10]. But in general little is known on the structure of the tip apex, and different procedures for tip preparation are applied. In our experiments the tips were electrochemically etched from polycristalline W wire ($\sim$ 2.5 V DC) and occasionally cleaned in-situ by Ne^+ ion sputtering. Prior to the experiments the tip was cleaned by field desorption/field evaporation in front of a dummy sample, mostly a gold specimen [8,11]. This produced a very stable tip with a rather flat apex, indicated by the low lateral resolution. Voltage pulses (7-9 V), applied during tunneling on the sample or the dummy surface, reproducibly transferred this tip into a high resolution state. Following this procedure atomic corrugation was resolved on different metal surfaces.

3. Clean Metal Surfaces

3.1. STM IMAGING OF CLEAN METAL SURFACES - BASIC PRINCIPLES

In STM measurements on clean metal surfaces tip and sample surface form the most simple case of a tunnel junction. Almost free electrons tunnel through a vacuum gap between two metal electrodes at almost identical potentials (small bias potentials). By analogy with one-dimensional descriptions for tunneling [12] an exponential decay of the tunnel current with increasing tip-surface separation was expected and experimentally verified in the first STM experiments [13].

In the meantime different theoretical concepts have been developed for a three-dimensional description of STM imaging of (metal) surfaces. Several papers in this volume give more details on this subject [14-16]. Most of these methods are based on the 'Transfer-Hamiltonian' formalism introduced by Bardeen, which neglects interactions between the electrodes and allows to calculate the tunnel current from the wave-functions of the separated electrodes [17]. Tip-surface interactions are explicitly taken into account in some of the most recent calculations [15,16,18]. Earlier attempts of describing the tunnel current by calculating the transmission coefficients for electrons incident on a corrugated barrier have not been pursued [19-21].

Tersoff and Hamann calculated the tunnel current between surface and tip by using the 'Transfer Hamiltonian' approximation. In these calculations they confined the tip to spherical symmetry with s-wave functions only [22,23]. They concluded that under these conditions the tip follows a line of constant charge density at E_F at the center of the tip. On metal surfaces the charge distribution at E_F can further be approximated by that of the total charge density. The exponential correlation between gap width and tunnel current was reproduced in these calculations. These authors also found that for a periodically corrugated surface the corrugation amplitude observed in the STM image decayed exponentially with increasing tip-surface separation. This result allows a very useful first estimate of the corrugation to be expected in STM images of metal surfaces. It was also pointed out that this corrugation is very similar to the corrugation of the scattering potential for He diffraction. Because of the larger distance between tip (center) and surface in STM measurements as compared to that of the turnaround point of scattered He atoms (~3Å to the plane of surface ion cores) the corrugation amplitudes in the STM image are expected to be lower than those calculated from He diffraction intensities.

The lateral resolution of STM measurements was estimated to $\sqrt{2(r_t+s)}$ (r_t: tip radius, s: distance between surface and tip perimeter), which leads to values around 4 to 5 Å for rather small tips and tip-surface distances [22,23]. These calculations also demonstrated the difficulty of extracting precise structural data from STM images. For tip-surface distances around 5 Å (s = 5 Å, r_t = 4 Å) it was shown to be almost impossible to decide on the position and even on the presence of low-lying surface atoms in the (1x3) reconstructed Au(110) surface [23].

Although STM experiments on clean metal surfaces have concentrated so far on structural properties, a number of different STM results on the electronic properties of metal surfaces was reported as well.

Measurements of the local tunnel barrier were performed either by measuring local current-distance curves or by recording dlnI/ds, via modulation techniques, while scanning the surface. On clean, flat surfaces typical values for the tunnel barrier are in the range of 2 eV to 4 eV which is significantly lower than the work function of the respective surfaces. (Varying results on the same surface indicate that the influence of the tip shape cannot be neglected). Two-dimensional images of the local tunnel barrier on structured surfaces show maxima in the tunnel barrier at the position of the corrugation maxima. For surface steps pronounced minima in the tunnel barrier are found on the lower terrace side of the step edge [24]. These results can be understood in the picture described above, in which the STM trace follows the line of constant charge density (at the center of the tip). The tendency of the charge density distribution to smear out with increasing distance ('Smoluchowski effect' [25]) results in a faster decay in charge density with increasing distance at the position of topography maxima, equivalent to a larger tunnel barrier at these positions. At topographical minima or at the lower terrace side of step edges the decay is slower which leads to a smaller value for the tunnel barrier. This behavior was reproduced in calculations on a stepped Ni surface [26]. It contrasts expectations based on the usual concepts of the work function [27]. It should be noted, however, that these measurements of the tunnel barrier differ strongly from work function measurements: The lateral variations in the tunnel barrier occur on a scale comparable to the vertical distance of the tip, whereas the work function by definition is measured far away from the surface, but at distances small compared to the lateral extension of the homogeneous surface plane [27].

Several possibilities were discussed to explain the rather low values determined for the tunnel barrier as compared to the respective work function. For contaminated surfaces, where extremely low barriers heights of a few tenths of an eV were observed, a mechanical contact between tip and surface was proposed [28]. In this case the conductivity of the insulating contamination layer depends on the pressure exerted by the tip [28]. On clean surfaces the lower values of the tunnel barrier were mostly attributed to image potential effects. The attractive image potential $V_{im} = -e^2/4r$ felt by a charged particle in front of a metal surface, at a distance r from the image plane, leads to considerable deviations from a square or trapezoidal barrier shape [29]. With decreasing gap width the 'effective' barrier height is reduced and finally collapses which is discussed in more detail in ref. 30. Image potential effects are reflected also by the resonances observed in the I/V-curves or (dI/dV)-V spectra, respectively, on various clean surfaces [31-34]. These resonances are due to the formation of bound states in the triangular potential in front of the surface and less dependent on the electronic structure of the surface [34,35].

Valence level spectroscopy - performed by evaluation of I/V-curves and their derivative or by direct measurement of dI/dV via modulation techniques - was rarely applied on clean metal surfaces so far and came to partly contradictory results. In STM measurements on Pd(111) and Au(111) Kaiser and Jaklevic could reproduce the spectroscopic features around E_F known from conventional valence-level spectroscopy [36]. In a later study Kuk et. al. similarly reported highly structured dI/dV curves for measurements on Au(100) [37]. Brodde et al. in contrast found no additional structure in I/V-measurements on Cu(111) and Au(111) [38]. It should also be noted here that for transition metal surfaces and for not too small gap widths the tunnel current is dominated by electrons tunneling into or out of sp-states rather than by d-electrons, since the latter are more localized close to the surface [26]. Consequently the pronounced structure in the density of states found in photoemission or inverse photoemission spectroscopy may not be resolved in STM spectroscopy, as far as it relates to d-state density. Because of the delocalized nature of the valence-level states also no spatial variation in the measured I/V-curves was found so far. Local spectroscopy performed on Cu(111) gave identical results over the entire surface [38].

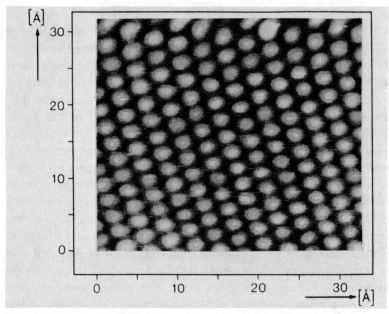

Fig. 1 Atomic resolution STM image of Al(111) (top-view, 34 Å x 34 Å, $V_t = -50$ mV, $I_t = 6$ nA, corrugation 0.3 Å, from ref. 8).

3.2. ATOMIC RESOLUTION IMAGING OF CLEAN METAL SURFACES

The resolution of individual atoms of the surface atomic lattice contrasts expectations from the Tersoff-Hamann model [22,23]. In the first report on atomic resolution STM images on (close-packed) metal surfaces, obtained on Au(111), Hallmark et al. attributed this phenomenon to the presence of a surface state on Au(111) near E_F [6]. Shortly later atomic resolution was achieved also on Al(111), a top-view image of the unreconstructed atomic lattice of that surface is reproduced in fig. 1 [8,9]. Aluminum is regarded as a model system for an (almost) free electron metal, the corrugation in this image is 0.3 Å, hence hardly subject to localized state effects in STM imaging. Self-consistent calculations show the corrugation in the total electron charge density to be negligible already 2 Å in front of the (111) surface [39]. Atomic resolution was reproducibly achieved using the tip preparation method described in the previous section. It was also shown that the transition into a high-resolution state of the tip is associated with an elongation of the tip by 10 to 20 Å. Different bias voltages between +1 eV and -1 eV had no measurable effects on the corrugation amplitude. Measurements of the tunnel current and the corrugation amplitude at different tip-surface separations revealed an exponential decay of both of these

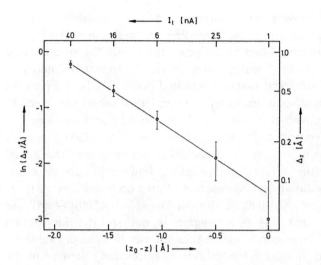

Fig. 2 Tunnel current and measured atomic corrugation of STM images on Al(111) at different tip-surface separations (V_t = -50 mV). The zero point for the gap width is arbitrary, atomic resolution was achieved over a range of 2 Å in gap width variation (from ref. 8).

quantities with increasing distance from the surface. This is illustrated in fig. 2. From these data a tunnel barrier of ϕ = 3.5 eV is evaluated, a maximum corrugation amplitude of 0.8 Å was found [8,9]. Although the absolute value of the gap width in these measurements is not known, it can be estimated from the known resistance for point contact (10-30 kΩ, s $\approx$ 2.5Å) [40] and the measured distance dependence of the tunnel current in fig. 2 to 4.5 Å at 40 nA. The atomic corrugation was resolved for different tip- surface separations over a range of 2 Å. Atomic resolution consequently was achieved at least 5 Å off from the surface, assuming almost point contact for the high current situation, or even 6.5 Å away if one uses the above estimate.

The physical origin for the observed atomic corrugation is still under discussion. As already mentioned it was attributed to the presence of a surface state on Au(111) [6]. On Al(111) a significant role of surface states was ruled out from the absence of any measurable bias dependence in the corrugation amplitude [8, 9]. In that study instead interactions between tip and surface were favored for explaining the high lateral resolution. First model calculations demonstrated that for small gap widths significant modifications of the surface electronic structure are induced by the presence of the tip ('tip induced states') [18] and that the tunnel current can no longer be approximated by the overlap of the unperturbed wave-functions of the two electrodes [16]. Interactions between tip and surface also result in forces between the two electrodes, and indeed Dürig et al. had reported forces in the nano-newton range acting between a metal tip and a metal surface under normal tunnel conditions [41,42]. Tip-induced, elastic deformations of the substrate had been identified as the origin of the giant corrugation amplitudes observed on graphite surfaces under certain conditions [43]. Such deformations are highly unlikely to occur on metal surfaces, but from observations during tip preparation Wintterlin et al. considered elastic deformations of the tip apex as a possible amplification mechanism [8,9].

Pursuing a different direction Chen investigated the possibility of increased resolution due to localized p- or d-states states on the frontmost tip atoms as compared to the extended, spherical s-states assumed previously [44]. From his calculations he found a significant increase in lateral resolution, sufficient to explain the resolution of individual atoms in a close-packed metal surface.

All of these effects can contribute to the increased lateral resolution necessary for resolution of individual atoms on close-packed metal surfaces. The actual mechanism and the role of the different contributions, however, is not yet clear. In the meantime atomic resolution has been achieved also on low-index surfaces of metals other than Au or Al such as Ru(0001) [45], Ni(100) [46,47] or NiAl(111) [48], indicating that this phenomenon is not specific for certain surfaces. According to our experiences high resolution was most easily achieved by using the tip preparation method described above, underlining the role of the specific state of the tip.

3.3. SURFACE RECONSTRUCTIONS

On several clean or adsorbate covered surfaces reconstructions lead to superstructures which result in a corrugation with a longer wavelength than that of the atomic lattice. Even for non-ideal tip conditions these structures are larger than the lateral resolution of the STM measurement and are therefore easily resolved by STM. Consequently a surface reconstruction was the first periodic structure observed in an STM image [1]. Structural details derived from STM images and the shape of the corrugation pattern predetermine structural models and can thus assist a structural analysis. This is particularly helpful for structures where large unit cells and a large number of atoms per unit cell complicate structure determination by diffraction methods or by ion scattering.

The clean surfaces of the (100) faces of Ir, Pt and Au and of Au(111) are known to be reconstructed. Current structural models propose the formation of a more densely packed, hexagonal topmost layer on a bulk-like substrate as a basic structural element for all of these reconstructions [49-53]. Because of the large unit cells, however, a quantitative structure determination of these reconstructed surfaces had not been possible so far. STM measurements on Au(100) [2,37], Pt(100) [54,55] and Au(111) [56,57] surfaces revealed a predominantly one-dimensional corrugation with a typical amplitude of 0.2 Å to 0.3 Å on all of these surfaces. The misfit between substrate and topmost layer results in slight vertical displacements of the atoms in that layer which is reflected by the corrugation seen by STM.

On the reconstructed (100) surfaces the hexagonal topmost layer is more closely packed than the square substrate layer and six rows of metal atoms reside on five rows of hollow sites along [1$\bar{1}$0]. The resulting periodicity of 14 Å orthogonal to these rows corresponds to the wavelength of the main corrugation in the STM images, and the amplitude reflects the vertical displacements due to the different 'adsorption' sites of the topmost metal atoms. On both of these surfaces the topmost layer is predicted to be slightly rotated with respect to the substrate lattice, induced by an additional contraction of the surface atoms also within the close-packed rows, normal to the corrugation. This rotation is reflected by the direction of the corrugation. The STM measurements could indeed identify the four different domain orientations expected from symmetry reasons. The rotation of the topmost layer also leads to a second, more subtle corrugation along the predominant corrugation lines and to periodic displacements within these lines. Both effects were resolved in STM studies of these surfaces [2,37,54]. The observation of these structural details allowed a quantitative description of the reconstructed layer even in the absence of atomic resolution.

On Au(100) [37] and Au(111) [56,57] atomic resolution images gave a more direct confirmation of the structural model and revealed additional structural details. For the clean, reconstructed Au(111) surface a uni-directional

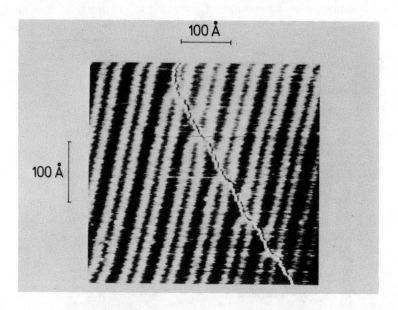

Fig. 3 Top-view STM image of the reconstructed Au(111) surface showing the transition regions (elevated lines) between fcc and hcp-stacking. The distance between line pairs in [1$\bar{1}$0] direction is 63 Å, the corrugation amplitude is 0.2 Å. Within the pairs the the lines are 19 Å apart, the depth of the smaller minimum (hcp region) is 0.12 Å (450 Å x 450 Å, from ref. 57).

contraction of the topmost layer was proposed, resulting in a stacking fault structure with periodic transitions from fcc to hcp stacking of the atoms in that layer [52,53]. The gradual transition between fcc and hcp stacking creates broad transition regions along [11$\bar{2}$] in which atoms are forced into less favorable, vertically displaced sites. In addition it involves a periodic lateral displacement of the surface atoms orthogonal to the [1$\bar{1}$0] contraction direction. Results of recent TEM, He diffraction and STM studies are in agreement with these ideas [52,53,56,57]. The fcc/hcp-transition regions show up as elevated lines in the STM image as seen in fig. 3. The paired arrangement of these lines with alternating, different spacings between neighboring lines reflects different stabilities of fcc and hcp regions. From a careful evaluation of the reconstruction pattern at step edges the wider domains can be identified as fcc regions [57]. Hence fcc stacking is slightly more favorable than hcp stacking, in agreement with the bulk geometry. The lateral displacement of the Au atoms within the unit cell is evident from high resolution images as the one shown in fig. 4. Without that displacement the Au atoms would be lined up along the [1$\bar{1}$0] direction which is indicated as a broken line in this figure. The absolute amount of the lateral displacement agrees well with the lateral distance between fcc and hcp adsorption site. Careful quantitative evaluation of the atomic positions in several atomic resolution images revealed that the contraction along the [1$\bar{1}$0] direction is not

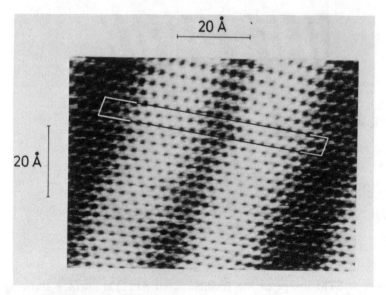

Fig. 4 Atomic resolution STM image (top-view, 80 Å x 60 Å, I_t = 630 nA, V_t = 69 mV) of the reconstructed Au(111) surface. The broken line in [110] direction clearly shows the lateral displacement of the individual Au atoms (white) between fcc and hcp-stacking regions (from ref. 57).

confined to the transition regions but acts rather uniformly. This contrasts predictions from the previously proposed soliton formalism for the fcc -> hcp stacking transition [53].

The (1x2) reconstructed (110) surface of Au and Pt represent a completely different type of structures. At the time of the first STM measurements on Au(110) [1] there was an ongoing controversy regarding their structure. These STM measurements showed no indication of any asymmetry in the surface corrugation as would have been expected for the 'saw-tooth' model [58], but they could not conclusively settle this debate. (In the meantime a 'missing row' structure with strong vertical and lateral distortions in deeper layers is generally accepted [59-61]). On these surfaces STM measurements turned out to be more important for the understanding of their defect structure and the physical origin of the reconstructed phases rather than of their ideal structure. Their LEED patterns often exhibit streaks in [01] direction, which indicates rather small domain sizes along the [001] axis [62]. Correspondingly Binnig and coworkers found a significant degree of disorder on Au(110) [1]. These results were interpreted in terms of a stabilization of the reconstructed surface by (111) microfacets [1]. But also surfaces with large domains of a well ordered (1x2) structure can be prepared reproducibly, as shown for Pt(110) in fig. 5 [63]. Hence there must be a specific stabilization of the (1x2) structure different than the formation of (111) microfacets [64-66]. The latter would equally allow also the existence of larger microfacets, equivalent to (1x3) or higher order structural subunits, which are absent in fig. 5. It is also concluded that the defect structure of these surfaces depends strongly on the respective procedures for sample preparation.

3.4. SURFACE DEFECTS

3.4.1 *Structural Defects.* Freshly prepared metal surfaces – after polishing to mirror finish – typically show a large number of rather smooth pits and hills of several hundred angstroms size. Only after careful and often tedious cleaning procedures extended, atomically flat terraces and mostly monoatomic steps begin to dominate the surface topography. On the different metal surfaces investigated in our group it had always been possible to finally reach terrace sizes of a few hundred angstroms. In a few cases, such as for Pt(111) [67], Ru(0001) [45,68] and Au(111) [57], terrace sizes in excess of 1000 Å were obtained. On most surfaces chemical and topographic quality of the surface seemed to be correlated, i.e. extended, flat terraces were achieved only after the surface was completely free of contaminants.

STM images of well prepared and clean metal surfaces resolve different kinds of structural features. Irregular structural elements include geometric or chemical defects such as steps, adsorbed foreign atoms/molecules or aggregations of these

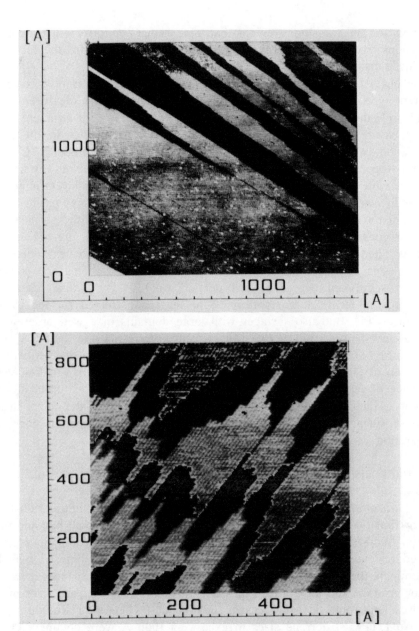

Fig. 5 Step-terrace topography of the (1x2) reconstructed Pt(110) surface. Terraces on different levels are alternately shaded dark and light (falling slope from foreground to background); top: area with predominant step direction along [1$\bar{1}$0], parallel to the close-packed rows of the surface (1450 Å x 1780 Å); bottom: area with predominant step direction along [001], orthogonal to the direction of the close-packed Pt rows (570 Å x 860 Å) (from ref.63).

adspecies. Advacancies or self-adsorbed, individual adatoms were not observed, presumably due to the high energies for two-dimensional condensation of metal atoms in a layer or at step edges. This differs from semiconductor surfaces where advacancies are found frequently.

Distribution, orientation and shape of surface steps as the main element of defect structures have long been subject of structural studies mainly by LEED [69]. In contrast to the averaged information obtained from that technique surface microscopy methods such as STM or low energy electron microscopy (LEEM) [70] have the advantage of giving a direct image of the local step structure. For example on low-index surfaces of high-melting materials such as Pt or Ru steps were often found to extend linearly over many hundred, even thousand angstroms and more [45,67,68]. In these cases the step edges are oriented along close-packed lattice directions. Other surfaces such as Al(111) [8,9] or Cu(110) [71-73] showed predominantly irregular, highly kinked steps without any significant orientational preference. For Cu(110) in fact there is good evidence that on the clean surface and at room temperature step edges are instable and can move over the surface by continuous two-dimensional evaporation and condensation of Cu atoms [73].

The step-terrace topography is influenced by the atomic structure of the surface and can subsequently exhibit strongly anisotropic features. This is illustrated by the STM images of the clean, (1x2) reconstructed (110) surfaces of Pt in fig. 5 [63]. These images were recorded at two orthogonal sides of an extended elevation of about 1μm diameter. They show a pronounced anisotropy in the step structure. The grey scale on these top-view images is adjusted such that terraces on successive levels are alternately shaded dark and light. In the top image we see terraces of hundred to several hundred angstroms width separated by monoatomic steps in [1$\bar{1}$0] direction. The kink density in these steps is very low, they often extend linearly over many hundred angstroms. In the bottom image the terraces equally stretch over the whole width of the image of ~600Å, but now the steps proceed in a zig-zag pattern. They consist of many small [1$\bar{1}$0] sections and a relatively high density of displacements in [001].

In these images a well-ordered, periodic corrugation with groves in [1$\bar{1}$0] direction is resolved, which relates to the 'missing row' (1x2) reconstruction of this surface [59-61]. The preferential formation of [1$\bar{1}$0] steps was attributed to an energetic stabilization of these steps by the presence of the (1x2) reconstruction [63]. As a consequence of the reconstruction the step edges form close-packed (111) microfacets, which had been proposed earlier as a stabilizing structural element for Au(110) [1]. The stabilizing effect of the (111) microfacets becomes directly apparent by comparison with the unreconstructed Pt(110) surface. On this surface, which can be obtained by CO adsorption, the pronounced preference for [1$\bar{1}$0] oriented steps is no longer observed, i.e. the stabilization of these steps is either absent or weaker [63].

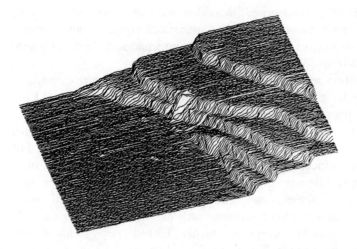

Fig. 6 STM image of a 550 Å x 490 Å area on a Au(111) surface with monoatomic steps. A vertical screw dislocation line emerges at the surface at the onset of two coincident steps on a flat terrace (from ref. 57).

In addition to reflecting surface properties the (local) step structure also contains information on the presence of bulk defects extending to the surface. Because of their very low density these defects are generally not accesible to integrating techniques. Nevertheless they are extremely important e.g. for bulk diffusion or segregation processes. Although no special study of (emerging) bulk defects has been reported yet, STM proves to be a unique tool for their investigation. The STM image in fig. 6 displays an area of a Au(111) surface in which a screw dislocation line emerges at the surface [57]. The bulk defect results in a vertical displacement within the layer crossed by the dislocation line. At the surface this is reflected by a step emerging on an otherwise flat terrace [74]. In the image in fig. 6 actually coincident steps emerges in the center of the image and thus mark the position of the (~vertical) dislocation line.

3.4.2. *Chemical Defects (Contaminants)*. Adsorbed foreign atoms or molecules (contaminants) represent another kind of irregular surface defect which is commonly observed by STM ('chemical defects'). Although STM measurements have no analytical capability in the sense of identifying the chemical nature of unknown adsorbed species, it is often possible to identify adsorbed species with a characteristic representation in the STM image by comparison with results of chemically sensitive measurements such as AES or XPS (X-ray photoelectron spectroscopy). After being identified once the species can be detected routinely by STM, down to very low concentrations far below common detection levels. This is demonstrated in the image in fig. 7, recorded on an Al(111) surface with traces of adsorbed carbon. In addition to the individual Al surface atoms characteristic extra-structures are resolved, which were attributed to carbon adatoms [75,76]. The carbon concentration in the surface area shown in fig. 7 is counted to 1.3% of a monolayer (ML). Clearly also much lower concentrations

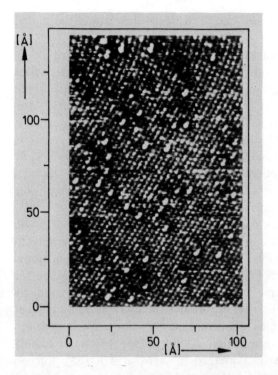

Fig. 7 STM image (top-view) of a 100 Å x 150 Å area on a Al(111) surface with traces of adsorbed carbon. Individual C adatoms on the Al substrate lattice are resolved as small protrusions (prominent white dots) (V_t = 90 mV, I_t = 100 nA).

would easily be detected, demonstrating that under certain circumstances STM measurements can serve as an extremely sensitive analytical tool.

4. Adsorbates and adsorbate covered surfaces

4.1 STM IMAGING OF ADSORBATES – BASIC PRINCIPLES

Adsorbates are characterized by their interaction with the substrate. This has consequences on their electronic structure and chemical nature, but also on their distribution over the surface. STM measurements on adsorbates and adsorbate covered surfaces allow both the observation and characterization of adsorbates. The (local) distribution and ordering of adsorbates within the adlayer or the influence of the substrate topography on these properties can be extracted from STM images, which can give insight also into dynamical surface processes such as adsorption/growth, diffusion or surface reactions.

In many of these problems useful contributions from STM measurements rely on the ability to resolve individual adsorbate species by STM, which had not been possible in earlier work on CO/Pt(100) [77,78] or CO/Ni(110) [79]. In recent STM studies on metal surfaces, however, adsorbed atoms and small molecules were resolved. The resolution of individual adsorbate species depends on different factors such as their mutual distance, the extent to which they affect the tunnel current and their mobility.

For a given lateral resolution individual adsorbate species are more easily resolved in low-coverage adlayers with a wider spacing than in close-packed adlayers or adlayer islands. Correspondingly in the first ordered adsorbate structure resolved by STM - a mixed C_6H_6/CO adlayer on Rh(111) [80,81] - the benzene molecules are separated by CO 'spacer' molecules. In the meantime individual adsorbates were resolved also in other ordered adlayers, such as p(2x2)O-Ru(0001) [68], (2x1)O-Cu(110) [71-73], (2x1)S-Mo(100) [7] or p(2x2)O-Ni(100) [46,47]. The resolution of individual CO molecules in the high coverage (2x1)p2mg structure on Pt(110) represents the first example for 'atomic' resolution in close-packed adlayers [82].

Measurements on ordered adlayers normally do not produce images of single particles but reflect the contours of an entire adsorbate layer comparable to images of close-packed substrate atoms. Imaging of individual, isolated adatoms requires these to be sufficiently immobile to remain on their adsorption site for the time of the STM measurement. In ordered layers this is achieved by adsorbate-adsorbate interactions, for single adsorbates the thermal mobility must often be reduced by cooling. Such measurements are very attractive in order to gain information on an individual adsorption complex (see below). Recently individual Xe adatoms were successfully imaged by STM at 4 K [83]. Carbon adatoms on Al(111) in contrast are sufficiently immobile on Al(111) and could be resolved also at room temperature [76].

The characterization of adsorbates by STM is based on their effect upon the tunnel current. The presence of an adsorbate leads to a modification in the electronic structure at and above the surface at the location of the adsorbate. This results from changes in the electronic structure of the substrate due to the chemical bond between substrate and adsorbate, and from internal adsorbate states. These modifications in the local electronic structure affect the tunnel current and thus the STM image. Hence the apparent size of the adsorption complex in the STM image, i.e. the local height variation in the STM trace at the position of the adsorbate, in general does not correspond to its binding geometry. Furthermore STM measurements characterize the entire adsorption complex which includes the neighboring substrate atoms and the chemical bond between adsorbate and substrate.

The correlation between electronic structure and tunnel current, and the impact of adsorbate induced modifications in the electronic structure is schematically illustrated in fig. 8 [84]. The applied bias voltage determines an energy window ΔE, and only states within this energy window can contribute to the tunnel current. Consequently only modifications in the density of states within this energy range have to be considered for the adsorbate induced change in tunnel current. Different from clean metal surfaces and comparable to

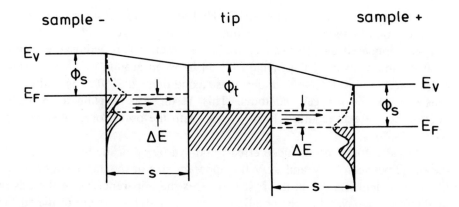

Fig. 8 Schematic description of the electronic effects in the tunnel current in an energy diagram: The left (right) junction corresponds to a negative (positive) bias on the sample, so as to permit tunneling from occupied states on the sample (tip) into empty states on the tip (sample). Only states within the energy window ΔE can contribute. The associated density of states $N_s(E)$ of the sample is sketched (solid line: occupied states, broken line: empty states), the arrows indicate the barrier-dependent tunneling probability. Φ_s, Φ_t: work function of sample and tip, respectively, E_F: Fermi level, E_v: Vacuum level, s: gap width (from ref. 84).

semiconductor surfaces the distribution of the charge density around E_F or within the energy gap ΔE is no longer equivalent to that of the total charge density, and the adstates are strictly localized on the adsorption site. Furthermore also the character and decay of the adstates into vacuum is important. States localized at the surface will contribute less than those that reach far into vacuum, such as metal s-states. Finally, as indicated in fig. 8, also the energetic position of the states within the energy window is of importance. States at the upper edge of the energy window will contribute most effectively to the tunnel current since they feel the smallest tunnel barrier and therefore also extend most into vacuum.

This qualitative picture was confirmed by first calculations. Lang modelled the tip-sample geometry by a Na atom (tip) and an adsorbate atom (sample) both adsorbed on a jellium background and then calculated the tunnel current between these electrodes by using a formalism comparable to the 'Transfer-Hamiltonian' [85]. He found that different adsorbates – S, He, or Na – will cause deviations in the STM traces which strongly differ from their geometric size [86]. Doyen and coworkers used a semiempirical method described in more detail in this volume [16] in order to calculate STM traces over oxygen adatoms on Ni(100) and Al(111) [87,88]. In that case the tip was modelled by a W atom adsorbed on a W(110) surface and the tunnel current was calculated exactly, i.e. tip-surface interactions were explicitly allowed. On Ni(100) they found a strong effect of the gap width on the apparent size. For tunnel distances larger than 6.5 Å the oxygen adatoms appeared as protrusions, while for smaller gap widths they lead to a local minimum in the STM image [87]. In between around 6.5 Å gap width the STM trace is not affected, under these conditions the STM is 'blind' to the presence of the oxygen adatoms. This result was explained in terms of interference effects resulting from overlap with the oscillatory, inner part of the wave functions.

Lang also found pronounced effects upon variation of the bias voltage [89]. In the range between −2 V and +2 V the apparent size of the adsorbates Na, S and Mo varied significantly. In the first two cases the apparent size of the adsorbate qualitatively behaved like the additional density of states in front of the surface at the position of the adsorbate, while in the latter case a strong maximum in the DOS was not reproduced in the STM trace. These tendencies can be understood from the different wave functions of the adsorbate. While on Na and S the additional DOS has mostly sp–character, the strong resonance around E_F on Mo is predominantly d–type and therefore more strongly localized directly in front of the surface.

First experiments showed indeed effects similar to those predicted by theory. STM measurements on O/Al(111) revealed a pronounced distance dependence in the apparent size of the O_{ad}. While for low tunnel currents and large gap widths the O_{ad} is represented by a minimum in the STM image, it results in a maximum for high currents. On the other hand for bias potentials in a range between +1 eV and −1 eV no measurable variation was found, provided the tip–surface distance was maintained [90]. Hence a flat DOS in this energy range at the location of the adsorbed oxygen atoms must be concluded.

4.2 LOCAL BOND EFFECTS

The modification of the electronic structure of the surrounding substrate atoms by the adsorbate reflects the chemical bond between substrate and adsorbate. STM images resolving individual atoms of both the substrate and the adsorbate can prove these effects and thus give an impression of the size and lateral extension of the electronic changes associated with the chemical bond. This is demonstrated in the two STM images in fig. 9 which depict (almost) the same area of an Al(111) surface with traces of adsorbed C_{ad} [76]. They only differ in the tunneling conditions used during recording (left: V_t = −20 mV, I_t = 41 nA, right: V_t = −70 mV, I_t = 41 nA). This is evident from the identical relative geometry of the two characteristic sets of triangularly arranged, slightly protruding Al atoms in the center part of the images. Each of these sets in fact surrounds the location of an adsorbed carbon atom, which is more obvious from the right image. In that case protrusions at the centers of the respective sets indicate the positions of the

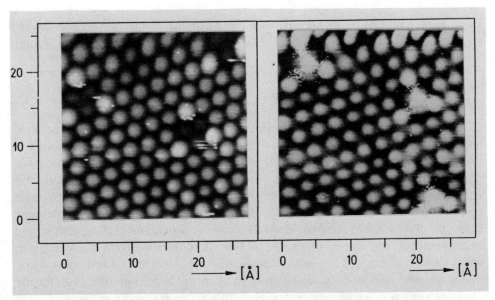

Fig. 9 High-resolution STM images recorded subsequently of (almost) the same Al(111) surface area, with two (three) carbon adatoms in the imaged area. Left: $V_t = -20$ mV, $I_t = 41$ nA; right: $V_t = -70$ mV, $I_t = 41$ nA (from ref. 76).

C atoms. The three nearest neighbor (nn) Al atoms directly bound to the C_{ad} appear to be displaced downwards (-0.3 Å), while the next-nearest neighbor (nnn) Al atoms are apparently shifted upwards (+0.2 Å). The apparent vertical displacements were interpreted in terms of modifications in the electronic structure of the substrate in the vicinity of the adsorbate rather than geometric displacements of the ion cores: The electronegative C atoms withdraw electronic charge from the neighboring Al atoms (which therefore appear depressed in the STM image). The Friedel-type oscillations of the charge density extending with decreasing amplitude from nearest neighbors (depletion) via next nearest neighbors (accumulation) to third-nearest neighbors (depletion) agree with calculations performed by Lang and Williams for electronegative adsorbates on a jellium surface [91]. This result also gives a direct measure of the range over which neighboring adsorbates may influence each other by indirect (i.e. mediated through the substrate) electronic interactions. A qualitatively similar result was obtained for a Si_{10} cluster adsorbed on Au(100) recently, although without resolution of the individual substrate atoms [37].

The data also give direct access to the adsorption site. They not only demonstrate that C atoms adsorb on the threefold-hollow site, from the orientation of the triangular sets of prominent nnn Al atoms it was also concluded that only one kind of threefold-hollow sites is occupied. Carbon adatoms near steps were found to be in registry with the Al atoms of the next-deeper layer. Hence the C_{ad} is solely adsorbed on hcp-sites with an Al atom of the second layer directly underneath [76].

4.3 SPECTROSCOPY OF ADSORBATES

STM spectroscopy of adsorbates principally resembles that of semiconductor surfaces and similar techniques may be applied [92]. The density of states is probed by current voltage-curves or their derivatives, and current distance-curves describe the decay of these states into vacuum. As mentioned before the contribution of different states to the total tunnel current is weighted by their respective tunnel barrier. Low-lying, occupied states thus will contribute very little to the tunnel current, equivalent to small effects in spectroscopic measurements for these states [93]. This is different from photoemission spectroscopy where this weighting factor is absent and energetically low-lying states are not systematically scaled down in intensity. In the opposite case, for the characterization of empty states, however, this argument works in favor of STM spectroscopy as compared to inverse photoemission spectroscopy. These effects are especially important for spectroscopy of adsorbates as shown below.

It should be noted that STM spectroscopy and measurements of the distance- or bias-dependent size of individual adsorbates reflect the same physical properties and contain the same physical information on the adsorption complex. The tunnel current depends on the lateral and vertical position (gap width) of the tip and the applied bias voltage as external parameter ($I_t = f(x,y,z,V_t)$) and these different measurements merely represent different cuts through a five-dimensional parameter space. For topography images ($z = f(x,y)_{V_t,I_t}$) x and y are varied keeping V_t and I_t constant, while for spatially resolved spectroscopic measurements the lateral position of the tip is maintained and either V_t or z are varied ($I_t = f(V_t)_{x,y,z}$, $I_t = f(z)_{x,y,V_t}$). Local I_t /V_t -curves or local I_t /z-curves can be constructed from a set of topograpy scans recorded at different distances or different bias voltages, respectively. Similarly topography images can be reconstructed from sets of spectroscopy measurements recorded at a variety of tip-surface distances, on the clean surface and close to or above an adsorption complex. So far spatially resolved spectroscopy has not been applied for the characterization of adsorbates as commonly as compared e.g. to STM studies on semiconductor surfaces. This is expected to change, however.

Spectroscopy also provides a means for detecting the presence of an adsorbate layer even if the individual adsorbed species are not resolved in STM topography images. This is the case e.g. for dilute, mobile adlayers. In dlnI_t/dV_t -curves recorded on a CO covered Ni(110) surface the presence of the adsorbate was clearly identified by a peak around 3.5 eV, which was attributed to the empty $2\pi^*$-state of adsorbed CO [79]. In contrast to the dominant feature at +3.5 eV the occupied states of CO_{ad}, the $5\sigma/1\pi$ states at –9.2 eV and and the 4σ state at –11.7 eV below E_F, were not detected in these measurements. Those measurements illustrate the restrictions imposed on STM spectroscopy of adsorbates: In contrast to semiconductor surfaces where the interest is centered predominantly on states around E_F, the electronic states of adsorbed species are often well below E_F, which renders their spectroscopic identification by STM difficult.

The characterization of the vibrational structure of adsorbates by inelastic tunneling spectroscopy is commonly performed for molecules adsorbed on solid tunnel junctions [94]. Except for one early report [95] inelastic tunneling measurements by STM had been successless so far, despite of extensive efforts [96]. Further details on this technique are found elsewhere [94,97,98].

4.4 LOCAL OBSERVATION OF ADSORBATES

The characterization of adlayers and adlayer processes by local observation of adsorbates is demonstrated in two examples. In the first one we will evaluate the mobility and local ordering in oxygen adlayers on different metal surfaces, the second one deals with STM observations on the growth mode and growth behavior during metal deposition on metal surfaces ('metal on metal systems').

Oxygen adsorption has been studied by STM on several metal surfaces – on Al(111) [9,75], on Ru(0001) [68] and on Ni(100) [46,47] – which allows a systematic comparison of these results. (On Ni(110) and Cu(110), where oxygen adsorption was equally investigated by STM, O_{ad} induces a surface reconstruction which will be dealt with in the following section). In all of these cases ordered structures are formed. On Al(111) individual oxygen adatoms are mobile at 300 K and were hardly observed by STM. At room temperature they instantaneously condense into small islands of 10 to 20 Å diameter as shown in fig. 10. In that image the individual Al substrate atoms are resolved, due to the electronic effects discussed above the oxygen islands show up as local depressions (dark) of 0.7 Å depth. The total coverage can be determined by counting the number of adsorption sites covered by the O_{ad} islands, it is estimated to $\Theta \approx 0.04$ ML in fig. 10. These immobile islands are metastable, at higher temperature they can be dissolved by two-dimensional evaporation and larger islands can grow on the expense of smaller ones. The instantaneous formation of (1x1) islands reflects strongly attractive interactions between oxygen adatoms on nearest neighbor adsorption sites. The O_{ad} islands are uniformly distributed over the surface, surface defects have no preferential activity for island formation.

On Ru(0001) the individual oxygen adatoms are also mobile at 300 K, but the mobility is sufficiently low that they can be observed by STM. At lower coverages small (2x2) clusters are continuously formed and dissolved again. At higher coverages $\Theta > 0.15$ stable (2x2) islands are formed, but there is still continuous condensation and evaporation at the perimeter of these islands going on (see fig.

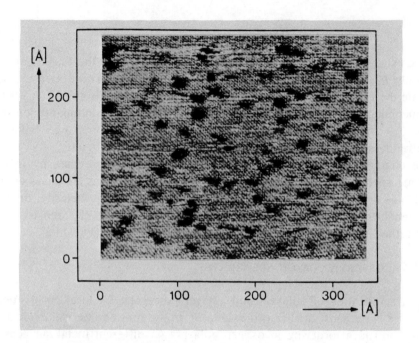

Fig. 10 STM image (top-view) of a 360 Å x 270 Å area on an Al(111) surface after exposure to 4 L O_2 at 300 K (V_t = -100 mV, I_t = 6 nA). The dark patches represent O_{ad} islands (coverage Θ = 0.04 ML), the individual Al substrate atoms are resolved (from ref. 75).

11). The attractive interaction between O adatoms on p(2x2) sites leading to that structure must be significantly weaker than that in the O adlayer on Al(111). This is not surprising because of the much larger distances between adatoms – 5.4 Å on Ru(0001) vs. 2.86 Å on Al(111). On Ru(0001) apparently repulsive interactions prevent the occupation of nearest or next-nearest (nnn) sites and support the formation of the ordered p(2x2) phase with increasing coverage.

The STM images indicate that also on this surface the mobility of the O adatoms is reduced by adsorbate-adsorbate interactions. Advacancies in an ordered island are sufficiently immobile that 'hopping' occurs on a timescale which is comparable to the frequency of STM images. Therefore individual hops of adatoms can be followed from image to image. This is illustrated by the three images in fig. 12 which were recorded subsequently on the same area of an oxygen covered Ru(0001) surface ($\Theta \approx 0.22$ ML). The advacancies in the characteristic quartett in the center of the images remain on their sites in the first two patterns. In the third pattern two of these vacancies have moved by one site. The vacancy at the bottom has changed its site from pattern to pattern. From extended series of such measurements a typical hopping frequency of 1-2 'hops' per vacancy and 100 sec was estimated. This can be converted into a diffusion

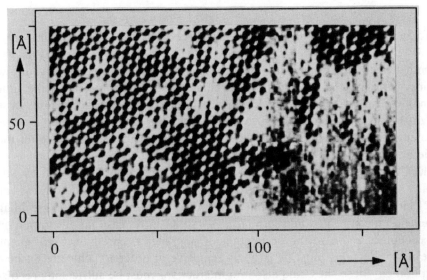

Fig. 11 STM image (top-view) of a 170 Å x 100 Å area on a Ru(0001) surface after exposure to 0.5 L O_2 at 500 K (V_t = 20 mV, I_t = 1.6 nA). Individual oxygen adatoms (dark) are resolved, which form an ordered p(2x2) structure with vacancies surrounded by a dilute lattice gas phase (from ref. 68).

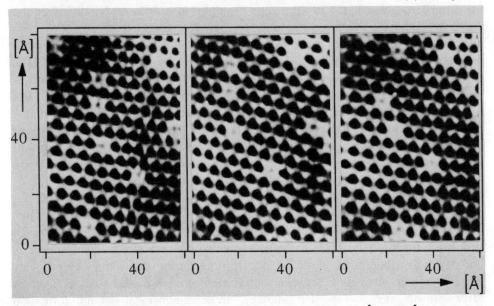

Fig. 12 Sequence of STM images of (almost) the same 60 Å x 80 Å area on an oxygen covered Ru(0001) surface, demonstrating the mobility of vacancies in the adlayer (Θ = 0.22 ML, V_t = 20 mV, I_t = 1.6 nA, from ref. 68).

rate for advacancies at this particular coverage by use of the Einstein formula [99].

Another method to determine surface diffusion rates by direct observation was described by Binnig et al., who had evaluated the 'flicker noise' in the tunnel current over an adsorbate covered surface [100]. Whenever an adatom passes through the surface area opposite to the tip apex the current is briefly modified. This method complements the one described above since it is applicable for more mobile adsorbates.

Dilute oxygen adlayers on Ni(100) − at coverages around 0.1 to 0.15 ML − similarly give no indication of island formation [47]. Individual adatoms or small p(2x2) clusters are rather uniformly distributed over the surface. These images showed that for annealed adlayers (T_{ann} = 450 K) in no case nn adsorption sites were occupied, and also the occupation of nnn sites occured very infrequently. Hence nn interactions must be strongly repulsive in order to completely rule out adsorption on these sites, and also nnn interactions must be sufficiently repulsive to make adsorption on those sites rather unfavorable. Interactions between third-nearest neighbor sites finally must be weak, otherwise no p(2x2) clusters (for repulsive interactions) or more pronounced island formation of the p(2x2) phase (for attractive interactions) would have been observed. The adlayer mobility differs from that in the two previous examples insofar as on Ni(100) even the isolated adatoms are almost immobile at 300 K and single 'hops' of oxygen adatoms can be followed from image to image. This observation agrees with reports that a well ordered p(2x2) phase is formed only after annealing [101]. The much higher mobility reported in ref. 100 referred to elevated temperatures.

These three examples demonstrated how information on the local order and mobility, gained from STM images, leads to conclusions on the adsorbate-adsorbate interactions. It mostly requires, however, 'atomic' resolution imaging of the respective adsorbates. Except for low-temperature STM's this kind of investigations is therefore limited to strongly interacting adsorbates which are sufficiently immobile and ordered already around room temperature.

STM studies on the initial stages of metal growth were concentrated so far on the semiconductor-metal interface. First examples for 'metal on metal' systems include STM studies on Au and Ag deposition on Au(111) [102,103] and on Cu deposition on Ru(0001) [45]. These measurements give a direct view on the size and layer distribution of metal islands formed during deposition. Hence they allow a definite distinction between the different growth modes such as layer by layer growth (Frank-van der Merwe), three-dimensional island formation (Volmer-Weber) or intermediate forms (Stranski-Krastanov) [104].

On close-packed metal surfaces the mobility of metallic adatoms is generally high [105]. In conjunction with the strongly attractive short-range interactions between these adatoms ('two-dimensional metallic cohesion') this explains the observed formation of extended islands in all of the above systems already at

300 K. These STM measurements also demonstrated that islands of the second or even higher layers are already formed (long) before the first layer is filled. For Cu/Ru(0001) their formation is clearly related to kinetic limitations. Apparently the islands in the second layer are too stable to dissolve again once they are formed by Cu atoms impinging on the first layer Cu islands, although Cu adatoms are more tightly bound in the first layer. Also the shapes and sizes of the metal islands are mostly determined by similar kinetic limitations rather than by their thermodynamics. All together kinetic limitations are often seen to dominate the growth behavior of thin metal films.

In some cases also the internal structure of the metal film can be resolved. For Cu/Ru(0001) pseudomorphic growth was observed in the first Cu layer [45], in agreement with earlier LEED results [105-107]. For the second layer, however, a one-dimensional corrugation was observed which was interpreted in terms of a uniaxial contraction of that layer, comparable to the structure of the topmost Au layer in the reconstructed Au(111) surface (see section 3.3). The two structures in fact result from similar situations: In both cases a hexagonal layer is placed on top of a hexagonal substrate, with the 'adlayer' atoms being more closely packed than those in the substrate lattice [45,57].

In this section we have neglected so far any effects caused by the STM measurement itself, both for the characterization and observation of adsorbates. For the latter case Eigler and Schweitzer had demonstrated that adsorbates can be repositioned by the electric field between tip and surface [83]. Hence care must be taken to rule out such effects in the respective experiment. Little is known on STM induced effects regarding the characterization of adsorbates by STM. First calculations on the electric field dependence of the electronic and vibrational states of adsorbed CO and CN by Bagus et al. [108,109] showed negligible effects for fields around 10^6 V/cm as they are typical for these STM measurements. Field effects had to be considered, however, for fields around 10^8 V/cm.

5. Surface reactions

Surface reactions lead to a chemical and/or structural modification of substrate or adsorbate and insofar are distinctly different from physical processes such as surface diffusion. Among these are reactions on the surface such as catalytic or decomposition processes, reactions between substrate and surface such as oxidation etc. and finally structural transformations of the surface. The different questions that can be answered by STM in this context relate to the initiation of the reaction process ('local activity', 'nucleation'), to its subsequent progress ('growth') and to the microscopic mechanism, i.e. the reaction mechanism on an atomic scale.

5.1 STRUCTURAL MODIFICATIONS OF THE SUBSTRATE

Structural modifications of metal surfaces due to the presence of an adsorbate or due to an ongoing (catalytic) reaction on the surface are frequently observed [110]. They range from adsorbate induced changes in the step–terrace topography via structural transformations in the topmost layer(s) ('adsorbate induced reconstructions') to larger scale modifications of the surface topography such as facetting.

Adsorbate induced changes in the step–terrace topography are most probable on surfaces where already for the clean surface steps are rather instable, as it was observed e.g. for Cu(110) [73] or Au(111) [111]. Correspondingly oxygen adsorption on Cu(110) was observed to lead to a significant rearrangement of the steps [73]. On Au(111) no similar effects were reported so far under vacuum conditions, but the dramatic increase in step mobility reported for Cl^- covered Au(111) surfaces under electrochemical conditions points into the same direction [112,113].

A number of structural transformations, both adsorbate induced surface reconstructions and the adsorbate induced removal of a reconstruction have been studied by STM. For the 'hex' -> (1x1) transition on Pt(100), which is initiated by different adsorbates such as CO, NO or C_2H_4, the 'nucleation and growth' mechanism proposed earlier [114] could be confirmed by direct observation [77,78]. Interestingly homogeneous nucleation was observed for adsorption of CO or NO, whereas for C_2H_4 adsorption heterogeneous nucleation was found to prevail [77]. This was interpreted in terms of a mechanism where in the first case the structural transition is rate limiting, while in the second case the decomposition of C_2H_4 as a preceding reaction step is rate limiting. Preferential decomposition at step edges results in the observed heterogeneous nucleation process for (1x1) formation. This proposed mechanism agrees well with macroscopic observations, where increased rates for hydrocarbon decomposition were found on stepped Pt surfaces [115]. The subsequent growth of the (1x1) phase exhibits a strong anisotropy, which the authors of ref. 77 related to strain effects in the clean, reconstructed surface layer of Pt atoms. In that study also a mechanism for the microscopic reaction path was proposed. Recent Monte–Carlo simulations based on this mechanism reached quantitative agreement with experimental observations [116,117].

The mechanistic understanding derived from STM investigations is demonstrated for the CO–induced (1x2) -> (1x1) transition on Pt(110). The clean Pt(110) surface exhibits a (1x2) 'missing row' reconstruction [60,61] which is removed by adsorption of CO. The two phases exhibit different densities of Pt atoms in the topmost layer, 0.5 ML in the (1x2) and 1.0 ML in the bulk-like

(1x1) phase. Hence the structural transformation must be accompanied by a change in atomic density in that layer and thus by transport of surface metal atoms. On the other hand this transformation was observed to occur already at temperatures around 250 K, which is far below the temperatures necessary for self-diffusion on this surface [58,118]. Different ideas were proposed to avoid this apparent contradiction, such as an order-disorder type transition [119] or the 'saw-tooth' structural model for the (1x2) surface [58]. But from the recent experimental results these ideas had to be discarded.

STM observations revealed that adsorption of CO at 300 K leads to the formation of characteristic square features on the surface [11,120]. This is reproduced in the STM images in fig. 13. These features are first seen at CO coverages around 0.2 ML. Further adsorption leads to an increase in their number, while their shape remains constant. At coverages around 0.5 ML the entire surface is covered by adjacent squares. The appearance of the square features must be associated with the structural transformation, i.e. the new (1x1) phase is formed within these squares. This leads to a 'nucleation and growth' type mechanism for that process. The uniform distribution of the squares over the surface points to homogeneous nucleation. Their constant size but increasing number indicates that the nucleation rate is much larger than the growth rate. In fact, each (1x1) nucleus stops growing once it has reached the size of a typical square feature.

The atomic scale mechanism for this transformation becomes obvious from the detailed image and the atomic model of one of these squares in fig. 14. It is made up by a small, bulk-like (1x1) area and a surrounding (1x1) rim. As demonstrated in the model these are formed by the disruption of a close-packed [1$\bar{1}$0] row of Pt atoms in the top layer of the reconstructed surface. Four Pt atoms of that row are displaced sideways into the troughs of the (1x2) surface, this way forming the square feature observed by STM. A two-step mechanism is derived from these observations: Exceeding a certain critical CO coverage ($\Theta_{crit} \approx 0.2$ ML) local clusters of CO can destabilize the close packed rows in the (1x2) phase, and in a first step a Pt atom from that row can 'hop' sideways into the adjacent trough. At this temperature the Pt atom possesses a limited mobility and can migrate, in a second step, one or two sites along the trough until it is trapped by two neighboring Pt atoms. The disrupture of the row leads to a destabilization of the terminal Pt atoms of that row, and those atoms can follow the same pattern until all terminal sites in the troughs are occupied. At that point the trapped Pt atoms stabilize the disrupted central row and further growth of the square (1x1) nucleus is inhibited.

The same sequence of steps can explain also STM observations at 350 K, which because of the strongly anisotropic growth of (1x1) islands under these

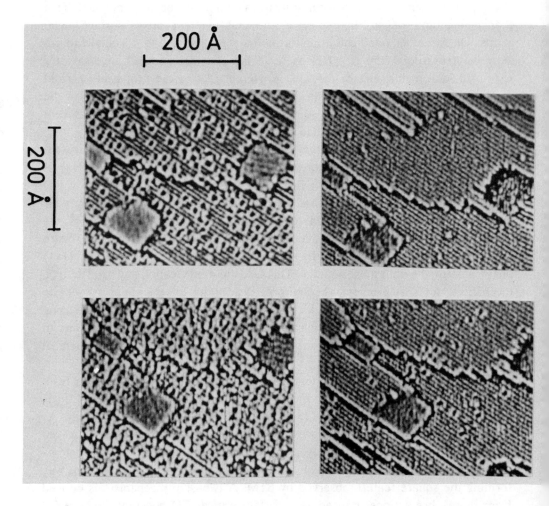

Fig.13 Series of STM images (top-view, ca. 350 Å x 430 Å) recorded on the same area of a (1x2) reconstructed Pt(110) surface during exposure to CO at 300 K (V_t = 225 mV, I_t = 0.4 nA) top left: clean surface; top right: 1.1 L CO; bottom left: 1.5 L CO; bottom right: 3 L CO (from ref. 11).

conditions seemed to follow an entirely different mechanism [11]. At this higher temperature the trapping of the Pt atoms in the troughs is much less efficient and the atoms can migrate further down along the troughs. This way the terminal atoms of the disrupted row are not stabilized and one Pt atom after another can

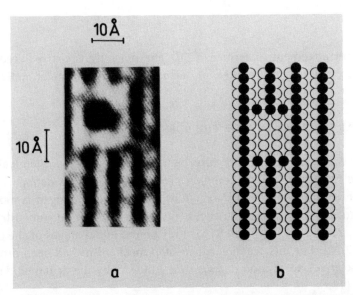

Fig.14 Structure of a (1x1) nucleus formed upon CO adsorption on (1x2) reconstructed Pt(110) at 300 K. a) High resolution STM image; b) atomic model, black: topmost layer Pt atoms, circles: second layer Pt atoms (from ref. 11).

move into the adjacent trough, which is energetically more favorable. The resulting local structure looks as if entire strings of Pt were displaced sideways. Similar structures were observed also by FIM for the reverse (1x1) -> (1x2) transition on the clean surface and were interpreted in terms of correlated jumps of such strings [121,122]. But from comparision with the low temperature results a sequential process as described above is more probable in the present case.

The oxygen induced (1x1) -> (2x1) reconstructions on Ni(110) and Cu(110) were subject of a number of STM investigations [71-73,123,124]. The (2x1)O phase was observed to grow in strongly anisotropic islands with preferential growth along [001], a 'nucleation and growth ' mechanism was concluded [71]. A recent STM study resulted in the proposal of a completely new reaction mechanism, based on an 'added row' model [73]. A mechanism analogous to a two-dimensional precipitation from a dilute, fluid phase was deduced for the formation of the (2x1)O structure. The fluid phase consists of individual Cu and O adatoms on the unreconstructed Cu substrate, and nuclei of the 'solid' (2x1)O phase are formed by condensation of adatoms into an 'adsorbed' -Cu-O-Cu-O-Cu- string [73]. The Cu adatoms 'evaporate' from step edges and migrate over the surface. As mentioned above the step edges on Cu(110) are very instable at room temperature, and the (2x1)O formation is accompanied by a significant rearrangement of terraces [73]. The mass transport required for that process thus occurs by dissolution of terraces and migration of individual Cu atoms.

5.2 CHEMICAL MODIFICATION OF THE SUBSTRATE

Reactions between substrate and adsorbate, which lead to a chemical modification of the substrate, include common processes such as oxidation, nitridation etc. These seem to be very simple reactions, but their mechanism is mostly very little understood. At least in some cases further insight can be gained from STM studies, as is demonstrated for an STM study on the initial stages of the oxidation of Al(111) [9,75]. For this reaction a two-step mechanism had been proposed in which (dissociative) adsorption of oxygen is followed by the actual oxidation in a second step [125]:

$$Al(111) \dashrightarrow Al(111) + O_{ad} \dashrightarrow Al(111) + AlO_n$$

Numerous subsequent studies, however, came to widely differing conclusions regarding the onset of oxidation. Spectroscopic evidence for the simultaneous existence of more than two oxygen species on the surface led to the proposal of a second kind of adsorbed species ('subsurface oxygen') [126].

From the STM measurements an island growth mechanism was derived for the adsorbed oxygen, with small O_{ad} islands equally distributed over the surface (see section 4.4). With increasing exposure mainly the number of islands grew, while their average diameter changed very slowly. These adsorption characteristics were maintained over a wide exposure range. At exposures around 50 to 60 L characteristic new features appeared in the STM images which is illustrated in the sequence of patterns in fig. 16. Small cone-like protrusions form at the step edges, which were not present previously. At this exposure Auger spectroscopy also revealed the onset of oxide formation. These features are strongly different from the adsorbed oxygen species found in the O_{ad} islands, which are still visible as irregular little 'holes' on the flat terraces of these images. The protusions were therefore associated with oxide nuclei [9].

The preferential formation of oxide nuclei at step edges ('heterogeneous nucleation') was reproduced in similar measurements. It contrasts the homogeneous nucleation mechanism for the formation of O_{ad} islands. But it is consistent with simple ideas on oxide formation since the penetration of the Al lattice is much easier conceivable at step edges as compared to unperturbed, flat terraces.

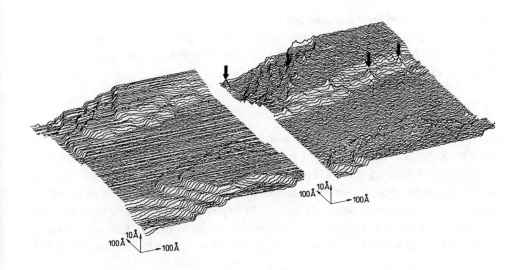

Fig.15 Subsequent STM images recorded on almost the same area of an oxygen covered Al(111) surface for different exposures (800 Å x 800 Å, 300 K). left: O_2 exposure 4 L; right: O_2 exposure 64 L (from ref. 9).

These observations were interpreted in a mechanistic picture which is very similar to the original one [9]. The oxidation indeed proceeds in a two-step mechanism, adsorption as a first step is followed by oxidation in a second step. This mechanism, however, is correct only on a local scale. Integrated over the surface, these two processes can occur simultaneously. Oxidation begins long before the surface is completely covered by O_{ad}, and the onset of oxidation depends strongly on the crystallographic properties, i.e. the concentration of steps and defects on the surface. This interpretation is supported by results of a previous study by Testoni and Stair, who reported an almost instantaneous onset of oxidation on a highly-stepped 4(111)x(111) Al surface, whereas on a flat Al(111) surface oxidation was observed only at exposures above 50 L [127]. The presence of different adsorbed species in spectroscopic measurements finally was explained by different electronic and vibrational properties of the adsorption complex depending on the number of O_{ad} bound to an underlying Al surface atom [9,128]. This number varies for edge atoms or atoms within an island, hence the average island size determines the ratio between the different species observed spectroscopically and variations in island size, caused e.g. by slight annealing, lead to changes in that ratio.

This reaction scheme, which requires only two oxygen species for a complete description of the oxidation process, is consistent with results of other techniques.

The strong variation in the minimum exposure for oxidation is mainly attributed to different defect densities on the respective samples. The temperature and coverage dependence in the population of the different spectroscopic states is related to different island sizes.

5.3 REACTIONS ON THE SUBSTRATE

Surface reactions of or between adsorbed species, which do not alter the substrate surface at all, such as an (ideal) catalytic reaction, are not very suitable for an investigation by STM. The time-scale of STM measurements differs too much from these mostly very rapid reactions. First STM studies were reported on decomposition reactions such as the STM study on C_2H_4 decomposition on Pt(100) [77]. Also STM induced decomposition reactions belong into this category [129]. This latter subject is covered in more detail by Quate in this volume [130].

6. Conclusions

The current state of STM investigations on clean and adsorbate covered metal surfaces was reviewed. STM investigations were seen to open new perspectives for the characterization of the defect structure of these surfaces; surface defects play an important role in the interaction with adsorbates and in surface reactions. The characterization of the local adsorption complex by STM gives access to details of the chemical bond between substrate and adsorbate. STM measurements on the local order, distribution and mobility of adsorbates allow conclusions on the effects of adsorbate-adsorbate interactions and can complement other techniques. The characterization of thin film growth, a very promising application of STM, belongs into this category. Finally the use of STM studies for revealing mechanistic details by direct observation was demonstrated for various surface reactions. In some cases completely new mechanistic concepts were derived.

Acknowledgements

I gratefully acknowledge contributions of J. Barth, H. Brune, Dr. D. Coulman, Dr. T. Gritsch, C. Günther, H. Höfer, E. Kopatzki, G. Pötschke, J. Wiechers and especially of Dr. J. Wintterlin. This work was supported by the Deutsche Forschungsgemeinschaft via SFB 128 and SFB 338.

References

1] G.Binnig, H.Rohrer, Ch.Gerber and E.Weibel,
Surface Sci. 131, L379 (1983).
2] G.Binnig, H.Rohrer, Ch.Gerber and E.Stoll,
Surface Sci. 144, 321 (1983).
3] G.Binnig, H.Rohrer, Ch.Gerber and E.Weibel,
Phys. Rev. Lett. 50, 120 (1983).
4] O.M.Magnussen, J.Hotlos, R.J.Nicols, D.M.Kolb and R.J.Behm, subm..
5] H.Siegenthaler and R.Christoph, this volume.
6] V.M.Hallmark, S.Chiang, J.F.Rabolt, J.D.Swalen and R.J.Wilson,
Phys. Rev. Lett. 59, 2879 (1987).
7] B.Marchon, P.Bernhardt, M.E.Bussel, G.A.Somorjai, M.Salmeron and
W.Siekhaus, Phys. Rev. Lett. 60, 1166 (1988).
8] J.Wintterlin, J.Wiechers, H.Brune, T.Gritsch, H.Höfer and R.J.Behm,
Phys.Rev. Lett. 62, 59 (1989).
9] J.Wintterlin, Ph.D. Thesis Univ. Berlin (1989).
10] Y.Kuk, P.J.Silverman and H.Q.Nguyen,
J. Vac. Sci. Technol. A6, 524 (1988).
11] T.Gritsch, D.Coulman, G.Ertl and R.J.Behm,
Phys. Rev. Lett. 63, 1086 (1989).
12] R.H.Fowler and L.Nordheim,
Proc. R. Soc. London Ser. A119, 173 (1928).
13] G.Binnig, H.Rohrer, Ch.Gerber and E.Weibel,
Appl. Phys. Lett. 40, 178 (1982).
14] J.Tersoff, this volume.
15] S.Ciraci, this volume.
16] G.Doyen, E.Koetter, J.Barth and D.Drakova, this volume.
17] J.Bardeen, Phys. Rev. Lett. 6, 57 (1961).
18] S.Ciraci, A.Baratoff and I.P.Batra, in press.
19] N.Garcia, C.Ocal and F.Flores, Phys. Rev. Lett. 50, 2002 (1983).
20] E.Stoll, A.Baratoff, A.Selloni and P.Carnevali,
J. Phys. C17, 3073 (1984).
21] N.Garcia and F.Flores, Physica 127B, 137 (1984).
22] J.Tersoff and D.Hamann, Phys. Rev. Lett. 50, 1998 (1983).
23] J.Tersoff and D.Hamann, Phys. Rev. B31, 805 (1985).
24] R.J.Behm, Scanning Microsc. Suppl. 1, 61 (1987).
25] R.Smoluchowski, Phys. Rev 60, 661 (1941).
26] G.Doyen and D.Drakova, Surface Sci. 178, 375 (1986).
27] R.Forbes, this volume
28] M.D.Pashley and J.B.Pethica, J. Vac. Sci. Technol. A3, 757 (1985).
29] J.G.Simmons, J. Appl. Physics 34, 1793 (1963).

30] G.Binnig, N.Garcia, H.Rohrer, J.M.Soler and F.Flores,
Phys. Rev B30, 4816 (1984).
31] R.S.Becker, J.A.Golovchenko and B.S.Swartzentruber,
Phys. Rev. Lett. 55, 987 (1985).
32] G.Binnig, K.H.Frank, H.Fuchs, N.Garcia, B.Reihl, H.Rohrer, F.Salvan and A.R.Williams, Phys. Rev. Lett. 55, 991 (1985).
33] F. Salvan, H.Fuchs, A.Baratoff and G.Binnig,
Surface Sci. 162, 634 (1985).
34] J.H.Coombs and J.K.Gimzewski, J.Microsc. 152, 841 (1988).
35] K.H.Gundlach, Solid State Electron. 9, 949 (1966).
36] W.J.Kaiser and R.C.Jaklevic, IBM J. Res. Develop. 30, 396 (1986).
37] Y.Kuk, P.J.Silverman and F.M.Chua,
J.Microsc. 152, 449 (1988).
38] A.Brodde, St.Tosch and H.Neddermeyer, J. Microsc. 152, 441 (1988).
39] K.Mednik and L.Kleinman, Phys. Rev. B22, 5768 (1980).
40] J.K.Gimzewkski and R.Möller, Phys. Rev. B36, 1284 (1987).
41] U.Dürig, J.K.Gimzewski and D.W.Pohl,
Phys. Rev. Lett. 57, 2403 (1986) .
42] U.Dürig, O.Züger and D.W.Pohl, J. Microsc. 152, 259 (1989).
43] J.M.Soler, A.M.Baro, N.Garcia and H.Rohrer,
Phys. Rev. Lett. 57, 444 (1986).
44] J.Chen, pers. commun..
45] G.Pötschke and R.J.Behm, subm..
46] A.Brodde, G.Wilhelmi, H.Wengelnik, D.Badt and H.Neddermeyer, pers. commun. and to be publ..
47] E.Kopatzki and R.J.Behm, subm..
48] S.Günther, Diploma thesis Univ. München (1990).
49] P.W.Palmberg and T.N.Rhodin, J. Chem. Phys. 49, 134 (1968).
50] K.Heinz, E.Lang, K.Strauss and K.Müller,
Appl. Surface Sci. 11/12, 611 (1981).
51] M.A.van Hove, R.J.Koestner, P.C.Stair, J.P.Biberian, L.L.Kesmodel, I.Bartos and G.A.Somorjai, Surface Sci. 103, 189 (1981).
52] K.Takayanagi and K.Yagi, Trans. Jpn. Inst. Met. 24, 337 (1983).
53] U.Harten, A.M.Lahee, J.P.Tönnies and Ch.Wöll,
Phys. Rev. Lett. 54, 2619 (1985).
54] R.J.Behm, W.Hösler, E.Ritter and G.Binnig,
Phys. Rev. Lett. 56, 228 (1986).
55] R.J.Behm, W.Hösler, E.Ritter and G.Binnig,
J. Vac. Sci. Technol. A4, 1330 (1986).
56] Ch.Wöll, S.Chiang, R.J.Wilson and P.H.Lippel,
Phys. Rev. B39, 7988 (1989).
57] J.V.Barth, H.Brune, G.Ertl and R.J.Behm, subm..

58] H.P.Bonzel and S.Ferrer, Surface Sci. 118, L263 (1982).
59] D.Wolf and W.Moritz, Surface Sci. 163, L655 (1985).
60] P.Fenter and T.Gustafson, Phys. Rev. B38, 10197 (1988).
61] P.Fery, W.Moritz and D.Wolf, Phys. Rev. B38, 7275 (1988).
62] D.Wolf, H.Jagodzinski and W.Moritz,
Surface Sci. 77, 265 (1978).
63] T.Gritsch, D.Coulman, R.J.Behm and G.Ertl, subm..
64] K.M.Ho and H.P.Bohnen, Phys. Rev. Lett. 57, 4452 (1986).
65] V.Heine and L.D.Marks, Surface Sci. 165, 65 (1986).
66] K.W.Jacobsen and J.K.Norskov, in 'The Structure of Surfaces II',
J.F. van der Veen and M.A.van Hove eds. (Springer, Berlin 1988).
67] B.Eisenhut and R.J.Behm, unpubl. data.
68] C.Günther and R.J. Behm, to be publ..
69] M.Henzler, Surface Sci 168, 744 (1986).
70] M.Mundschau, E.Telieps, E.Bauer and W. Swiech,
Surface Sci. 223, 413 (1988).
71] F.M.Chua, Y.Kuk and P.J.Silverman, Phys. Rev. Lett. 63, 386 (1989).
72] F.M.Chua, Y.Kuk and P.J.Silverman, J. Vac. Sci. Technol.(in press).
73] D.Coulman, J.Wintterlin, R.J.Behm and G.Ertl, subm..
74] H.Bethge, M.Krohn and H.Stenzel, in 'Electron Microscopy in Solid State
Physics', H.Bethge and J.Heydenreich eds. (Elsevier, Amsterdam, 1987).
75] J.Wintterlin, H.Brune, H.Höfer and R.J.Behm,
Appl. Phys. A47, 99 (1988).
76] H.Brune, J.Wintterlin, G.Ertl and R.J.Behm, subm..
77] E.Ritter, R.J.Behm, G.Pötschke and J.Wintterlin,
Surface Sci. 181, 403 (1987).
78] W.Hösler, E.Ritter and R.J.Behm,
Ber. Bunsenges. Phys. Chem. 90, 205 (1986).
79] C.Ettl and R.J.Behm, unpubl. data.
80] H.Ohtani, R.J.Wilson, S.Chiang and C.M.Mate,
Phys. Rev Lett. 60, 2398 (1988).
81] S.Chiang, R.J.Wilson, C.M.Mate and H.Ohtani,
J. Microsc. 152, 567 (1988).
82] T.Gritsch, G. Ertl and R.J.Behm, to be publ..
83] D.M.Eigler and E.K.Schweitzer, Nature (in press).
84] R.J.Behm and W.Hösler, in 'Chemistry and Physics of Solid Surfaces VI',
R.Vanselow and R.Howe eds. (Springer, Berlin, 1986), p.361.
85] N.D.Lang, Phys. Rev. Lett. 55, 230 (1985).
86] N.D.Lang, Phys. Rev. Lett. 56, 1164 (1986).
87] G.Doyen, D.Drakova, E.Kopatzki and R.J.Behm,
J. Vac. Sci Technol. A6, 327 (1988).

88] E.Kopatzki, G.Doyen, D.Drakova and R.J.Behm,
     J. Microsc. 152, 687 (1989).
89] N.D.Lang, Phys. Rev. Lett. 58, 45 (1987).
90] H.Brune, J.Wintterlin, G.Ertl and R.J.Behm, to be publ..
91] N.D.Lang and A.R.Williams, Phys. Rev. B18, 616 (1978).
92] R.M.Feenstra, this volume.
93] R.M.Feenstra, J.A.Stroscio and A.P.Fein, Surface Sci. 181, 295 (1986)
94] 'Tunneling Spectroscopy', P.K.Hansma ed. (Plenum, New York, 1982);
     E.L.Wolf, 'Principles of Electron Tunneling Spectroscopy'
     (Oxford Univ. Press, Oxford, 1985).
95] D.P.E.Smith, M.D.Kirk and C.F.Quate, J. Chem. Phys. 86, 6034 (1987).
96] D.M.Eigler, pers. commun..
97] B.N.J.Persson and A.Baratoff, Phys. Rev. Lett. 59, 339 (1987).
98] B.N.J.Persson and J.Demuth, Solid State Commun. 57, 769 (1986).
99] G.Ehrlich, in 'Chemistry and Physics of Solid Surfaces V',
     R.Vanselow and R.F.Howe eds. (Springer, Berlin, 1984).
100] G.Binnig, H.Fuchs and E.Stoll, Surface Sci. 169, 295 (1986)
101] W.D.Wang, N.Wu and P.A.Thiel, J. Chem. Phys. 92, 2025 (1990).
102] M.M.Dovek, C.A.Lang, J.Nogami and C.F.Quate,
     Phys. Rev. B40, 11973 (1989).
103] J.Nogami C.A.Lang, M.M.Dovek and C.F Quate,
     Surface Sci. 224, L947 (1989).
104] J.H.van der Merwe, in 'Chemistry and Physics of Solid Surfaces V',
     R. Vanselow and R.F. Howe eds. (Springer, Berlin, 1984).
105] K.Christmann, G.Ertl and H.Shimizu,
     J. Catalysis, 61, 397 (1980).
106] J.E.Houston, C.H.F.Peden, D.S.Blair and D.W.Goodman,
     Surface Sci. 167, 427 (1986).
107] C.Park, E.Bauer and H.Poppa, Surface Sci. 187, 86 (1987).
108] P.S.Bagus, C.J.Nelin, W.Müller, M.R.Philpott and H.Seki,
     Phys. Rev. Lett. 58, 559 (1987).
109] P.S.Bagus, C.J.Nelin, K.Hermann and M.R.Philpott,
     Phys. Rev. B36, 8169 (1987).
110] G.A.Somorjai and M.A.van Hove, Progr. Surface Sci., in press.
111] R.C.Jaklevic and L.Elie, Phys. Rev. Lett. 60, 120 (1988).
112] J.Wiechers, T.Twomey, D.M.Kolb and R.J.Behm,
     J. Electroanalyt. Chem. 248, 451 (1988).
113] D.Trevor, C.E.D.Chidsey and D.N.Loiacono,
     Phys. Rev. Lett. 62, 929 (1989).
114] R.J.Behm, P.A.Thiel, P.R.Norton and G.Ertl,
     J. Chem. Phys. 78, 7437 (1983).

P.A.Thiel, R.J.Behm, P.R.Norton and G.Ertl,
J. Chem. Phys. 78, 7448 (1983).

115] S.M.Davies and G.A.Somorjai, in 'The Chemical Physics of Solid Surfaces and Heterogeneous Catalysis', D.A.King and D.P. Woodruff eds. (Elsevier, Amsterdam, 1982).

116] R.J.Behm, in 'Diffusion at Interfaces: Microscopic Concepts', M. Grunze, H.J.Kreutzer and J.J.Weimer eds. (Springer, Berlin, 1988) p. 93.

117] A.E.Reynolds, D.Kaletta, G.Ertl and R.J.Behm, Surface Sci. 218, 452 (1989).

118] D.W.Basset, and P.R.Webber, Surface Sci. 70, 520 (1978)

119] J.C.Campuzzano, A.M.Lahee and G.Jennings, Surface Sci. 152/153, 68 (1985).

120] T.Gritsch, D.Coulman, G.Ertl and R.J.Behm, Appl. Phys. A49, 403 (1989).

121] G.L.Kellog, Phys. Rev. Lett. 55, 2168 (1985).

122] Q.Gao and T.T.Tsong, Phys. Rev. Lett. 57, 452 (1986).

123] A.M.Baro, G.Binnig, H.Rohrer, Chr. Gerber, E.Stoll, A.Baratoff and F.Salvan, Phys. Rev. Lett. 52, 1304 (1985).

124] E.Ritter and R.J.Behm, in 'The Structure of Surfaces II', J.F. van der Veen and M.A.van Hove eds. (Springer, Berlin, 1988) p. 261.

125] I.P.Batra and L.Kleinman, J. Electron Spectrosc. Rel. Phenomen. 33, 175 (1984).

126] J.L.Erskine and R.L.Strong, Phys. Rev. B25, 5547 (1982).

127] A.L.Testoni and P.C.Stair, Surface Sci. 171, L491 (1986).

128] C.F.Mc Conville and D.P.Woodruff, Surface Sci. 171, L447 (1986).

129] E.E.Ehrichs, S.Yoon and A.L.de Lozanne, Appl. Phys. Lett. 53, 2287 (1988).

130] C.F.Quate, this volume.

SCANNING TUNNELING MICROSCOPY: SEMICONDUCTOR SURFACES, ADSORPTION, AND EPITAXY

R. M. FEENSTRA
IBM Research Division, T. J. Watson Research Center
Yorktown Heights, New York, USA 10598

ABSTRACT. The application of the scanning tunneling microscope (STM) to the study of clean and adsorbate covered semiconductor surfaces is discussed. Various imaging and spectroscopic methods are described, and the methods are illustrated with examples of experimental results. Emphasis is placed on the relationship between the electronic and structural properties of a surface, and the role that this relationship plays in the interpretation of STM images.

1. Introduction

Since its inception in 1982, the scanning tunneling microscope (STM) has proven to be a powerful tool in the study of surfaces.[1-5] Ordered arrays of atoms, and disordered atomic features, have been observed on many metal and semiconductor surfaces. Clean surfaces, as well as isolated adsorbates and thin overlayers, have been studied. The STM has been used in a variety of environments including ultra-high vacuum (UHV), air, and various liquids, and at temperatures ranging from liquid helium to above room temperature.

The power of the tunneling microscope lies in its ability to spatially and energetically resolve the electronic states on a surface. Spatially, the states can be observed with atomic resolution; 5 Å lateral resolution is routinely achieved, and features on the 3 Å scale can be resolved under favorable circumstances. Energetically, states which lie within a few electron-volts on either side of the Fermi-level can be observed with an energy resolution of a few kT. Details of the geometric arrangement of atoms on the surface are reflected in the spatial distribution of electronic states, and the STM thus provides a probe of the atomic structure of surfaces. The connection between the electronic states and the atomic structure depends on the type of system: For metals, the states generally follow the atoms in a uniform fashion, and the STM thus provides a direct "topographic" view of the atoms. For semiconductors, the electronic states often reflect "non-topographic" details of the surface. In that case, the interpretation of STM images in terms of their relationship to the geometric structure of the surface is a nontrivial task, and a major part of this article is devoted to a description of how such an interpretation can be accomplished.

In this article, we mainly discuss STM data obtained from clean, ordered semiconductor surfaces in ultra-high-vacuum. This topic is just one of the many areas in which the STM has been applied. Aside from semiconductors, the surfaces of metals, semi-metals (graphite), superconductors, insulators, organics, and other types of materials have been studied. One powerful aspect of the tunneling microscope is its ability to access a large number of variables, including current, voltage, tip-sample separation (z-position), and lateral (x,y) position on the sample. This ability leads to a large number of variations in the techniques used in acquiring the data, and these different techniques often emphasize different aspects in the interpretations of the data. No single acquisition method can be expected to solve all STM problems — the range of problems which can be addressed with the STM is just too broad to be encompassed by one single method of data acquisition and analysis. In this article, we illustrate many of the methods which have been developed and used to date.

Following this introductory section, we present in Section 2 a brief description of the essential mechanical and instrumental aspects of the scanning tunneling microscope. In Section 3 we discuss the theoretical basis for tunneling, and we review work pertaining to the interpretation of STM images. The remainder of the paper is devoted to describing the various STM methods which have been applied to semiconductor systems. The methods are divided into three broad categories: microscopy, spectroscopy, and spatially resolved spectroscopy. Much overlap exists between these categories. The microscopic imaging methods discussed in Section 4 include constant-current imaging and voltage-dependent imaging. Interpretation of the images in terms of surface structural features is emphasized. The basic spectroscopic techniques discussed in Section 5 include constant-current spectroscopy and constant-separation spectroscopy. The role of the tunneling transmission term in the spectroscopic measurements is described, including the various normalization schemes which have been developed for dealing with this term. The methods which have been developed for spatially resolved spectroscopy, discussed in Section 6, include acquisition of spectra at a few spatial locations, or at many spatial locations. When current-voltage curves are acquired at every pixel in an image, one can form spectroscopic images of the surface. The interpretation of such spectroscopic images is discussed.

2. Basic Principles

In Figure 1 we display a schematic view of the STM, taken from the original work of Binnig, Rohrer, Gerber, and Weibel.[1] In operation, the probe tip is brought to within 5 – 10 Å of the sample surface. Fine motion of the tip is accomplished with the piezodrives P_x, P_y, and P_z. A voltage V_T is applied between the tip and sample, producing the tunneling current J_T. The tunneling current is extremely sensitive to the separation between tip and sample, varying typically by one order-of-magnitude for each Å change in tip-sample separation. Thus, by scanning the tip across the sample, one can obtain an image of the sample topography. In practice, to avoid tip-sample contact, the scan is accomplished by continuously adjusting the tip height

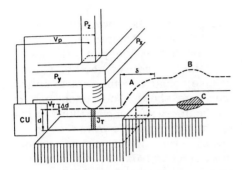

Fig. 1. Schematic view of the scanning tunneling microscope. The piezodrive P_x and P_y scan the metal tip over a surface. The control unit (CU) applies the appropriate voltage V_p to the piezodrive P_z for a constant tunnel current J_T at tunnel voltage V_T. The broken line indicates the z displacement in a y scan at (A) a surface step and (B) a contamination spot, C, with lower work function. [From Ref. 1].

in order to maintain a constant tunneling current. The tip height as a function of lateral position, $z(x,y)$, thus constitutes a constant-current image of the surface. In the simplest case, such images reflect the topography of the sample, as illustrated in Fig. 1 where the tip moves up over a surface step. More realistically, the tunnel current is also affected by local electronic effects on the surface. For example, a contamination spot may affect the work function of the surface, and thus will change the tunnel current and subsequent tip height, as pictured in Fig. 1. STM images thus contain a mixture of topographic and electronic information. In many cases, these types of information can be separated by acquiring images at a variety of tip-sample voltages.

Figure 2, also taken from the work of Binnig and Rohrer,[3] illustrates a number of other aspects of the STM. We see the probe tip mounted on a piezo-tripod, directed towards a sample which is mounted on an electrostatic walker or "louse". This walker, described in Ref. 2, provides the coarse motion in the microscope, bringing the sample up to the tip within the scan range of the z-piezo (about 1 μm). The apex of the tip, in close proximity to the sample, is shown in two magnified views. At the highest magnification, dotted lines are used to indicate contours of constant electron density (state density) for the tip and sample. The tunnel current is determined by the overlap of these wave-functions from the tip and sample, and the exponential dependence of the current comes from the exponential fall-off of the wave-functions into the vacuum region. Because of this exponential dependence of the tunnel current, only one or a few atoms right at the apex of the probe-tip contribute significantly to the total current. The precise arrangement of atoms on the end of the tip affects the resolution of the microscope; this phenomena is routinely observed by any operator of an STM as discrete changes in the resolution during the course of acquiring images. This dependence of the images on the sharpness and shape of the tip must always be kept in mind when analyzing STM images. Features which could result from asymmetric or multiple tips must be reproduced many times before they can be confidently identified as a property of the sample itself. Nevertheless, many features of the images (eg. the size and orientation of the unit cell) are relatively independent of tip structure, and thus form a most reliable database for structural determination with the STM.

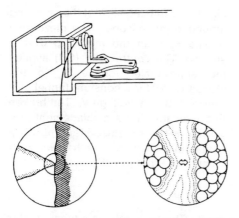

Fig. 2. Schematic view of the physical principle and technical realization of the STM. In the upper part of the figure, the three piezodrives, the tip, and the sample holder with the sample on the louse are shown. The bottom part of the figure shows amplifications of the tip-sample region down to the atomic level. The dotted curves denote electron charge-density contours on a logarithmic scale. [From Ref. 3].

We mention here a few additional technical features of the STM which are not illustrated in Figs. 1 or 2. First, there is the issue of vibration isolation. Clearly the mechanical stability between probe-tip and sample must be very high to avoid vibrational noise in the tunnel current. Typical stability levels are in the range $0.01 - 0.1$ Å. In the original STM designs, this stability was accomplished using the simple, but effective, method of building the microscope as compact as possible, and then suspending it on soft springs. These springs damp out vibrations with frequency greater than their resonance frequency ($\simeq 1$ Hz). The resonant motion of the springs themselves can be suppressed by eddy current damping using permanent magnets. This original design of the STM remains today one of the very best methods for designing and building the microscopes. Nevertheless, a number of other designs have been developed which simplify and improve on various aspects of the instrument. A review of these designs is presented elsewhere.[6]

Another important technical feature of the STM has to do with unintentional spatial drift between tip and sample. This drift typically amounts to a few Å per minute movement of the tip relative to the sample, in the x, y, or z direction. At this rate, the drift often amounts to a significant fraction of an image size in the time required to acquire the image. Sources of the drift include thermal expansion, and creep of the piezoelectric elements. The drift can be minimized by operating at low temperature, although this option is generally not available. The drift between tip and sample is a major factor in acquiring and analyzing data from the microscope. In the simplest case, the images themselves will be distorted. This distortion can be subsequently corrected either by measuring the drift using several successive images, or by using the known size of a unit cell on the surface. A more difficult problem arises when performing spectroscopic measurements at selected spatial locations. Typically it is difficult, or impossible, to go back precisely to a specified location and measure a spectrum after an image has been acquired. One way around this problem is to acquire the spectral information simultaneously with the image.

3. Theory

Constant-current imaging is the original imaging method of the STM. This method has already been described in Section 2; essentially, the probe-tip is scanned over the surface, with the z-position of the tip adjusted to maintain the tunnel current constant. The resultant image, z(x,y) for constant tip-sample voltage V and current I, is often called a "topograph" of the surface. In this section we review the work which addresses the question of what exactly is measured in these constant-current images.

The tunnel current is determined by the overlap of wave-functions between the tip and sample, as illustrated in Fig. 3. This dependence is most clearly seen in the Bardeen tunneling formalism, where the tunneling current takes the form[7,8]

$$I = \frac{2\pi e}{\hbar} \sum_{\mu\nu} f(E_\mu)[1 - f(E_\nu - eV)]\delta(E_\mu - E_\nu) |M_{\mu\nu}|^2 \qquad (1)$$

with

$$M_{\mu\nu} = \frac{\hbar^2}{2m} \int d\vec{S} \cdot (\psi_\mu^* \vec{\nabla}\psi_\nu - \psi_\nu \vec{\nabla}\psi_\mu^*). \qquad (2)$$

Here, $f(E_\mu)$ and $f(E_\nu - eV)$ are Fermi-Dirac occupation factors at the energies E_μ and $E_\nu - eV$ of the left and right-hand electrodes respectively. The delta-function ensures energy conservation. $M_{\mu\nu}$ is called the "matrix-element" for the process, and ψ_μ and ψ_ν are the wave-functions of the left and right-hand electrodes respectively. The matrix-element is evaluated over any surface lying entirely within the vacuum region separating the two electrodes.

Equations (1) and (2) can be used directly to illustrate some simple aspects of the tunneling problem. We consider a 1-dimensional problem, where the electric field in the junction is neglected, the same work-function ϕ is assumed for both electrodes, and all barrier rounding effects are neglected. The wave-functions for the resulting rectangular-barrier problem are simple decaying exponentials; denoting the separation between electrodes as s, we have

$$\psi_\mu = \psi_\mu^0 \exp(-\kappa z) \qquad (3)$$

and

$$\psi_\nu = \psi_\nu^0 \exp[-\kappa(s-z)] \qquad (4)$$

where

$$\kappa = \sqrt{\frac{2m\phi}{\hbar^2}} \qquad (5)$$

is the decay constant of the wave-functions. For a typical work-function of $\phi = 4.5$ eV, the decay constant has a value $\kappa = 1.1$ Å^{-1}. Inserting Eqs. (3) and (4) into (2), we have

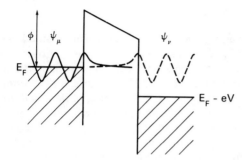

Fig. 3. Energy diagram illustrating tunneling between two electrodes.

$$M_{\mu\nu} = \frac{\hbar^2}{2m} \int dS\, 2\kappa (\psi_\mu^0)^* \psi_\nu^0 \exp(-\kappa s) \ . \tag{6}$$

Substituting into Eq. (1) we find that

$$I \propto \sum |\psi_\mu^0|^2 |\psi_\nu^0|^2 \exp(-2\kappa s) \ . \tag{7}$$

Here, using a shorthand notation, the summation sign denotes the sum over all relevant states which have appropriate energies and occupations to participate in the tunneling process. Note that in Eqs. (6) and (7) the explicit dependence on z has dropped out — that is, the current can be evaluated along any surface separating the two electrodes.

Equation (7) illustrates several important features of the tunneling process. First, the factor $\exp(-2\kappa s)$ gives the well-known exponential dependence of the current on separation. Using the above value of $\kappa = 1.1\,\text{Å}^{-1}$, this exponential dependence produces an order-of-magnitude change in the tunneling current for each Å change in the tip-sample separation. Second, within our simplified picture of a perfectly rectangular barrier, the quantities $|\psi_\mu^0|^2$ and $|\psi_\nu^0|^2$ give the state-density (probability density) at the surface of the electrodes. In this picture then, we can consider the tunnel current to be a probe of the surface state-density, $\rho \equiv |\psi^0|^2$, and the exponential factor is regarded as the matrix-element for the process.

Equation (7) was well known before the advent of the STM. In his treatise on tunneling, Duke arrives at essentially an identical formula displaying the proportionality between the tunneling current and the state-density at the surface of the electrode(s).[9] However, it was then stated that "the [use of] plane wave basis functions, the neglect of irregularities in the actual (as opposed to the average) barrier potential, and the inapplicability of the WKB approximation to an abrupt junction conspire to strip this formal result of any simple physical interpretation".[9] Clearly, the situation changed after the STM was invented! The atomic-resolution images of surfaces,[1] and the oscillations of the tunneling current seen in the field emission range,[2] beautifully demonstrate the spatial localization and quantum coherence of the tunneling current in the STM. The observation of surface states in spectroscopic measurements with the STM demonstrate conclusively the close relationship between the tunnel current and the surface state-density.[10-13]

To understand the spatial resolution of the STM, it is necessary to go beyond the simple 1-dimensional barrier problem discussed above, and extend the discussion to 3-dimensions. In terms of the Bardeen formalism, this extension was accomplished by Tersoff and Hamann[8], and by Baratoff.[14] Tersoff and Hamann assume the probe-tip to be a uniform sphere of radius R. In the limit $R \to 0$ the tip is replaced by a point-probe, located at a position $\vec{r}_0$. For the case of small voltage and temperatures (so that the relevant energy of the tunneling electrons is the Fermi energy E_F) they find that

$$I \propto \sum_\nu |\psi_\nu(\vec{r}_0)|^2 \delta(E_\nu - E_F) \ . \tag{8}$$

Thus the tunnel current is proportional to the local state-density of the sample, evaluated at the point $\vec{r}_0$. A constant-current STM image then simply corresponds to a surface of constant state-density. Tersoff and Hamann have shown that this same result can also be obtained for real probe-tips with nonzero radius. In that case the point $\vec{r}_0$ corresponds to the center of radius of curvature of the tip. This point is located a distance $R + d$ from the sample, where R is the tip radius and d is the distance from the tip apex to the sample. As the probe tip becomes blunter, R increases, and the STM effectively measures the state-density at increasing distances away from the sample. As shown below, atomic features in the state-density are attenuated as one moves away from the surface, and in this way the resolution of the STM decreases as the probe tips become blunter.

For metal surfaces, the local state-density at a surface roughly follows the corrugation of the surface atoms. For semiconductor surfaces the situation is more complicated, since, at a given energy, the surface states tend to be localized on some atoms or bonds but not on others. In both cases, metals and semiconductors, we must also consider the changes in the state-density as the distance from the surface increases. Generally, as the distance increases, contours of constant state-density tend to smooth out, with the higher order Fourier components being more rapidly attenuated. This phenomena is illustrated in Fig. 4, where we show schematically the wave-function ψ, state-density $\psi^*\psi$, and the variation in state-density with increasing distance z from the surface.[15] Fig. 4(a) illustrates the situation for a row of atoms on the surface, where the wave-function is assumed to alternate between maximum and minimum values on subsequent atoms. The state-density then has a maximum value on half the atoms and a minimum value on the other half. As the distance from the surface increases, the state-density smooths out, until eventually we are left with only a single sinusoidal corrugation component. Another situation is pictured in Fig. 4(b), in which we consider an isolated pair of atoms on the surface with the wave-function concentrated more on one atom than the other. Now, as the separation from the surface increases, the state-density will smooth out until it reaches a limiting form with a single maximum located somewhere between the two surface atoms. In both of these examples, the contours of constant state-density far from the surface

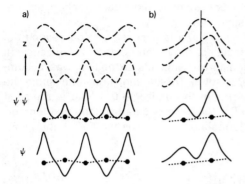

Fig. 4. Schematic illustration of the wave-function (ψ), the state-density ($\psi^*\psi$), and the decay of the state-density with distance (z) from the surface. [From Ref. 15].

cannot be interpreted in one-to-one correspondence with the surface atoms, and thus, in general, one must exercise caution in the interpretation of STM images.

The smoothing out of the state-density with increasing separation from the surface can be expressed in a simple analytic form, as derived by Stoll.[16] If we consider a Fourier component of the corrugation with period a and amplitude at the surface of h_s, then, for a distance d between surface and tip apex the observed corrugation amplitude, Δd, is given by

$$\frac{\Delta d}{h_s} = \exp\left\{\frac{-\pi^2(R+d)}{\kappa a^2}\right\} \tag{9}$$

where R is the tip radius and κ is the electron decay constant defined above. This equation shows that, as the corrugation period a decreases, the attenuation of the corrugation amplitude rapidly increases. The effective resolution of the STM can be seen by plotting numerical values from Eq. (9), for fixed $R + d$, as shown in Fig. 5. Smaller values for $\Delta d/h_s$ imply a reduced resolution of the microscope. We see that larger probe-tip radii, or greater tip-sample separations, both tend to reduce the resolution. This effect of the tip radius has been experimentally confirmed.[17] In practice, the ultimate resolution of the microscope is limited by the vibrational noise level in the instrument, below which a corrugation cannot be observed. For some value of the surface corrugation amplitude h_s, this noise limit corresponds to a particular value of $\Delta d/h_s$. Typically, corrugations which are attenuated by more than a factor of 0.1 will be difficult to observe. This cutoff value is shown by the dashed line in Fig. 5, and the intersection of the various curves with this line specify the effective resolution of the microscope. For the very small $R + d$ value of 4 Å, we find a resolution of 3 – 4 Å, which is about the best that can be expected with the STM. A more realistic value for $R + d$ of 20 Å produces a resolution in the range 6 – 10 Å, which is easily attainable with the STM. These values for the tip radius refer to the *microscopic* radius at the apex of the probe-tip; the *macroscopic* tip radius can be much larger than this value. Probe-tips prepared by the usual methods of electro-chemical etching apparently tend to have small "mini-tips" along their surface.[18]

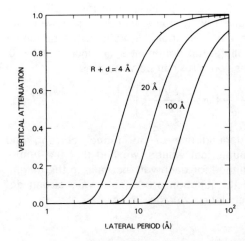

Fig. 5. The ratio of observed corrugation amplitude to actual corrugation amplitude, as a function of the corrugation period. Results are shown for fixed values of $R + d$, where R is the probe-tip radius and d is the distance from tip apex to sample surface.

We emphasize here that values for the STM resolution given by Eq. (9) are only applicable to smoothly varying sinusoidal surface corrugation. For semiconductor (or semi-metal) surfaces in which electronic effects can be large, the resolution can be considerably enhanced over these values. A dramatic example of this effect has been presented by Tersoff for the surface of graphite,[19] in which it is argued that zeroes of the surface wave-function prevent the decay of the corrugation amplitude with increasing tip-sample separation. In that case, one may expect a resolution equal to the hexagonal spacing of graphite, which is 2.46 Å. Effects of this sort may also occur on semiconductor surfaces when only a few states are being sampled by the tunneling current.

For the understanding of spectroscopic measurement with the STM, it is necessary to consider the effects of a nonzero voltage V between tip and sample. These effects are usually incorporated into an expression for the tunnel current by writing

$$I \propto \int_0^{eV} \rho_1(E)\rho_2(E - eV)T(E,V)dE \tag{10}$$

where $\rho_1(E)$ and $\rho_2(E - eV)$ are the surface state-densities of sample and tip respectively at energies relative to their Fermi-levels, and $T(E,V)$ is the transmission term for the tunneling electrons. A number of approximations are made in writing this equation, including neglecting the effect of the electric-field on the surface state-densities and neglecting the dependence of the transmission term on parallel momentum, but nevertheless the expression does illustrate some qualitative features of the tunneling process. In the WKB approximation, the transmission term has the form

$$T(E,V) = \exp\left\{-2s\left[\frac{2m}{\hbar^2}\left(\phi - E + \frac{eV}{2}\right)\right]^{1/2}\right\}. \tag{11}$$

Let us consider some aspects of Eqs. (10) and (11). First, note the dependence of the transmission term on energy E. Highest-lying states, with largest E, will contribute most strongly to the tunneling process. For positive voltage the dominant state occurs at $E = eV$, and for negative voltage the dominant state occurs at $E = 0$. In *both* cases, the maximum in the transmission term can be written as

$$T_{max}(V) = \exp\left\{-2s\left[\frac{2m}{\hbar^2}\left(\phi - \frac{|eV|}{2}\right)\right]^{1/2}\right\}. \quad (12)$$

This equation gives the lowest order dependence of the tunnel current, and conductivity, on voltage. Substituting in numerical values, we find that the tunnel current varies by about one order-of-magnitude for each volt increase in the magnitude of V. Thus, the electric-field across the junction produces a very strong dependence of the tunnel current and conductivity on voltage.

4. Microscopy

4.1 Constant-Current Imaging

The basic STM imaging technique — constant-current imaging — has already been described in Sections 2 and 3. In general the images can be understood in terms of contours of constant state-density, and for semiconductor surfaces the state-density can often be interpreted in terms of dangling bonds at the surface. The dangling bonds themselves are attached to surface atoms, so the surface geometry can be deduced by measuring the location of the dangling bonds. In this section we consider cases where the surface geometry can be deduced directly from a single STM image of the surface. The occupation of a bond (empty, filled, or partially filled). determines whether or not it is seen at some particular applied voltage between tip and sample. In the following section we consider cases where the voltage dependence of the images is important in understanding the surface geometry.

A classic example of direct geometric information in an STM image is the adatom array seen on the Si(111)7x7 surface.[20] Fig. 6 shows an STM image of that surface. The arrangement of 12 adatoms per unit cell is clearly seen in the image, and provides immediate and direct structural information about the surface. Each adatom ties up three dangling bonds from the substrate, resulting in a single dangling bond on the adatom; this reduction in dangling bond density provides the driving force for this type of surface reconstruction. It should be noted that, for the adatoms, electronic contributions to the image are still significant.[21] Also, convolution of the images with the tip shape can occur and can distort the images,[22] although this effect can be fairly readily evaluated by imaging the system with several different probe tips. Adatoms have been observed on a number of other clean semiconductor surfaces, including the Si(111)2x2, 5x5, and 9x9,[21] as well as the Ge(111)c2x8 surface.[23] This latter case provides an example of local symmetry which cannot be seen with diffraction methods, in which the c2x8 reconstruction on the surface was seen to consist of an alternating arrangement of 2x2 and c4x2 unit cells.

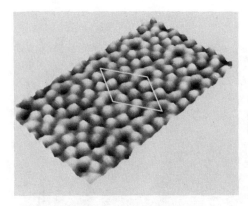

Fig. 6. STM image of the Si(111)7x7 reconstruction. A 7x7 unit cell is shown outlined by white lines. Twelve adatoms appear within the unit cell.

In contrast to the case of adatoms, vacancies may occur in semiconductor surface reconstructions. These missing atoms create large features in the images which also can be directly interpreted in terms of geometric structure. One example of this are the missing dimers seen on GaAs(001),[24] which form an ordered 2x4 arrangement on the surface. Missing dimer defects have also been studied on the Si(001) surface.[25] In that case they form an ordered 2x8 arrangement, stabilized by very small concentrations of Ni on the surface.

Steps on semiconductor surface are readily imaged with the STM. The relative occurrence of monoatomic or biatomic steps on Si(001) and Ge(001) has been studied,[26] and has some relevance to epitaxial growth on these surface. Detailed structure of the step edge has been investigated for Si(111)7x7,[27] Si(001),[28] and also for the Si(111)2x1 surface.[29] In Fig. 7 we show an STM image of the Si(111)2x1 surface, showing a step on the surface. The step contains two different types of structures along the step edge. Although the existence of the step is, of course, directly seen in a single image, the detailed interpretation of the structure along the step edge turns out to require voltage-dependent STM imaging.[29]

In addition to clean semiconductor surfaces, the STM has been used to study the local structure of many surfaces covered with adsorbates or overlayers. For example, adatoms have been found to form on Si(111) faces covered with Al,[30] In,[31] or Ga.[32] Depending on the amount of metal coverage, many different phases form for these overlayer systems. Adatoms have also been observed in one case on a non-(111) semiconductor face, namely, Sn on GaAs(110), as illustrated in Fig. 8. In this case, the Sn forms a double layer, with the first Sn layer roughly occupied sites which are an extension of the GaAs substrate, and the second Sn layer forming adatoms on top of the first layer.[33] In the [1$\bar{1}$0] direction, the spacing between the Sb atoms is 6 Å, corresponding to 1.5 unit cell lengths in that direction. Considerable disorder occurs in the top Sn layer because of two energetically equivalent stacking sequences, so that this 1.5x local symmetry is only short range. Thus, this arrangement of Sn atoms, which immediately reveals the structure of the overlayer, is not visible with diffraction methods.

In many cases, STM images may reveal two (or more) different structural phases of the surface in the same image. In such cases, a very important property of the

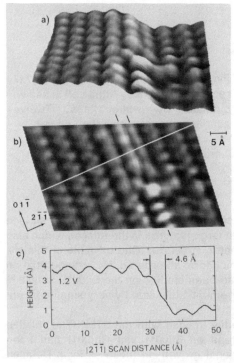

Fig. 7. STM image of a monatomic step on Si(111), acquired at a sample voltage of +1.2 V; (a) perspective view, (b) top view of the same data, and (c) cross-sectional cut along the line indicated in (b). The step edge is identified by tic marks at the border of the image in (b). [From Ref. 29].

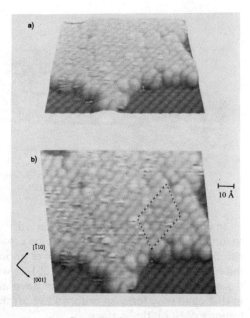

Fig. 8. Perspective view (a) and top view (b) of an STM image of a cleaved p-type GaAs(110) surface covered with 1 ML of Sn. The image is taken at sample bias of −1.5 V. A locally ordered region of the Sn overlayer is highlighted. The surface height is shown with a grey-scale, ranging from 0 Å (black) to 9 Å (white). [From Ref. 33].

image is the *registration*, that is, the relative orientation between the surface phases. Given the structure of one of the surface phases, the relative orientation of the other can lead to a determination of its structure. Application of this method usually requires some computer graphics which allows a grid to be fit to selected areas of the image, and then extended over the part of the surface with unknown structure. For example, in Fig. 9 we show a determination of the local binding site of Au atoms on the GaAs(110) surface.[34] On the clean surface, the corrugation maxima correspond to the location of As atoms (this relationship is discussed in Section 4.2 below). The small white dots in the image are fit to the position of those maxima. The Au atoms on the surface, seen as the large white maxima, are then seen to be located equidistant from four As atoms. This location places the Au atom right beside a surface Ga atom, as shown in Fig. 9(c). The method of registration was used in a

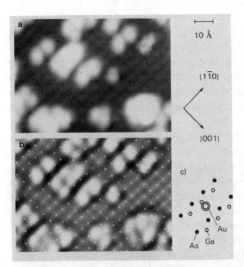

Fig. 9. 90 × 60 Å² STM image of 0.25 ML Au on the GaAs(110) surface, acquired at a sample voltage of −2.5 V. The grey-scale corresponds to (a) surface height, and (b) surface curvature. A top view of the atomic positions is pictured in (c), with a 2.2x expanded lateral scale. [From Ref. 34].

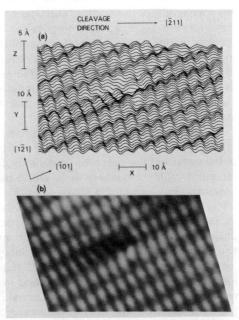

Fig. 10. Constant-current STM image of the Si(111)2x1 surface. (a) Line scans of the surface. (b) Grey-scale image with drift correction. [From Ref. 15].

study of the structure of the annealed Ag/Si(111) system[35-37] (although, in this case, the registration was apparently not sufficient to completely determine the structure). A particularly simple registration of metal atoms on Si(111) occurs at lower temperatures, in which the metal adsorbates mainly occupy one half of the 7x7 unit cell.[38,39] The registration method has also recently been used in the structural determination of Sb overlayers on the GaAs(110) surface.[40]

4.2 Voltage-Dependent Imaging

We now turn to cases of STM imaging where the voltage-dependence of the images is important in building up a total picture of the surface geometric structure. The significance of the voltage-dependence was first recognized for the Si(111)2x1 surface.[41] Figure 10 shows an STM image of that surface. Fig. 10(a) shows the direct line scans of the data, and these are assembled into a top-view in Fig. 10(b). In the top-view a drift-correction has been made to the data to achieve the known dimensions of the 2x1 unit cell, 6.65 x 3.84 Å. A small defect is seen near the center of that image. Such defects are relatively common on cleaved Si surfaces, although their exact nature is not known.

The image of Fig. 10 contains a single topographic maximum per 2x1 unit cell, and the magnitude of the vertical corrugation which gives rise to the maximum is about 0.5 Å Both the corrugation magnitude, and the shape of the topographic maxima (oblong in Fig. 10) are highly dependent on the shape of the probe-tip. As discussed in Section 3, a blunt tip produces a reduced corrugation amplitude, and asymmetry in the shape of the tip can give rise to asymmetry (circular or oblong) in the shape of the topographic maxima. This dependence on tip geometry thus makes the vertical corrugation amplitude itself an unsuitable quantity to use in a determination of the surface structure

Given the unsuitability of the corrugation amplitude in determining the surface structure, it is desirable to find some property of the surface which is independent of the details of the probe-tip and yields a simple, qualitative measure of the surface structure. Such a quantity turns out to be given by the *difference* in lateral position between empty and filled states, as seen in the voltage-dependence of the STM images. Figures 11(a) and (b) show two STM images of the Si(111)2x1 surface, acquired at sample voltages of 1 V and − 1 V respectively.[41] The images are similar, each consisting of an array of topographic maxima with one maximum per unit cell. Cross-hairs are superimposed on the images, located at identically the same surface locations in each image. Referring to the intersection of the cross-hairs, we see in Fig. 11(a) a topographic maximum, and in Fig. 11(b) we find a saddle-point in the topography. Thus, the topographic maxima have shifted by half a unit cell in the $[01\bar{1}]$ direction. This shift in the maxima occur between surface states which are normally empty or filled, as seen in the positive or negative voltage images respectively.

The shifts seen in the images of Fig. 11 can be quantitatively assessed by looking at cross-sectional cuts, as shown in Fig. 12. In the $[01\bar{1}]$ direction we find a shift of half a unit cell between the line cuts, as shown in Fig. 12(a). In the $[2\bar{1}\bar{1}]$ direction we observe a small shift in the corrugation, as shown in the cross-sections of Fig. 12(b). To produce more quantitative results, containing information from the entire image, the images can be fitted to a surface consisting of the sum of two sinusoids, with adjustable periods, amplitudes, and phases. This procedure yields corrugation shifts between empty and filled states for the images of Fig. 11 of about 2.0 ± 0.3 Å for the $[01\bar{1}]$ direction, and 0.8 ± 0.2 Å for the $[2\bar{1}\bar{1}]$ direction.[15]

In most cases on semiconductor surfaces, the topographic maxima observed in the STM images can be associated with dangling bonds at the surface. The dangling bonds, in turn, are attached to surface atoms, so that by imaging all of the dangling bonds we can build up a structural model for the surface. Energies of the dangling bonds may vary, so that certain bonds (atoms) may be seen at some tip-sample voltages but not at others. Thus, to build up a complete picture of the surface structure it is necessary to image the surface over a range of voltages. In the above example of the Si(111)2x1 surface, different dangling bonds were seen at positive and negative voltages. In this case the interpretation of the energy dependence is particularly simple, as described below. However, in general, we will not know in detail the energetics of the surface states, and thus it is not possible to predict which bonds will be imaged at particular voltages. Experimentally, one simply acquires images

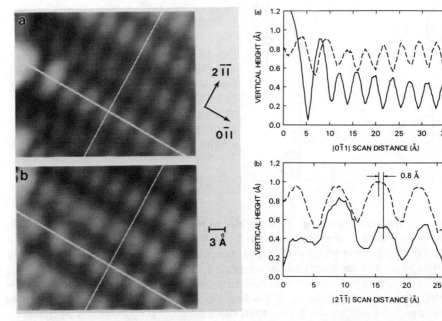

Fig. 11. Two STM images of the Si(111)2x1 surface, acquired simultaneously at sample voltages of (a) +1 V and (b) −1 V. [From Ref. 15].

Fig. 12. Cross-sections of the images in Fig. 11, along (a) the [0$\bar{1}$1] direction, and (b) the [2$\bar{1}\bar{1}$] direction. Absolute height is arbitrary here. For the [0$\bar{1}$1] direction the corrugation shifts by half a unit cell (1.92 Å) when the voltage changes from −1 to +1 V, and for the [2$\bar{1}\bar{1}$] direction the corrugation shifts by about 0.8 Å. [From Ref. 15].

at as many different voltages as possible, and then attempts to construct a structural model on the basis of that data set.

For the Si(111)2x1 surface, and several other surfaces with two atoms per unit cell, the spatial shift between empty and filled states provides a simple, qualitative way of understanding the surface structure. To understand the difference between empty and filled surface states, we introduce the concept of "buckling" on a semiconductor surface. We use this term here to refer to charge transfer between equivalent or inequivalent atoms on a semiconductor surface, as illustrated in Fig. 13. We consider two atoms on the surface, which may be two isolated atoms (eg. forming a dimer) or two members of a chain of atoms on the surface. We suppose that in the nonrelaxed surface each of these atoms has a dangling bond, and each bond is occupied by a single electron, as shown in Fig. 13(a). A 2-dimensional set of such bonds will form a half-filled band of states on the surface, as pictured in Fig. 13(b). On a semiconductor surface this band generally ends up lying somewhere within the bulk band

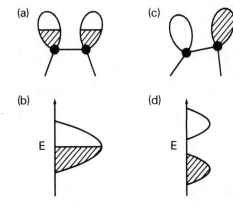

Fig. 13. Schematic view of buckling on a semiconductor surface. Surface atoms with half-filled bonds, (a), form a half-filled band of surface states, (b). Charge transfer from one bond to the other, (c), splits this band of states into a filled band and an empty band, (d).

gap. We use the term "buckling" here to refer to the transfer of an electron from one of the dangling bonds to the other, creating one filled bond and one empty bond, as shown in Fig. 13(c). This transfer of charge results in a splitting of the band of states into a filled band and an empty band, separated by a band-gap, as pictured in Fig. 13(d). Accompanying the charge transfer there will be some relaxation in the position of the atoms on the surface. If the two atoms of the surface are initially inequivalent (*eg.* a Ga and an As atom), then the energy of their bonds is necessarily different, and some buckling of the surface always occurs. If the two atoms are equivalent (*eg.* two Si atoms in a symmetric dimer on the Si(100) surface), then buckling may or may not occur depending on the energetics of the situation. It is important to note that buckling does *not* require the transfer of an *entire* electron from one dangling bond to the other. In reality, all the states on the surface are composed of a linear combination of both dangling bonds. For the unbuckled surface this combination will include equal contributions from both bonds, whereas on a buckled surface the spectral weight of one bond will be concentrated in the filled states and the other bond will be preferred in the empty states.

We return now to the Si(111)2x1 surface, and examine various models for the surface structure. Figure 14(a) shows our results for the dangling bond positions imaged at positive and negative voltages. In Fig. 14(b) - (d) we show the atomic positions of various structural models of the surface. Each model has two surface atoms per unit cell, with one half-filled dangling bond per surface atom. The atoms are structurally inequivalent, so that the dangling bonds necessary have slightly different energies. Thus, some buckling will occur, and one dangling bond will be seen at negative voltages and the other at positive voltages. In Fig. 14(b) - (d) we show one surface atom as a solid circle (filled bond) and the other as an open circle (empty bond). The models then can be directly compared with the measured positions of the dangling bonds. We see that the only model consistent with the data is the π-bonded chain model, Fig. 14(b). In this way, we determine the structure of the surface.

A second example of voltage-dependent imaging is provided by the GaAs(110) surface. That surface consists of of equal numbers of Ga and As atoms, with one

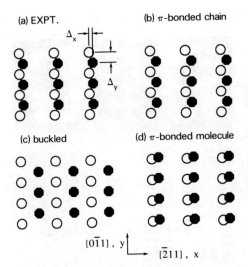

Fig. 14. Schematic top view of the spatial location of filled states (solid circles) and empty states (open circles) for the STM experiment on Si(111)2x1 and various structural models. The difference in lateral position between filled and empty states is denoted by Δ_x and Δ_y. [From Ref. 15].

dangling bond per surface atom. As mentioned in the discussion of buckling above, charge transfer will occur from the Ga to the As atoms, leaving the Ga dangling bond somewhat more empty, and the As dangling bond somewhat more full. Thus, when this surface is imaged with the STM one expects to see Ga or As atoms in the empty or filled states respectively. This reversal in the images was predicted theoretically early on,[8] and was later confirmed experimentally.[42] The experimental results are shown in Fig. 15. We see that the shift between empty and filled states is quite close to what is expected for the shift between Ga and As surface atoms. For this surface, the observed voltage dependence has compared in detail with theory, yielding an estimate for the amount of buckling at the surface.[42]

As discussed above for Si(111)2x1 and GaAs(110), the voltage dependence of the STM images arises simply from a two-band model for the surface states. Other examples of surfaces which can be understood in this way include Si(001)2x1,[43] Ge(001)2x1,[44] and Sb/GaAs(110).[40] This two-band model is not applicable to more complicated surfaces, although even in those cases the voltage dependence can often provide simple and direct structural information for the surface. One example which has been extensively studied is the stacking fault of the Si(111)7x7 surface. Experimentally, it is found that the two halves of the 7x7 unit cell appear to have different heights at different voltages.[11,45,46] This voltage dependence immediately indicates some structural inequivalence between the halves of the unit cell. One of the structural models for this surface, the dimer-adatom-stacking fault (DAS) model, predicts just such a structural inequivalence in the form of a stacking fault between the two halves of the unit cell. The observed voltage dependence is thus consistent with the idea of a stacking fault, and, together with the adatoms seen directly in the STM images, provides strong support for the correctness of the DAS model.

An extensive study of DAS and non-DAS related structures, using the voltage-dependent method, has been recently performed by Becker and co-workers.[21] Many

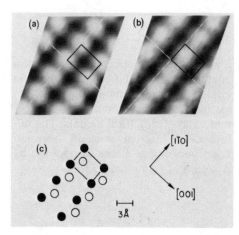

Fig. 15. STM images of the GaAs(110) surface, acquired at sample voltages of (a) +1.9 and (b) −1.9 V. The surface height is given by a grey-scale, ranging from 0 (black) to (a) 0.83 and (b) 0.65 Å (white). (c) Top view of the surface atoms. Arsenic atoms are represented by open circles and Ga atoms by closed circles. The rectangle indicates a unit cell, whose position is the same in all three figures. [From Ref. 42].

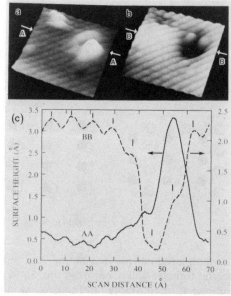

Fig. 16. STM images of an oxygen defect, acquired simultaneously at sample voltages of (a) −2.5 V and (b) +1.5 V. (c) Surface height contours along the lines AA (solid line) and BB (dashed line) shown in (a) and (b), respectively. (Absolute height is arbitrary). Oxygen exposure is 120 Langmuirs. The tick marks indicate the position of the Ga sublattice. [From Ref. 47].

structures were investigated, including Si(111)2x2, 5x5, 7x7, 9x9, and Ge(111)c2x8. The main purpose of the study was to determine the structure of the Ge(111)c2x8 surface, for which one DAS model and one non-DAS model had been proposed. Using voltage-dependent imaging, it was found that for DAS type structures, Si(111)5x5, 7x7, and 9x9, topographic maxima of empty and filled states are spatially coincident (ie. no shift occurs between empty and filled states). The topographic maxima are attributed to dangling bonds on the adatoms which are partially filled, and thus contribute to both empty and filled states. However, it was found for the non-DAS Si(111)2x2 structure that a shift occurred between empty and filled states. The empty states are attributed to an empty adatom band, and the filled states to a filled rest atom band. For the case of the Ge(111)c2x8 structure, a clear shift was also observed between empty and filled states, thus indicating that this surface also has a non-DAS type structure.

The voltage-dependence of STM images can be used to reveal other properties of the surface in addition to geometric structure. In the example of GaAs(110) above, the differentiation between Ga and As surface atoms provided a crude type of chemical identification of the surface atoms. Another property which can be seen in the images is the charge state of adsorbates, as demonstrated in studies of oxygen on the GaAs(110) surface.[47] Fig. 16 shows images of an oxygen adsorbate, acquired at sample voltages of -2.5 and $+1.5$ V. At negative voltage the adsorbate appears as a surface protrusion, and at positive voltage it appears as a surface depression. The voltage-dependence in this case was interpreted in terms of scattering of the current carriers inside the semiconductor itself. The electronegative oxygen adsorbate is taken to be negatively charged. It is then repulsive to electrons in the conduction band (positive voltage) leading to a decrease in the current, whereas it is attractive to holes in the valence band (negative voltage) leading to an increase in the tunnel current.[47]

4.3 Other Imaging Methods

The method of constant-current imaging, discussed above, is usually performed in a "slow scan" mode at a rate slower than the response of the STM feedback loop. Alternatively, one may use a "fast scan" method, in which the raster scanning of the piezos is performed faster than the response time of the feedback loop.[48] In the simplest case one then approaches a situation where the tip height z is a constant, and the image is formed by measuring the variations in tunnel current across the surface. Such a "current image" is then related to a constant-current image simply by the exponential relationship between current and tip height. The fast-scan method can only be applied to relatively flat surfaces, since the feedback loop cannot respond fast enough to prevent tip-sample contact on a rough surface. Nevertheless, this method appears to be advantageous for scanning surfaces in air, since the fast data acquisition provides an effective means of noise rejection.

A second method exists for current imaging, that of "current-imaging tunneling spectroscopy" (CITS).[13] This method is essentially a spatially resolved spectroscopic measurement, and we reserve discussion of the technique until Section 6 below. Similarly, the method of conductivity imaging at constant-current is also discussed in Section 6.

We conclude this section by mentioning the technique of work-function imaging with the STM, which is one of the techniques developed early on in the field.[1] In this method one modulates the tip-sample separation, thus measuring a quantity proportional to the square-root of the work function between tip and sample. The work function provides a type of average of the electronic properties at the surface. This method has not been widely applied to the study of semiconductor surfaces, although it may be applicable in some cases. One example of a possible application of this method would be to look for work function changes arising from charge transfer between adsorbates and the semiconductor surface.

5. Spectroscopy

5.1 Constant-Current Spectroscopy

In addition to constructing images of a surface, the tunneling microscope can be used to directly probe the spectrum of surface states. On semiconductors, such investigations are generally limited in energy to a few eV on either side of the Fermi-level; at voltages higher than this, field-emission resonances dominate the spectrum.[2,49,50] The first method used for measuring tunneling spectra with the STM was to measure conductivity, as a function of voltage, at constant tunnel current. Surface states on the (111) face of clean and Au covered Si were measured with this method.[10,11] The major disadvantage of this constant-current technique is that one cannot scan through zero volts, since at at zero volts and constant current, contact between the tip and sample will occur. For a metal, $dI/dV = I/V$ at low voltage, so that at constant current dI/dV diverges like $1/V$ as $V \to 0$. At high voltages however, the constant-current method is very effective at measuring the spectrum, since the large dependence of conductivity on voltage (discussed in Section 3) is automatically nulled out by the constant-current operation.

5.2 Constant-Separation Spectroscopy

Practically all semiconductor surfaces posses an energy band gap separating bands of filled and empty surface states. This surface gap may be larger or smaller than the bulk band gap. To simultaneously measure the empty and filled states on either side of the gap, it is necessary to scan though zero voltage. In that case the constant-current feedback loop cannot be used, since operation at zero volts would cause the probe-tip to touch the sample surface. To overcome this problem, an "interrupted feedback" method was developed in which probe-tip is frozen at any particular (x,y,z) coordinate using a sample-and-hold circuit, and the current versus voltage $I(V)$ characteristic is measured.[12,51] To build up a large dynamic range in the current, the measurement can be repeated for several different values of the tip-sample separation.

Figure 17 shows a series of $I(V)$ characteristics obtained from the Si(111)2x1 surface.[41] Each $I(V)$ curve is measured at a constant value of the tip-sample separation, and the separation is increased from curve $a - m$. A plot of the separation versus voltage $s(V)$, for a constant current of 1 nA, is shown in the lower part of the figure. At high voltages, above 4 V, the $s(V)$ curve displays oscillations from the barrier resonances in the vacuum region between tip and sample.[2] At lower voltages, between -4 and 4 V, surface state-density features show up as the various kinks and bumps occurring the the $I(V)$ curves.

The state-density features seen in the $I(V)$ curves are obscured by the fact that the tunneling current depends exponentially on both separation and applied voltage, as discussed in Section 3. This dependence can be partially removed by plotting the ratio of differential to total conductivity, $(dI/dV)/(I/V)$. Note that taking the ratio of conductivity to current, $(dI/dV)/I$, is equivalent to measuring the spectrum at

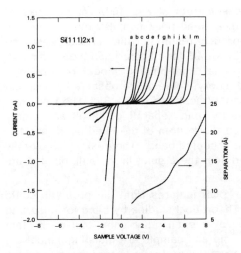

Fig. 17. Tunneling current versus voltage for a tungsten probe-tip and Si(111)2x1 sample, at tip-sample separations of 7.8, 8.7, 9.3, 9.9, 10.3, 10.8, 11.3, 12.3 14.1, 15.1, 16.0, 17.7, and 19.5 Å for the curves labelled $a-m$, respectively. These separations are obtained from a measurement of separation versus voltage, at 1 nA constant current, shown in the lower part of the figure. [From Ref. 41].

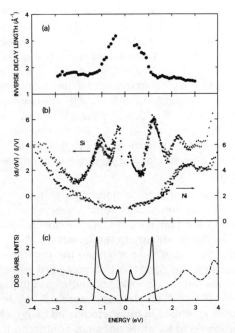

Fig. 18. (a) Inverse decay length of the tunneling current as a function of energy (relative to the surface Fermi-level). (b) Ratio of differential to total conductivity for silicon and for nickel. The different symbols refer to different tip-sample separations. (c) Theoretical DOS for the bulk valence band and conduction band of silicon (dashed curve, taken from Ref. 54), and the DOS from a 1-dimensional tight-binding model of the π-bonded chains (solid line). [From Ref. 41].

constant-current. Multiplying by V, $V \cdot (dI/dV)/I \equiv (dI/dV)/(I/V)$, then cancels out the low-voltage divergence of the constant-current method. The resultant "normalized conductivity" provides a relatively direct measure of the surface density-of-states (at least for a metal).[52,53] The results of such an analysis, applied to the data of Fig. 17, are shown in Fig. 18(b). There, each type of data symbol refers to a different curve (ie. a different tip-sample separation) from Fig. 17. The measurements at different separations fall on top of each other in Fig. 18, demonstrating that the quantity $(dI/dV)/(I/V)$ provides a convenient "invariant" quantity.

A number of peaks are evident in the spectrum of Fig. 18(b), at energies of −1.1, −0.3, 0.2, 1.2 and 2.3 eV, along with a small peak near 3.2 eV. To identify those peaks, Fig. 18(c) shows a plot of theoretical results for the surface density-of-states (DOS). The dashed curve gives the bulk DOS for silicon,[54] and the solid curve gives the surface DOS for a one-dimensional tight-binding model of π-bonded chains.[55] By comparison of experiment and theory, some of the observed peaks can be immediately identified with known electronic features of the Si(111)2x1 surface: The peaks of −0.3 and 0.2 eV encompass the gap separating the fill and empty states. The peak at −1.1 eV corresponds to the bottom of the filled band, and the peak at 1.2 eV corresponds to the top of the occupied band. The peak at 2.3 eV may arise from the lowest lying conduction band near the L point in the bulk band structure.

It is important to note that, in principle, DOS features from the probe-tip could contribute to spectroscopic observations. Theoretical studies by Lang for adsorbed atoms on two surfaces which are scanned over each other show that each atom contributes equally to the spectrum (ie. the tip and sample are indistinguishable in spectroscopy).[53] To evaluate the contribution of the probe-tip to the above observations on the Si(111)2x1 surface, the same tips were used to perform spectroscopic measurements on evaporated nickel films. The resulting spectrum in also shown in Fig. 18(b). A single peak, near 2.7 eV, is seen in the spectrum, and this peak has been identified in other studies as being a Ni-related surface state. Thus, the tungsten probe tip apparently does not contribute to the observed spectra in Figs. 17 and 18.

Another interesting quantity can be derived from the data of Fig. 17. The relative separation of each $I(V)$ curve can be obtained by selecting the appropriate values from the observed $s(V)$ curve, thus constituting a measurement of current versus separation at various voltages. The current is found to decay exponentially with separation, and the measured decay constants are plotted in Fig. 18(a). These decay constants correspond to the quantity 2κ as used above. In the simplest approximation we expect a decay constant of $2\kappa = 2.2$ Å^{-1} for a work-function of $\phi = 4.5$ eV. This value is close to what is observed in Fig. 18(a) at energies with magnitudes greater than 1 eV. At energies closer to zero, the decay constant is observed to increase sharply. This increase is attributed to tunneling through states with large values of parallel wave-vector. Current arising from a state with parallel wave-vector $k_\parallel$ will decay into the vacuum according to[8]

$$\kappa = \sqrt{\frac{2m\phi}{\hbar^2} + k_\parallel^2} \ . \tag{13}$$

Using this formula, a value of $k_\parallel \simeq 1.1$ Å^{-1} is found for the states near zero energy in Fig. 18(a). This value is in agreement with the surface state band structure of the Si(111)2x1 surface, for which states near zero energy arise from the edge of the surface Brillouin zone, with maximum wave vector of 0.94 Å^{-1}.

5.3 Variable-Separation Spectroscopy

As discussed above, the major advantage of constant-separation spectroscopy over the constant-current method is that one is able to scan through zero voltage. On the other hand, the constant-current method has an intrinsic advantage in terms of the dynamic range: since the current in constant, the conductivity always maintains large, measurable values throughout the entire voltage scan. A large dynamic range is achievable in the constant-separation method only by performing repeated scans in various values of the tip-sample separation (or by signal averaging at one particular tip-sample separation). This procedure can be tedious for routine measurements, since the resultant spectrum always must be "pasted together" from a number of discrete segments.

To permit acquisition of spectra with a large dynamic range, in a single scan, a method was developed which falls somewhere between the constant-current and constant-separation techniques. The method was primarily developed for measurements on wide band-gap materials, in which large conductivity outside the gap and small conductivities inside the gap must be simultaneously measured. In the method,[40,56] the tip-sample separation is continuously varied while the bias voltage is scanned, using a linear ramp in the tip-sample separation. The separation is reduced as the voltage decreases from some starting value down to zero, and the separation is then increased as the voltage increases away from zero. The total range of the tip-sample separation is typically 4 Å, thus providing an additional 4 orders-of-magnitude of dynamic range in the measurement. A modulation method is used to measure the conductivity dI/dV, with this derivative evaluated at *constant* tip-sample separation at each point in the scan.

Given the current and conductivity along some contour of tip-sample separation, one then can perform the conventional $(dI/dV)/(I/V)$ normalization of the data. A significant problem arises here for the case of wide band-gap materials.[40] Near the edge of the band-gap, the quantity I/V approaches zero faster than dI/dV, so that their ratio diverges. This problem is a failure of the normalization method to provide a meaningful estimate of the surface state-density. This failure can be illustrated by writing an approximate expression for the surface state-density as

$$\rho(V) \simeq \frac{dI/dV}{\tilde{T}(V)} \tag{14}$$

where $\tilde{T}(V)$ is an approximate form for the tunneling transmission term (*eg.* as given by Eq. (12)). In the conventional normalization, the total conductivity I/V provides an estimate for $T(V)$. Such an estimate is meaningful for a metal, in which I/V is nonzero at all voltages, but this estimate is completely invalid for a semiconductor, in which I/V is zero inside the band gap but the transmission term itself is nonzero. To overcome this problem, an approximate normalization method has been used in which I/V is broadened over the band gap region, with the resultant quantity denoted $\overline{I/V}$. The normalized conductivity is then given by the ratio $(dI/dV)/(\overline{I/V})$.

In Fig. 19 we show results for the normalized conductivity of clean and metal covered GaAs(110) surfaces.[57] For the clean surface, a band gap is seen with width

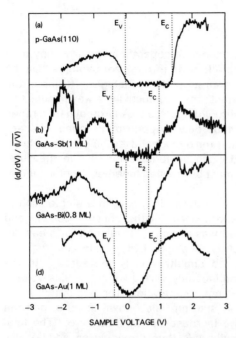
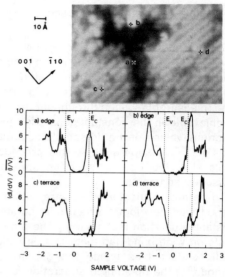

Fig. 19. Normalized conductivity versus voltage, for (a) clean p-type GaAs(110), and for monolayer films of (b) Sb, (c) Bi, and (d) Au. The horizontal lines separating the spectra give the zero level for the conductivity. [From Ref. 57].

Fig. 20. Normalized conductivity versus voltage, measured on top and on the edge of an Sb terrace on GaAs(110). The associated 120 × 80 Å^2 STM image is shown, with checkered markers denoting the locations at which the spectra were acquired. [From Ref. 59].

close to 1.4 eV, corresponding to the bulk band gap of GaAs. Band gaps are also seen for the Sb and Bi covered surfaces, with widths of 1.4 and 0.7 eV respectively. In these two cases the metal overlayers form ordered structures on the surface, and these ordered structures themselves have gaps in their energy spectrum. For the case of Au on GaAs(110), the Au layer is somewhat disordered, and a well-defined band gap does not form in the spectrum.

6. Spatially Resolved Spectroscopy

6.1 Spectroscopy at a Few Points

A very powerful aspect of the STM lies in its ability to perform *both* spatial imaging and electronic spectroscopy. The combination of these types of measurements is referred to here as spatially resolved spectroscopy. The major technical hurdle involved in these types of measurements arises from the thermal and/or piezo-electric drift of the tunneling microscope. As discussed in Section 2, it is normally difficult to acquire an STM image and then afterwards return to some specific location on the surface to measure a spectrum. Thus, various methods have been developed to simultaneously measure the spatial topography and the spectroscopy.

The earliest example of spatially resolved spectroscopy was performed on the Si(111)7x7 surface.[11] There, constant-current spectra were measured on the two different halves of the surface unit cell, and a particular surface state was found associated with the faulted half of the unit cell. In addition, this result was reproduced on the SnGe(111)7x7, GeSi(111)5x5, and Si(111)9x9 surfaces.[58] The measurement method in this particular case involved acquiring very small images of the surface, at sufficiently high rate so that instantaneous position of the probe-tip relative to the two halves of the unit cell was known. When the tip was seen to be well positioned over the center of either side of the cell, a spectrum was acquired. This method produced fine results in this case, but it may be difficult to apply to cases where higher spatial resolution is required, or when nonperiodic features on a surface are to be investigated.

An alternate method of acquiring a few spectra in an image consists of interrupting the raster scan, at specified locations, during acquisition of the image.[56] This can be done in real-time using a "hot-key" on the computer. Scanning of the image in this case is done quite slowly, and the image is plotted point-by-point, so that the raster scan can be interrupted at any desired spatial position. An example of this type of data in shown in Fig. 20, with the spectra acquired using the variable-separation method. The surface consists of a monolayer of Sb on GaAs(110); the grey region in the image consists of the ordered Sb overlayer, and the black area is the GaAs substrate.[59] Spectra (c) and (d) were acquired on the overlayer, and display a band gap of about 1.4 eV. Spectra (a) and (b) were acquired at the edge of the Sb terrace, and show additional states appearing inside the band gap region. These types of states have been observed for a number of metals on GaAs(110), and are significant in terms of understanding the formation of Schottky barriers on these systems.[57]

6.2 Spectroscopy at Many Points

For both of the above examples, modulation methods were used to measure the tunneling conductivity. Such methods provide good dynamic range, but they are limited in speed by the modulation frequency and/or lock-in amplifier time constant used in the measurement. Spectra can be acquired faster, but with less dynamic range, simply by measuring current versus voltage $I(V)$ directly, and later using numerical differentiation to get the conductivity. In that case many $I(V)$ curves can be acquired in the STM image. One can predefine some grid of pixels in the image at which $I(V)$ curves will be acquired, or, ultimately, one can acquire an $I(V)$ curve at every pixel in the image.

The acquisition of rapid $I(V)$ curves is done using the interrupted-feedback method[12] described in Section 5.2 above. Application of this method to every pixel in an image was given the name "current-imaging tunneling spectroscopy" (CITS).[13] In this method, one starts by arranging to scan the surface in the constant-current

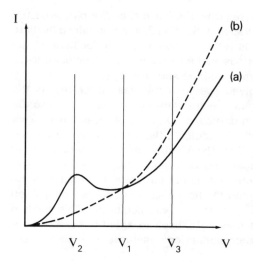

Fig. 21. Schematic illustration of current-voltage characteristics at two different spatial locations (a) and (b) on a surface. The currents are equal at the voltage V_1. At V_2, a surface state feature appears at location (a), and it can be imaged at that voltage. An image formed at the voltage V_3 will show a maximum at spatial location (b), associated with the varying background level of the current.

mode, as some tip-sample voltage V_1. Then, at each pixel in the scan, a current-voltage $I(V)$ curve is measured at constant tip-sample separation. The $I(V)$ curves can then later be examined to reveal structure due to surface states. This method was used to reveal the states on the Si(111)7x7 surface associated with various structural features: adatoms, rest atoms, and stacking fault.[13] NH_3 on the Si(111)7x7 surface has also been studied using this method.[60]

6.3 Spectroscopic Imaging

In addition to a method of spatially resolved spectroscopy, the CITS technique can also be used as an imaging method. One can form images by plotting the current (or conductivity) at some voltage V_2, which in general is different than the scanning voltage V_1. In this case, the images represent the *difference* in tunnel current measured at V_1 and V_2. Such images can be very powerful in revealing the electronic structure of a surface. However, since the images represent a current difference, they suffer from some difficulties in interpretation, as described below.

The general problem encountered in the interpretation of current (or conductivity) images is illustrated in Fig. 21. We plot in that figure two possible current versus voltage curves, assumed to exist at two different lateral positions (a) and (b) on the surface and at some particular values of tip-sample separation used in the measurement. If these curves were measured with the CITS method, then the currents are constrained to be equal at the feedback voltage V_1, as shown in Fig. 21. In both of the curves, we show the current increasing rapidly with voltage. This increasing "background" on the current is always present in real data, and arises from the transmission term for the electrons through the vacuum, as discussed in Section 3. In curve (a) of Fig. 21 we show a possible surface-state peak near the voltage V_2, which does not appear in curve (b). Now let us consider forming images of the current. At the voltage V_2 we will see a maximum in the image near the spatial position (a), and this maximum can correctly be interpreted as a surface-state related feature. However, at the voltage V_3, the images will show a maximum near the spatial position (b). In this case the maximum arises purely from varying background on the current, and is not surface-state related.

As illustrated in the above example, one must exercise caution in interpreting the features seen in current images.[61,62] Similar comments apply for differences between current images, and for conductivity images measured at constant current.[63] In general, a complete spectrum of the surface-states must first be examined before one can assign specific topographic features to individual surface-states. These problems of the "background" are more severe for current images measured by the CITS method than for conventional constant-current images because, as remarked above, the CITS images actually represent a current *difference* and as such they are more sensitive to changes in the background level of the tunnel current. These problems in interpretation of CITS images have been discussed in detail for the Si(111)2x1 surface,[61] and for the Si(111)7x7 surface.[62]

7. Summary

In this article, we have discussed the interpretation of images obtained with the scanning tunneling microscope. For semiconductor surfaces, these images are dominated by the influence of dangling bonds on the surface. An understanding of the images thus translates into an understanding of the surface wave-functions, expressed in terms of the state-density $\rho \equiv |\psi|^2$. The state-density may be peaked on certain atoms or bonds, but not on others, so that a one-to-one interpretation of the STM images in terms of surface atoms is not possible. However, by acquiring images at enough tip-sample bias voltages, one may succeed in imaging all of the surface dangling bonds and in this way build up a complete picture of the surface structure. In measurements of tunnel current versus voltage, the STM obtains an entire spectrum of the surface density-of-states. Effects of the tunneling matrix element can be partially, but not completely, removed from this spectrum. To lowest order the matrix element simply depends exponentially on tip-sample separation, with decay constant κ. Matrix element effects can be expressed in terms of increasingly more accurate approximations for κ, including the dependence of κ on energy, tip-sample voltage, and parallel wave-vector.

References

1. G. Binnig, H. Rohrer, Ch. Gerber, and E. Weibel, Phys. Rev. Lett. **49**, 57 (1982).
2. G. Binnig and H. Rohrer, Helvetica Physica Acta **55**, 726 (1982).
3. G. Binnig and H. Rohrer, Surface Science ***152/153***, 17 (1985).
4. G. Binnig and H. Rohrer, IBM J. Res. Develop. **30**, 355 (1986).
5. P. K. Hansma and J. Tersoff, J. Appl. Phys. **61**, R1 (1987).

6. Y. Kuk and P. J. Silverman, Rev. Sci. Instrum. **60**, 165 (1989).
7. J. Bardeen, Phys. Rev. Lett. **6**, 57 (1961).
8. J. Tersoff and D. R. Hamann, Phys. Rev. Lett. **50**, 1998 (1983).
9. C. B. Duke, *Tunneling in Solids* (Academic Press, New York, 1969), p. 253.
10. F. Salvan, H. Fuchs, A. Baratoff, and G. Binnig, Surf. Sci. **162**, 634 (1985).
11. R. S. Becker, J. A. Golovchenko, D. R. Hamann, and B. S. Swartzentruber, Phys. Rev. Lett. **55**, 2032 (1985).
12. R. M. Feenstra, W. A. Thompson, and A. P. Fein, Phys. Rev. Lett. **56**, 608 (1986).
13. R. J. Hamers, R. M. Tromp, and J. E. Demuth, Phys. Rev. Lett. **56**, 1972 (1986).
14. A. Baratoff, Physica **127B**, 143 (1984).
15. R. M. Feenstra and J. A. Stroscio, Physica Scripta **T19**, 55 (1987).
16. E. Stoll, Surf. Sci. **143**, L411 (1984).
17. Y. Kuk and P. J. Silverman, Appl. Phys. Lett. **48**, 1597 (1986).
18. C. F. Quate, Physics Today **39**, 26 (August, 1986).
19. J. Tersoff, Phys. Rev. Lett. **57**, 440 (1986).
20. G. Binnig, H. Rohrer, Ch. Gerber, and E. Weibel, Phys. Rev. Lett. **50**, 120 (1983).
21. R. S. Becker, B. S. Swartzentruber, J. S. Vickers, and T. Klitsner, Phys. Rev. B**39**, 1633 (1989).
22. S. Park, J. Nogami, and C. F. Quate, Phys. Rev. B**36**, 2863 (1987).
23. R. S. Becker, J. A. Golovchenko, and B. S. Swartzentruber, Phys. Rev. Lett. **54**, 2678 (1985).
24. M. D. Pashley, K. W. Haberen, W. Friday, J. M. Woodall, and P. D. Kirchner, Phys. Rev. Lett. **60**, 2176 (1988).
25. H. Niehus, U. K. Köhler, M. Copel, and J. E. Demuth, J. Microscopy **152**, to be published.
26. J. E. Griffith, J. A. Kubby, P. E. Wierenga, R. S. Becker, and J. S. Vickers, J. Vac. Sci. Technol. A **6**, 493 (1988).
27. R. S. Becker, J. A. Golovchenko, E. G. McRae, and B. S. Swartzentruber, Phys. Rev. Lett. **55**, 2028 (1985).
28. P. E. Wierenga, J. A. Kubby, and J. E. Griffith, Phys. Rev. Lett. **59**, 2169 (1987).
29. R. M. Feenstra and J. A. Stroscio, Phys. Rev. Lett. **59**, 2173 (1987).
30. R. J. Hamers, J. Vac. Sci. Technol. B **6**, 1462 (1988).
31. J. Nogami, S. Park, and C. F. Quate, J. Vac. Sci. Technol. B **6**, 1479 (1988).
32. J. Nogami, S. Park, and C. F. Quate, Surf. Sci. **203**, L631 (1988).
33. C. K. Shih, E. Kaxiras, R. M. Feenstra, and K. C. Pandey, to be published.
34. R. M. Feenstra, J. Vac. Sci. Technol. B **7**, Jul/Aug 1989.
35. R. J. Wilson and S. Chiang, Phys. Rev. Lett. **58**, 369 (1987).

36. E. J. van Loenen, J. E. Demuth, R. M. Tromp, and R. J. Hamers, Phys. Rev. Lett. **58**, 373 (1987).

37. R. J. Wilson and S. Chiang, Phys. Rev. Lett. **59**, 2329 (1987).

38. U. K. Köhler, J. E. Demuth, and R. J. Hamers, Phys. Rev. Lett. **60**, 2499 (1988).

39. St. Tosch and H. Neddermeyer, Phys. Rev. Lett. **61**, 349 (1988).

40. P. Mårtensson and R. M. Feenstra, Phys. Rev. B**39**, 7744 (1989); J. Microscopy **152**, to be published.

41. J. A. Stroscio, R. M. Feenstra, and A. P. Fein, Phys. Rev. Lett. **57**, 2579 (1986).

42. R. M. Feenstra, J. A. Stroscio, J. Tersoff, and A. P. Fein, Phys. Rev. Lett. **58**, 1192 (1987).

43. R. J. Hamers, R. M. Tromp, and J. E. Demuth, Surf. Sci. **181**, 346 (1987).

44. J. A. Kubby, J. E. Griffith, R. S. Becker, and J. S. Vickers, Phys. Rev. B**36**, 6079 (1987).

45. G. Binnig, H. Rohrer, F. Salvan, Ch. Gerber, and A. Baro, Surf. Sci. **157**, L373 (1985).

46. R. M. Tromp, R. J. Hamers, and J. E. Demuth, Phys. Rev. B**34**, 1388 (1986).

47. J. A. Stroscio, R. M. Feenstra, and A. P. Fein, Phys. Rev. Lett. **58**, 1668 (1987).

48. A. Bryant, D. P. E. Smith, and C. F. Quate, Appl. Phys. Lett. **48**, 832 (1986).

49. G. Binnig, K. H. Frank, H. Fuchs, N Garcia, B. Reihl, H. Rohrer, F. Salvan, and A. R. Williams, Phys. Rev. Lett. **55**, 991 (1985).

50. R. S. Becker, J. A. Golovchenko, and B. S. Swartzentruber, Phys. Rev. Lett. **55**, 987 (1985).

51. A. P. Fein, J. R. Kirtley, and R. M. Feenstra, Rev. Sci. Instrum. **58**, 1806 (1987).

52. R. M. Feenstra, J. A. Stroscio, and A. P. Fein, Surf. Sci. **181**, 295 (1987).

53. N. D. Lang, Phys. Rev. B**34**, 5947 (1986).

54. J. R. Chelikowsky and M. L. Cohen, Phys. Rev. B**10**, 5095 (1974).

55. R. Del Sole and A. Selloni, Phys. Rev. B**30**, 883 (1984).

56. R. M. Feenstra and P. Mårtensson, Phys. Rev. Lett. **61**, 447 (1988).

57. R. M. Feenstra, P. Mårtensson, and R. Ludeke, *Proceedings of the Materials Research Society, Symposium on High Resolution Microscopy of Materials*, to be published.

58. R. S. Becker, B. S. Swartzentruber, and J. S. Vickers, J. Vac. Sci. Technol. A **6**, 472 (1988).

59. R. M. Feenstra and P. Mårtensson, *Proceedings of the 19th International Conference on the Physics of Semiconductors*, DHN Ltd. (Warsaw, 1988).

60. R. Wolkow and Ph. Avouris, Phys. Rev. Lett. **60**, 1049 (1988).

61. J. A. Stroscio, R. M. Feenstra, D. M. Newns, and A. P. Fein, J. Vac. Sci. Technol. A **6**, 499 (1988).

62. Th. Berghaus, A. Brodde, H. Neddermeyer, and St. Tosch, Surf. Sci. **193**, 235 (1988).
63. G. Binnig and H. Rohrer, IBM J. Res. Devel. **30**, 355 (1986).

SPECTROSCOPY USING CONDUCTION ELECTRONS.

H. VAN KEMPEN
Research Institute for Materials
University of Nijmegen
Toernooiveld
NL – 6525 ED Nijmegen,
The Netherlands

ABSTRACT. These notes describe two spectroscopic methods which use conduction electrons as spectroscopic probes. Mainly low temperature experiments will be discussed.

1. General Introduction

In this series of three lectures I will focus on low temperature experiments which can be performed with small metallic points. Two regimes will be discussed: 1) The point makes a metalic (small) contact with the sample; 2) There is only a tunneling contact. In the first regime electrons can be ballistically injected from the point to the sample where they arrive with an excess energy equal to the potential difference present over the contact. In the last regime the scanning tunneling microscopes are operated. In the other regime one usually does not perform microscopy, although sometimes STM-like devices are used to make the contact in a controlled way. For both regimes a number of examples will be discussed.

2. Point contact spectroscopy

2.1. INTRODUCTION

Point contacts between normal metals show non-linearities in the current (I)-voltage (V) relation from which detailed information about the electron scattering lifetime can be obtained. It is well known that for bulk normal metals, Ohm's law, $V = RI$, is obeyed for a very wide range of currents and voltages. The validity of this linear I $\sim$V relationship is due to the fact that the electrons cannot gain enough energy between scattering events to make the energy dependence of the scattering time significant. For most pure metals current densities of $10^9 - 10^{10} A/cm^2$ are needed to accelerate the electrons between collisions to near the Debye energy. While it is clear that these current densities cannot be obtained in bulk metal, they can readily be obtained in point contacts between two pieces of metal where the contact size is small compared with the mean free path l. When a voltage V is applied over the point contact, an electron going from the one metal to the other gains an energy eV, and nonlinear effects can be observed.

For the geometry shown in Fig. 1a, one can consider two limiting cases. The limit of $l \ll a$, where a is the diameter of the contact, has been treated by Maxwell [1] by solving

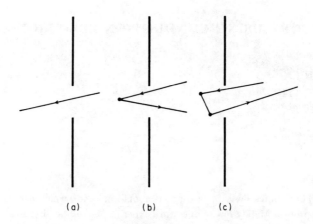

Figure 1: Schematic view of the successive iterative contributions to the current through a point contact: zeroth-order Sharvin current (a), first-order single collision back flow (b), and second-order double-collision back flow (c).

the Poison equation. One finds

$$R = \rho/2a \tag{1}$$

The other limit, $l \gg a$, has ben studied by Sharvin [2]. This limit is often called the Knudsen limit in analogy to a similar problem in kinetic gas theory. In this case the electrons are in a strong potential gradient when they pass from one metal to the other. This results in a resistance

$$R = 4\rho l/3\pi a^2 \tag{2}$$

which is essentially independent of l because ρl is a material constant.

Between the two limits Wexler [3] has derived a formula interpolating between (1) and (2):

$$R = \frac{4\rho l}{3\pi a^2} + \Gamma(K)\frac{\rho}{2a} \tag{3}$$

with $\Gamma(K) \simeq 1$; K is the Knudsen number l/a. Since ρl is a constant, only the second term can contribute to the voltage dependence of R:

$$\frac{dR}{dV} = \frac{\Gamma(K)\rho l}{2a}\frac{d1/l}{dV} \tag{4}$$

If one assumes the validity of Matthiessen's rule then one can separate the contributions of different scatterings processes to $1/l$. If, for example, besides the elastic electron-impurity scattering only inelastic electron-phonon scattering is present then

$$1/l = 1/l_i + 1/l_{ep}$$

$$\frac{d1/l}{dV} = \frac{d1/l_{ep}}{dV} = \frac{1}{v_F}\frac{d\tau_{ep}^{-1}}{dV} \tag{5}$$

where v_F is the Fermi velocity, and τ_{ep} is the electron-phonon scattering time. In the limit of $T = 0$, τ_{ep} for an electron with energy ϵ above the Fermi energy is given by:

$$\tau_{ep}^{-1}(\epsilon) = 2\pi \int_0^{\epsilon/\hbar} d\omega \alpha^2 F(\omega) \tag{6}$$

where $\alpha^2 F(\omega)$ is the well known Eliashberg function. Combining (4), (5) and (6) produces a simple relation between dR/dV and $\alpha^2 F(\omega)$:

$$\frac{dR(V)}{dV} = \frac{\pi \Gamma(K) \rho l e}{a \hbar v_F} \alpha^2 F(eV) \tag{7}$$

A more fundamental discussion has been given by Kulik et al. [4-6] and by van Gelder [7,8] based on an iterative solution of the Boltzmann equation. The zero-order solution, corresponding to the situation depicted in Fig. 1a, gives the energy independent Sharvin resistance. The first-order solution, in the Knudsen limit (Fig. 1b), leads to nonlinearities in the current voltage relation very similar to (7). The main difference is that instead of $\alpha^2 F(\omega)$ an $\alpha^2 F_p(\omega)$ is found which includes an efficiency factor $(1-\theta/\tan\theta)$ [8]. This factor, which shows up when more attention is payed to the geometry of the paths of the back scattered electrons, is similar to the efficiency factor $1-\cos\theta$ which plays an important role in the DC transport theory of metals. Higher order processes as in Fig. 1c can also be calculated and are sometimes observed experimentally.

2.2. EXPERIMENTAL TECHNIQUES

The first experiments on point contact spectroscopy were performed by Yanson [9,10] using tunnel junction technology. A point contact between two evaporated metal films separated by an oxide layer, is made by puncturing the oxide layer by electric break down or by mechanical pressure. An other technique was used by Jansen et al. [11]. A pointed metal wire (the "spear") is pressed against the surface of a flat piece of metal (the "anvil"). By carefully adjusting the pressure by means of some mechanical device the contact resistance can be adjusted to values up to 50 Ω which corresponds to contact diameters of the order of 25 Å. A third method consists of pressing together two sharp edges of two bulk pieces of metal. The last two mentioned methods have the advantage that they can be applied to single crystals and specially the last method is suitable for research into the anisotropy of the samples.

The electrical measurements use the standard modulation method, which is well known from tunneling experiments, to find the first and second derivatives.

2.3. SOME EXAMPLES OF EXPERIMENTAL RESULTS

2.3.1. Electron-Phonon Interaction. The most well known application of point contact spectroscopy has been the determination of the electron-phonon interaction. Examples are: the noble metals (Fig. 2) [12], the alkali metals [14] (Fig. 3), very anisotropic metals like zinc (Fig. 4) [15].

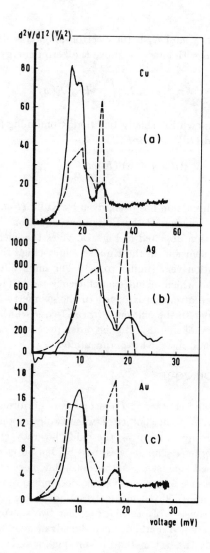

Figure 2: Measured d^2V/dI^2 of the noble metals. The broken curves are the phonon density of states obtained from neutron scattering experiments [13].

For the alkali metals a comparison with theoretical calculations can be made. For the other metals however the $\alpha^2 F_p(\omega)$ is not known in detail although the phonon density of states is known from neutron scattering experiments. For superconducting metals

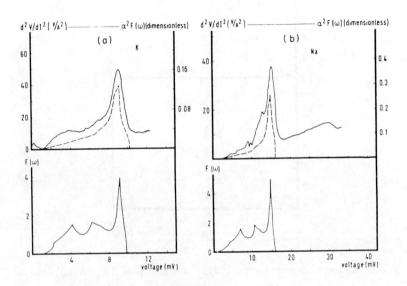

Figure 3: Measured d^2V/dI^2 for (a) potassium, $T = 1.2$ K and (b) sodium, $T = 1.5$ K. The broken curves are theoretical $\alpha^2 F$ functions [22]. The bottom curves give the phonon density of states from neutron scattering experiments [16,17].

however it is possible to obtain $\alpha^2 F(\omega)$ from the $I versus V$ relation of tunnel junctions [18]. Therefore it is interesting to apply point contact spectroscopy to superconductors and compare the $\alpha^2 F_p(\omega)$ determined with point contact spectroscopy with the $\alpha^2 F(\omega)$ from the tunneling observations. This comparison has been made by Yanson et al. [9] and good agreement has been found (Fig. 5).

Occasionally the higher order processes indicated in Fig. 1c, like the double process and even the tripple process, can be observed (Fig. 3,6) [19].

There are two weak points in this method for $\alpha^2 F_p(\omega)$ determination: For low energies the method is not very sensitive, because the signal is very small and other, stronger, effects are often present (zero bias anomaly). A second point is that an average over the Fermi surface is measured. For very anisotropic metals like zinc the anisotropy shows up in the $\alpha^2 F_p(\omega)$ measurements as we have seen (Fig. 4) but for metals like the noble metals, were also quite large variations in the electron scattering probability occur, only a weak effect can be observed.

Indeed one should like to have the ability to select certain k-space orbits or even k-space points to get the complete picture. Both points can be improved upon (at least in

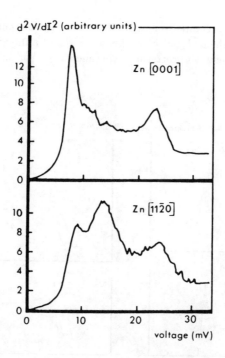

Figure 4: Measured anisotropy in d^2V/dI^2 for Zn (after Yanson and Batrak [15]).

principle) by a double point contact method to be described later.

2.3.2. Crystal Field Levels. Akimenko et al. [20] performed point contact experiments on $PrNi_5$ and observed singularities in the second derivatives which they could ascribe to Pr^{3+} transitions from the ground state to the excited states.

Also Frankowsky and Wachter found crystal field excitations in their experiments on TmSe [21].

For more examples see also Ref. [22].

2.4. DOUBLE POINT CONTACT SPECTROSCOPIC EXPERIMENTS

As mentioned earlier in section 2.3 weak points of the point contact spectroscopy are that the method does not work for low energies and gives only some average over the Fermi surface. However with the double point transverse focusing method, introduced for the firs time by Tsoi [23], both limitations can be overcome. Tsoi demonstrated that the electron-surface interaction can be studied by a double point contact arrangement. The basic set-up is given in Fig. 7. Two point contacts are made on the surface of a pure metal single crystal. Electrons which are injected in the crystal by one of the the contacts, the emitter, are focused by a magnetic field on the surface. When the field has the right value the electrons are focused on the second point contact, the collector, and a signal is observed. For fields which are multiples of this focusing fields, signals will also be present at the collector if the

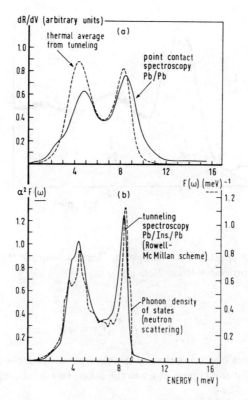

Figure 5: (a) Comparison of point contact and superconducting tunneling results. Full curve: dR/dV of a lead micro contact in a tunnel junction (after Yanson [9]). Broken curve: thermal average of $\alpha^2 F$ for a bath temperature of 2 K. (b) Comparison of phonon density of states $F(\omega)$ (broken curve) and Eliashberg function $\alpha^2 F(\omega)$ (full curve) obtained by superconducting tunneling [18].

electrons are specularly reflected. An example is shown in Fig. 8 [24]. One can see that, besides the electron orbits, also the hole orbits can be observed by this method. Generally, every closed orbit can be studied by a proper choice of orientation of the direction of the crystal, the field and the line connecting the points.

A mechanical device allowing all these alignments and also the controlled placements of the point contacts at low temperature has been described by Hoevers *et al.* [25]. In this device two point contacts can be placed independently of each other on the sample. The sample can be rotated and shifted in the horizontal plane. In addition also the magnetic field can be rotated in the horizontal plane.

When the injector point contact is a Sharvin contact one can inject the electrons with a known energy. The signal at the detector contact becomes then a function of injector voltage and contains information about the energy dependence of the mean free path:

$$V_{det}(eV) = V_{det}(0) exp(-t/\tau)$$

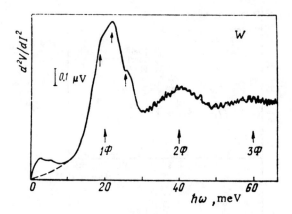

Figure 6: Point contact spectrum for Tungsten. Arrows indicate the position of the one, two and three phonon processes (from Ref. [19]).

and, at $T = 0K$,

$$\frac{1}{\tau(\epsilon)} = 2\pi \int_0^{\epsilon/\hbar} \alpha^2 F(\omega) d\omega$$

τ is the energy-dependent scattering time of the electrons and t is the time needed to travel over half a cyclotron orbit from one contact to the other, $t = \pi m/eB$, where B is the focusing field.

For low energies the Eliashberg function can be approximated by $b\omega^2$ which gives $1/\tau \sim b\epsilon^3$. From a decrease of the focusing peak hight with increasing electron energy, the coefficient b can be found. The value found is an average for the selected orbit. The results can be compared with radio frequency size effect experiments, where, from the temperature dependence in the same low energy limit, the parameter b can be determined also. In Fig. 9 [26] results are shown for silver for two different orbits. The measured curves are compared with the calculation following the above formula.

An interesting feature of the curves in Fig. 9 is that for higher emitter voltages the signal rises again. This is due to the electrons which have lost their energy close to the emitter. Those electrons can be considered to come again from a point source. The smaller the source of these electrons is, the narrower and higher their focusing peak will be. Also this high energy signal is dependent on the configuration chosen. The electrons scattered close to the emitter are focused at the collector only if they are scattered in a specific direction, namely the starting direction of the focusing orbit. In this sense this experiment gives information on the scattering probabilities into a certain point in k-space.

The high sensitivity in the low energy range can also be obtained with a single point contact although without the k-space selectivity, by using a "focusing mirror" which sends the electrons back again to the emitter. Such a mirror does indeed exist but several experimental limitations prevent a practical use of this method. More details will be given in the next section.

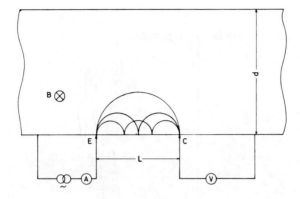

Figure 7: Double point contact geometry for observation of electron focusing. The electrons are emitted into the metal at E and are focused by the field B on the collector contact C. For higher fields the electrons can reach C after several reflections as also shown in this figure.

3. Tunneling Spectroscopy With Point Contact Junctions

3.1. INTRODUCTION

The essential difference between the spectroscopic capabilities of metallic point contacts and point contact tunnel junctions can be understood by considering Fig. 1b. Fig. 1b shows the diagram of a metallic point contact. An incoming electron (from the right) passes from the high potential side to the low potential side from where it can be scattered back by a scattering process like, for example, phonon emission. The probability that the electron indeed scatters back through the point contact is, in the ideal case, only dependent on geometrical factors as indicated schematically in Fig. 1b. Also in the case of a tunnel junction the incoming electron can be scattered back after tunneling in the direction where it came from. However the probability to cross the junction barrier again, and in doing so contributing to the resistance, is in practical cases extremely small. From these considerations it follows that the tunnel junctions are generally not sensitive to electron-excitation scattering processes. However the tunneling current is dependent on the (local) density of states of the electrodes. So tunnel junctions are very well suited to study density of states by I-V measurements. (Also the tunneling probability can be energy dependent due to inelastic tunneling allowing another spectroscopic method which will be discussed in section 4.)

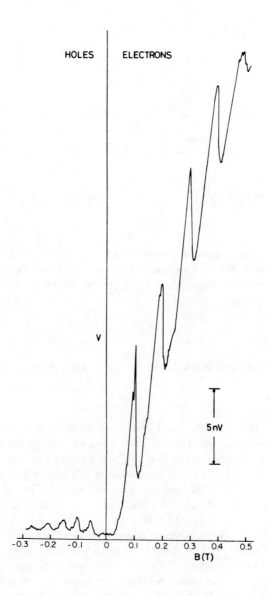

Figure 8: Reflections of electrons and holes from the (100) surface of a silver crystal.

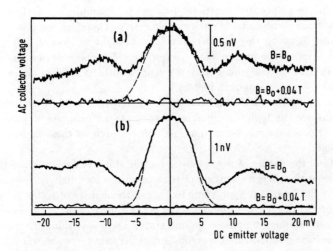

Figure 9: (a) Collector voltage *versus* dc emitter voltage for Ag(100). The magnetic field is directed along [001] and the line connecting the point contacts is perpendicular to it. Traces are for the focusing field, B=B_o, and for B = B_o+0.04T. The dashed line is a fit with b = 60 eV^{-2}. (b) Same as (a), but with the magnetic field rotated over ≈35°. Traces are for the focusing field and for B = B_o+0.04T. The dashed line is a fit with b = 75 eV^{-2}.

3.2. FIXED POINTS

In this part experiments will be described were the mechanical principle of the STM plays only a minor role. This means that, although STM-like devices are used to make the point contact junctions, imaging is not pursued.

Evaporated tunnel junctions have been used for a long time to study Density Of State (DOS), especially of superconductors. However there exist quite some materials for which it is difficult to make evaporated junctions because, for example, the surfaces are to rough or to reactive or because no suitable native oxide exists. Therefore one has tried to develop vacuum junctions in which the two electrodes are brought in close vicinity by some mechanical means. Among the first to succeed were Leo [27] whose device was used for DOS measurements in TTF-TCNQ and Poppe [28] who used his STM-like device to determine the DOS of heavy Fermion superconductors.

One class of materials which has been studied extensively lately with mechanical point contact junctions are the high T_c superconductors. As an example of what can be done and what kind of problems can be encountered the research on these superconductors will be described next.

To understand the nature of the superconducting state, one needs quite basic information like: How does the energy spectrum of the excitations look like, is there an energy gap and if so, what is the ratio of the gap value to T_c, can the density of states be described by a Bardeen, Cooper, Schriever (BCS) model, what is the nature of the ground state, and is the ground state a zero momentum paired ground state. These questions are specially relevant for the high transition temperature superconductors because there a detailed understanding is still missing.

Tunneling is one of the most used methods to study the density of states around the Fermi surface in superconductors. The results published sofar on high T_c superconductors show a considerable scattering in the obtained results. Several effects can contribute to this situation. In this section we will concentrate on the following points:
1) Coulomb gap.
2) Incremental charging.
3) Pressure effects.
4) Coherence length, Andreev reflection.

3.2.1. Coulomb Gap. The effect to be discussed in this section is not directly related with superconductivity, but is an effect generally present when the capacitance of a tunnel junction is very small. For very small tunnel junctions the charging energy $e^2/2C$ (e is the electron charge and C is the capacitance) is large compared to k_BT. In that case it is energetically unfavourable to tunnel as long as the capacitor is charged to less than $e/2$ ("Coulomb blockade"). This leads to tunneling of the electrons one by one at regular time interfals: Single Electron Tunneling (SET) oscillations induced by the Coulomb blockade. This effect strongly influences the I *versus* V curves [29,30]. Around the origin there is a quadratic voltage dependence while for large voltages the curve approaches an asymptote displaced from the origin with $\Delta V = e/2C$ (Fig. 10). The full curve can be calculated with a semiclassical Monte Carlo simulation [31,32]).

Fig. 11 shows two results obtained for a small junction formed between a tungsten

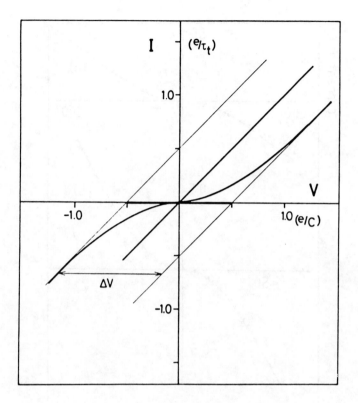

Figure 10: Expected I-V curve for SET. The current is expressed in units of e/RC.

needle and a ceramic YBa$_2$Cu$_3$O$_{7-\delta}$ sample and a stainless steel sample. It is clear from the curves that SET does not have anything to do with superconductivity.

As is obvious from the figures it is possible to confuse the SET curves with Superconductor-Insulator-Normal metal (SIN) junction curves. A distinction however is that the SET characteristics have a continuous rising derivative. Also, the apparent gap for the SET curves can be increased at will by lowering of the capacitance (increasing the tip-sample distance). Of course at the same time the tunneling resistance chances too. As the resitance has an exponential distance dependence and the capacitance a 1/distance dependence, a linear relation between $1/C$ and log$R$ should be found as is indeed the case (Fig. 12). In this figure the R and C values deduced from Fig. 13 are used.

A question is what the combined effect will be, when both the Coulomb blockade effect and a superconductivity gap are present at the same time. Monte Carlo simulations have been performed by van Bentum [33]. In Fig. 14 the calculated tunneling characteristics are plotted for three representative values of the capacitance, assuming a conventional BCS density of states for the superconductor. From these calculations one can conclude that the gap structure is strongly depressed by the SET oscillations.

This is even more clear from the calculations of the derivative dI/dV which are shown in Fig. 15 for the same set of capacitances. As a rule of thumb, the peak in the derivative is

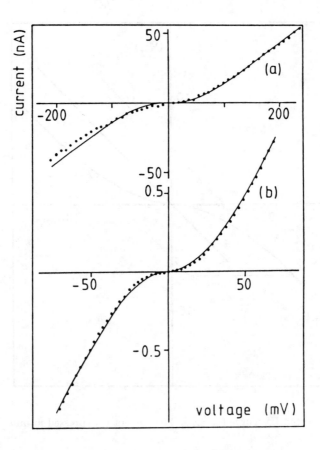

Figure 11: I-V characteristic of a tunnel junction between a tungsten tip and a stainless steel surface (solid dots), the continuous curve is a calculation from Monte Carlo simulations (a). I-V characteristic of a junction between a tungsten tip and an $YBa_2Cu_3O_{7-\delta}$ surface, compared with a Monte Carlo simulation (b).

broadened by approximately e/C. More cumbersome however is the fact that the maximum in the derivative shifts by approximately $e/2C$. For very small capacitance junctions it is therefore not correct to take the position of the maximum in dI/dV as a measure for the energy gap. As is clear from Fig. 17, this may lead to an overestimate by as much as 25 to 50%.

3.2.2. Incremental Charging. Another effect, related to the SET effect discussed above, shows up when a small particle is situated between tip and sample [29,30,32]. Then there are two small capacitances in series and the tunnel current can only flow if sufficient voltage is present to charge the particle with one or more electrons. In that case, the total charge on the common electrode is no longer a classic variable, but becomes quantized in units of the electron charge. If the net charge of the grain is zero, then the potential is a geometric mean of the potential difference between tip and sample. In addition, the potential of the grain increases stepwise if the number of additional charge carriers increases. The maxi-

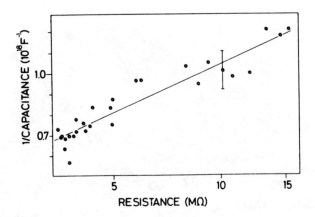

Figure 12: Inverse capacitance vs $\ln R_t$. The linear behavior indicates a simple d^{-1} law for the capacitance.

mum charge on the particle is limited by the applied voltage on the tip, and the tunnel current increases stepwise if an additional electron can be accommodated on the grain. This characteristic shape of the I versus V curve is generally referred to as the Coulomb staircase. More in particular, no current will flow until a sharply defined threshold voltage is reached. This threshold voltage is determined mainly by the larger capacitance of the two series junctions, and may be misinterpreted as an energy gap in the superconductor. A more extended treatment can be found in van Bentum et al. [34].

The charging effect has been observed in very small evaporated junctions with $C \approx 10^{-15}$ F at very low temperatures by Fulton et al. [35]. Using small isolated particles one can create much smaller capacitances and the effect can be seen at liquid helium temperature. Figs. 16 shows results of a measurement on a small Aluminium particle. Very similar results have been obtained for $YBa_2Cu_3O_{7-\delta}$ [34]. Note that the I-V curves around the origin show a resemblance to the BCS curve of a superconducting–superconducting tunnel junction. A clear discrimination between the charging effect and a superconducting multiple gap structure can be obtained again by observing the distance (and so the capacitance) dependence of the I versus V relation: In the case of incremental charging the apparent gaps increases for larger distances.

3.2.3. Pressure Effect. The values for the order parameter, measured at different positions on the same sample of $La_{1.85}Sr_{0.15}CuO_4$ and for different samples of the same composition were found to vary from 5 to 10 meV. The resistance measurements, however, did not indicate an associated variation of T_c for these samples. Therefore it was proposed that a large portion of the scattering is due to the local pressure exerted by the tip on the surface. Although the force on the surface is very small, the corresponding microscopic contact area does lead to a non-negligible pressure. Examination of the tip after the experiment indicates that the deformation pressure of tungsten whiskers, of order of 100 kBars [36], can

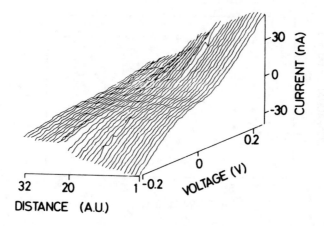

Figure 13: I-V characteristics of a tunnel junction between a tungsten tip and a stainless-steel surface. The distance between tip and surface is reduced between successive I-V scans.

easily be reached.

In Fig. 17 a set of I–V characteristics is displayed where deliberately the pressure P is increased at the junction area. The whisker was formed in an S-shape, with an experimentally determined spring constant of $8.6 \pm 0.3 \times 10^{-7}$ N/nm. The piezo voltage is directly proportional to the pressure at the contact area. Although normal parallel tunneling prohibits a detailed comparison with theory, a clear gap structure is visible which shifts to higher voltages with increasing pressure. The inset shows the width of the gap as a function of the piezo displacement.

At low pressures a linear increase of the gap was found. At very high pressure, a saturation takes place, which is probably due to the deformation of the tungsten probe. Assuming a deformation limit for the tungsten whiskers of $P_d = 100$ kBar, we obtain a contact area of 0.07 μm^2, which is reasonable for the probes used. In this way one can make an estimation of the pressure dependence of T_c, yielding $dT_c/dP \approx 0.7$ K/kBar, in good agreement with the results of Chu et al. [37]. It would be interesting to measure superconducting properties of these samples at high temperatures to check whether there is a related increase of the critical temperature. However the set-up used was not well suited for this purpose due to thermal drifts.

3.2.4. *Coherence Length, Andreev Reflection.* It is known from many experiments that the high temperature superconductors have an extremely small coherence length. Deutscher and Müller [38] have shown that this can lead to a strongly decreased value of the gap at the surface. They conclude that in that case tunneling gives no information on the bulk properties. A possible alternative to find the bulk value of the delta is to measure the energy dependence of the Andreev reflection.

Andreev reflection works as follows. When an electron, with an energy smaller than

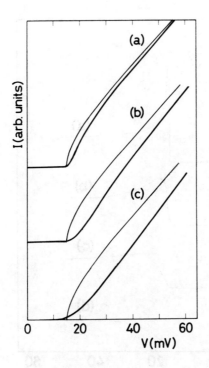

Figure 14: Calculation of the I–V characteristics of a SIN junction for various values of the capacitances. a:$C=3\times 10^{-17}$ F, b: $C=1\times 10^{-17}$ F and c: $C=5\times 10^{-18}$ F. The energy gap is 15 meV. Thin lines: ideal BCS characteristics.

the gap energy Δ, reaches a superconductor-normal metal interface, it cannot enter the superconductor as a quasiparticle, since there are no states below $E = \Delta$. Instead, in the "classical" superconductors, it can condense, together with a second electron of opposite spin and momentum from the normal metal, to form a Cooper pair in the superconductor. The hole, or missing electron, that results in the normal metal will move back in exactly the direction where the incident electron came from because of energy and momentum conservation (Andreev reflection [39]). So the observation of the Andreev retro-reflection is an indication of the presence of momentum compensated pairs in the superconducting ground state. To observe the energy dependence of the Andreev reflection, we use a Sharvin point contact geometry. The point contact is used to inject electrons into the normal metal layer. When Andreev reflection is present holes will travel back through the point contact and will give rise to the so called excess current. In other words, for electron energies below Δ, one expects an increase in the current and so a decrease of the point contact resistance. With the Sharvin contact electrons can be injected with energies up to eV (where V is the voltage over the point contact) and the energy dependence of the Andreev reflection can be determined and so the gap value can be found (Fig. 18). The Andreev reflection has been measured using both single [40,41] as well as double [42,43] point contact techniques

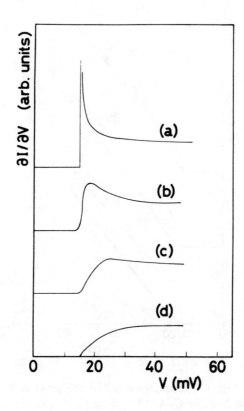

Figure 15: Calculation of dI/dV of same set of tunnel characteristics as in Fig. 14, (b,c,and d); a: ideal BCS case.

on "classical" superconductors like Pb. Typical results for high T_c superconductors are displayed in the upper part of Fig. 19. The point-contact resistance at zero voltage is higher than that at low bias voltages; at voltages higher than approximately 10 mV the resistance starts to increase and finally becomes more or less constant above 15 mV. For comparison, we have also plotted a typical experimental result obtained for a Ag-Pb bilayer (lower curve).

There are two mechanisms which can contribute to the observed energy dependence, the energy dependence of the Andreev reflection itself and the energy dependence of the electron-phonon scattering in the normal metal. Measurements of the energy-dependent electron-phonon coupling constant $\alpha^2 F(\omega)$ of Ag indicate that the electron mean free path will decrease rapidly just in the region of the expected gap energy of the $YBa_2Cu_3O_{7-\delta}$ [11]. When the mean free path decreases drastically the excess current becomes smaller and this effect can dominate the energy dependence of the Andreev reflection. Therefore, the results obtained from Fig. 21 give information on the superconductive gap in $YBa_2Cu_3O_{7-\delta}$ or the electron-phonon interaction in the normal metal or a mixture of both. Making the normal metal layer thicker, it is possible, at least in principle, to make the electron-phonon

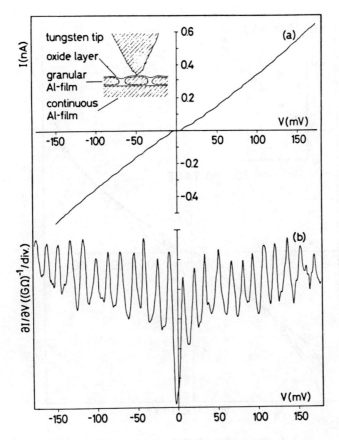

Figure 16: Incremental charging observed between a Tungsten tip and an Aluminium (a) sample; (b) gives the derivative.

scattering the dominant feature and by tuning the thickness one can make the experiment sensitive in the higher or lower energy range.

Note that the Andreev reflection supplies us with the "electron focusing mirror" we introduced in the discussion in section 2.4.

3.3 SPECTROSCOPIC IMAGING

By making images with different tunneling voltages one gets very detailed information on the surface properties as has been spectacularly shown for semiconductors. This will be fully described in a chapter on that subject (Feenstra, this issue). In this section some examples of the low temperature spectroscopic imaging of superconductors will be given.

3.3.1. Technical Details. Many different constructions have been developed. Presently most are based on a tube scanner combined with a (differential) micrometer screw for

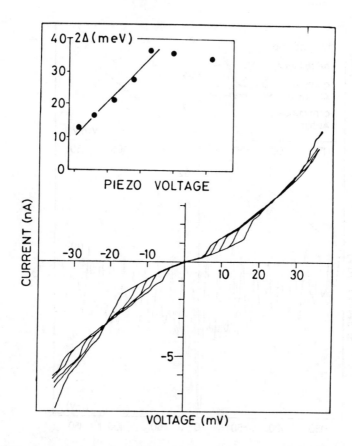

Figure 17: I-V curves of a tunnel junction of a W tip and a La- compound superconducting sample. The curves are measured with different piezo-voltages on the z-piezo. The inset shows the piezo-voltage dependence of the gap.

coarse adjustment. Generally, to maintain simplicity in the mechanical construction, no coarse adjustment in the lateral direction is present.

One of the bigger problems of low temperature STM is the limited possibility to clean the surfaces. Breakage or cleavage of the samples can be the solution for some materials. But also extended set-ups have been build combining UHV with low temperature and all the usual cleaning and characterization tools, for example de Lozanne et al. [44]. But even then problems can arise because some materials, like the high temperature superconductors, loose the oxygen when cleaned in the usual way.

3.3.2. Superconductors. Several attempts have been made to image the position dependence of the gap energy, for example of Nb_3Sn [45], NbN [46] and $YBa_2Cu_3O_{7-\delta}$ [47].

Large variations over the surface have been found probably due to irregular surface structure. Also when imaging one has to take into account of course the possible influence of the charging and Coulomb blockade effects discussed in section 3.2.

A very nice example is the direct observation by Hess et al. [48] of the Abrikosov flux

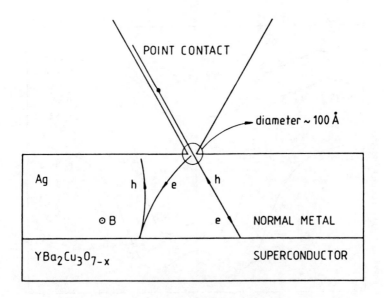

Figure 18: Arrangement of single point Andreev experiment.

line pattern in NbSe$_2$, Fig. 20. The observations are made by scanning with an energy such that a large current difference between the superconductive regions and the normal vortex cores exists. This experiment showed also an unexpected and unununderstood current-voltage dependence in the vortex core which indicated that the electron states of the core are clearly different from the bulk normal state. (An explanation has been put forward recently by Overhauser et al. [49].)

4. Inelastic Tunneling Spectroscopy

Jaklevic and Lambe [50] found in 1966 spectralike structures in the d^2I/dV^2 versus V plots of their tunnel junctions. These spectra could be assigned to the vibrational modes of molecules present in the junction barrier. Since the discovery of this Inelastic Electron Tunneling Spectroscopy, or shortly IETS, this technique has become a commonly used tool for the study of adsorbed molecules (see Hansma's review ref. [51]).

If a molecule, which has a vibrational mode with frequency ω, is present in the barrier, an electron can tunnel and at the same time excite this vibrational mode. Hence the electron will loose an energy $\hbar\omega$. This is only possible if the applied voltage V satisfies the relation $\hbar\omega \geq eV$, since otherwise there are no empty states available for the electron in the second electrode. So at a certain voltage V_0 ($V_0 = \hbar\omega/$ e) an inelastic tunnel channel is switched on resulting in an increase in current. Above V_0 the tunnel current will continue to increase since more and more final states come into play. Thus if we sweep the voltage

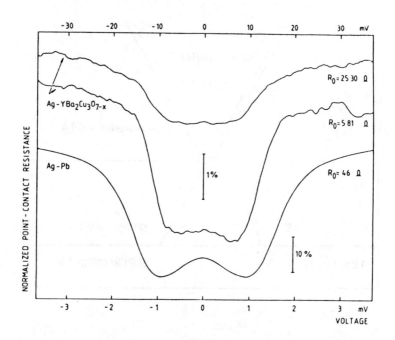

Figure 19: Differential resistance of a point contact *versus* voltage. Geometry as in Fig. 18.; Ag: 0.25 μm thick, YBa$_2$Cu$_3$O$_{7-\delta}$: 1 μm thick. The upper curves show the result for two different point contact resistances. The lower curve shows for comparison the result for a Ag-Pb bilayer.

across the junction we will find an increase in the slope of the I-V curve above V_0. In the first derivative steps of order of .1 to 1% result. The second derivative d^2I/dV^2 *versus* V gives a spectra like structure with a peak at V_0.

For performing IETS experiments a junction is created by evaporating a thin metal film ($\approx$ 2000Å), allowing this film to oxidize, exposing this oxide layer to a vapor of the substance one wants to study and finally evaporate a second metal on top to complete the junction. To study the current-voltage relationship in detail and detect the small changes in current, one uses a modulation technique. This is done by applying a dc voltage sweep plus an ac modulation of frequency ω to the junction. A lock-in amplifier then detects the signal at frequency 2ω. This gives the desired second derivative signal which is plotted against the dc voltage on a recorder. In this way spectra like the one for benzoic acid on aluminum, shown in Fig. 21 [51], are obtained. This real spectrum has a number of peaks because there are a number of vibrational modes in such an organic molecule. If one could do the same with a STM all kinds of interesting possibilities would be created. For instance one could study the vibrational modes of parts of larger molecules. However, in order to detect the small changes in current due to inelastic processes the tunnel current and hence the tip sample distance has to be very constant (up to 10^{-4}-10^{-5} Å). To resolve the vibrational modes the STM should also be operated at liquid helium temperatures. Surprisingly these

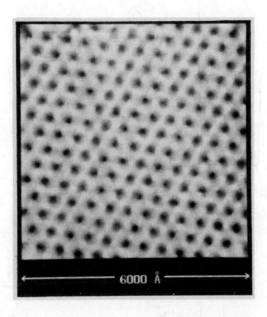

Figure 20: Flux line pattern observed by Hess et al. [48] in NbSe$_2$.

precautions seem to be superfluous in some cases.

Fig. 22 shows dI/dV *versus* V plots recorded with a STM, with constant lateral tip position, at room temperature, on a stepped Ni(111) surface [52]. Interesting is the double peak structure at 250 meV with negatively biased tip. When the voltage was fixed at a peak position the lateral spread of the peak could be estimated: it was of the order of inter atomic distances! This peak may be related to an inelastic process involving the C-O stretching vibration, which is known from other techniques to give a double peak at 250 meV [53]. Note that the peaks appear in the first in stead of the second derivative, only at one polarity and are resolved at room temperature. A similar result was obtained by Smith *et al.* [54] for sorbic acid on graphite at low temperatures. Probably the inelastic effects are enhanced when electrons tunnel from a tip, which in an ideal case could mean that the current is supported by one atom or even one atomic orbital. Also resonant tunneling effects can play an important role. The much higher signal strength could perhaps be partly explained by the direct interaction model (as opposed to the long range models which describe standard IETS results) of Kirtley and Soven [55]. This does not explain the occurrence in the first instead of the second derivative. Until now however there is no satisfactory explanation for these observations.

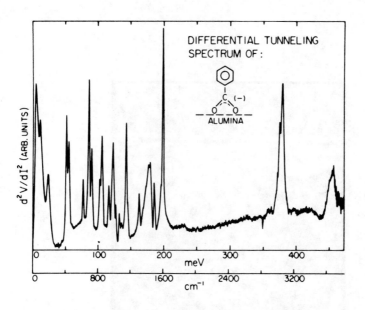

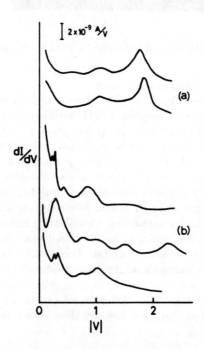

Figure 21: Upper panel: Differential tunnel spectrum of benzoic acid on aluminium (from ref. [51]); lower panel: Spectra from the Ni-W tunneljunction. a: tip positive; b: tip negative.

5. References

1. Maxwell, J.C. (1904) 'A treatise on electricity and magnetism', Oxford, Clarendon.

2. Sharvin, Yu N. (1965), Zh. Eksp. Teor. Fiz. 48, 984. Engl. Transl. Sov. Phys. JETP 21, 665.

3. Wexler, G. (1966), Proc. Phys. Soc. London 89, 927.

4. Kuluk, I.O., Omel'yanchuk, A.N., and Shekhter, R.I. (1977), Fiz. Nizk. Temp. 3, 1543. Engl. Transl. Sov. J. Low Temp. Phys. 3, 740.

5. Kulik, I.O., Shekhter, R.I., and Omel'yanshuk, A.N. (1977), Solid State Commun. 23, 301.

6. Kulik, I.O., and Yanson, I.K. (1978), Fiz. Nizk. Temp. 4, 1267. Engl. Transl. Sov. J. Low Temp. Phys. 4, 596.

7. Van Gelder, A.P. (1978), Solid State Commun. 25, 1097.

8. Van Gelder, A.P. (1980), Solid State Commun. 35, 19.

9. Yanson, I.K. (1974), Zh. Eksp. Teor. Fiz. 66, 1035. Engl. Transl. Sov. Phys. JETP Lett 27, 197.

10. Yanson, I.K., and Shalov, Yu.N. (1976), Zh. Eksp. Teor. Fiz. 71, 286. Engl. Transl. Sov. Phys. JETP 44, 148.

11. Jansen, A.G.M., Mueller, F.M., and Wyder, P. (1976) in Proc. 2nd Rochester Conf. Superconductivity in d and f-band Metals, ed. D.H. Douglass, Plenum, New York.

12. Jansen, A.G.M., Mueller, F.M., and Wyder, P. (1977), Phys. Rev. B 16, 1325.

13. Lynn, J.W., Smith, H.G., and Nicklow, R.M. (1973), Phys. Rev. B 8, 3493.

14. Jansen, A.G.M., van den Bosch, J.H., van Kempen, H., Ribot, J.H.J.M., Smeets, P.H.H., and Wyder, P. (1980), J. Phys. F 10, 265.

15. Yanson, I.K. and Batrak A.G.(1978), Zh. Eksp. Teor. Fiz. Pis'ma Red 27, 212. Engl. Transl. Sov. Phys. JETP 39, 506.

16. Cowley, R.A., Woods, A.D.B., and Dolling, G. (1966), Phys. Rev. 150, 487.

17. Gilat, G., and Raubenheimer, L.J. (1966), Phys. Rev. 144, 390.

18. Rowell, J.M., McMillan, W.L., and Dynes, R.A. (1973) 'A tabulation of the electron-phonon interaction in superconducting metals and alloys'. Part I. Bell Telephone Laboratories, Murray Hill, NJ. Unpublished.

19. Yanson, I.K., and Shklyarevskii, O.I. (1983), Fiz. Nizk. Temp. 9, 676. Engl. Transl. Sov. J. Low Temp. Phys. 9, 343.

20. Akimenko, A.I., Ponomarenko, N.M., Yanson, I.K., Janos, S., and Reiffers, M. (1984a), Fiz. Tverd. Tela (Leningrad) 26, 2264. Engl. Transl. Sov. Phys. Solid State 26, 1374.

21. Frankowski, I., and Wachter, P. (1982), Solid State Commun. 41, 577.

22. Duif, A.M. (1987), Thesis, University of Nijmegen.

23. Tsoi, V.S. (1974), Zh. Eksp. Teor. Fiz. Pis'ma Red. 19, 114. Engl. Transl. JETP Lett. 19, 70.

24. Benistant, P. (1980), Doctoral Thesis, University of Nijmegen, Nijmegen.

25. Hoever, H.F.C., Hermsen, J.G.H., and van Kempen, H. (1989), Rev. Sci. Instrum. 60, 1316.

26. van Son, P.C., van kempen, H., and Wyder, P. (1987a), Phys. Rev. Lett. 58, 1567.

27. Leo, V., (1981), Solid State Commun. 40, 509.

28. Poppe, U., (1985), J. Magn. Magn. Mat. 52, 157.

29. Averin, D.V., and Likharev, K.K. (1986), J. Low Temp. Phys. 62, 345.

30. Ben-Jacob, E., and Geffen, Y. (1985), Phys. Lett. 108A, 289.

31. van Bentum, P.J.M., van Kempen, H., van de Leemput, and Teunissen, P.A.A. (1988), Phys. Rev. Lett. 60, 369.

32. Mullen, K., Ben-Jacob, E., Jacklevic, R.C., and Schuss, Z. (1988), Phys. Rev. B37, 98.

33. van Bentum, P.J.M., Hoevers, H.F.C., van de Leemput, L.E.C. and van Kempen, H. (1988) J. magn. magn. mat. 76+77, 561.

34. van Bentum, P.J.M., Smokers, R.T.M., and van Kempen, H. (1988), Phys. Rev. Lett. 60, 2543.

35. Fulton, T.A., and Dolan, G.J. (1987), Phys. Rev. Lett. 59, 109.

36. Pethica, J.B., private commun.

37. Chu, C.W., Hor, P.H., Meng, R.L., Gao, L., Huang, Z.G., and Wang, Y.Q. (1987), Phys. Rev. Lett. 58, 405.

38. Deutscher, G., and Müller, K.A., (1987), Phys. Rev. Lett. 59, 1745.

39. Andreev, A.F. (1964), Zh. Eksp. Fiz. 46, 1823. Engl. Transl. Sov. Phys. JETP 19, 1228.

40. Benistant, P.A.M., van Gelder, A.P., van Kempen, H. and Wyder, P. (1985), Phys. Rev. B 32, 3351.

41. van Son, P.C., van Kempen, H. and Wyder, P. (1987), Phys. Rev. Lett. 59 2261.

42. Bokzo, S.I., Tsoi, V.S. and Yavkolev, S.E. (1982), Pis'ma Zh. Eksp. Teor. Fiz. 36, 123. Engl. Transl. JETP Lett. 36, 152.

43. Benistant, P.A.M., van Kempen, H. and Wyder, P. (1983), Phys. Rev. Lett. 51, 817.

44. de Lozanne, A.L., Ng, K.W., Pan, S., Silver, R.M., and Berezin, A. (1988), J. of Microsc. 152, 117.

45. de Lozanne, A.L., Elrod, S.A., and Quate, C.F., (1985), Phys. Rev. Lett. 54, 2433.

46. Kirtley, J.R., Raider, S.I., Feenstra, R.M., and Fein, A.P., (1987), Appl. Phys. Lett. 50, 1607.

47. Volodin, A.P., and Khaikin, M.S. (1987), Pis'ma Zh. Eksp. Teor. Fiz. 46, 466. Engl. Transl. JETP Lett. 46,588.

48. Hess, H.F., Robinson, R.B., Dynes, R.C., Valles,Jr., J.M. and Waszicak, J.V. (1989), Phys. Rev. Lett. 62, 214.

49. Overhauser, A.W. and Daemen, L.L. (1989), Phys. Rev. Lett. 65, 1619.

50. Jaklevic, R.C., Lambe, J. (1966), Phys. Rev. Lett. 17, 1139.

51. Hansma, P.K. (1982) 'Tunneling Spectroscopy', Plenum Press, new York and London.

52. van de Walle, G.F.A., van Kempen, H., Wyder, P. (1987), Surf. Sci. 181, 27.

53. Erley, W., Ibach, H, Lehwald, S. and Wagner, H. (1979), Surf. Sci. 83, 585.

54. Smith, D.P.E., Kirk, M.D., Quate, C.F., (1987), J. Chem. Phys. 86, 6034.

55. Kirtley, J. Soven, P. (1979), Phys. Rev. B 19, 1812.

SCANNING TUNNELING OPTICAL MICROSCOPY (STOM) OF SILVER NANOSTRUCTURES

R. BERNDT, A. BARATOFF and J. K. GIMZEWSKI
IBM Research Division
Zurich Research Laboratory
CH-8803 Rüschlikon
Switzerland

ABSTRACT. Local photon emission properties in the visible range of granular and island silver films have been characterized using STM. The experimental data are discussed in the light of model calculations. The observed fluorescence spectra and their dependence on tunnel voltage are interpreted theoretically in terms of radiative decay of local surface plasmons excited dominantly via inelastic tunneling. Spatial mapping of the resulting integrated light intensity (photon maps) reveals a close relation to topographic features. The sharp spatial contrast, lack of significant polarity dependence, and the detection of field emission resonances in isochromat photon spectra favor inelastic tunneling as the dominant process.

1. Introduction

In a scanning tunneling microscopy (STM) experiment, the tunnel current flowing between tip and sample is usually the prime source of information (Binnig et al., 1986). From its variation with tip position and voltage, data pertaining to the local density of states, topography and apparent tunnel barrier height can be derived. The tunnel current is generally dominated by the elastic channel. To observe inelastic contributions to the tunneling conductance, extreme requirements on gap stability, measurement time and current density will have to be fulfilled (Adkins et al., 1985).

Reflecting this state of affairs, almost all theoretical studies concerning STM have so far concentrated on elastic, non-dissipative channels. The mechanisms whereby an injected hot electron thermalizes or an inelastic tunneling channel operates, although crucial to the ultimate fate of tunneling electrons, have received little attention. An experimental possibility to obtain such information is to observe signals arising from purely inelastic events. Photon emission processes represent just such a channel: elastic events create zero emission so that, despite the low signal intensities, useful signal-to-background ratios can be achieved even at sub-nA currents on sub-nm areas.

In this contribution, we discuss light emission from small silver islands and polycrystalline silver films excited by tunneling electrons in an STM. For silver films, it is well known that electrons can excite long-lived surface plasmon modes with energies below 3.8 eV, i.e. in the tunneling range below the work function, a feature not accessible by conventional electron bombardment. These modes can then couple to light via a breakdown of translational symmetry, a subject recently reviewed by Raether (1988). Such a scenario was proposed by McCarthy and Lambe (1977) soon after their discovery of light emission from metal-oxide-metal sandwich tunnel junctions with gold or silver top electrodes. They observed enhanced emission for silver particles formed above such junctions. The voltage range over which light emission could be observed was limited by breakdown of the MOM junction. Recently, using an optical analogue of STM (scanning near-field optical microscopy) (Pohl 1989), local plasmons in gold films were investigated with a lateral resolution of ~20-50 nm (Fischer and Pohl, 1989).

One unique feature of the STM is the opportunity to excite locally plasmon modes in for example crystallites, isolated particles, other small structures or molecules, using a wide range of tunnel voltages V_T. In addition to the surface topography, the variations in light intensity can be recorded simultaneously. Consequently, local emission spectra were also measured and compared with tunneling dynamic conductance spectra (Coombs et al., 1988). These possibly not only represent a unique method to locally study inelastic and electron thermalization tunneling processes in mesoscopic systems with dimensions much smaller than the characteristic wavelength of the emitted light but, as will be shown, we have discovered that a contrast mechanism for variations of light emission intensity exists on an atomic scale. In STM experiments, the tunneling current from a single atom or a small cluster of atoms at the apex of the tip determines the "point" of excitation. Maintaining the total current at a constant value results in photon maps that reflect the local relative probability of a tunneling electron to generate an excitation which finally results in light emission. If all emitted light were to be collected, the measured signal would be the integrated photon yield. For convenience, we use that term for a collection of wavelengths in the visible spectrum from $\lambda = 350$ to 830 nm (3.5 to 1.5 eV).

In the following, we discuss recent experiments and combine them with theoretical model calculations to convey the basic physics involved.

2. Experimental Details

The experiments described here were all performed in ultrahigh vacuum (UHV) ($p < 1 \times 10^{-10}$ torr) using samples of thermally cleaned Si(111) crystals upon which Ag was condensed in vacuo at room temperature. The *average* coverage was monitored using a quartz crystal thickness monitor. The general arrangement and details of the STM and the UHV chamber have been described elsewhere (Gimzewski et al., 1987). Photons were collected at an angle of 45° using an externally mounted condenser lens focussed onto a photon detector. The solid angle at the detector was approximately 0.1 steradian. The STM was usually operated at a tunneling voltage that is higher than the lowest detectable photon energy (approximately 1.5 eV; ~830 nm). The upper detection cutoff is approximately 3.5 eV.

Spatial maps of the integrated photon signal which we call *photon maps* were recorded simultaneously with topography in a constant tunnel current mode using an externally mounted photomultiplier (Hamamatsu R268, peak quantum efficiency 30% at 400 nm). Peltier cooling enabled a photon-counting mode to be achieved with a dark count rate lower than 30 cps. The photon yield at a tunneling current of $i_T \sim 1$ nA and $V_T = -3.5$ V was sufficient to operate the STM at scan speeds of

~10 ms/pixel with peak count rates of 2×10^4 cps. Spectral distributions of the photons emitted for a predefined tip position above the surface were measured via a fiber-optic cable coupled to a grating spectrometer and an optical multichannel analyzer (EG&G, Model 1420R-1024HQ) interfaced to a computer. Typical acquisition times at $i_T \sim 300$ nA for photon spectra of ~60 s were required to obtain adequate statistics.

3. Results of Measurements

3.1. POLYCRYSTALLINE SILVER FILMS

In Fig. 1, the topography and photon intensity map of a thick ($d \simeq 100$ nm) silver film condensed at $T = 300$ K are shown. The structural details of such films have been reported by Gimzewski et al. (1985). The topography displays a preferred (111) texture with grain sizes in the 20-50 nm range. The corresponding photon map shows clearly contrasting regions relating to individual grains; some of them emit light strongly, while no light was detected from others. (Note that the relation between photon signal and gray level in all figures is strictly linear.)

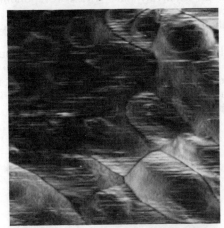
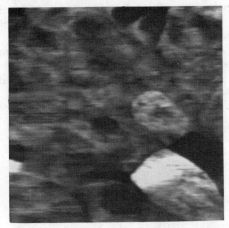

Figure 1. Conventional (edge-enhanced) topography and photon intensity image of a 1000×1000 Å² area of a thick (1000 Å) Ag film on Si(111) taken at a tip voltage of $V_T = -3.5$ V and a tunnel current of $i_T = 1$ nA. Above left, a top view of the topography is shown (gray scale extending over 50 Å), and at the bottom left a pseudo-3D illuminated picture. The simultaneously recorded photon intensity is shown on the top right (gray scale extending from the dark count rate to 5000 cps).

These sharp changes in the yield resulting from sub-nanometer lateral displacements in the tip position are particularly interesting: Although the emission area is unknown *a priori*, variations of the photon yield on a scale much smaller than the wavelength of the emitted light are compatible with changes in the inelastic excitation probability on a near-atomic scale. Therefore we find it convenient to define emitters as excitation *points* which give rise to a high photon yield.

3.2. SILVER ISLANDS ON Si(111)

Silver is accepted to grow in the Stranski-Krastanow mode on Si(111) (Venables et al., 1980). Consequently, isolated three-dimensional silver crystallites can be obtained by condensing small amounts of silver and subsequently annealing to temperatures > 490 K. An example of a topograph and accompanying photon map of a 600×600 Å^2 area of such a film is shown in Fig. 2. As in the other figures shown, the fast scan direction is from left to right with the detector facing the upper left-hand corner. The principal feature observed is an isolated oblate-spherical silver grain roughly 200 Å in diameter and 50 Å in height against a relatively flat background. Parts of two other grains are also evident at the top edge of the picture; the rest of the surface appears to be relatively flat. The corresponding photon map has three main characteristics:

i) The islands produce considerably higher photon signals than the flat surface areas.

ii) The yield from flat areas is significantly higher than the dark count rate of our detection system.

iii) The yield shows distinct intra-island intensity variations; in this particular example no emission appears to occur at their edges. In other cases islands apparently emit preferentially from the periphery.

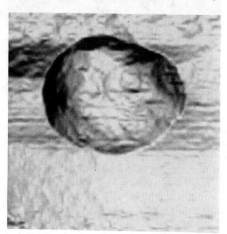 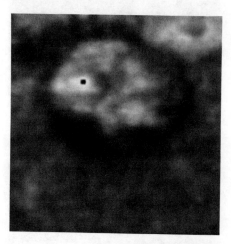

Figure 2. Topography (edge enhanced) (left) and photon intensity map (right) of an ~300-Å thick Ag film on Si(111) annealed at 490 K. Scan area: 350×350 Å^2, $V_T = -3.4$ V, $i_T = 1.5$ nA, height gray scale over 50 Å, light intensity scale ranges from dark count rate to 3000 cps.

3.3. STEP STRUCTURES

In certain cases an atomically flat area of the same film was observed, here with two atomic steps (Fig. 3). Closer inspection reveals some small-scale structure within the steps which we tentatively attribute to preferential residual gas adsorption. Surprisingly, the flat areas yield high intensity in the photon map (Fig. 3b) while the steps provide no detectable photon signal. To confirm that the variation in light intensity was not due to variations in i_T, we simultaneously recorded maps of the tunneling current. They displayed no significant correlation with the photon maps and appeared structureless.

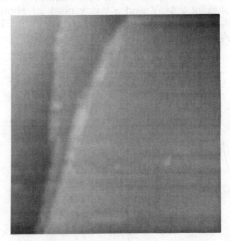

Figure 3. Topography (top view and pseudo-3D image, left) and photon map (right) of a stepped area (250×250 Å^2) of the film shown in Fig. 2. $V_T = -3.5$ V, $i_T = 1.5$ nA, light intensity scale: 3000 cps, height scale: 5 Å.

3.4. FLUORESCENCE SPECTRA

Up to now, we have presented the integrated photon signal. Spectral distributions of emitted light at a series of voltages between tip and sample from fixed regions as a function of the tunnel voltage were previously reported (Gimzewski et al., 1989). They showed that a resonant enhancement in photon emission at $h\nu \simeq 1.9$ and 2.4 eV could be obtained. Due to our primitive experimental geometry, acceptable signal-to-noise ratios were achieved using higher tunnel currents (300 nA). Figure 4 shows a series of spectra representing averages over areas with a linear dimension of ~20 Å scan length. The spectra contain three clearly resolved features whose relative heights depend on the tip voltage. Their positions are well defined and vary slightly between the individual spectra in the ranges of 400-450 nm, 530-600 nm and 720-830 nm (3.1-2.8 eV, 2.3-2.1 eV, 1.7-1.5 eV). This variability can be due to a direct dependency on V_T, but it also stems from changes in lateral tip position, as discussed in Section 5. A considerable number of the spectra, of which we present only two here ($V_T = -3.4$ V, $V_T = +2.5$ V) show apparent emissivities enhanced by a factor of ~10. We found that light emission is observed with comparable intensities for both tip polarities. Earlier measurements under the same conditions on similar samples (Gimzewski et al., 1989) displayed a strong peak at 2.4 eV and a weak peak or shoulder at 1.9 eV, with strengths growing with V_T beyond the quantum threshold, $eV_T = \hbar\omega$, which then decreased above $V_T = -3.5$ eV.

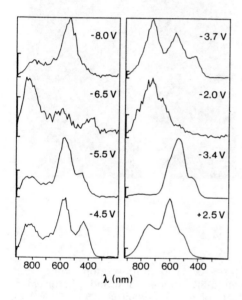

Figure 4. Spectral distributions of photon yield recorded for the sample of Fig. 1. The tip voltage for each spectrum is given. Vertical bars correspond to a count rate of 100 cps, except in the spectra for $V_T = -3.4$ V and $+2.5$ V where they represent 1000 cps. Light with a wavelength shorter than 350 nm was not detected because of absorption by an intervening Pyrex window.

4. Theoretical Estimates

4.1. MODEL

Stimulated by the initial experiments on polycrystalline Ag films (Coombs et al., 1988; Gimzewski et al., 1989), Persson and Baratoff (1989) took a fresh look at mechanisms for photon emission in STM. They considered tunneling from an s-like orbital at the apex of the tip to a similar orbital on the surface of a spherical free-electron-like metallic particle. The radius R of the particle is assumed to be small compared to the photon wavelength, but large compared to the typical wavelength of an injected electron. Both conditions are fulfilled for grains similar to those observed experimentally (Figs. 1 and 2). Within this model, a clear picture of all possible processes was provided and their branching ratios, including the dependence on R and V_T, were estimated. Maximum emission is predicted to occur for

$R \sim 200$ Å with a yield $\gtrsim 10^{-3}$ photons per injected electron, the dominant mechanism being the radiative decay of the dipolar plasmon excited by inelastic tunneling. This yield is comparable to that estimated earlier by Coombs et al. (1988), but is at least one order of magnitude larger than the highest yield previously achieved in light emission from tunnel junctions covered with Ag particles (Bloemer et al., 1987). The model does explain the dependence of the *integrated yield P* on V_T, but is too simple to predict the spatial, spectral and angular distributions of the emitted light.

4.2. DISCRETE SURFACE PLASMONS

Surface plasmons are electromagnetic normal modes with fields confined close to the boundaries of a single conducting body or several such bodies. In the electrostatic approximation, which is valid if $kR < 0.1$, where $k = \omega/c$ is the photon wave vector, the frequency of the fundamental dipolar mode is given by

$$1 + [\mathcal{R}e\, \varepsilon(\Omega) - 1]A = 0 \tag{1}$$

(Meier et al., 1985). Here ε is the complex dielectric function of the material and A the depolarization factor of the body. The corresponding field distribution is depicted in Fig. 5(a) for a sphere ($A = 1/3$). Multipole modes of order $\ell > 1$ need not be considered because the radiation efficiency is proportional to $(kR)^{2\ell}$. The corresponding fields exhibit 2^ℓ reversals around the body, so that for $\ell \gg 1$, their frequencies approach that of short-wavelength surface plasmons on a planar boundary, as illustrated in Fig. 5(b). This limiting frequency ω_s is given by

$$\mathcal{R}e\, \varepsilon(\omega_s) + 1 = 0. \tag{2}$$

For free electrons, $\Omega = \omega_p/\sqrt{3}$ and $\omega_s = \omega_p/\sqrt{2}$, ω_p being the bulk plasma frequency such that $\mathcal{R}e\, \varepsilon(\omega_p) = 0$. In silver, however, the onset of strong transitions from the d-bands to the Fermi level at $\hbar\omega_d = 3.8$ eV causes a steep increase of $\mathcal{R}e\, \varepsilon(\omega)$ for $\omega < \omega_d$. As a result, ω_s and Ω are close to ω_d. Thus the observation of several emission peaks at lower frequencies cannot be explained without invoking several dipolar modes, which are red-shifted by electromagnetic coupling to neighboring particles and to the tip itself. The field distribution of a mode which will be excited by a tunneling electron is sketched in Fig. 5(c). It is analogous to that of the "slow" mode in tunnel junctions (Mills et al., 1982). The main shift is probably due to the proximity of the tip and is present even in the case of well-separated particles.

A related situation (metallic spheres close to a conducting half-plane) has been considered by Rendell and Scalapino (1981). They found a distribution of dipole-active modes with broad maxima at 1.9 and 2.5 eV for an Ag sphere of diameter comparable to the lateral dimensions of the grains in our films. Note further that for a collection of spheres, electromagnetic coupling merely redistributes the strengths of the multipoles which would be induced in individual spheres by an arbitrary external field (Rojas and Claro, 1986). In estimating the integrated yield from all dipolar modes, it is therefore convenient to consider an isolated free-electron-like particle with R equal to the radius of curvature of the grain closest to the

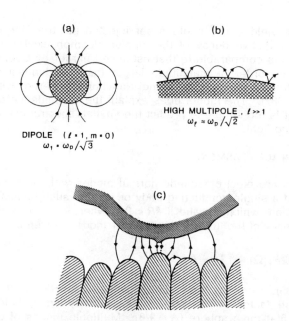

Figure 5. Instantaneous electric field lines for surface plasmon modes: (a) lowest and (b) high multipole modes of metallic sphere; (c) local slow mode between STM tip and granular sample (dotted lines indicate the much more localized tunneling current).

tip and assume that it has a single dipole resonance at the frequency $\hbar\Omega$ of the main peak (2.4 eV) observed in the emission spectra below the field emission range ($|V_T| < 5$ V).

4.3. PLASMON EXCITATION AND DECAY

Possible processes involving the intermediate excitation of the dipolar surface plasmon and their estimated branching ratios at each step are summarized in Fig. 6. Transition probabilities were calculated to the lowest order in the interaction of an electron with the barrier (elastic tunneling) or with the potential of the surface plasmon in the electrostatic approximation. Since this potential has different forms outside and inside the particle, it is convenient to distinguish between excitation via inelastic tunneling (a) and hot-electron decay (b). If one channel dominates, the other channel and its interference with the former may be neglected.

Once excited, dipole resonances can decay into electron-hole pairs (Zaremba and Persson, 1987) or via photon emission with a net yield P given by the product of the relevant branching ratios. The inverse lifetime due to the nonradiative process is $\gamma_{eh} \simeq v_F/R$, as expected on dimensional grounds, whereas that due to radiation is

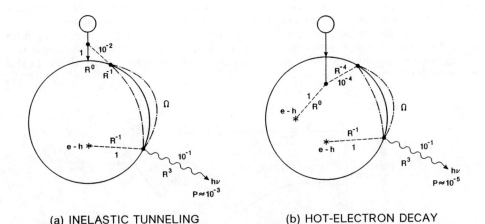

(a) INELASTIC TUNNELING (b) HOT-ELECTRON DECAY

Figure 6. Branching ratios for processes involving the excitation of a surface plasmon mode of frequency Ω (schematically indicated by dash-dotted lines) on a spherical metallic particle of radius R. The R-dependence, typical estimates for $R \sim 200$ Å and the final photon yield P are indicated for (a) inelastic tunneling and (b) elastic tunneling followed by hot-electron decay. Competing nonradiative channels involving electron-hole pair creation are denoted by e-h.

$$\gamma_{ph} = \frac{2}{3}\left(\frac{R}{c}\right)^3 \Omega^4. \tag{3}$$

This is simply the power radiated by the dynamic dipole $p_\Omega = (\hbar\Omega R^3/2)^{1/2}$, obtained by equating its classical field energy to $\hbar\Omega$. If $kR > 0.2$, the radiative damping becomes dominant, and retardation effects must also be considered. Judging from results in that range (Meier et al., 1985), we expect p_Ω and P to reach a maximum and then to decrease with increasing R.

Dipolar plasmon excitation by an elastically injected hot electron was calculated in a manner similar to γ_{eh}. However, it is extremely unlikely compared to direct decay into electron-hole pairs. While it is relatively improbable compared to elastic tunneling, dipolar plasmon excitation via inelastic tunneling clearly dominates. It is actually better calculated as a coherent resonant process, which gives

$$P_{ph} = \frac{e^2 \Omega^5 R^2}{6\hbar \gamma c^3} \frac{2ms^2}{\phi}\left(1 - \frac{\hbar\Omega}{eV_T}\right), \tag{4}$$

where $\gamma = \gamma_{eh} + \gamma_{ph}$, and ϕ and s are the mean barrier height and minimum thickness, respectively. This integrated photon yield P_{ph}, calculated consistently within the electrostatic approximation, i.e. with $\gamma = \gamma_{eh}$, is compared with that from hot-electron decay in Fig. 7. Its dependence on the tunnel voltage, contained in the last factor of Eq. (4), is apparent for $eV_T < \hbar\omega_s$ in Fig. 8. Once this threshold is exceeded, nonradiative surface plasmons with short wavelengths $\sim s$ are efficiently excited with a probability of $P_s \simeq C(1 - \hbar\omega_s/eV_T)$ and steal intensity from other processes. The constant C calculated from an earlier theory (Persson and Baratoff, 1988) is ~5 times smaller than the value required to fit the data shown in Fig. 8 for $eV_T > \hbar\omega_s$. An additional mechanism, e.g. relative enhancement of the elastic tunneling channel by the first field emission resonance, appears to contribute in that energy range.

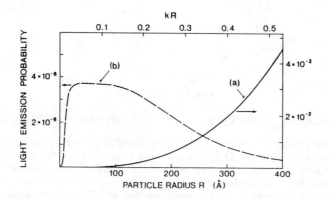

Figure 7. Dependence of the integrated photon yield on particle radius for processes (a) and (b) of Fig. 6 (note the different scales) in the electrostatic approximation. Radiation damping and retardation effects would cause the yield to reach a maximum at $kR \sim 0.25$.

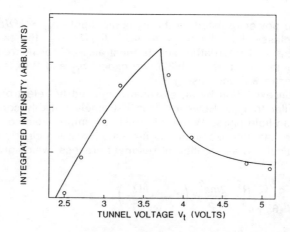

Figure 8. Circles: integrated photon signal deduced from the data in Fig. 1 of Coombs et al. (1989); solid curve: calculated dependence of the integrated photon yield on tunnel voltage assuming a dipolar mode with energy $\hbar\Omega = 2.4$ eV and a limiting surface plasmon energy $\hbar\omega_s = 3.7$ eV. The overall scale and the relative strength of the nonradiative excitation which sets in for $eV_T > \hbar\omega_s$ have been adjusted.

5. Discussion

In accordance with the theoretical picture outlined, we observe strong light emission associated with nearly spherically shaped silver islands. Evidence supporting inelastic tunneling as the dominant process is the sharpness of the contrast in the photon yield with spatial position. In a hot-electron picture, electrons with energies in the eV range have inelastic mean free paths in the range of several hundred Å for silver (Kirtley et al., 1983). Consequently, one would expect a resolution in that range if the plasmons were created by electron scattering alone. Secondly, the previously reported observation of field emission resonances in the photon signal (Coombs et al., 1988) is compatible with inelastic tunneling into resonance states. It is unlikely that hot electrons once injected into the bulk will interact with these states, which are confined between the tunnel barrier and the surface.

The overall agreement between experimental data and the theoretical description is encouraging and provokes further questions. First some topographically similar silver grains emit light. The differences in the photon signal due to variations in the electromagnetic coupling to the surrounding material require further consideration. Second, the photon yield is not distributed uniformly even over a single emitting grain, an effect not caused by variations in the total tunneling current. Whereas in the theoretical model, the ratio of inelastic excitation and elastic tunneling probabilities depends only on V_T and R, local variations of that ratio can arise in the experiments for the following reasons:

i) An electron which loses energy while tunneling sees a higher effective barrier and a different local density of final states. This can give rise to atomic-scale variations because, e.g. the total tunneling current (mainly elastic) has a different dependence on the tip-sample separation.

ii) The field distribution of the local surface plasmon(s) and their excitation probability depend on the position of the tip relative to the nearest grain(s).

Finally, the chemical identity of the tip apex is not known and a silver end layer may be necessary for enhanced light emission to occur. This would reduce the damping of dipolar plasmon modes between tip and sample. Coupled modes have been observed using TEM for small Al particles in close proximity (Batson, 1982). Correspondingly, a nonuniform coating could result in contrasts because the excitation region changes as the tip is scanned over a surface protrusion. A more detailed understanding of the observed emission properties, in particular a detailed correlation between spectral and topographic information, would be desirable to disentangle these factors. We are actively pursuing this and other aspects of local photon emission such as luminescence of molecules and semiconductors. The results obtained to date form a bridge between photoemission-related studies and STM, and prompt a rethinking of optical properties affected by nanometer-scale structures.

6. Acknowledgement

We wish to thank Heini Rohrer for supporting and encouraging this work. Reto Schlittler provided invaluable experimental assistance. Jürgen Sass, Bruno Reihl, Jim Coombs, Dieter Pohl, and Bo Persson were all actively involved in different aspects of this concept of photon emission on a local scale.

7. References

Adkins, C. J., and Phillips, W. A. (1985) 'Inelastic electron tunneling spectroscopy', J. Phys. C: Solid State Phys. 18, 1313-1346.

Batson, P. E. (1982) 'Surface plasmon coupling in clusters of small spheres', Phys. Rev. Lett. 49, 936-940.

Binnig, G., and Rohrer, H. (1986) 'Scanning tunneling microscopy', IBM J. Res. Develop. 30, 355-369.

Bloemer, M. J., Mantovani, J. G., Goudonnet, J. P., James, D. R., Warmack, R. J., and Ferrell, T. L. (1987) 'Observation of driven surface-plasmon modes in metal particulates above tunnel junctions', Phys. Rev. B 35, 5947-5953.

Coombs, J. H., Gimzewski, J. K., Reihl, B., Sass, J. K., and Schlittler, R. R. (1988) 'Photon emission experiments with the scanning tunneling microscope', J. Microscopy 152, 325-336.

Fischer, U. Ch., and Pohl, D. W. (1989) 'Observation of single-particle plasmons by near-field optical microscopy', Phys. Rev. Lett. 62, 458-461.

Gimzewski, J. K., Humbert, A., Bednorz, J. G., Reihl, B. (1985) 'Silver films condensed at 300 and 90 K: scanning tunneling microscopy of their surface topography', Phys. Rev. Lett. 55, 951-954.

Gimzewski, J. K., Sass, J. K., Schlittler, R. R., and Schott, J. (1989) 'Enhanced photon emission in scanning tunnelling microscopy', Europhys. Lett. 8, 435-440.

Gimzewski, J. K., Stoll, E., and Schlittler, R. R. (1987) 'Scanning tunneling microscopy of individual molecules of copper phthalocyanine adsorbed on polycrystalline silver surfaces', Surface Science 181, 267-277.

Kirtley, J. R., Theis, T. N., Tsang, J. C., and DiMaria, D. J. (1983) 'Hot-electron picture of light emission from tunnel junctions', Phys. Rev. B 27, 4601-4611.

Lambe, John and McCarthy, S. L. (1976) 'Light emission from inelastic electron tunneling', Phys. Rev. Lett. 37, 923-925.

McCarthy, S. L., and Lambe, John (1977) 'Enhancement of light emission from metal-insulator-metal tunnel junctions', Appl. Phys. Lett. 30, 427-429.

Meier, M., Wokaun, A., and Liao, P. F. (1985) 'Enhanced fields on rough surfaces: dipolar interactions among particles of sizes exceeding the Rayleigh limit', J. Opt. Soc. Am. B 2, 931-949.

Mills, D. L., Weber, M., and Laks, Bernardo (1982) 'Light emission from tunnel junctions' in P. K. Hansma (ed.), Tunneling Spectroscopy, Plenum Press, New York, pp. 121-152.

Persson, B. N. J., and Baratoff, A. (1988) 'Self-consistent dynamic image potential in tunneling', Phys. Rev. B 38, 9616-9627.

Persson, B. N. J., and Baratoff, A. (1989) 'Theory of photon emission in electron tunneling to metallic particles' (in preparation).

Pohl, D. W. (1989) 'Scanning near-field optical microscopy (SNOM)' in Advances in Optical and Electron Microscopy Academic Press Ltd., London (to be published).

Raether, H. (1988) 'Surface Plasmons on Smooth and Rough Surfaces and on Gratings', Springer Tracts in Modern Physics 111, Springer-Verlag, Berlin.

Rendell, R. W., and Scalapino, D. J. (1981) 'Surface plasmons confined by microstructures on tunnel junctions', Phys. Rev. B 24, 3276-3294.

Rojas, R., and Claro, F. (1986) 'Electromagnetic response of an array of particles: Normal-mode theory', Phys. Rev. B 34, 3730-3736.

Venables, J. A., Derrien, J., and Janssen, A. P. (1980) 'Direct observation of the nucleation and grown modes of Ag/Si(111)', Surface Science 95, 411-430.

Zaremba, E., and Persson, B. N. J. (1987) 'Dynamic polarizability of small metal particles', Phys. Rev. B 35, 596-606.

SURFACE MODIFICATION WITH THE STM AND THE AFM

C. F. QUATE
Department of Applied Physics
Stanford University
Stanford, California 94305
USA

ABSTRACT. The Tunneling and Force Microscopes are superb instruments for imaging atomic and molecular structures. From the beginning it has been clear that they can be used to modify surfaces in various ways. In this article we review the conventional methods and compare these with potential for surface modification with the STM and AFM. We reach the conclusion that the new instruments can be used for microfabrication of structures on solid substrates with a resolution that is improved by "a factor of ten beyond the present capabilities."

1. Introduction

The Scanning Tunneling Microscope introduced by IBM-Zurich has now evolved into a family of units which fit under a term coined by Bob Melcher - the SXM's. This family includes the Scanning Tunneling Microscope (STM) [1], the Atomic Force Microscope (AFM) [2], the Magnetic Force Microscope (MFM) [3], the Capacitive Microscope [4,40], and the Microscope for Electrochemistry [5]. The range of topics and the range of scale continues to expand.

In this review we will discuss the role for these instruments in the area of surface modification. Surface modification is divided into four categories:

(1) Reconstruction of annealed surfaces of semiconductors amd metals

(2) Modification through surface chemistry of annealed surfaces of semiconductors and metals

(3) Lithography for patterning surfaces on a fine scale, and

(4) Creation of memory sites for storage of digital information.

In the first two entries, the entire surface is modified with a controlled change in the environment. In the latter two, selected points on the surface are modified with the manipulation of the scanning tip.

Surface reconstruction is covered in this issue by Feenstra and Behm. Surface chemistry is exemplified by the work of Avouris and Wolkow [6] and Schardt et al.[7]. Lithography via the STM is discussed by Ehrichs and de Lozanne [8].

We first remind ourselves of what is available with conventional technology. We are confident that the diameter of a cell for bit storage will be shrunk to 0.35 microns by 1993. This translates into a density of 10^9 bits/cm^2. The technology will be based on magnetic systems. For line widths we turn to one of the most advanced activities in the US. Sematech, in Texas, is designed to fabricate the most advanced designs in semiconductor technology [9]. In that organization, the line widths at present are 0.8 microns, in 1992 they project a width of 0.5 microns, and in 1993 it will be reduced to 0.35 microns.

Exploratory work on the fabrication of lines less than 0.1 micron in width has been proceeding for the past decade. Nanolithography is the term used to describe this area. Nanolithography relies on the combination of high resolution electron microscopes and PMMA - an electron resist introduced by Hatzakis [10]. Broers et al. [11] used this combination in 1978 to fabricate lines 250 Å in width. In a following paper [12] he discussed the resolution limits with PMMA resist. Howard and the group at Bell [13], have routinely fabricated lines with widths of 100 Å. Muray with his colleagues at Cornell [14] have exploited AlF$_2$ films to produce feature sizes as small as 40 Å on 80 Å centers. Newman at Stanford [15], in prize winning work, has written a full page of text with minimum feature size of 150 Å. All of this has been summarized in an article by Mackie and Beaumont [16]. They point out that nanolithography "...may open...a new field of study for solid state physics - an entirely unexplored collection of effects for device engineers to work with."

What remains for the SXM's after these pioneering studies with the high voltage beams from commercial electron microscopes? What remains is solid substrates. The best work with the STM's has been carried out with patterns on thin membranes, usually silicon nitride. Howard et al. [17] discuss these effects and compare lithographies on solid silicon substrates with those on membranes. High energy electrons penetrate several microns into solid substrates and create showers of secondaries which 'bloom' into a spherical volume much larger than the diameter of the primary beam. Membranes are used to circumvent the problem of 'blooming'. The STM's with the narrow electron beam from the scanning tips changes this relationship. The penetration depth of low energy electrons is small and 'blooming' is not encountered. The SXM's extend the horizons even further.

What are the possibilities with the new instrumentation ? We are encouraged by the words of Alex Broers when he discusses lithography. In Oxford at STM '88 [18] he wrote "The STM has been used to modify structures with an apparent resolution of a few tenths of a nanometer and therefore offers the possibility of extending the resolution of microfabrication a factor of ten beyond present capability." In an article on miniaturization Keyes [19] discusses the limits on feature size imposed by physical laws. He writes "It is seen that the fundamental limit of about 10^2 atoms will be reached early in the next century if present rates of miniaturization continue." His projection versus time is shown in Fig. 1.

The decrease in size of a site for a single bit is illustrated. In the year 2008 the curve indicates that 1000 atoms will be needed to store one bit. We note that 1000 atoms are contained in an area measuring 100 Å on a side.

Lines written with the STM are now approaching 100 Å in width [8]. In a later section we report that a single bit, 40 Å across, can be stored in an area that measures 100 Å on a side. If this size can be sustained in a real device it will mean that the density of bit storage will be increased to 10^{12} bits/cm^2 as compared to the density projected with conventional technology of 10^9 bits/cm^2.

The concepts are fascinating. The potential gain is large. It warrants a sustained and intensive effort.

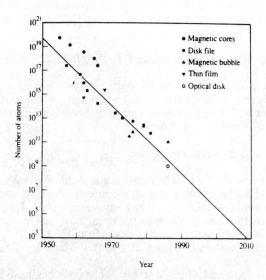

Figure 1. The number of atoms required to store one bit of digital information in discrete magnetic entities. (Courtesy of R. Keyes, IBM)

2. Surface Modification with the STM - Initial Trials

Surface modification with the STM was evident in the talks presented at the conference in Lech, Austria [20]. The reconstruction of silicon surfaces and the small indents in a gold substrate were described. This early work indicated that the path into lithography followed by conventional microscopes was open for the STM. Abraham et al. [21] have shown that it is a simple matter to indent the surface of gold and immediately image the indent by scanning the tip over the area. This work was a precursor to later work directed toward the creation of memory sites at predetermined points on smooth surfaces.

Ringger et al. [22] illustrate the fabrication of patterns 10 nanometers in width with a moving tip. They marked the 'contamination resist', a hydrocarbon layer, that forms in a natural way on surfaces exposed to the oil from diffusion pumps. The surface was a glassy substrate, $Pd_{81}Si_{19}$. These authors believe that the electrons in the tunneling polymerize the hydrocarbon and produce lines of carbon that show up in their image.

This was followed with an experiment by McCord and Pease [23]. They used the STM in a field emission mode where the tip was withdrawn to the point where the electrons were extracted from the tip via field emission into the vacuum. In the initial work the electrons from the tip were used to expose the hydrocarbon covering the substrate. Their later work, where they exposed metal halide films to form various patterns, is treated in the chapter in this volume by Ehrichs.

A fourth experiment was carried out by Becker et al. [24]. They reported on the atomic-scale modification of a germanium crystal. They did this by holding their tip stationary over the selected site and increasing the tip voltage to 4 volts with 20 pA of current. When they imaged the selected area with the tip voltage reduced to 1 volt they observed a surface protrusion with a height of 1 Å and a diameter of 8 Å. The authors believe that the tip in the normal course of imaging the surface picks up an atom of Ge. In the process of modifying the surface the Ge atom is transferred from the tip to the substrate.

3. Conventional Technologies related to Surface Modification

In this section we will address a variety of topics relevant to surface modification. We are concerned with the creation of memory sites for storage and the writing of fine lines. The new scanning probes will, in our opinion, permit us to advance beyond the limits projected with the conventional systems treated here.

A. DEPOSITION OF ATOMS WITH ION BEAMS

In his work with liquids, Taylor [25] has shown that a column in the form of a cone can be raised from a liquid surface with a strong electric field. Bell and Swanson [26] have carried this over to liquid metal cones. They have demonstrated that a strong E-field will form a cone with a sharp apex. At the apex the strength of the electric field can be large - 1 V/Angstrom. This is sufficient to extract atoms from the tip through a process known as field evaporation. The system is known as a Liquid Metal Ion Source (LMIS). It is an extraordinary source of ions.

Why does the cone form and why is it stable ? The cone begins to form when the E-field attracts the liquid surface toward the opposing electrode. The liquid cone is shaped by the countering stresses of surface tension and electrostatic pressure. The surface tension - tending to flatten the cone - is normal to the cone face and proportional to $1/r$, where r is the distance along the cone from the apex.

The countering stress due to the E-field tends to sharpen the cone. This stress is proportional to $E^2=V^2/r^s$. The parameter, s, is a function of the cone angle. For a cone angle of 49.3° it is equal to unity and the two components of stress are balanced over the entire surface of the cone. Provided, of course, that the magnitude of the voltage is adjusted properly. This relation was worked out by G. I. Taylor and these special liquid cones carry his name [25].

The cone once formed is remarkably stable. The movement of the apex over long periods of time is negligible. This stability makes it possible to use the focused ion beams in several applications, such as mask repair, maskless implantation of ions and micromachining. In the commercial systems the ion beam is 500 Å in diameter.

The liquid metal cone as a source of ions is an elegant component for the STM. It is natural to consider writing fine lines with an LMIS mounted as a tip. This is what Ben Assayag et al. have done [27]. They found that it was straightforward to write lines 100 nm in width and 10 microns in length. They moved the liquid metal tip close to the silicon substrate, formed the Taylor cone and deposited the lines of Gallium on the silicon. They used the field emission mode with several nanoamps of current and 50 volts on the tip. The cone was very stable. This stability is a good omen for the future as we try to write finer and finer lines.

B. ABLATIVE SYSTEMS

An archival, read-only, system with important characteristics is based on the ablation of thin films to form physical holes for the memory cell. Jipson and Jones [28] discuss the thermal properties of metal films and compare them with organic films of PMMA and hydroxy squarylium (HOSq). These authors point out that the organic films, PMMA and

HOSq, have low melting points and small values of thermal conductivity and diffusivity. These films should be useful for the formation of holes through ablation. They demonstrate that the energy required to burn one micron dots in HOSq films, 30 nm thick, is similar to that required for burning holes in Te films described below.

In a paper by Lou et al. [29] they discuss the micromachining of physical holes in thin films of Te based alloys deposited on PMMA. The melting point of Te is one of the lowest for metals. Jipson and Jones [28] point out that the thermal conductivity and diffusivity of Tellurium is nearly two orders of magnitude below that of aluminum. The alloyed films are heated with a focused laser beam. It is believed that the laser beam produces localized melting. The radial conductivity is negligible in films with a thickness less than 1000 Å and for heating times less than 1 microsecond. The ablated holes formed in this manner are less than 1 micron in diameter. They are stable for periods in excess of two years.

We can speculate on the sensitivity of these films to heating with current from tunneling tips. We extrapolate from laser beam heating, admittedly a very large step but, nonetheless, an interesting exercise. Films 200 Å thick require an energy density of 0.15 nJ/micron^2. This translates to an energy of 10^{-4} joules for melting a spot 100 Å in diameter. If we use 10 volts on the tip of the STM and 10 nA of current, we can deposit this amount of energy onto the film in 10^{-7} seconds, a reasonable time for writing bits. It suggests that the ablation of thin films is an area worthy of investigation.

C. MATERIALS WITH A PHASE TRANSITION

The read-only system with ablation, soon evolved into a read-write system [30]. Two methods, phase change systems and magneto-optic systems, can be used for the read-write cycle. The phase change system, a derivative of the work on ablating amorphous films exploits the transition between the crystalline and amorphous states. The film initially in the crystalline state is reduced to the amorphous state when a localized region is heated above the melting temperature with a focused laser beam. The localized region is allowed to cool quickly and thermal quenching produces the amorphous state. Erasure is accomplished by annealing the film with a temperature above the glass transition temperature and below the melting temperature. The time required for erasure is longer than that for writing. In a typical case, a laser beam is used to heat a spot 1 micron in diameter with a 10 mW beam for 100 nS. The erase cycle requires 2 mW for 10 microseconds.

This system could be adapted to the STM if the heating cycle was initiated with the tunneling current. The change of state from crystalline to amorphous would produce large changes in the tunneling current. No evidence for this effect has yet been published.

D. MAGNETO-OPTICAL DEVICES

The second system for read-write memory is based on thermal heating of magneto-optical material. We will include a discussion of writing with optical beams in magneto-optical material even though a system for writing in this material has not evolved. Reading the patterns follows from the imaging on magnetic patterns with the MFM.

Magneto-optical devices are discussed in an article by Bloomberg and Connell [31]. Magneto-optics relies on the reversal of magnetization in the region of the domain. The information is recorded thermomagnetically by increasing the temperature of the film to the point where an externally applied magnetic field will reverse the magnetization in the region being heated. In the usual case, a focused laser beam is used to heat the region of the domain. Erasure is done with the same thermal process with the direction of the external magnetic field reversed. The ideal materials are amorphous films of heavy rare earth-transition metal (Fe, Co) alloys.

In Fig. 2 we present an illustration of writing in magneto-optic material as taken from the work of Rugar [38].

Figure 2. Magnetic Domains in $Tb_{19}Fe_{81}$. These domains were written with a laser beam in this magneto-optic material with write power decreasing from left to right. Smallest domain diameter shown on the right is approximately 0.2 μm. (Courtesy of D. Rugar, IBM Almaden)

E. MAGNETIC IMAGING WITH THE MAGNETIC FORCE MICROSCOPE (MFM)

The imaging of domains is certainly feasible with the SXM and this will be the basis of reading when a system for writing is introduced. We will present the work on imaging in this section. Imaging is done with a modified version of the Atomic Force Microscope (AFM). In the AFM the tip from an STM is combined with the cantilever of a stylus profilometer. It is used to map features on insulating substrates in a scanning mode. The force between the sharp tip and the surface of the sample deflects the cantilever in proportion to the force between the tip and the sample. As the sample is scanned beneath the tip, the varying force is sensed by monitoring the deflection of the cantilever. The detecting element, sensitive to sub-angstrom deflections, is based on several technologies [32].

With a magnetic tip this device is used to image magnetic fields. The first trials were with materials where the magnetization was normal to the plane of the film [3]. This was followed with a demonstration that magnetization in the plane of the film could be imaged with equal ease [33]. These patterns were deliberately written but naturally occurring patterns can also be imaged [34]. Recent work by Mamin et al. [35] demonstrates that this is a powerful technique for studying the structure of magnetic domains. An illustration taken from their work is shown in Fig. 3.

Other forms of microscopy such as Lorentz microscopy [36] require special and tedious methods for preparing the sample. This is not the case with the MFM. In a study of samples used in thin film magnetic recording Mamin et al. [37] have demonstrated that they can image magnetic structure independent of topography. They argue that the information collected with the MFM is comparable to that found in the images from the SEM.

Magnetic domains are important. Microdomains can be very small. In thin films of $Tb_{19}Fe_{81}$ stable domains can be as small as 500 Å in diameter [38].

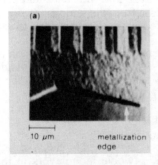 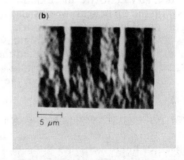

Figure 3. Images from the MFM of magnetization domains in a recording material. (Reproduced with permission.)

F. FERROELECTRIC DOMAINS

Polarization domains in ferroelectric materials are close in analogy with the domains of magnetization in magnetic materials. The technology for storage based on polarization has not kept pace with that in magnetics. Historically their use in storage devices was inhibited by the slow data rates. That has now changed. Thin films of Potassium Nitrate (KNO_3) have switching speeds that are competitive with magnetic devices.

The recent analysis of switching times in the ferroelectric films of KNO_3 by Araujo et al. [39] indicates that the switching times are less than 50 ns. This for a cell deposited on glass, 25x25 microns in area, 75 nm in thickness, with a switching voltage of 6.3 volts.

It is likely that domains of polarization can be created in ferroelectric thin films with the E-field from a scanning tip of the capacitive microscope. The polarized domain can be imaged with a low voltage on the tip. Electrostatic forces, such as those that arise from the reversed polarization, are easily measured by applying a voltage between the tip of a force microscope and the sample. This mode is sometimes known as the Capacitive Microscope. Early work on this system was done by Martin et al. [4] and by Erlandsson et al [41]. It would be a powerful technique for modifying surfaces in a reversible manner.

G. ELECTROCHEMISTRY

The STM adapted for the purposes of Electrochemistry where the tip is immersed in an electrolyte is covered in the lectures by Siegenthaler [42]. We are interested in this mode as a method for surface modification. The reports on this topic are divided between those who exploit ionic currents and those who suppress ionic currents to exploit electronic currents. Operation of the STM in liquids was demonstrated in Santa Barbara by Sonnenfeld and Hansma [43].

In the group led by Bard in Texas, [44] the ionic, or faradic, currents are to modify surfaces. They find that a tip, scanning over a film of ionic conductors can induce the transport of metal ions through the ionic conductor to the surface being modified. With this method they have written lines less than 0.2 microns in width with silver and copper.

A review by Dovek et al. [45] describes the microscope adapted for electrochemistry. Surface modification with electrochemistry started with the work of Lin et al. [46, 5] when they etched the surface of GaAs with ionic currents from a moving tip. The GaAs was immersed in an electrolyte and illuminated with light. The line widths were as small as 0.3 microns with the tip held at -4 volts and spaced 1 micron from the sample.

Schneir and Hansma [47] have used tips insulated with glass coatings to mark gold surfaces covered with fluorocarbon greases.

Itaya and Tomita [48] have studied the electrochemical deposition of Ag on pyrolytic graphite with the graphite surface held at -0.1 volt. After deposition they found that the silver could be easily stripped from the graphite by scanning the area with the graphite surface at +0.25 volts.

H. NEGATIVE RESISTANCE IN MONOMOLECULAR FILMS

A proposal for modification of molecular films has been advanced by Sakai et al. [49]. He shows that multilayers of squalilium dyes "SOAZ", twelve monolayers in thickness, exhibits a switching and memory phenomena.

The device used by Sakai was a planar structure started with a gold film evaporated on glass. The "SOAZ" film was deposited on the gold and covered with aluminum. After 'forming', with the application of 6 volts, the device could be switched from the high impedance state to the low impedance state with voltages as low as 1 volt. The switching times were in the range of 10-50 ns. Sakai suggests that the switching is associated with field-induced intramolecular structural changes. The switching can, in principle, be induced with the intense field from the tip. Once the film is switched frm the high impedance to the low impedance, the state could be read with the scanning probe at a low voltage. This phenomena is important for the scanning probes.

I. SURFACE MODIFICATION WITH MOLECULAR FILMS

Surfaces are easily modified by attaching organic molecules to electrode surfaces. The interest in these modifications has been described in detail by the electrochemists [50]. These surfaces can be manipulated and further modified with the scanning probes. Organized monolayers are laid down with two methods. In the first method, the molecules are placed on a liquid surface and subsequently transferred to a solid substrate with the Langmuir-Blodgett (LB) technique [51]. In the second method, the molecules are organized with molecular self assembly techniques as described by Sabatani et al. [52]. Early on, high energy beams [53] were used to modify the monomolecular films. More recently, the STM has been used to alter molecular LB films through energy transfer from the tunneling electrons [54].

Another approach to the study and manipulation of molecular films was used by Foster et al. [55].They determined that selected molecules show up in the images when they are pinned to graphite.

Molecules of DMP were placed on the graphite substrate. Initially, the molecules do not appear in the image. They are diffusing freely throughout the liquid unless they are pinned with a chemical bond to the substrate. The molecules can be pinned to the substrate with a sharp voltage pulse applied to the tip. The amplitude of the pulse is in excess of 3.5 volts and it lasts for 0.1 microseconds. Presumably energy sufficient to form a chemical bond to the carbon surface is transferred from the tunneling electrons to the molecules. The attached molecules appear in the image as bright objects. This hypothesis is reinforced with the observation that a similar voltage pulse will detach the molecule if the voltage pulse is applied to the molecule. The proposed mechanism for contrast relates to the polarization of the molecules by the tip E-field [56].

The attachment and detachment of molecules in this fashion should allow us to form an ordered pattern over the entire surface of the substrate.

In a second important case, Foster's group studied the surface of graphite while it was covered with a liquid-crystal [57]. In the image they observed long range molecular order at the liquid-solid interface. They repeated the tip pulsing procedure described above and found that the area under the tip increased in brightness. They attribute this to the formation of a chemical bond between the substrate and the molecule under the tip.

This system should also be capable of writing ordered patterns across the entire substrate.

J. METALLIC GLASS

A unique approach for modifying surfaces is employed by the group in Basel with Güntherodt [58]. They note that the amorphous metallic glasses, $Rh_{25}Zr_{75}$, have low values of thermal conductivity and diffusivity similar to those for the films used in the ablative systems discussed earlier. In these special materials, the tunneling current can melt small localized regions on the glass substrate. The group at Basel, found that metallic glass - smoothed to atomic flatness with Ar^+ ion milling - is a suitable material for experiments with writing. They use the current from the STM tip to locally melt a small region on the glass surface. The E-field on the tip pulls the pool of liquid toward the tip to form a Taylor cone. The cone freezes in place when the current is reduced to moderate the heat. The microstructures produced in this fashion, with a height of several nm's, measure 10 nm in diameter. This process involves thermal melting and the writing time is a limiting parameter. As it stands, it is an intriguing idea, but it will require further work to characterize this as a writing tool.

K. SURFACE CHARGING WITH THE SCANNING TIP

The group with Rugar at IBM have modified surfaces by placing charge on the surfaces. They use an AFM with a voltage between the tip and the sample, the Capacitive Microscope [4]. Electrostatic forces are easily measured in this instrument. Stern et al. [59] used this system to deposit charge on insulating surfaces. They pulsed the tip to 100 V for 25 ms. The tip was held 1000 Å from the surface and the charge was deposited over a region 2 microns in diameter. The source of ions is the corona discharge near the tip. Stern found that it was possible to deposit a "reproducible amount of charge in a predefined location." An illustration of surface charging is shown in Fig. 4.

The presence of the charged surface was detected as a change in the electrostatic force on the tip when the microscope was used in the imaging mode with a small voltage on the tip.

They estimate that the sensitivity is sufficient to detect 100 electrons with the tip 1000 Å from the surface. For this reason, surface charging will be interesting for read-write storage if a method for stabilizing the charge is developed.

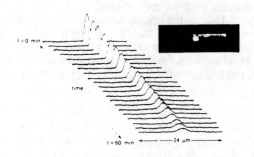

Figure 4. The contours of charge deposited on PMMA as monitored with the Force Microscope. The traces are spaced at 3-min intervals to show the decay of charge. Inset: Gray-scale image (24x5 microns) of the negatively charged region. (Reproduced with permission.)

4. Surface Modification on Gold

Marking of the gold surface has been extensively studied for it is easily indented with the STM. The pioneering effort from the Berkeley group [21] stimulated several workers to continue the investigation. It is a special case that merits elaboration.

Gold is useful in this work because it is inert in air with atomically flat surfaces that are easily prepared. Jaklevic and Elie [60], in their UHV studies of self diffusion on metals, prepared their surface by annealing a single crystal of gold. Hsu [61] found that a gold wire forms into a molten ball when it is melted in the flame of an oxy-acetylene torch. These balls upon cooling form facets on the (111) face which extend, with atomic flatness, over 1000 Å. A more practical system, epitaxial gold films grown on mica, has evolved from the work of Hallmark et al. [62]. Epi-films prepared in this way were studied by Emch et al., both as a media for marking and as a substrate for deposition of molecules [63]. They verified that it was an easy matter to form an array of craters in the surfaces with the STM tip operating in air. As mentioned above, Schneir and Hansma [47] have reported on the depressions formed by touching the STM tip to gold surfaces covered with non-polar liquids.

The interest in gold indents has waned because of the high diffusion rates for gold atoms on the gold surface [57, 60]. The diffusion rates in gold are such that depressions created with tip touching disappear within several hours. Terraces move, and fill in, on a time scale that is easily observed in the sequential images from the STM.

Gold is an informative surface for the study of self diffusion, but it is not a viable surface for lithography.

5. Surface Modification with Molecular Films

The LB technique is used for the deposition of thin layers of amphiphilic molecules on solid substrates. These films are promising candidates for applications in microelectronics

[64]. Thin polymer films have been deposited on graphite and the smoothness of this substrate makes it possible to study the films with the STM and the AFM. The polymer of choice is poly (octadecylacrylate) (PODA) where the interaction between side chains is strong enough to produce ordered structures in bulk crystals [57]. Both the AFM and the STM are used to image the structure of PODA films on graphite. The films can be modified with the STM by applying a voltage pulse to the tip. With a 0.1 microsecond voltage pulse greater than 4 volts, it is possible to cut the polymer fibrils and mark the substrate on a scale of 100 Å [65].

The voltage pulsing strategy for marking molecular films is not applicable to the AFM, but nonetheless, Albrecht [66] has shown that this instrument can be used to mark mica substrates covered with monomolecular films. The films on mica were prepared by Sagiv [67]. Two different films were studied, octadecyltrichlorosilane (OTS) and octadecyl phosphate ($C_{18}P$). Both films are robust and chemically inert. The OTS molecules ionically bonded to the substrate. Both films appear in the AFM images with a hexagonal close packed array where the molecules are spaced 5-8 Å. This agrees with the vertical orientation of the molecules as suggested by Sagiv. The OTS films, which polymerize in the lateral plane, were surprisingly sturdy. It was not possible to mark, or damage, these films with the tip of the AFM.

The results were quite different with the $C_{18}P$ films. They do not polymerize in the lateral planes and it is rather easy to remove the molecules from the substrate with a large force on the cantilever. The marking process is quite repeatable. The boundaries at the edge of the pattern are distinct and sharp.

Lines have written in $C_{18}P$ with a width of 100 Å. These initial experiments indicate that this system might be used for marking a mica substrate in a controlled manner.

6. Surface Modification with the STM - Recent Results

Graphite is a layered compound where the layers are held together with weak Van der Waals forces. The bonds in the layer between the carbon atoms are strong and a single layer is exceedingly robust. It is an inert material that is easily cleaned by cleaving. It will sustain abusive treatment without impairing the atomic smoothness of the surface. Graphite has been studied with the STM in a wide variety of environments. It is more, or less, the prototype material for the scanning microscopes.

A surprising effect has been discovered by Albrecht et al., in the course of their studies of graphite coated with a thin layer of liquid crystal molecules. D.P.E. Smith [68] demonstrated that with a voltage pulse longer than 10 microseconds small holes could be created in a molecular monolayer placed on graphite. When Albrecht et al. [69] set out to reproduce this result, they found that molecules were not necessary. The holes could be created on freshly cleaved graphite without the molecular coating. Furthermore, they were able to show that carbon atoms were removed from the topmost layer of the graphite sample. This removal creates holes with a depth of approximately 4 Å which corresponds to the interlayer spacing of the graphite. They produced holes with a voltage pulse 3-8 volts in amplitude and 10 microseconds in duration. The holes are shown in Fig. 5. The process can be used to reliably produce craters about 40 Å in diameter.

Patterns written automatically with the instrument under computer control are displayed in Fig. 6. It is a demonstration of the present state of the writing technique.

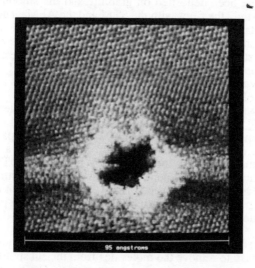

Figure 5. A single hole written on the surface of graphite. The cross-section through the hole is shown in the line trace to the left of the image. The hole is 40 Å in diameter and 7 Å in depth. The graphite lattice is clearly visible around the hole.

Figure 6. Nearly 500 individual holes were written to record this message on graphite. 99.6% of the intended dots were successfully written. The letters are about 300 Å tall.

In our experiments, it was necessary to have water molecules present when the holes were created. If the sample is placed in a vacuum chamber the holes do not appear. When water vapor was introduced into the chamber with a pressure exceeding 20 Torr the holes reappeared. This is consistent with hole creation that was originally discovered with the sample in air. It is known that water vapor condenses into a thin layer on the surface of graphite [70, 49, 52]. From these experiments we now know that water molecules, or at least vapors with hydrogen components, are necessary if the process is to proceed. Penner [71] has found that holes are created with the instrument immersed in water. It does not occur with an atmosphere of oxygen, or nitrogen, surrounding the sample.

One explanation for this has been put forward by Siekhaus [72]. He refers to the work on molecular beams and the degradation of graphite surfaces bombarded with water molecules and ions [73]. The process is called ion enhanced etching. It was determined that water molecules react with the graphite surface to form CH and H_2 molecules. The reaction removes carbon atoms from the graphite surface. They found that the reaction does not occur with other molecules; atomic hydrogen is the exception for it does react with the graphite. Furthermore, the reaction - at least for hydrogen - is enhanced with electrons. Ashby [74] has studied the production of methane (CH_4) when the surface of fresh graphite is bombarded simultaneously with hydrogen atoms and low energy electrons. She found that the production of methane was increased with the electron bombardment, an example of electron enhanced etching.

Siekhaus points out that surface reactions of this type are enhanced with electrons, ions, or photons. With this in mind, he proposes that the electrons tunneling from the tip serve to enhance the reaction between the molecules of water and the carbon atoms in the surface layer. He argues that the carbon atoms, removed from the upper layer, combine chemically with the molecules of water to form H_2 and CH.

If this proves to be correct, it would explain the negative results with the graphite in vacuum and atmospheres of nitrogen and oxygen and it should lead to new studies of electron enhanced chemical etching on a local scale. This could be important, as important as lithography.

References

1. Binnig, G. and Rohrer, H. (1986) 'Scanning tunneling microscopy', IBM J. Research and Development 30, 355-369.
2. Binnig, G., Quate, C. F., and Gerber, Ch. (1986) 'Atomic force microscope', Physical Review Letters 56, 930-933.
3. Martin, Y., Rugar, D., and Wickramasinghe, H. K. (1988) 'High-resolution magnetic imaging of domains in TbFe by force microscopy', Applied Physics Letters 52, 244-246.
4. Martin, Y., Abraham, D. W., and Wickramasinghe, H. K. (1988) 'High-resolution capacitance measurement and potentiometry by force microscopy', Applied Physics Letters 52, 1103-1105.
5. Bard, A. J., Fan, F-R. F., and Kwala, J. (1989) 'Scanning electrochemical microscopy, introduction and principles', Analytical Chemistry 61, 132-138.
6. Avouris, Ph. and Wolkow, R. (1989) 'Atom-resolved surface chemistry studied by scanning tunneling microscopy and spectroscopy', Physical Review B 39, 5091-5100.

7. Schardt, B. C., Yau, S-L., and Rinaldi, F. (1989) 'Atomic resolution imaging of adsorbates on metal surfaces in air: iodine adsorption on Pt(111)', Science 243, 1050-1053.
8. Ehrichs, E. E. and de Lozanne, A. L. This volume.
9. The New York Times, May 24,1989.
10. Hatzakis, M. (1969) 'Electron resists for microcircuit and mask production', J. Electrochemical Society; Electrochemical Technology 116, 1033-1037.
11. Broers, A. N., Harper, J. M. E., and Molzen, W. W. (1978) '250-Å linewidths with PMMA electron resist', Applied Physics Letters 33, 392-394.
12. Broers, A. N. (1981) 'Resolution limits of PMMA resist for exposure with 50 kV electrons', J. Electrochemical Society 128, 166.
13. Howard, R. E., Liao, P. F., Skocpol, W. J., Jackel, L. D., and Craighead, H. G. (1983) 'Microfabrication as a scientific tool', Science 221, 117-121.
14. Muray, A., Isaacson, M., and Adesida, I. (1984) 'AlF$_3$ - a new very high resolution electron beam resist', Applied Physics Letters 45, 589-591.
15. Newman, T. H., Williams, K. E., and Pease, R. F.W. (1987) 'High resolution patterning system with a single bore objective lens', J. Vacuum Science & Technology B 5, 88-91.
16. Mackie, S. and Beaumont, S. P. (1985) 'Materials and processes for nanometer lithography', Solid State Technology 28, 117-122.
17. Howard, R. E., Hu, E. L., and Jackel, L. D. (1981) 'Multilevel resist for lithography below 100 nm', IEEE Transactions on Electron Devices ED-28, 1378-1381.
18. Broers, A. N. (1988) 'Resolution limits for lithography', Proceedings, Third International Conference on Scanning Tunneling Microscopy, Oxford, 4-8 July, 1988, p.S6, Royal Microscopical Society 23, Part 3, Supplement.
19. Keyes, R. W. (1988) 'Miniaturization of electronics and its limits', IBM Journal of Research and Development 32, 24-28.
20. See, IBM J. Research and Development 30, July and September 1986.
21. Abraham, D. W., Mamin, H. J., Ganz, E., and Clarke, J. (1986) 'Surface modification with the scanning tunneling microscope', IBM J. Research and Development 30, 492-499.
22. Ringger, M., Hidber, H. R., Schlögl, R., Oelhafen, P., and Güntherodt, H.-J. (1985) 'Nanometer lithography with the scanning tunneling microscope', Applied Physics Letters 46, 832-834.
23. McCord, M. A. and Pease, R. F. W. (1986) 'Lithography with the Scanning Tunneling Microscope', J. Vacuum Science & Technology B 4, 86-88.
24. Becker, R. S., Golovchenko, J. A., and Swartzentruber, B. S. (1987) 'Atomic-scale surface modifications using a tunnelling microscope', Nature 325, 419-421.
25. Taylor, G. I. (1964) 'Disintegration of water drops in an electric field', Proceedings Royal Society London A280, 383-397.
26. Bell, A. E. and Swanson, L. W. (1986) 'The influence of substrate geometry on the emission properties of a liquid metal ion source', Applied Physics A 41, 335-346.
27. Assayag, G. Ben, Sudraud, P., and Swanson, L. W. (1987) 'Close-spaced ion emission from gold and gallium liquid metal ion source', Surface Science 181, 362-369.
28. Jipson, V. B. and Jones, C. R. (1981) 'Infrared dyes for optical storage', J. Vacuum Science & Technology 18, 105-109.
29. Lou, D. Y., Blom, G. M., and Kenney, G. C. (1981) 'Bit oriented optical storage with thin tellurium films', J. Vacuum Science & Technology 18, 78-86.

30. Connell, G. A. N. (1986) 'Problems and opportunities in erasable optical recording', in L. J. Brillson (ed.), Frontiers in Electronic Materials & Processing, American Institute of Physics Conference Proceedings No. 138, New York, pp. 29-39.
31. Bloomberg, D. S. and Connell, G. A. N. (1985) 'Prospects for magneto-optic recording', IEEE Computer Society Press, pp. 32-38.
32. Martin, Y., Williams C. C., and Wickramasinghe, H. K. (1988) 'Tip-techniques for microcharacterization of materials', Scanning Microscopy 2, 3-8; Martin, Y., Williams, C. C., and Wickramasinghe, H. K. (1986) "Atomic force microscope - force mapping and profiling on a sub 100-Å scale', J. Applied Physics 61, 4723-4729.
33. Abraham, D. W., Williams, C. C., and Wickramasinghe, H. K. (1988) 'Measurement of in-plane magnetization by force microscopy', Applied Physics Letters 53, 1446-1448.
34. Grütter, P., Meyer, E., Heinzelmann, H., Rosenthaler, L., Hidber, H.-R., and Güntherodt, H.-J. (1988) 'Applications of atomic force microscopy to magnetic materials', J. Vacuum Science & Technology A 6, 279-282; Grütter, P., Wadas, A., Meyer, E., Heinzelmann, H., Hidber, H.-R., and Güntherodt, H.-J. (1989) 'Magnetic force microscopy of a CoCr thin film', submitted to J. Applied Physics.
35. Mamin, H. J., Rugar, D., Stern, J. E., Terris, B. D., and Lambert, S. E. (1988) 'Force microscopy of magnetization patterns in longitudinal recording media', Applied Physics Letters 53, 1563-1565.
36. Reimer, L. (1984) 'Transmission electron microscopy' in, Springer Series in Optical Science, vol. 36, Springer, New York; Celotta, R. J. and Pierce, D. T. (1986) 'Polarized electron probes of magnetic surfaces', Science 234, 333-340.
37. Mamin, H. J., Rugar, D., Stern, J. E., Fontana, R. E. Jr., and Kasiraj, P. (1989) 'Magnetic force microscopy of thin permalloy films', submitted to Applied Physics Letters.
38. Rugar, D., Lin, C.-J., and Geiss, R. (1987) 'Submicron domains for high density magneto-optic data storage', IEEE Transactions on Magnetics MAG-23, 2263-2265.
39. Araujo, C., Scott, J. F., Godfrey, R. B., and McMillan, L. (1986) 'Analysis of switching transients in KNO_3 ferroelectric memories', Applied Physics Letters 48, 1439-1440; Dimmler, K., Parris, M., Butler, D., Eaton, S., Pouligny, B., Scott, J. F., and Ishibashi, Y. (1987) 'Switching kinetics in KNO_3 ferroelectric thin-film memories', J. Applied Physics 61, 5467-5470; and Bondurant, D. and Gnadinger, F. (1989) 'Ferroelectrics for nonvolatile RAMs', IEEE Spectrum 26, 30-33.
40. J. R. Matey and J. Blanc, (1985) 'Scanning capacitance microscopy', J. Applied Physics. vol. 57, p. 1437-1444.
41. Erlandsson, R., McClelland, G. M., Mate, C. M., and Chiang, S. (1988) 'Atomic force microscopy using optical interferometry', J. Vacuum Science & Technology A 6, 266-270.
42. Siegenthaler, H. This volume.
43. Sonnenfeld, R. and Hansma, P. K. (1986) 'Atomic resolution microscopy in water', Science 232, 211-213; Drake, B., Sonnenfeld, R., Schneir, J., and Hansma, P. K. (1987) 'Scanning tunneling microscopy of processes at liquid-solid interfaces', Surface Science 181, 92-97. Interestingly, the same group were the first to immerse the AFM in water - see Drake, B., Prater, C. B., Weisenhorn, A. L., Gould, S. A. C., Albrecht, T. R., Quate, C. F., Cannell, D. S., Hansma, H. G.,

and Hansma, P. K. (1989) 'Imaging crystals, polymers, and processes in water with the atomic force microscope', Science 243, 1586-1589.
44. Bard, A. J., Fan, F.-R. F., Kwak, J., and Lev, O. (1989) 'Scanning electrochemical microscopy. Introduction and principles', Analytical Chemistry, 61, 132-138.
45. Dovek, M. M., Heben, M. J., Lewis, N. S., Penner, R. M., and Quate, C. F. (1988) 'Applications of scanning tunneling microscopy to electrochemistry', in M. P. Soriaga (ed.), Electrochemical Surface Science, ACS Symposium Series 378, American Chemical Society, Washington, D.C. Chap 13, pp. 174-201.
46. Lin, C. W., Fan, F-R. F., and Bard, A. J. (1987) 'High resolution photoelectrochemistry etching of n-GaAs with the scanning electrochemistry and tunneling microscope', J. Electrochemical Society 134, 1038-1039.
47. Schneir, J. and Hansma, P. K. (1987) 'Scanning tunneling microscopy and lithography of solid surfaces covered with nonpolar liquids', Langmuir 3, 1025-1027. For a more complete description see, Sonnenfeld, R., Schneir, J., and Hansma, P. K. (1989) 'Scanning tunneling microscopy: a natural for electrochemistry', in J. O'M. Bockris, B. Conway and R. E. White (eds.), invited chapter, to appear in Modern Aspects of Electrochemistry, 21st volume, Plenum Press, New York.
48. Itaya, K. and Tomita, E. (1988) 'Scanning tunneling microscope for electrochemistry - a new concept for the in situ scanning tunneling microscope in electrolyte solutions', Surface Science 201, L507-L512.
49. Sakai, K., Matsuda, H., Kawada, H., Eguchi, K., and Nakagiri, T. (1988) 'Switching and memory phenomena in Langmuir-Blodgett films', Applied Physics Letters 53, 1274-1276.
50. Murray, R. W. (1984), in A. J. Bard (ed.), Electroanalytical Chemistry, vol. 13, Marcel Dekker, New York, pp. 191-368.
51. Kuhn, H. (1983) 'Functionalized monolayer assembly manipulation', Thin Solid Films 99, 1-16; Mann, B. and Kuhn, H. (1970) 'Tunneling through fatty acid salt monolayers', J. Applied Physics 42, 4398-4405.
52. Sabatani, E., Rubinstein, I., Maoz, R., and Sagiv, J. (1987) 'Organized self-assembling monolayers on electrodes', J. Electroanalytical Chemistry 219, 365-371.
53. Zingsheim, H. P. (1977) 'STEM as a tool in the construction of two-dimensional molecular assemblies', in O. Johari (ed.), Scanning Electron Microscopy, vol. 1, IIT Research Institute.
54. Albrecht, T. R., Dovek, M. M., Lang, C. A., Grütter, P., Quate, C. F., Kuan, S. W. J., Frank, C. W., and Pease, R. F. W. (1988) 'Imaging and modification of polymers by scanning tunneling and atomic force microscopy', J. Applied Physics 64, 1178-1184.
55. Foster, J. S., Frommer, J. E., and Arnett, P. C. (1988) 'Molecular manipulation using a tunnelling microscope', Nature 331, 324-326.
56. Spong, J. K., Mizes, H. A., La Comb, L. J. Jr., Dovek, M. M., Frommer, J. E., and Foster, J. S. (1989) 'Contrast mechanism for resolving organic molecules with tunneling microscopy', Nature 338, 137-139.
57. Foster, J. S. and Frommer, J. E. (1988) 'Imaging of liquid crystals using a tunnelling microscope', Nature 333, 542-545.
58. Staufer, U., Wiesendanger, R., Eng, L., Rosenthaler, L., Hidber, H.-R., Güntherodt, H.-J., and Garcia, N. (1988) 'Surface modification in the nanometer range by the scanning tunneling microscope', J. Vacuum Science & Technology A 6, 537-539.

59. Stern, J. E., Terris, B. D., Mamin, H. J., and Rugar, D. (1988) 'Deposition and imaging of localized charge on insulator surfaces using a force microscope', Applied Physics Letters 53, 2717-2719.
60. Jaklevic, R. C. and Elie, L. (1988) 'Scanning-tunneling-microscope observation of surface diffusion on an atomic scale: Au on Au(111)', Physical Review Letters 60, 120-123.
61. Hsu, T. (1983) Ultramicroscopy 11, 167.
62. Hallmark, V. M., Chiang, S., Rabolt, J. F., Swalen, J. D., and Wilson, R. J. (1987) 'Observation of atomic corrugation on Au(111) by Scanning Tunneling Microscopy', Physical Review Letters 59, 2879-2882.
63. Emch, R., Nogami, J., Dovek, M. M., Lang, C. A., and Quate, C. F. (1989) 'Characterization of gold surfaces for use as substrates in scanning tunneling microscopy studies', J. Applied Physics 65, 79-84.
64. Roberts, G. G., Vincett, P. S., and Barlow, W. A. (1981) 'Technological applications of Langmuir-Blodgett films', Physics Technology 12, 69-75.
65. Dovek, M. M., Albrecht, T. R., Kuan, S. W. J., Lang, C. A., Emch, R., Grütter, P., Frank, C. W., Pease, R. F. W., and Quate, C. F. (1988) 'Observation and manipulation of polymers by scanning tunneling and atomic force microscopy', J. Microscopy 152, Pt. 1, 229-236.
66. Albrecht, T. R. (1989) 'Advances in atomic force microscopy and scanning tunneling microscopy', Chap 6, Thesis, Stanford University.
67. Weizmann Institute of Science, Rehovot, Israel.
68. Smith, D. P. E. IBM Physikgruppe, Munich, private communication.
69. Albrecht, T. R., Dovek, M. M., Kirk, M. D., Lang, C. A., Quate, C. F., and Smith, D. P. E. (1989) 'Nanometer-scale hole formation on graphite using a scanning tunneling microscope', submitted to Applied Physics Letters.
70. Pohl, D. W. (1987) Proceedings of the Adriatico Research Conference on Scanning Tunneling Microscopy - Fundamentals and Theoretical Progress, Trieste, Italy, July 28-31, 1987.
71. Penner, R. California Institute of Technology, private communication.
72. Siekhaus, W. J. Lawrence Livermore Research Laboratory, private communication.
73. Olander, D. R., Acharya, T. R., and Ullman, A. Z. (1977) 'Reactions of modulated molecular beams with pyrolytic graphite IV. Water vapor', J. Chemical Physics 67, 3549-3562.
74. Ashby, C. I. H. (1982) 'Temperature dependence of electron bombardment enhanced reactivity of different types of graphite', J. Nuclear Materials 111 & 112, 750-756; Balooch, M. and Olander, D. R. (1975) 'Reactions of modulated molecular beams with pyrolytic graphite III. Hydrogen', J. Chemical Physics 63, 4772-4786.

SCANNING PROBE MICROSCOPY OF LIQUID-SOLID INTERFACES

P.K. Hansma, R. Sonnenfeld,† J. Schneir,‡ O. Marti,* S.A.C. Gould, C.B. Prater, A.L. Weisenhorn, B. Drake, H. Hansma, G. Slough,** W.W. McNairy,** and R.V. Coleman,** Department of Physics, University of California, Santa Barbara, CA 93106

ABSTRACT. Four scanning probe microscopes, the scanning tunneling microscope (STM),[1] the atomic force microscope (AFM),[2] the scanning electrochemical microscope (SEM)[3] and the scanning ion-conductance microscope (SICM)[4] have all been used to image liquid-solid interfaces. Images in this report illustrate the variety of systems that can be studied: from iron corroding in salt water to selenium atoms in liquid nitrogen and from graphite covered with vacuum grease to proteins covered with oil. The technological and biological importance of liquid-solid interfaces has driven and will ensure rapid growth in this field.

1. Introduction

There are three ways to keep a surface clean: 1) keep the sample in ultra-high vacuum, 2) cover it with another solid material or 3) cover it with a clean liquid. Most surface science has been done on the first class of interfaces: surfaces in ultra-high vacuum. Here, many powerful experimental tools can be used such as low energy electron diffraction, Auger electron spectroscopy and many, many more. The second class of interfaces is at the heart of modern semiconductor technology. Critical interfaces are solid-solid interfaces buried far enough away from the surface of the device to be protected from contamination. The third class of interfaces will be treated here. Liquid-solid interfaces are key to understanding diverse phenomenon: from lubrication to corrosion, from adhesion to life. Scanning probe microscopes can give images of unprecedented resolution of these important interfaces.

This report will be divided into four sections by the electrical conductivity of the liquid and of the solid. In order, they will be 1) nonconducting liquid on a conducting solid, 2) conducting liquid on a conducting solid, 3) nonconducting liquid on a nonconducting solid and 4) conducting liquid on a nonconducting solid. These are arranged in order of difficulty, though now that the techniques are known and the appropriate scanning probe microscopes are built all can be studied without real problems.

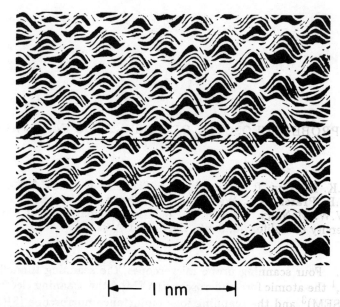

Figure 1. STM image of atoms on the surface of 2H-TaSe$_2$ covered with liquid nitrogen. Note the single atom vacancies which demonstrate that individual atoms are being imaged. Figure from Ref. 5.

1.1. Nonconducting Liquid on a Conducting Solid

The first atomic resolution image of a solid covered with a liquid was of a cleaved surface of 2H-TaSe$_2$ covered with liquid nitrogen.[6] An early image of a related compound, 2H-TaSe$_2$ revealed not only the atoms on the surface but also single atom vacancies (Figure 1). Since that time these and other transition metal chalcogenides covered with liquid nitrogen and helium have been studied in detail in Professor Coleman's lab.[5,7-10] Imaginative workers in Professor Clarke's lab imaged a transition metal chalcogenide under liquid pentane, which permitted them to study the temperature dependence of charge density waves.[11] Even individual molecules of sorbic acid have been imaged on a graphite surface covered with liquid helium.[12]

It is also possible to image solid surfaces covered with nonconducting liquids at room temperature. Even relatively viscous materials such as greases can be used as the liquid phase (Figure 2). Advantages of covering a solid with a nonconducting liquid are that 1) it can replace an unknown and inhomogeneous layer of contaminants and water with a known, homogeneous, clean liquid for fundamental studies, 2) it can protect air sensitive materials like GaAs[13] and 3) it can permit large enough voltages to be applied to the tip for lithography.[13,15-18]

Foster and Frommer showed that it is possible to see an adsorbed liquid crystal array with enough resolution to see functional groups within individual molecules

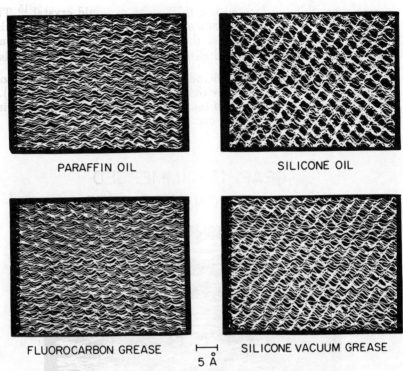

Figure 2. STM images of atoms on a graphite surface that was covered with different oils and greases. Note that the spacing of the atoms is comparable in all of the images suggesting that viscous drag is a negligible factor-even for grease. Figure from Ref. 13.

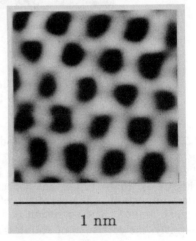

Figure 3. AFM image of highly oriented pyrolytic graphite that is covered with paraffin oil. This image is an orthogonalized version of the original image in Ref. 14.

by covering a graphite surface with a thick drop of liquid crystal.[15] Thus they imaged the array at the interface between the solid and the rest of the liquid crystal drop. Together with Arnett, they also showed that it is possible to write and erase molecular scale features on a graphite surface covered with organic fluids.[19]

The final image in this section demonstrates that the AFM[2] can also be used to image conducting solids covered with nonconducting liquids (Figure 3). In fact, the AFM can now image every one of the four classes. It is the only scanning-probe microscope currently available that can.

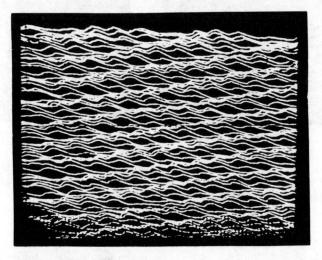

Figure 4. STM image of a sample of graphite covered with water. This image demonstrated the potential of the STM for giving electrochemists atomic resolution images of electrodes without removing them from an electrochemical cell. The scale bar is 3Å. Figure from Ref. **20**.

1.2. Conducting Liquid on a Conducting Solid

The first atomic resolution image in this class was of graphite covered with water (Figure 4). Many important processes occur at this class of interfaces. One of the most important is the corrosion of metals by aqueous solution such as sea-water. Figure 5 shows the corrosion of an iron film in contact with salt water. More recently, the corrosion of stainless steel has been studied by Bard's group.[21]

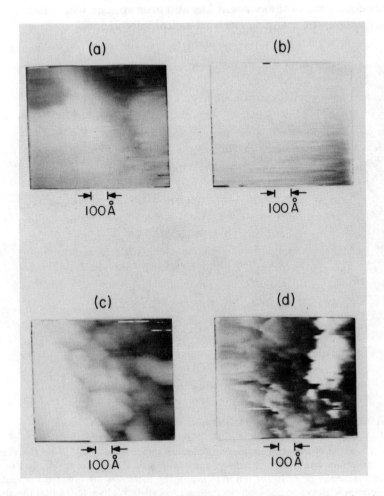

Figure 5. STM images of the corrosion of a thin iron film covered with salt water, 2mg of sodium chloride per 150 mℓ. (a) before immersion, (b) after immersion, (c) 30 seconds after image b, and (d) 150 seconds after image c. Black to white on these images corresponds to a height increase of approximately 10 nm. Figure from Ref. 22.

This class has so many important applications, especially in electrochemistry, that it has been the most studied of the four classes in this article.[23] Ariva was one of the first to discuss, in detail, the opportunities for STM work in electrochemistry.[24] Since that time images of GaAs[23,25], platinum[26-28] and nickel[29] electrodes, covered with electrolytes have been obtained. A breakthrough has been an atomic resolution image of a graphite surface under potential control during imaging.[30]

Scanning probe microscopes can also image processes that occur on electrodes.

As examples, the electrodeposition of silver[31,32] and gold[33] have been imaged. The electrodeposition of these metals has also been studied with another scanning probe microscope, the scanning electrochemical microscope.[34]

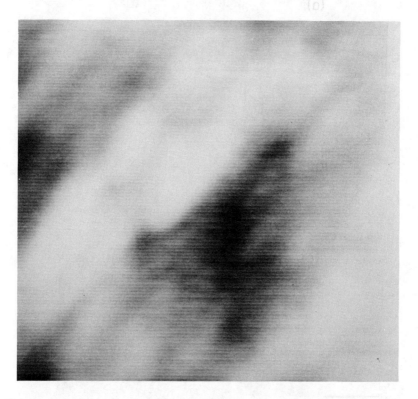

Figure 6. (a) AFM image of fibrinogen molecules under paraffin oil. Note the characteristic molecules composed of 3 balls in a row with a total length of ≈ 50 nm that are oriented diagonally in a. In (b) (on next page), the molecules are oriented horizontally because the scan direction has been electronically rotated. We have found that the ability to electronically rotate scan directions is of great benefit in imaging soft materials such as proteins. The image area is 90 × 90 nm.

1.3 Nonconducting Liquid on a Nonconducting Solid

Figure 6 shows the blood protein molecule, fibrinogen, imaged under paraffin oil. This image is, to our knowledge, the first of this class to be presented anywhere. We will dwell only briefly on this class, because it is less important than the following case and because not much has been done yet. There are, however, applications to lubrication and catalysis waiting to be explored.

Figure 6(b). Caption on previous page.

1.4 Nonconducting Liquid on a Nonconducting Solid

This class includes many important systems in biology, medicine and technology; from mitochondria in cytoplasm to painted ships in seawater. The AFM can obtain images fast enough (a few seconds per image) to observe many biological and chemical processes in real time. For example, Figure 7 shows the corrosion of aluminum in salt water. Though aluminum itself is a conductor, surfaces that have been exposed to air or water become nonconducting at their surface through the formation of aluminum oxide. Thus the study of the corrosion of aluminum requires a microscope that can image nonconducting surfaces. Another important example in this class is mica covered with aqueous solutions (Figure 8).

Though the STM cannot work with a bulk nonconductor, it can image thin nonconducting layers – even if they are covered with a conducting liquid. As

Figure 7. (a) AFM images of an aluminum surface corroding in seawater. The images, 7(a) on this page and 7(b,c) on the next page, were taken approximately $1\frac{1}{2}$ minutes apart. The scan area was 450 × 450 nm. Further microscope development is being focussed on lowering the drift so that processes can be observed over smaller scan areas for times up to an hour.

examples, DNA layers covered with water[39,40] and the outer sheaths of cell walls covered with water[41] have been imaged.

The final images in this paper are taken with a new scanning probe microscope, the scanning ion-conductance microscope (SICM).[8] The probe of the SICM is an electrolyte-filled micropipette. The flow of ions through the opening of the pipette is blocked at short distances between the probe and the surface, thus limiting the ion conductance. A feedback mechanism can be used to maintain a given conductance and in turn determine the distance above the surface. Figure 9 is one of the first images from this microscope. The SICM can not only image topography but also the local ion currents above surfaces. It has been used, for example, for imaging ion currents through the 0.8 micrometer pores in a Nuclepore[42] membrane filter (Figure 10).

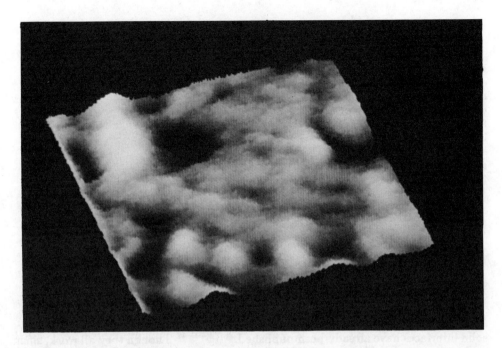

Figure 7(b). Caption on previous page.

Figure 7(c). Caption on previous page.

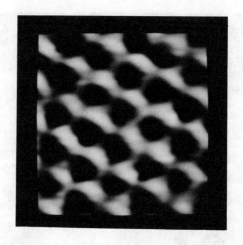

Figure 8. AFM image of mica covered with an aqueous solution. Note the white rings of atoms that surround each black hole. The holes form a nearly hexagonal array that is characteristic of cleaved mica. The image area is 2.6 × 2.6 nm. This image is a top view of some data from Ref. 39.

1.5 Instruments

Many designs for scanning probe microscopes suitable for use in imaging liquid-solid interfaces have already been published.[20,22,40-47] Though they all work, such rapid advances have been made that designs can be obsolete before they ever appear in print. Therefore we will not discuss designs here except to note that we are no longer running any STMs in our laboratory because optical-lever AFMs[48-50] can be easily adapted to image liquid-solid interfaces[39] of all four classes. They offer superior imaging performance, especially with aqueous solutions, since there is no tip insulation to deteriorate with time. The other three scanning probe microscopes remain useful for local electrical measurements.

1.6 Summary

Scanning probe microscopes can image every class of liquid-solid interfaces. Because of the importance of liquid-solid interfaces in technology and biology it is a safe prediction that the use of scanning probe microscopes for exploring these interfaces will continue to grow rapidly.

This is a certainty over the short term; this camera-ready report references only a representative selection of articles that have already (March 1989) been published. The large number of high quality preprints that we have seen guarantees growth, at least in the number of published articles, over the short term. Review articles have already been written[51,52] and should appear soon.

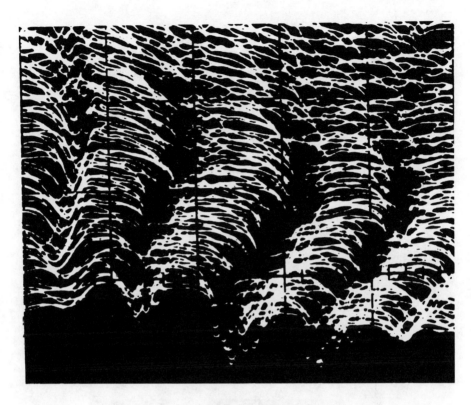

Figure 9. SICM image of a plastic replica diffraction grating covered with 0.5 M NaCl. The measured line spacing of 2.0±0.3 µm is consistent with the grating line spacing of 1.9 µm. The dc voltage between an electrode inside the micropipette and the bath was approximately 0.3 V.

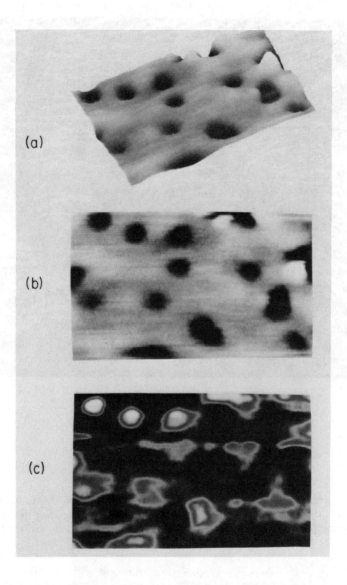

Figure 10. (A) A SICM topographic image of the 0.8-μm diameter pores in a Nuclepore membrane filter.[38] (B) The same image presented in a top view. (C) A SICM image of the ion currents coming out through the pores. The gray scale goes from black at the background level of current, 8 nA, up to white at the maximum level of $\approx$ 40 pA above the background. The imaged area is 7.8 μm by 4.5 μm for all three images. Figure from Ref. 4.

1.7 Acknowledgements

We thank G. Binnig, H. Rohrer, Ch. Gerber, and E. Weibel for starting such an interesting and useful field; and G. Somorjai for pointing out the importance of studying liquid-solid interfaces. Our work is supported by the National Science Foundation-Solid State Physics-Grant DMR8613486 (PKH,SACG,ALW,HH), U.S. Department of Energy Grant No. DE-FG05-84ER45072 (RVC,GS,WWM) and the Office of Naval Research (PKH,CBP,BD).

1.8 References

†Present address: IBM Almaden Research Center, 650 Harry Road, San Jose, CA 95120-6099.

‡ Present address: National Institute for Standards and Technology, Metrology, A107, Gaithersburg, MD 20899.

*Present address: Institut für Quantenelectronik, ETH Hönggerberg, HPT, CH-8093 Zürich, Switzerland.

**Department of Physics, University of Virginia, Charlottesville, VA 22901.

1. G. Binnig, H. Rohrer, Ch. Gerber, E. Weibel, Phys. Rev. Lett. **49**, 57 (1982).

2. G. Binnig, C. F. Quate, Ch. Gerber, Phys. Rev. Lett. **56**, 930 (1986).

3. Scanning electrochemical microscope. A.J. Bard, F-R.F. Fan, J. Kwak, and O. Lev, Anal. Chem. **61**, 132 (1989).

4. P.K. Hansma, B. Drake, O. Marti, S.A.C. Gould, and C.B. Prater, Science **243**, 641 (1989).

5. C. G. Slough, W. W. McNairy, R. V. Coleman, B. Drake, and P. K. Hansma, Phys. Rev. B **34**, 994 (1986).

6. R.V. Coleman, B. Drake, P.K. Hansma and G. Slough, Phys. Rev. Lett. **55**, 394 (1985).

7. B. Giambattista, A. Johnson, R.V. Coleman, B. Drake and P.K. Hansma, Phys. Rev. B **37**, 2741 (1988).

8. R.V. Coleman, W.W. McNairy,. C.G. Slough, P.K. Hansma, B. Drake, Surf. Sci. **181**, 112 (1987).

9. B. Giambattista, W.W. McNairy, C.G. Slough, A. Johnson, L.D. Bell, R.V. Coleman, J. Schneir, R. Sonnenfeld, B. Drake and P.K. Hansma, Proc. Natl. Acad. Sci. USA **84**, 467 (1987).

10. B. Giambattista, A. Johnson, and R.V. Coleman, Phys. Rev. B **37**, 2741 (1988).

11. R.E. Thomson, U. Walter, E. Ganz, J. Clarke, A. Zettl, P. Rauch, F.J. DiSalvo, *Phys. Rev. B* **38**, 10734 (1988).

12. D.P.E. Smith, M.D. Kirk and C.F. Quate, *J. Chem. Phys.* **86**, 6034 (1987).

13. R. Sonnenfeld, J. Schneir, B. Drake, P. K. Hansma, and D. E. Aspnes, *Appl. Phys. Lett.* **50**, 1742 (1987).

14. O. Marti, B. Drake, P. K. Hansma, *Appl. Phys. Lett.* **51**, 484 (1987).

15. J.S. Foster and J.E. Frommer, *Nature* **333**, 542 (1988).

16. J. Schneir, O. Marti, G. Remmers, D. Gläser, R. Sonnenfeld, B. Drake, P. K. Hansma, and V. Elings, *J. Vac. Sci. Technol. A* **6**, 283 (1988).

17. J. Schneir, P.K. Hansma, V. Elings, J. Gurley, K. Wickramasinghe, R. Sonnenfeld, *Scanning Microscopy Technologies and Applications*, ed. E.C. Teague, *Proc. SPIE* **16** (1988).

18. J. Schneir, R. Sonnenfeld, O. Marti, P. K. Hansma, J. E. Demuth, and R. J. Hamers, *J. Appl. Phys.* **63**, 717 (1988).

19. J. S. Foster, J. E. Frommer, and P. C. Arnett, *Nature* **331**, 324 (1988).

20. R. Sonnenfeld and P. K. Hansma, *Science* **233**, 211 (1986).

21. Fu-Ren, F. Fan and A.J. Bard, *J. Electrochem. Soc.* **136**, 166 (1989).

22. B. Drake, R. Sonnenfeld, J. Schneir, P.K. Hansma, *Surf. Sci.* **181**, 92 (1987).

23. R. Sonnenfeld, J. Schneir, B. Drake, P. K. Hansma, and D. E. Aspnes, *Appl. Phys. Lett.* **50**, 1742 (1987).

24. A.J. Arvia, *Surf. Sci.* **181**, 78 (1987).

25. C.W. Lin, F.-R.F. Fan and A.J. Bard, *J. Electrochem. Soc.* **134**, 1038 (1987).

26. H.Y. Liu, F.-R. Fan, C.W. Lin, and A.J. Bard, *J. Am. Chem. Soc.* **108**, 3838 (1986).

27. Fu-Ren F. Fan and Allen J. Bard, *Analytical Chemistry* **60**, 751 (1988).

28. K. Itaya, K. Higaki, and S. Sugawara, *Chem. Lett.*, 421 (1988).

29. O. Lev, F.R. Fan and A.J. Bard, *J. Electrochem. Soc.* **135**, 783 (1988).

30. A.A. Gewirth and A.J. Bard, *J. Phys. Chem.* **92**, 5563 (1988).

31. R. Sonnenfeld and B. C. Schardt, *Appl. Phys. Lett.* **49**, 1172 (1986).

32. D.H. Craston, C.W. Lin and A.J. Bard, *J. Electrochem. Soc.* **135**, 785 (1988).

33. J. Schneir, V. Elings and P.K. Hansma, *J. Electrochem. Soc.* **135**, 2774 (1988).

34. O.E. Hüsser, D.H. Craston and A.J. Bard, *J. Vac. Sci. Technol.* **B6**, 1873 (1988).

35. S. M. Lindsay and B. Barris, *J. Vac. Sci. Technol. A* **6**, 544 (1988).

36. B. Barris, U. Knipping, S.M. Lindsay, L. Nagahara and X.X. Thundat, *BioPolymers* **27**, 1691 (1988).

37. D.C. Dahn, M.O. Watanabe, B.L. Blackford, M.H. Jericho, *J. Vac. Sci. Technol. A* **6**, 548 (1988).

38. Nuclepore Corporation, Pleasanton, California.

39. B. Drake, C.B. Prater, A.L. Weisenhorn, S.A.C. Gould, T.R. Albrecht, C.F. Quate, D.S. Cannell, H.G. Hansma, P.K. Hansma, *Science*, March 24, (1989).

40. B. Drake, R. Sonnenfeld, J. Schneir, P. K. Hansma, G. Slough, and R. V. Coleman, *Rev. Sci. Instrum.* **57**, 441 (1986).

41. P.K. Hansma, *J. Vac. Sci. Technol.* **A6**, 2089 (1988).

42. D.P.E. Smith and G. Binnig *Rev. Sci. Instrum.* **57**, 2630 (1986).

43. S. Morita, I. Otsuka, T. Okada, H. Yokoyama, T. Iwasaki, and N. Mikoshiba, *Jpn. J. Appl. Phys.* **26**, L1853 (1987).

44. K. Itaya and S. Sugawara, *Chem. Lett.*, 1927 (1987).

45. M.D. Kirk, T.R. Albrecht, C.F. Quate, *Rev. Sci. Instrum.* **59**, 833 (1988).

46. K. Itaya and E. Tomita *Surf. Sci.* **201**, L507 (1988).

47. S. Morita, I. Otsuka, and N. Mikoshiba, *Phys. Scripta* **38**, 277 (1988).

48. N.M. Amer and G. Meyer, *Bull. Am. Phys. Soc.* **33**, 319 (1988).

49. G. Meyer and N.M. Amer, *Appl. Phys. Lett.* **53**, 1045 (1988).

50. S. Alexander, L. Hellemans, O. Marti, J. Schneir, V. Elings P.K. Hansma, M. Longmire and J. Gurley, *J. Appl. Phys.* **65**, 164 (1989).

51. P.K. Hansma, J. Schneir and R. Sonnenfeld, *Modern Aspects of Electrochemistry*, 21st Vol., ed. R.E. White, (Plenum Publ., NY), to appear.

52. Moris M. Dovek, Michael J. Heben, Nathan S. Lewis, Reginald M. Penner, and Calvin F. Quate, to appear in an ACS symposium series volume *Molecular Phenomena at Electrode Surfaces*, ed. M. Soriaga.

In-situ Scanning Tunneling Microscopy in Electrochemistry

H. Siegenthaler
Institut für Anorganische, Analytische und Physikalische Chemie, Universität Bern
CH-3012 Bern (Switzerland)

R. Christoph
IBM Research Division, Zurich Research Laboratory
CH-8803 Rüschlikon (Switzerland)

ABSTRACT: We discuss principal aspects and experimental concepts of STM at potential-controlled electrodes in electrolytic environment and illustrate them with typical electrochemical applications.

1. Introduction

Electrochemical systems and reactions are encountered in a wide field of scientific and technological topics, such as electrochemical synthesis and trace analysis, galvanic metal deposition and refining, and electrochemical energy production and storage. In addition, electrochemical processes play a fundamental role in metal corrosion as well as in its prevention by appropriate protection techniques.

In most electrochemical systems of practical interest, the electrode substrates in contact with the electrolyte exhibit marked deviations from structural and chemical homogeneity, such as
- step and screw dislocations and monolayers and submonolayers of foreign adsorbates,
- complex surface morphology and polycrystallinity,
- chemical phase heterogeneity and passive layer formation.

These heterogeneities can influence the local mechanism of electrochemical reactions in a wide resolution range between sub-nm and µm-dimensions, e.g. by partial blockade of the charge transfer and local inhibition or enhancement of surface diffusion, and the establishment of non-homogeneous transport regions of electrochemically active species in the electrolyte. Although previously developed idealized concepts of the interphase electrode/electrolyte have been in many cases a valuable basis for electrochemical thermodynamics and kinetics [1], a better knowledge of the structural properties of electrode surfaces and their interaction with electrochemical mechanisms is required for improving the understanding and control of electrochemical processes at real electrodes. In the last few years, electrochemical research has thus been focussed on the implementation of inhomogeneous charge transfer and transport models into kinetic measurement techniques (in particular electrochemical impedance analysis [2]), and on the combination of the electrochemical system response with structural electrode properties investigated by non-electrochemical methods [3]. STM-imaging of electrode surfaces has first been applied ex-situ in air by *Garcia, Baro, Arvia et al.* [4] for investigating - partially in combination with SEM - the influence of electrochemical surface pretreatment techniques upon the surface morphology of Pt and Au electrodes.

An important aspect of combined electrochemical/structure-sensitive studies is the *conservation of the electrolytic interfacial environment* during the structural investigation. This requires, ideally, the application of structure-sensitive in-situ techniques, which can be directly applied in electrolyte solutions. Besides in-situ methods such as IR-spectroscopy, X-ray diffraction and EXAFS [5], and recently developed non-linear optical techniques [6], STM is considered a highly valuable tool for measurements in an electrolytic environment, since it combines *real-time imaging* in a wide resolution range with the possibility of *local monitoring* of substrate structure and morphology.

In this presentation, we discuss some principal aspects and experimental concepts of *in-situ STM in electrolytic systems* and illustrate them with typical electrochemical applications. Beyond the mentioned scope of electrochemistry at real electrodes, electrochemical aspects of in-situ STM apply also to STM-imaging of biological compounds in water and electrolyte solutions. In addition, in-situ STM may prove an interesting new method for modelling electrolytic tunneling concepts associated with electrochemical charge transfer, as suggested by *Sass and Gimzewski* [7].

2. STM at Electrode Surfaces in Electrolytic Environment

2.1. PRINCIPAL ASPECTS

Fig.1 shows schematically the special situation encountered in electrolytic STM: In most electrolytic systems, differently to STM in UHV or in an inert gas atmosphere, direct tunneling between substrate and tip (pathway A in Fig.1) will be superposed by electrochemical processes, i.e. heterogeneous charge transfer reactions with electrolytic species at both the electrolyte/substrate and the (laterally exposed) electrolyte/tip interface. Without appropriate control, these reactions can affect the results and reproducibility of STM-imaging, as well as the quantitative interpretation of STM parameters, in various ways:

a) At the *tip*, the superposition of time-dependent electrochemical current components will generally deteriorate the performance of the STM control system by fluctuations and enhanced noise in the total tip current. In addition, the electrochemical reactions at the tip

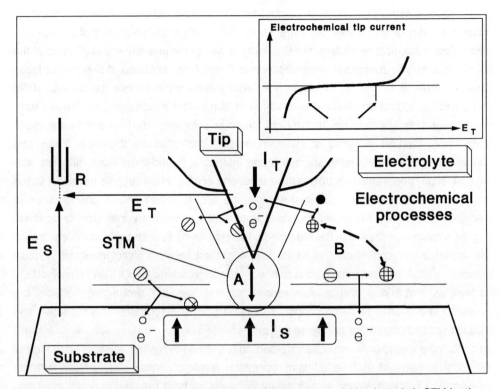

Fig.1: Schematic presentation of the special situation encountered in electrolytic STM by the superposition of direct tunneling (pathway A) and electrochemical charge transfer reactions at substrate and tip. The broken arrow B indicates the possibility of linkage between electrochemical reactions at substrate and tip via electrolytic transport of dissolved reactants. E_T, E_S = potential of tunneling tip and substrate vs. reference electrode R. I_T, I_S = current at tip and substrate. *Inset:* Schematic example of the correlation between the electrochemical component of the tip current and E_T for a tip/electrolyte combination with ideally polarizable potential range (arrow marks).

may cause alterations of the tip geometry by phase deposition/dissolution, or changes in the interfacial properties by specific monolayer or submonolayer adsorption/desorption of electrolyte components (including trace contaminants).

b) *Uncontrolled electrochemical reactions at the substrate* can disturb the reproducibility of STM-imaging by arbitrary changes in the morphology and/or chemical composition of the substrate surface. Practically relevant examples are metal surfaces in undeaerated aqueous electrolytes (including undeaerated water), or in acid solution, where the combination of oxidative metal dissolution with O_2- and/or H^+-reduction will cause corrosion of base metals (including Ag) also under zero net current. Another example is the well-known electrochemical interaction of carbonaceous substrates (especially glassy carbon) with dissolved oxygen, leading to the formation of strongly-bound oxygen coverages [8].

c) A specially complex feature of electrolytic STM configurations, which is sometimes observed under unfavourable experimental conditions [9] is the possibility of linkage between electrochemical processes at the substrate and at the laterally exposed parts of the tip via *electrolytic transport of electrochemical reactants* between the two interfaces (broken arrow B in Fig.1). In electrolytes with ionic concentrations above ca. 10^{-2}M (achieved by adding an electrochemically inert supporting electrolyte), electrolytic transport beyond distances of ca. 10 nm from the tip or substrate interface can be treated in good approximation in terms of *diffusion models* controlled by the local fluxes and concentrations of the diffusing species at the interfacial boundaries. Such diffusive coupling of locally separated electrochemical reactions across an electrolytic transport space may extend over distances above the µm-range and is a well known phenomenon in electrochemistry [10]. At smaller distances from the tip or substrate interface, or on lowering the ionic concentration in the electrolyte, the increasing field strength in the electrolyte will enhance the contribution of ionic field migration to electrolytic transport. Typical electrochemical reactants involved in transport coupling include H^+, OH^- and dissolved O_2, as well as metal ions and soluble redox couples. Specially pronounced effects are expected in presence of redox couples that undergo fast electrochemical reduction and oxidation at both substrate and tip (e.g. Fe^{2+}/Fe^{3+} at Pt or Au). In this case, already small potential differences between substrate and tip (i.e. small tunneling voltages) can lead to the establishment of marked diffusion-controlled electrochemical steady-state currents between the two electrodes, whose extent depends on both reactant concentration and geometry of the STM-configuration. The prevention of such transport-linked electrochemical interaction between substrate and tip by an appropriate experimental STM-concept is thus an essential prerequisite for electrolytic STM studies, in particular for the STM-imaging of *controlled electrochemical changes at the substrate*.

It is obvious, that the uncontrolled interference of electrochemical processes according to a) - c) should be reduced as far as possible to achieve optimum STM performance. Ideally, this implies
- diminuation of electrochemical charge transfer at the lateral parts of the tip without affecting the direct tunneling between tip and substrate, combined with
- full extrinsic control of the electrochemical processes at the substrate.

These conditions can be approached in a qualitative concept [11-15] presuming that the electrochemical reactions at substrate and tip follow the mechanisms of electrochemical kinetics and thermodynamics without noticeable interference by the direct tunneling transfer in the STM domain A. In this case, they can be controlled by conventional potentiostatic techniques, i.e. by independent adjustment of the potential differences E_S and E_T between substrate and tip, respectively, and a reference electrode R serving as electrolytic potential probe. As shown schematically in the inset of Fig.1, optimum conditions for diminishing the electrochemical components of the tip current are obtained with a tip/electrolyte combination, whose electrochemical behaviour exhibits a range with purely capacitive behaviour (arrow-marked range in the electrochemical current-E_T-curve), and by maintaining E_T within this (ideally polarizable) range during the STM experiment. The electrochemical reactions or reaction sequences at the substrate can be controlled under well-defined conditions by setting E_S. Although this potentiostatic control scheme, implying a strict local delimitation between electrochemical charge transfer and direct tunneling tip/substrate, relies on an idealized picture of a real electrolytic STM configuration, the results presented and cited in the present lecture show that it is a valuable basis for in-situ STM.

Independently on the electrochemical reactions, the high capacitance of electrolytic tunneling gaps, ranging around 20 - 100 μFcm^{-2}, can disturb STM measurements by external noise pick-up and by capacitive current components during voltage and distance modulation experiments. As first shown by *Sonnenfeld and Hansma* [16], these effects can be reduced significantly by using laterally coated tips with highly diminished exposed surface area. In this context, it is also essential that low-noise voltage sources are used for applying the potential differences E_S and E_T.

2.2. EXPERIMENTAL CONCEPTS

Fig. 2 shows different experimental concepts for electrolytic STM. The initial STM-measurements in water and electrolyte solutions by *Sonnenfeld, Schneir, Hansma et al.* [16,17] were carried out using a conventional setup (Fig. 2a) without independent potential control of substrate and tip. Besides imaging HOPG in water and aqueous electrolytes, these authors investigated the morphological properties of GaAs, of Fe during corrosion in NaCl, and of electrochemically deposited Ag and Au on HOPG and Au(111). *Itaya et al.* [18] studied HOPG substrates before and after electrochemical Pt deposition, as well as Pt substrates between various electrochemical oxidation/reduction cycles. In all these studies, the electrochemical deposition or substrate pretreatment sequences were performed separately by means of a conventional potentiostatic 3-electrode circuitry, which was disconnected from the substrate during the actual STM experiment.

Independent potentiostatic control of substrate and tip according to the principal considerations discussed in part 2.1 is achieved in the two experimental concepts of Fig. 2b,c, applied in our own STM investigations [11-13]. In both cases, the potentials E_S and E_T of substrate and tip are adjusted independently vs. a reference electrode by low-impedance low-noise voltage sources V_S and V_T, thus defining the *tunneling voltage* $U_T = E_T - E_S$. While the simpler experimental setup of Fig. 2b [11] is based on a current-carrying reference

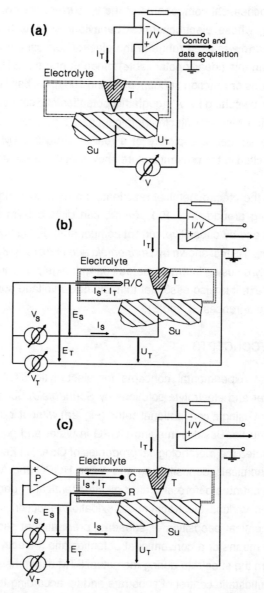

Fig.2: Different experimental concepts for STM in electrolyte solutions: *(a):* Conventional concept [16,17] without independent potential control of substrate Su and tip T vs. the electrolyte. Voltage source V for tunneling voltage U_T. Current/voltage transducer I/V for measuring the tip current I_T. *(b): Potentiostatic 3-electrode concept* [11] with additional current-carrying reference electrode C/R and voltage sources V_S and V_T for independent control of the potentials E_S of the substrate and E_T of the tip vs. the reference electrode. Current I_S at the substrate. *(c): Potentiostatic 4-electrode concept* [12,13] using a potentiostat P with current-carrying counter electrode C and currentless reference electrode R for controlling the potentials E_S and E_T with the voltage sources V_S and V_T. In both configurations (b) and (c), E_T should be kept within an ideally polarizable potential range (see inset of Fig. 1.)

electrode R/C, accommodating the total electrolytic current $I_S + I_T$, the 4-electrode assembly of Fig. 2c [12,13] features a true potentiostatic circuitry with potentiostat P, currentless reference electrode R and current-carrying counter electrode C. In both cases, the potential E_T of the tip vs. the reference electrode is adjusted to an ideally polarizable potential range, according to the schematic presentation in the inset of Fig. 1. A small Ag/AgCl electrode was used as current-carrying or currentless reference electrode in both configurations of Fig. 2a,b. The electrode is confined in contact with a 0.1M NaCl solution (pH 2.5) in a special electrolyte compartment which is separated from the main electrolyte of the STM cell by a ZrO_2 frit. This arrangement allows thus the use of different (also Cl⁻-free) electrolyte solutions in the STM cell, as long as the diffusion potential gradient across the ZrO_2 frit remains constant. The leakage of Cl⁻ ions from the reference electrode compartment remains at a negligible level over a time scale of > 3h. In the configuration of Fig. 2a, the Ag/AgCl reference electrode is sufficiently unpolarizable to remain at near-equilibrium also under a flow of current $I_S + I_T$ in the order of magnitude of μA, and deviations in the potentiostatic control caused by the uncompensated electrolyte resistance across the ZrO_2 frit (ca. 10^3 Ω) and by faradaic Cl⁻-concentration changes in the contacting NaCl-electrolyte can be neglected. Similar concepts as in Fig. 2c have been designed by *Wiechers, Behm et al.* [14], who used a hydrogen-charged Pd wire in equilibrium with H^+ as reference electrode, and by *Itaya et al.* [19], where the reference electrode consisted of an Ag wire in equilibrium with the Ag^+ ions in the electrolyte solution of the STM cell. *Bard et al.* [15] chose a slightly different approach by controlling the voltage between the substrate and a Pt counter electrode and measuring the potential difference between the substrate and a currentless reference electrode. A similar circuitry has been used later by *Trevor et al.* [20]. Potentiostatic 4-electrode concepts have also been applied by *Otsuka et al.* [21], *Green, Richter et al.* [22], and recently by *Uosaki and Kita* [23].

In order to keep the electrochemical components of I_T as low as possible, and to avoid electrochemical changes at the tip surface, it is highly desirable that E_T remains within an ideally polarizable range during the entire STM experiment. This is of special relevance in the case of *STM-imaging at variable substrate potential*, since any shift of E_S at constant tunneling voltage U_T will cause a concurrent shift of E_T. Depending on the electrochemical properties of the tip/electrolyte system, potentiostatic STM-investigations can thus be performed in two different techniques:

a) If the electrochemical current - E_T -curve exhibits a sufficiently wide potential range with ideally polarizable behaviour, a *constant tunneling voltage U_T* may be maintained during STM-imaging at variable E_S, as long as E_T remains well within the ideally polarizable potential region.

b) Practical experience shows, however, that residual faradaic reactions (e.g. reduction of O_2 in insufficiently deaerated solutions) or specific adsorption of electrolytic compounds (including contaminants) at the tip may severly limit the applicable range of E_T. In this case, it is preferable to maintain E_T at a constant optimum value, thus allowing for a shift of the tunneling voltage U_T with changing E_S, which, evidently, is to be considered in the evaluation of the STM results. Most of our STM-investigations, as well as other studies [14,21] have been carried out by this method.

If the electrochemical components of the tip current cannot be reduced to the low pA levels for special experimental reasons, the resulting disturbance of the STM control circuitry can usually be diminished by setting accordingly higher values for I_T, i.e. by applying higher tunneling currents. However, since the electrochemical reactions at the tip proceed in the STM mode, they may still induce chemical or morphological changes at the tip, that can affect the reproducibility of the measurements.

From the principal aspects discussed in part 2.1 it is clear, that reliable and reproducible in-situ STM studies imply due consideration of the electrochemical properties of the chosen tip/electrolyte combination. In aqueous electrolytes, valve metals, e.g. Ta, exhibit a specially wide potential range with vanishing faradaic current density, due to the formation of passivating surface layers. Our initial experiments [11] were carried out with Ta tips etched in HF, using the experimental configuration of Fig. 2b, and show indeed that faradaic current densities at these tips remain between ca. 40nA/cm^2 and -20nA/cm^2 in a potential range between 0 and -500 mV vs. SCE. If the exposed area of these tips is reduced by epoxy-coating to $< 2 \cdot 10^{-3}$ cm^2, these low current densities yield only small residual faradaic currents between ca. 5pA and -3 pA within a considerably wide potential range of 500 mV. However, the STM results suggest that tunneling at these tips occurs through a (possibly semiconducting) nonmetallic surface layer [24], whose electronic and morphological properties require further investigation.

Most of the STM investigations presented in the next chapter have been carried out with Au tunneling tips [10,11]. These tips were prepared from 50μm Au wire by electrochemical etching in HCl and subsequent mounting in a thin ceramic or glass tube. The laterally exposed parts of the tip were then coated with epoxy resin, leaving only ca. 10μm of the tip end exposed. In undeaerated solution, the residual faradaic currents at these tips range well below 10pA in a potential interval between ca. +250 and -100 mV vs. SCE. Although the applicable range for E_T is considerably smaller than at the Ta tips, their STM performance appears more reproducible, since they are free of passivating surface layers within the mentioned potential range. *Wiechers, Behm et al.* [14] have used W tips coated with Apiezon wax, whereas electrochemically etched Pt tips sealed in glass capillaries were applied by *Bard et al.*[15] and by *Itaya et al.* [18,19]. Other STM studies were performed with Pt-Ir tips sealed in glass capillaries [16,17,21,23], or coated with fingernail polish [22] or with a silicon polymer [20]. In a recent paper by *Gewirth et al.* [25], a simple dip-coating technique is presented for preparing epoxy-covered Pt-Ir tips with very small exposed tip areas. *Heben* et al. [26] showed a technique for producing Pt-Ir tips with ultra-small exposed areas, by melt-coating the tip with glass or poly(α-methylstyrene) followed by approaching the tip to an HOPG substrate in an air-operated STM at high bias voltages. Systematic evaluations of different tip materials and preparation techniques, as well as studies of the long-time tip behaviour under electrolytic STM operation, are essential requirements for the further improvement of in-situ STM. Besides comparative in-situ and ex-situ STM studies of the tip performance, other high-resolution imaging techniques (e.g. high-resolution TEM [27]) and electrochemical methods based on recently developed microelectrode concepts [28] are considered valuable additional tools for investigating the chemical and morphological surface properties of the tips. First examples of the electrochemical tip characterization in terms of a microelectrode have been presented recently [25,26].

3. Examples of STM investigations at potential controlled electrodes

3.1. IN-SITU STM AT CHEMICALLY POLISHED Ag(100)-SUBSTRATES

An important aspect of the electrochemical deposition of foreign metal phases at solid (in particular non-alloying) substrates is the preceding formation of metal monolayer coverages in the underpotential range (i.e. at substrate potentials positive of the equilibrium potential of the pertaining bulk metal phase). The formation of such underpotential deposits has been observed with many metal/substrate combinations and has been investigated mostly at monocrystalline substrates of different low-index orientations, using a variety of voltammetric and related electrochemical methods [29]. Although underpotential deposition has been assigned in many systems to the adsorptive formation of more or less ordered surface layers at thermodynamic equilibrium, there is experimental evidence in some cases [30-32] for non-equilibrium phenomena that are explained qualitatively by slow structural transformations within the adsorbate/substrate system, depending on the structural and chemical properties of the substrate/electrolyte interface. For example, while Pb and Tl underpotential adsorbates show marked time-dependent changes of the charge and coverage isotherms at chemically polished Ag(111), the corresponding isotherms exhibit equilibrium behaviour at chemically polished Ag(100) substrates, at least within the experimental limits of the voltammetric investigation method [28]. In addition, studies with electrolytically grown Ag(111) substrates with variable step densities suggest that the rate of the observed nonequilibrium phenomena increases with increasing step density of the substrate [31]. It is thus evident, that a better insight in the surface morphology of chemically or electrochemically processed nonideal single-crystal substrates is an important prerequisite for the structure-related investigation of metal adsorption and phase deposition.

As a first example with relatively well-known electrochemical behaviour, we have therefore studied chemically polished Ag(100) substrates with and without Pb underpotential and phase deposits [12,13]. The substrates consisted of commercial melt-grown single crystals oriented to the Ag(100) face. As in the previous electrochemical studies of Pb and Tl adsorption, the surface of the electrode was first polished mechanically in a series of diamond polishing sequences, followed by an established [12] chemical chromate polishing technique and a test of the electrochemical response of Pb adsorption/desorption in a conventional electrochemical cell. All STM measurements were performed in 0.5M $NaClO_4$ solution at pH 2.5 with and without 10^{-3}M Pb^{2+}, using a potentiostatic 4-electrode STM-assembly based on the concept of Fig. 2c, with a detachable electrochemical cell unit and a Ag/AgCl reference electrode. The deaeration of the electrolyte was restricted to previous degassing of the stock solution with N_2 in a separate cell, before filling the STM cell unit. The rate of concentration increase of dissolved oxygen in the STM cell and of the leakage of Cl^- ions from the reference electrode compartment were tested electrochemically in special experiments by recording cyclic voltammograms of Pb adsorption/desorption after different time intervals. These tests indicate that changes in the composition of the electrolyte are sufficiently slow to enable reproducible STM conditions over a time scale of several hours. The substrate transfer between different cells for surface renewal and electrochemical testing, as well as to and from the STM cell, was carried out in a special transfer unit under protection by an electrolyte cover [12]. During an STM-measurement sequence, the tip potential E_T was kept at a constant optimum value $E_T > -25$ mV vs. SCE, where the residual electrochemical current was usually below 2pA, and where Pb-Adsorption at the lateral parts of the tip can be neglected. Since the substrate potential E_S covered a range between ca. -200 and ca. -950 mV vs. SCE, tunneling occurred in all cases from substrate to tip.

Typical examples of the different surface features found on the chemically polished, *lead-free substrates* are shown in the STM images of Fig.3, recorded in absence of Pb^{2+} with $I_T \leq 1$ nA at substrate potentials E_S within the range of electrochemical stability of Ag. Fig.3a displays an area of ca. 70 x 80 nm with relatively flat terraces of 5 - 20 nm width, that are separated by steps of variable height. in the central part, mono- and double-atomic steps of ca. 0.3 and 0.6 nm height are resolved (arrow marks). It is typical for these chemically polished substrates, that the atomically flat terraces are restricted to smaller dimensions than in the case of flame-annealed Au(111) surfaces [14]. At all Ag(100) substrates we noticed that flat areas of the type of Fig.3a are interspersed by regions with dome-shaped features (Fig.3b) extending in height up to > 5 nm. These domains exhibit steep lateral parts where no monoatomic terraces can be resolved. A similar dome-shaped morphology has also been found by ex-situ STM at emersed electrochemically pretreated Pt and Au substrates [4] and in a recent gas-phase and electrolytic STM study of electrochemically grown Ag(100) substrates [33]. As shown below, the dome-shaped parts of the substrate appear to be more inert towards Pb phase deposition than the flatter domains. We assume therefore that the formation of these marked heterogeneities, which seems to be typical for chemically or electrochemically processed substrates, is influenced by

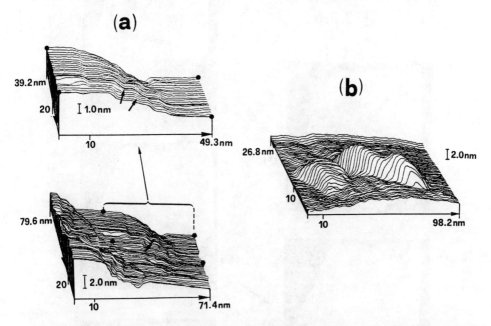

Fig.3: In-situ STM images of chemically polished Ag(100) substrates in 0.5M NaClO$_4$ (pH 2.5). *(a):* Relatively flat substrate area (70 x 80 nm) with small atomically flat domains and mono- and double-atomic steps (arrow marks). The substrate part in the lower figure marked by dots is displayed in enlarged scale in the upper figure. E_S = -195 mV vs. SCE; E_T = 65 mV vs. SCE; I_T = 500 pA; scan rate = 80 nms^{-1}. *(b):* Substrate area (ca. 100 x 27 nm) with dome-shaped domains. E_S = -255 mV vs. SCE; E_T = 45 mV vs. SCE; I_T = 1 nA; scan rate = 100 nms^{-1}. From [13], © Pergamon Press 1988.

adsorbed trace impurities. Additional images recorded in absence of Pb^{2+} at different substrate potentials E_S within a range between -195 and -955 mV vs. SCE indicate [13], that the topographic features are not significantly affected by the considerable change of the tunneling voltage U_T imposed by shifting E_S at constant tip potential E_T. This result is of relevance in view of the Pb adsorption and phase deposition experiments discussed below, which are induced by shifting E_S within the same potential range. The decrease of the tunneling current from the nA range to 50 pA at constant U_T and E_S yields only negligible effects upon the STM imaging, as expected for nonperiodic features and conventional tunneling barriers [34]. In addition, it is found that I_T follows in good approximation an exponential dependence on the tip-sample separation [13], as observed also in electrolytic STM at Au(111) and Ni substrates [14,15]. We therefore conclude that electrochemical contributions to the tip current are negligible.

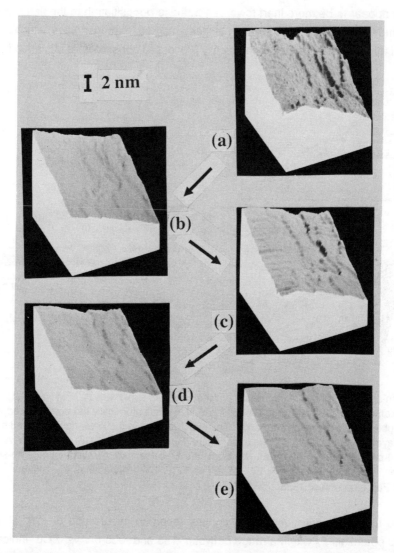

Fig.4: Potentiostatic STM imaging sequence of repetitive Pb underpotential adsorption/desorption at chemically polished Ag(100) in 0.5M $NaClO_4$ + 10^{-3}M Pb^{2+} (pH 2.5) [12,13]. Window size 20x20 nm. The images are processed by simulated shading. The lateral drift of the substrate with regard to the scan window during the imaging sequence is negligible. The images are alternatingly recorded on the adsorbate-free substrate at E_S = -255 mV vs. SCE (a,c,e), and on the same substrate window at E_S = -470 mV vs. SCE with a full adsorbate coverage (b,d). While shifting E_S, the tip is maintained scanning over the entire window. I_T = 500 pA; E_T = -5 mV vs. SCE; scan rate = 40 nms^{-1}; the equilibrium potential of a Pb bulk substrate in the same electrolyte is -480 mv vs. SCE.

The effect of *underpotential Pb adsorption* at Ag(100) is presented in Fig.4 in a sequence of 5 processed STM images of the same substrate window (ca. 20x20 nm), recorded alternatingly at a E_S = -255 mV vs. SCE on a substrate surface without lead coverage (Figs.4a,c,e), and at E_S = -470 mV vs. SCE (Figs.4b,d), where the establishment of a full lead adsorbate coverage of ca. $1.6 \cdot 10^{-9}$ Moles/cm^2 has been found by electrochemical methods [30]. While shifting E_S, the tip was left scanning over the entire STM window, in order to decrease diffusional shielding of the substrate in the surroundings of the tip. During the complete imaging sequence of Fig.4, the lateral drift of the substrate relative to the scan window remained negligible. Fig.4a shows the Pb-free original substrate with distinct stepped features up to 1 nm height on the right-hand half of the window, and with a relatively "rough" appearance of the simulation-shaded surface. After the formation of the adsorbate coverage at E_S = -470 mV vs. SCE (Fig.4b), the substrate appears smoother and shows softer, partially displaced contour lines of the stepped features. After desorbing the Pb adsorbate (Fig.4c), the electrode surface retains a smoother appearance than the initial Pb-free substrate of Fig.4a, with minor, but noticeable changes in the step contours. These effects continue in a subsequent adsorption/desorption sequence (Figs.4d,e), resulting in a flatter Pb-free surface (Fig.4e) than the original substrate. It should be pointed out that no comparable effects are observed in corresponding substrate polarization sequences in absence of Pb^{2+}. We believe therefore that the observed effects are truely due to Pb adsorption/desorption, even though we cannot observe superstructure-like lateral patterns under our present STM imaging conditions. Additional evidence for the formation of a chemically different substrate in the images of Figs.4b,d is found in modulation experiments of the tip/substrate separation, where a negative shift of E_S to the range of the full adsorbate coverage is accompanied by an increase of the inverse decay length [13]. The adsorption/desorption effects displayed with simulated shading in the images of Fig.4 are observed in the original constant-current line scans as corresponding displacement effects of the stepped features, combined with changes in the noise amplitudes. The smoothening of the Pb-free substrates (Figs.4a,c) by repetitive Pb adsorption/desorption might be caused by an irreversible desorption of originally formed anion adsorbates (e.g. residual adsorbates of anionic trace components from the chromate polishing solutions [12] and/or ClO_4^--ions adsorbed preferentially at the step edges [31], as well as an enhancement of Ag surface diffusion by interaction with the adsorbed Pb coverage.

Even though there are no thermodynamically stable binary Pb/Ag alloy phases under the given experimental conditions, very pronounced morphological changes of the nonideal substrate are induced by the *deposition and dissolution of bulk Pb coverages*. Fig.5 shows a typical example of the effect of repetitive Pb phase deposition/dissolution in a series of consecutive images recorded alternatingly on the Pb-free substrate at E_S = - 155 mV vs. SCE (Figs.5a,c,e,g), and in the potential range of Pb phase deposition at E_S = -955 mV vs. SCE (Figs.5b,d,f). Fig.5a shows a 20x20 nm window of the Pb-free substrate,

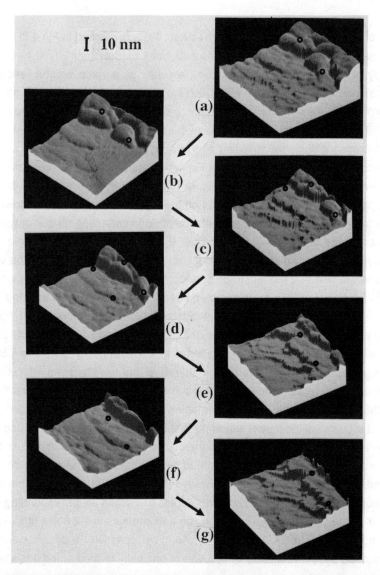

Fig.5: STM imaging of repetitive Pb phase deposition/dissolution on chemically (chromate-) polished Ag(100) in 0.5 M NaClO$_4$ + 10^{-3}M Pb^{2+} (pH 2.5) [12,13]. Window size 20x20 nm. The images are processed by simulated shading. The circles mark reference points of the substrate showing a slight drift relative to the scan window. The STM pictures represent consecutive recordings alternating between the lead-free substrate at E$_S$ = -155 mV vs. SCE (*a,c,e,g*) and the substrate with a Pb deposit of ca. 10-15 monolayers at E$_S$ = -955 mV vs. SCE (*b,d,f*). During Pb deposition and dissolution, the tip is kept scanning. In the course of repeated Pb deposition/dissolution, the formation of pronounced steep steps with heights up to about 8 nm is observed in the flat parts of the underlaying substrate. E$_T$ = -5 mV vs. SCE; I$_T$ = 80 pA; scan rate = 50 nms^{-1}.

which consists mostly of relatively flat terraces with heights up to ca. 3 nm, bordering on a region with accumulated dome-shaped patterns of the type shown in Fig.3b. After depositing a thin Pb phase coverage, estimated to comprise about 10-15 monolayers (Fig.5b), the flat substrate region appears markedly smoother, whereas the upper parts of the domed features remain practically unchanged. After anodic dissolution of the entire Pb coverage (including the underlaying Pb adsorbate), the formation of pronounced steep steps is observed in the flat parts of the underlaying substrate (Fig.5c). These steps increase markedly in the course of further Pb deposition/dissolution cycles (Figs.5d-g), reaching eventually heights of up to 8 nm. The formation of the steps appears to be restricted to certain parts of the substrate, while the domed domains remain practically inert to both Pb deposition and morphological changes. We interpret these nm-scale phenomena as recrystallization of certain substrate domains which may be induced by the formation of metastable alloy phases or by exchange processes within the range of several substrate and deposit layers, occurring preferentially at substrate parts with high dislocation densities (grain or subgrain boundaries) or inhomogeneous adsorption of trace contaminants.

It is interesting to notice that, at the investigated Ag(100) substrates, both the smoothening of the substrate by repetitive Pb adsorption/desorption and the marked step formation by repetitive Pb phase deposition/dissolution are accompanied by only minor changes in the charge and coverage isotherms of Pb adsorption. Additional STM-studies of underpotential adsorption, in particular at substrate/adsorbate systems with pronounced nonequilibrium behaviour of the isotherms [30-32], should give further insight for correlating the electrochemical system response of nonideal electrodes with their STM features in the nm- and sub-nm scale.

3.2. OTHER INVESTIGATIONS AND OUTLOOK

Green et al. [22] studied lead adsorption at vapour-deposited Au(111) layers on mica and observed also the smoothing of terrace surfaces during the build-up of the Pb adsorbate, combined with lateral shifts of terrace boundaries and levelling of pits. A marked enhancement of the surface diffusion of substrate atoms by the adsorption of anions has been found by *Wiechers et al.* [14] on flame-annealed Au(111) substrates in 0.05 M H_2SO_4 + $5 \cdot 10^{-3}$M NaCl, where repetitive adsorption/desorption of Cl⁻ induces pronounced irreversable shifts of terrace ledges. The authors observed also, that the establishment of the adsorbed chloride coverage is accompanied by increasing noise in the STM imaging patterns. The increase of Au surface diffusion in presence of adsorbed chloride at Au(111) can also be deduced from the results of *Trevor et al.* [20], who observed a faster potentiostatic "annealing" of anodically roughened vapour-deposited Au(111) layers in presence of chloride adsorbates.

The studies mentioned above, as well as our investigations of Pb adsorption on Ag(100), suggest that the STM images of adsorption in the atomic resolution range cannot be interpreted on a purely topographic basis, but are strongly determined by the electronic structure of the chemically heterogeneous adsorbate/substrate system and may also be affected by time-averaged dynamic processes of lateral surface diffusion and exchange reactions at the electrolytic interface.

In the *resolution range >1nm*, the topographic interpretation of electrolytic STM features is more straightforward. This range covers presumably the main field of electrochemical applications of practical interest, where potentiostatic STM can provide highly valuable *local* information on the morphological properties of electrode surfaces and their change during electrochemical reactions, which is not directly accessible by classical electrochemical techniques. Of special relevance in this context is the influence of multilayer or bulk deposition/dissolution upon the nm-scale morphology of the substrate, such as the irreversible morphological changes presented in Fig.5, or slow relaxation phenomena of the substrate, as observed at electrochemically polished Au(111) surfaces after deposition and dissolution of several monolayers of Pb [27]. Another application in the nm- to μm-resolution range is the field of *conducting organic polymers* such as polypyrrole or polyaniline [35], which are considered as electrochemical energy storage devices. The electrochemical behaviour of these systems depends not only on their electronic properties (e.g. the extent of oxidative doping), but are strongly affected by their complex inhomogeneous 3D morphology, which is believed to change during emersion from the electrolyte. Test results of the potentiostatic STM imaging of polypyrrole films on HOPG are shown in Fig.6 for a thin electropolymerized film on HOPG, with an average thickness of ca. 1.8 nm [36]. The film morphology is characterized by marked 3D islands, which are interspersed on a relatively flat film area. Other experiments show that the 3D inhomogeneity increases at higher average film thickness. An important aspect is the marked influence of the tunneling current at constant tunneling voltage: Upon increasing I_T from 100 pA (Fig.6a) to 1 nA (Fig.6b), the surface appears more ripped, presumably due to strong tip/substrate interactions, or even to direct contact. We believe that future STM investigations of these systems can give valuable contributions on their electronic and morphological properties.

First potentiostatic in-situ STM studies of *semiconductor substrates* have been reported recently [37,38]. *Thundat et al.* [37] applied the principle of potentiostatic STM to image surface modifications by three important electrolytic reaction categories, i.e. electrochemical phase deposition (Ni on Ge), photoinduced electrochemical deposition (Au on GaAs), and photocorrosion (GaAs). *Itaya and Tomita* [38] demonstrated the role of band bending for electron tunneling from the conduction band of n-TiO_2 to the tip in KCl solution by shifting the substrate potential. Especially the latter example demonstrates new application potentials given by the freely adjustable parameter E_S.

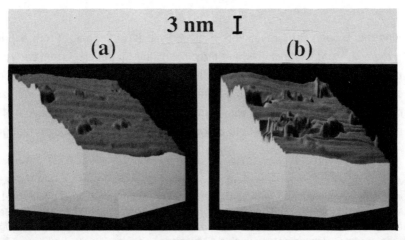

Fig.6: In-situ STM-images of an electropolymerized polypyrrole film on a HOPG substrate in 0.1M $NaNO_3$ (pH 2). Window size 100x100 nm. $E_S = -55$ mV vs. SCE; $E_T = 6$ mV vs. SCE ; scan rate = 200 nms^{-1}. The average film thickness is estimated from coulometric evaluation of the electropolymerization to ca. 1.8 nm. The images display the same substrate window, imaged at different set tunneling current $I_T = 100$ nA (a) and $I_T = 1$ nA (b). The images are processed by simulated shading. From [36].

Beyond the presented application field of electrolytic STM for surface imaging, it is important to gain more insight into the *structure of the electrolytic interface and the mechanism of tunneling through electrolytic junctions*. This requires systematic quantitative measurements of the dependence of I_T on gap width and tunneling voltage at different substrate potentials, and the correlation of these data with sufficiently sound structural models of the electrolytic junction. Such attempts are still in their initial phase, since the experimental precision for these measurements has only just reached an acceptable standard. From a recent measurement series of the distance dependence of I_T [27], performed on Au(111) surfaces with Au and Pt/Ir tips, we still observe considerable variations of the inverse decay length between measurements with different tips of the same material. We assign this behaviour, which is also often observed in the gas phase, to differences in the surface geometry and /or chemical surface composition of the tip, originating presumably from the tip preparation process, as well as from contamination during air contact or during penetration across the air/electrolyte interface. An additional factor to be considered especially at nonideal electrode substrates is the influence of surface corrugation upon the experimentally measured decay lenghts. In spite of the remaining experimental error range, it appears now certain that the inverse decay lengths at Au and Ag substrates are considerably lower in an aqueous electrolytic junction than in a vacuum barrier. This behaviour, which can be explained qualitatively by the contribution of the solvent dipoles to the dipolar moment at the metal/aqueous interface [39] may be an important factor for determining the experimental resolution of electrolytic STM imaging. From experiments including the point-contact regime, where the discontinuity in the tip current can reach two orders of magnitude, we estimate that tunneling under usual experimental conditions occurs over distances up to ca. 1.5nm. Further experiments along this line, including studies in different solvents, combined with continuous improvement of the involved experimental procedures for tip and substrate preparation and handling are important goals for future work.

The STM experiments involving the authors were partially supported by the Swiss National Foundation for Scientific Research and were performed in collaboration with P. Lustenberger, H. Rohrer and H. Wiese (IBM), and with M. Binggeli, D, Carnal and R. Nyffenegger (University of Bern). We also thank F. Niederhauser for the assistance in the preparation of STM-tips, and U. Dürig, Ch. Schönenberger and G. Travaglini for valuable discussions.

References
1. See for example A.J. Bard, L.R. Faulkner, "Electrochemical Methods", John Wiley & Sons, New York (1980).
2. See for example M. Sluyters-Rehbach and J.H. Sluyters, "A.C. Techniques", in Comprehensive Treatise of Electrochemistry (Editors: E. Yeager et al.), Vol.9, pp.177-292, Plenum Press, New York (1984).
3. See for example Ber.Bunsenges.Phys.Chem. **92**, pp.1169-1444 (1988).
4. L. Vazquez, A.M. Baro, N. Garcia, A.J. Arvia et al., J.Am.Chem.Soc. **109**, 1730 (1987); Surf.Sci. **181**, 98 (1987). J. Gomez, A.M. Baro, A.J. Arvia et al., J.Electroanal.Chem. **240**, 77 (1988).
5. B. Bittins-Cattaneo, S. Wilhelm, E. Cattaneo, H.W. Buschmann and W. Vielstich, Ber.Bunsenges.Phys.Chem. **92**, 1210 (1988); L. Blum, H.D. Abruna, J. White, J.G. Gordon, G.I. Borges, M.G. Samant and O.R. Melroy, J.Chem.Phys. 85, 6732 (1986); M.G. Samant, M.F. Toney, G.L.Borges, L. Blum and O.R. Melroy, J.Phys.Chem. **92**, 220 (1988).
6. V.L. Shanon, D.A. Koos and G.I. Richmond, J.Phys.Chem. **91**, 5548 (1987).
7. J.K. Sass and J.K. Gimzewski, J. Electroanal. Chem. **251**, 241 (1988).
8. D. Laser and M. Ariel, J.Electroanal.Chem. **52**, 291 (1974); R.E. Panzer, Electrochim.Acta **20**, 635 (1975).
9. R. Christoph, unpublished results.
10. See for example A.T. Hubbard and F.C. Anson, "The Theory and Practice of Electrochemistry with Thin Layer Cells", in Electroanalytical Chemistry (Editor: A.J. Bard), Vol.4, pp.129-214, Marcel Dekker, New York (1970); F.E. Woodard and C.N. Reilley, "Thin Layer Cell Techniques", in Comprehensive Treatise of Electrochemistry (Editors: E. Yeager et al.), Vol.9, pp.353-392, Plenum Press, New York (1985).
11. P. Lustenberger, H. Rohrer, R. Christoph and H. Siegenthaler, J.Electroanal.Chem. **243**, 225 (1988).
12. R. Christoph, Ph.D. thesis, Bern (1989).
13. R. Christoph, H. Siegenthaler, H. Rohrer and H. Wiese, Electrochim.Acta **34**, 1011 (1989).

14. J. Wiechers, T. Twomey, D.M. Kolb and R.J. Behm, J. Electroanal. Chem. **248**, 451 (1988).
15. O. Lev, Fu-Ren Fan and A.J. Bard, J.Electrochem.Soc. **135**, 783 (1988).
16. R. Sonnenfeld and P.K. Hansma, Science **232**, 211 (1986).
17. R. Sonnenfeld, B.C. Schardt, Appl. Phys. Lett. **49**, 1172 (1986); B. Drake, R. Sonnenfeld, J. Schneir and P.K. Hansma, Surf. Sci. **181**, 92 (1987); R. Sonnenfeld, J. Schneir, B. Drake, P.K. Hansma and D.E. Aspnes, Appl. Phys. Lett. **50**, 1742 (1987); J. Schneir, V. Elings and P.K. Hansma, J. Electrochem.Soc. **135**, 2774 (1988).
18. K. Itaya and S. Sugawara, Chem. Lett. 1927 (1987); K. Itaya, K. Higaki and S. Sugawara, Chem. Lett. 421 (1988); K. Itaya, S. Sugawara and K. Higaki, J.Phys.Chem., in press.
19. K. Itaya and E. Tomita, Surf.Sci. **201**, L507 (1988).
20. D.J. Trevor, C.E. Chidsey and D.N. Loiacono, Phys.Rev.Lett. **62**, 929 (1989).
21. I. Otsuka and T. Iwasaki, J. Microscopy **152**, 289 (1988).
22. M.P. Green, M. Richter, X. Xing, D. Scherson, K.J. Hanson, P.N. Ross, R. Carr and I. Lindau, J.Microscopy **152**, 823 (1988); J.Phys.Chem. **93**, 2181 (1989).
23. K. Uosaki, H. Kita, J. Electroanal.Chem. **259**, 301 (1989).
24. V. Macagno and J.W. Schultze, J. Electroanal.Chem. **180**, 157 (1984).
25. A.A. Gewirth, D.H. Craston and A.J. Bard, J.Electroanal.Chem. **261**, 477 (1989).
26. M.J. Heben, M.M. Dovek, N.S. Lewis, R.M. Penner and C.F. Quate, J. Microscopy **152**, 651 (1988).
27. M. Binggeli, R. Nyffenegger, D. Carnal, R. Christoph, H. Rohrer and H. Siegenthaler, to be published; R. Christoph, H. Siegenthaler and H. Rohrer, to be published.
28. R.M. Wightman and D.O. Wipf, "Voltammetry at Ultramicroelectrodes", in Electroanalytical Chemistry (Editor: A.J. Bard), Vol.15, pp.267-353, Marcel Dekker, New York (1989)
29. D.M. Kolb, "Physical and Electrochemical Properties of Metal Monolayers on Metallic Substrates", in Advances in Electrochemistry and Electrochemical Engineering (Editors: H. Gerischer and C.W. Tobias), Vol.11, pp.125-271, John Wiley & Sons, New York (1978).
30. E. Schmidt and H. Siegenthaler, J. Electroanal.Chem. **150**, 59 (1983), and references cited therein.
31. T. Vitanov, A. Popov, G. Staikov, E. Budevski, W.J. Lorenz and E. Schmidt, Electrochim.Acta **31**, 981 (1986).
32. H.O. Beckmann, H. Gerischer, D.M. Kolb and G. Lehmpfuhl, paper presented at the 12th Faraday Symp.Chem.Soc., Southampton (1977).
33. M.H.J. Hottenhuis, Ph.D. thesis, Nijmegen (1988).
34. N. Lang, Phys.Rev.Lett. **58**, 45 (1987).
35. See for example G.K.Chandler and D.Pletcher, in Electrochemistry, Specialist Periodic Report, Vol.**10**, pp. 117 ff, The Royal Society of Chemistry, London (1985).
36. F. Büchi, Ph.D. thesis, Bern (1989).
37. T. Thundat, L.A. Nagahara and S.M. Lindsay, J.Vac.Sci.Technol., in press.
38. K. Itaya and E. Tomita, Chem.Lett. 285 (1989).
39. W. Schmickler, lecture presented at the IBM Europe Institute on Ultramicroscopy, Garmisch-Partenkirchen, August 1989.

IMAGING AND CONDUCTIVITY OF BIOLOGICAL AND ORGANIC MATERIAL

G. Travaglini,* M. Amrein,** B. Michel* and H. Gross**
*IBM Research Division
Zurich Research Laboratory
CH-8803 Rüschlikon
Switzerland

**Institute of Cell Biology
Swiss Federal Institute of Technology
CH-8093 Zurich
Switzerland

ABSTRACT. I review the state of the art of scanning tunneling microscopy in biology, and present STM images of coated and uncoated biological macromolecules. I further discuss the electron-transfer mechanisms which allow STM imaging of mesoscopic organic objects.

1. Introduction

Scanning tunneling microscopy (STM) in the biological fields includes a wide range of topics. The most crucial questions in this field are whether organic molecules or macromolecules satisfy the STM imaging requirements. If so, what does imaging mean? How must we prepare biological samples and which substrates do we have to use for a reproducible STM investigation in this field? To visualize biological specimens at high resolution in biological solutions, it will be necessary to operate the STM in liquids. How important then are electrochemical processes at the tip-liquid-object interface for STM imaging?

The following six lectures will focus on the principal arguments spanning this field with the aim of answering some of the questions raised above. The lectures are divided into an experimental and a theoretical category. In the first category, I present a chronological overview of the pioneering efforts, problems and significant results which have established this powerful technique in biology. The two experimental contributions presented by R. Emch and A. Cricenti will permit us to consider specific STM applications in biology in more detail. The second category of lectures is mainly devoted to the STM imaging mechanisms of organic matter. The electronic conducting mechanism is in itself an extremely interesting and important question. Local probing by STM may turn out to be a significant technique to understand the electronic structure of organic systems. In the central contribution of the theoretical lectures, C. Joachim will present a quantum-chemical treatment of the enigmatic electron-transfer processes in organic molecules. Electron transfer through long molecular nonresonant bonds is considered as a possible mechanism for the STM imaging of long upright-oriented saturated chains. The imaging mechanism of biopolymers proposed by R. Garcia is that of a conductivity model based on disorder and relaxation effects on the electronic configuration of organic polymers. The lecture by H. Siegenthaler focuses on the liquid-biopolymer interface and possible extrinsic electrochemical kinetic processes occurring in the hydration shell of biopolymers, which may be involved in the imaging mechanism.

2. Imaging of Biological Molecules

Biological molecules are, unlike simple elementary organic molecules, huge macromolecules with a complicated three-dimensional structure. Such molecules have a diameter or thickness of at least 2 nm. Therefore they represent mesoscopic objects, which cannot be detected simply through a variation of the tunnel transmission as in the case of singly-lying, elementary molecules. To allow STM imaging, a not-so-obvious electron transfer carrying the STM current has to occur through or along the surface of the macromolecule. Aside from this exciting physical challenge, the wish to improve the resolution of biological specimens in a close to native conformation at the submolecular level or even to observe their functional activity renders STM a very appealing technique. In this spirit, STM investigations of biological objects began six years ago.

The first STM experiment in biology [1] was performed in vacuum on naked double-stranded DNA adsorbed on a carbon film. Long filamentary structures have been observed as depressions on the support in constant-current mode and interpreted as images of DNA. Note that the images obtained in constant-current mode are referred to here as "STM images" or "topographies." A significant STM result has been obtained in air on a 65 nm thick sample, namely on a naked virus, the bacteriophage Φ29 [2], adsorbed on graphite support. The images revealed the structures of the virus head, the capsid, as well as the virus-collar, as expected from air-dried samples. The first experiments performed in air on uncoated complexes of DNA and recA protein, a long filamentous well-characterized helix, adsorbed on graphite [3] permitted the reproduction of air-dried segments of this filamentary structure with the expected width and pitch of the recA protein spiral around the DNA filament. Other experiments performed in air on a naked purple membrane on graphite and porin membrane on C-film [4] showed that the tip can be scanned over such objects, reproducing the expected thickness at membrane steps. Similarly, Langmuir-Blodgett films of "insulating" long aliphatic chains adsorbed on silicon [5] and graphite [6] have been imaged. Other features in STM images have been obtained from experiments in liquid on DNA adsorbed in situ on gold supports [7]. All these STM experiments of exploratory nature on uncoated biological structures could however not resolve structural details but only showed rough contours of the samples. In spite of the poor resolution, these experiments quite unexpectedly established that STM imaging of thick naked organic material is possible.

First, attention was directed to an appropriate support for STM imaging, which − as in any kind of microscopy − is of crucial importance. The STM requirements for the support are manifold. It has to be mechanically rigid and of sufficient conductivity so that the voltage drop in the support is negligible with respect to the applied voltage. The surface corrugation of the support should be small on the scale of the features to be imaged, and ultimately atomically smooth. Undulations with a large "wavelength" compared to the lateral dimensions of the objects are not disturbing. The interaction of the support with the biological material has to be strong enough to immobilize the sample and to guarantee good electrical contact, without however modifying the conformation of the object. Finally, the support has to allow efficient adsorption of biological material. Standard transmission electron microscopy (TEM) supports such as carbon (C) or platinum-carbon (Pt-C) films evaporated on cleaved mica platelets have been tested. C and Pt-C films have good adsorbing properties and low corrugation. The Pt-C film, for example, has a granular structure with a typical grain size of about 15-20 Å laterally and 6 Å vertically. The tunneling current on Pt-C film is quite stable and not as noisy as on a pure C support. Cleaved highly oriented pyrolithic graphite (HOPG) and gold (111) surfaces have been also used. HOPG can be flat on an atomic scale over thousands of Å [2]; it is an ideal support as far as STM imaging is concerned, although biological sample preparation with a reasonable coverage becomes problematic. In addition, the interpretation of features measured on graphite is very dangerous, since all kinds of artifacts have been observed on graphite itself. Atomically flat gold supports can be achieved by controlled gold evaporation on heated mica platelets [8] or by melting and quenching gold wires in air, which subsequently form balls with have atomically flat (111) facets on the surface [9]. For the adsorption of biological samples, which normally have hydrophilic outer surfaces, the supports must also be rendered hydrophilic, e.g., by exposition to glow discharge. Adsorption on gold by attracting the molecules against the support via an applied electrical field has been also tested [7].

Several tip materials have been used in these experiments: W, Pt-Ir, and Au. These tips can either be obtained by etching a wire of W in KOH, of Pt-Ir in KCN, and of Au in concentrated HCl solution or, except for Au, by mechanical sharpening. Tunneling in air is restricted to voltages between -1.5 and $+1.5$ V. The typical STM current in these applications ranges from 50 pA to 2 nA.

To overcome the resolution difficulties, which have been attributed to the insufficient "conductance" of organic samples, STM experiments on replicated or metal-coated objects have begun. These relatively easy experiments had the advantages of not being subject to possible physical limitations related to the intrinsic nature of STM, and we could use of the conventional sample preparations like for TEM. Since the STM imaging requirements are fulfilled, such STM applications should be able to quickly provide convincing three-dimensional images. In addition, at this initial stage of STM applications on biological macromolecules, an important aspect is the correlation of the three-dimensional STM topography of the coated objects and the image of an uncoated sample. Comparison between the vertical dimensions observed on coated and bare samples prepared in parallel will help interpret the "topographies" obtained on the naked material. Moreover, even if the STM applications on coated objects do not represent a challenge for STM, they will provide a potential method for extending the limits of TEM on coated or replicated objects, and for obtaining useful biological information.

In this context, the first problem to solve is the "metallic" coating or the replica film. The requirements for such a film are three-dimensional stability when exposed to atmospheric conditions and small granularity to allow high-resolution images. The standard shadowing films used in TEM, such as tantalum-tungsten (Ta-W) or Pt-C films, suffer from oxidation processes or metal clustering on air contact. A carbon film must be additionally evaporated onto these films to preserve their high-resolution performance for air experiments [10]. This process does not in any way affect the contrasts in TEM, but does lead to tunnel instability in STM and will smear the interesting structural details since the measured topography will correspond to a twice-coated sample. A new coating film which satisfies the imaging requirements has been developed: Pt-Ir (iridium)-C [11]. A link between STM and TEM has been established for biological material by applying STM on freeze-dried recA-DNA complexes [12] adsorbed on standard TEM supports and coated with an evaporated "10 Å" thick Pt-Ir-C film, see Fig. 1.

Note that freeze-drying is an established TEM approach to prevent the biological samples from collapsing as the result of air-drying processes: this method permits structural studies of dehydrated biological specimens. The three-dimensional topography of the coated recA-DNA complexes reveals the right-handed single helix, the pitch and, on some strands, the position of the recA monomers. The structural information obtained is similar to that from a three-dimensional reconstructed TEM image. Other significant STM results have been obtained on replicated [13] and on coated biomembranes [14]. These results prove that in STM a three-dimensional relief data at macromolecular level can be achieved directly without averaging, whereas in TEM a three-dimensional image must be reconstructed from several averaged projections of the sample. When averaging is difficult and the relief corrugation cannot be reconstructed from projections, as in the case of flat objects with a very smooth surface relief, STM will be the only way to reveal the three-dimensional structure. STM images obtained on freeze-dried and coated shell protein of a polyhead-mutant of bacteriophage T4 confirm this [15]. Its relief is expected to have a maximal vertical corrugation of less than 10 Å, and thus be hardly explorable with the TEM technique. The STM images reveal a honeycomb lattice of proteins with a resolved sixfold symmetric relief having a maximal corrugation of 10 Å. The resolution limit of STM on coated or replicated material is limited by the grain size of the respective coating or replica film. On strongly corrugated samples, the obtained STM topography will also reflect features due to the tip geometry itself [4]. An averaging of images produced by different tips will minimize this artifact [15]. After these successes on coated materials, efforts are increasingly addressing the STM challenge in biology, that is the imaging of uncoated hydrated samples. Approaches using several biological samples have been based on standard TEM preparations. For example, experiments on filamentary structures such as double-stranded DNA or recA-DNA complexes have been

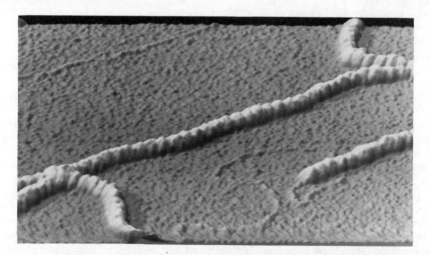

Figure 1. STM image with simulated shading of freeze-dried recA-DNA complexes coated with a Pt-C-Ir film. The area shown is 236 by 192 nm. Thick filaments correspond to recA-DNA complexes, tiny filaments to free DNA. From [12], © AAAS.

performed on HOPG and Pt-C film supports. Following the TEM preparation procedure, the samples were always "cleaned" after adsorption by extensive washing. On HOPG for example it was possible to trace only a couple of scans on "clean" singly-lying filamentary structures; afterwards the filaments were simply dragged along by the moving tip. Only DNA filaments trapped by graphite steps or recA-DNA filaments aligned in parallel yielded consistent images [3,16], see Fig. 2.

Experiments on "clean" DNA and recA-DNA complexes adsorbed on 30 to 1000 Å thick Pt-C films [16] presented the same difficulties encountered on HOPG. Although DNA or recA-DNA complexes were densely adsorbed on Pt-C film, the STM images only reproduced the Pt-C granularity. Nevertheless, the rare image obtained on air-dried recA-DNA filament aggregation on HOPG shows collapsed recA-DNA strands with resolved topographic details. This was an indication that the possibility of obtaining high-resolution images of naked material depends very strongly on several preparation parameters of the sample, which so far represent a stringent limitation to STM imaging. Images of dI/ds, that is the change in the tunneling current I on fast modulation of the tip-sample separation s recorded simultaneously with the topographies, are intrinsically more sensitive to small local variations of the current induced by the sample than I itself: therefore they should achieve a superior lateral resolution, and indirectly reveal local chemical properties as well as the elastic nature of the object. Indeed, the dI/ds images of the above-mentioned recA-DNA strands are more richly structured than the topography, revealing a laterally resolved substructure on recA monomers at the scale of 7 Å [16] as shown in Fig. 3.

Note that the dI/ds signal is only proportional to the square-root of the tunnel barrier Φ when measured on rigid surfaces and when on such surfaces the tunnel barrier Φ varies slowly as a function of the tip-surface separation s. Two-dimensional objects such as membranes should be more stably adsorbed to the support

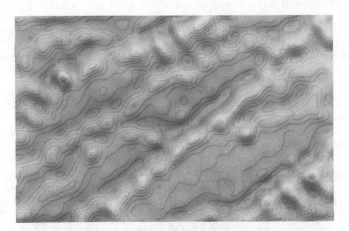

Figure 2. Top view with simulated shading of uncoated air-dried double-stranded DNA molecules adsorbed on graphite and trapped at graphite steps. The filaments exhibit a width of 2 nm and a periodicity between 33-35 Å. The image shows an area of 30.7 by 46.4 nm. From [17].

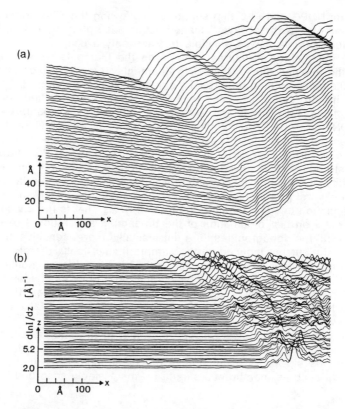

Figure 3. Unprocessed STM image of uncoated air-dried strands of recA-DNA complexes adsorbed on graphite (a), which were aligned in parallel, and corresponding dI/ds traces (b). The scans are intentionally displaced in the z-direction to emphasize the three-dimensional structure. The area corresponds to 72 by 10 nm. From [16], © 1988 Royal Swedish Academy of Sciences

than filamentary structures, and are thus less easily displaced by the moving tip. Indeed, it was possible to image membranes without difficulty even in a multilayer configuration [Fig. 4(a)]. The images obtained for example of air-dried purple membrane show an apparently low corrugated surface. In purple membrane, proteins are arranged in a rhombohedral lattice with a lattice constant of 62 Å. Detecting structural details of the surface of purple membrane is a challenging exercise requiring optimal preparation conditions, since its corrugation is expected to be less than 5 Å. In some regions, the membrane images present rows arranged in parallel and separated by a distance of 62 Å, and typical air-drying cracks occurring along characteristic lattice directions [Fig. 4(b)]. The lack of detailed structural information in the STM images can be attributed to a modification of the surface caused by air-drying or to interference effects in the current due either to the bulk of the membrane itself or to the underlying membranes. On singly-adsorbed phospholipid membrane, the resolution reaches the molecular level, where individual lipids are resolved [18].

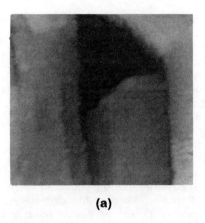 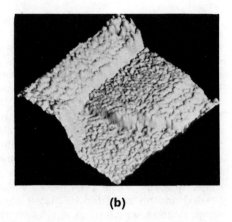

(a)   (b)

Figure 4. (a) Top view of uncoated air-dried multilayer configuration of purple membranes adsorbed on Pt-C support. (b) Three-dimensional representation of uncoated air-dried purple membranes lying across each other. The step height at the membrane edges is about 4.5 nm for a single step. The topmost membrane presents rows arranged in parallel and separated by about one lattice distance.

A very interesting point is that images of purple membranes can be obtained by applying a bias of only 50 mV. The fact that purple membranes are quite flat in large areas and that imaging can occur at very low bias suggested the use of purple membranes as a support for biological objects. Adsorption of organic material on an organic support should provide adsorption conditions more compatible to biological material. In fact, it was possible to image many uncoated single recA-DNA filaments adsorbed on the surface of multilayered purple membranes. The images of recA-DNA complexes show "dirty" filaments, as they were embedded in something I would attempt to define as a "humid salt" or the product of some electrochemical reaction. Even in TEM micrographs, filaments prepared as described seemed slightly embedded in "salt". Washing the preparation to remove the "salt" from recA-DNA complexes destroyed the preparation (nevertheless preparations on organic supports should be attempted again). The fact that preparations on membranes feature a relatively large amount of "salt" and that "dirty" recA-DNA filaments were reproducibly imaged suggested that STM experiments be performed on "dirty" recA-DNA complexes directly adsorbed on standard support to elucidate the importance of salt for imaging. On such supports, it should be possible to dilute the "salt" enough without diminishing the density of the adsorbed objects so that structural details of the sample can appear. Pt-C supports were then treated with magnesium chloride or magnesium acetate. In contrast to experiments on "clean" complexes, recA-DNA complexes adsorbed from solution onto a salt-treated Pt-C support and kept humid during measuring yield high-quality STM images [19]. Periodically, a drop of bidistilled water was put on the sample, and excess water was removed with blotting paper. (In the case of recA-DNA adsorbed on purple membranes, it is possible that the membranes, by virtue of their hydrophilic surface, maintain the necessary humidity for STM imaging in the presence of "salt.")

The correlation between the images of the uncoated hydrated recA-DNA filaments and the STM images obtained on coated filament is so good that it allows the "naked" image to be interpreted as the true topography of the sample. The measured filament height corresponds to the expected value of about 100 Å. Every striation of the hydrated complex directly shows a three to four-part structure, see Fig. 5. This substructure not only reveals the position but also the shape of the recA monomers on the upper side of the filament with no averaging. Unfortunately this primitive preparation did not allow imaging of the filament in the more sensitive dI/ds mode, which would facilitate resolving additional substructures beyond the topography. Similarly, a salt treatment has been used in the experiment on double-stranded DNA adsorbed on graphite [20]. Processed images show structures in a filamentary-like segment which have been interpreted as the major and minor grooves of the DNA helix. Subsequent STM experiments on uncoated DNA [21-24] and on uncoated HPI layer sheet [25] yield clearly interpretable high-resolution images, which once again confirm the applicability as well as the potential of STM in biology.

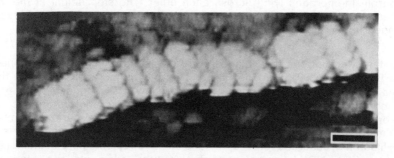

Figure 5. STM top-view image of an uncoated hydrated recA-DNA complex adsorbed on a magnesium-acetate-treated Pt-C film. Every striation, corresponding to one helical turn, is composed of three to four repeating parts. Such substructures not only show the position but also the shape of recA monomers without averaging. Scale bar: 10 nm. From [19], © AAAS.

3. Introduction to Imaging Mechanisms

This paragraph will focus on some of the physical aspects of the STM imaging mechanisms of organic macromolecules, and is an introduction to the theoretical lectures. Consider first the imaging of uncoated hydrated recA-DNA complexes. Which elements of the entire system are probed by the STM tip? The recA-DNA complex consists of a spiral of protein monomers around a double-stranded DNA. Even the position of DNA inside of the spiral is not clearly determined: therefore the physical interaction between the recA monomers and DNA is not yet quantifiable. Electron-transfer processes intrinsic to recA monomers involve intra- as well as interchain interactions.

Polypeptide chains could be regarded as pendant group polymers, with residues acting as pendant group of the polypeptide backbone. The electronic structure of pendant group polymers is quite similar to that of the single molecular subunits constituting the polymer. In terms of a semiconductor picture, the empty or the filled states, separated by an energy gap, are composed of discrete energy levels instead of energy bands. Carriers injected into empty states are therefore localized either by disorder as in the Anderson model or by correlation forming radicals. When pendant groups are charged, strong relaxation will shift the electronic energy states (HOMO-LUMO) to the middle of the gap [26,27]. Electron transfer through these relaxed electronic levels can occur over mesoscopic distances and therefore contribute to STM imaging. A quantitative discussion in such terms is presented in the lecture of R. Garcia.

In addition, proteins are highly structured macromolecules: their specific interchain interaction will also influence the quantum paths for electron-transfer processes. But, as proven experimentally, elements extrinsic to the recA-DNA macromolecules such as water and salt have been revealed as being determinant for imaging. The entire system therefore consists of the two electrodes tip and support, of recA-DNA complexes and of an electrolyte whose electrochemical potential is not defined, since no reference electrode was used in the experiment. Imaging was performed within a bias window of -300 to $+300$ mV. No leakage current has been observed within this window, which is a sign that the Fermi levels of the two electrodes are not resonant with the redox levels of the added ions. It also is practically impossible to explain the achieved lateral resolution of 20 Å in terms of an ionic migration from the tip via the molecule and toward the support. Do ions then act as dopants of the macromolecule or do they diffuse in the macromolecule, thereby relaxing the electronic terms of the molecule? Significant extrinsic imaging mechanisms will surely occur directly in the hydration shell or even in the Helmholtz layer of the macromolecule. Such strong polar regions contain solvated ions or specific adsorbed ions. Do fluctuations of the polarization coordinate promote an electron transfer (tunneling) between solvated ions or even between intrinsic macromolecular sites? If the macromolecule alone allowed sufficient electron transfer for imaging, then the role of the added salt and water would simply be to establish a suitable electrochemical potential reducing the resistance of the molecule–support-interface experienced in the experiments on "clean" complexes below the threshold value for imaging. Such electrochemical processes are discussed in more detail in the lectures of H. Siegenthaler.

Too many channels are involved in the imaging of recA-DNA complexes so that a quantitative discussion of intrinsic electronic properties of the measured molecule necessitates a series of experiments specifically testing the various channels. A possibility to investigate the physics of STM imaging of mesoscopic organic systems is to perform STM experiments on systems whose intrinsic physical properties are quantifiable and whose imaging is based on better defined parameters. Therefore, molecules consisting solely of a linear chain of C-C and C-H sigma bonds have been chosen in subsequent STM investigations [28,29]. These substances, called n-alkanes, are saturated hydrocarbons chains in which each carbon atom is bonded to four other atoms. Alkanes, also known as paraffin, are chemically inert substances and are hydrophobic because of the absence of polar and charged molecular groups. Alkanes with a chain length of up to 17 C are liquids at room temperature, those with longer chains are solids. Despite their high molecular weight, alkanes have low melting temperatures, showing that the structure of

Figure 6. Top view with simulated shading of uncoated paraffin-oil crystallized on Pt-C support. The area corresponds to 200 by 200 nm.

condensed alkanes is dominated by intramolecular forces. STM experiments performed under atmospheric conditions demonstrate that solid layers (Fig. 6) and crystalline-like structures of alkanes of various chain length up to a thickness of 100 nm can be imaged on a variety of supports.

This is surprising in view of the excellent insulating nature of the paraffins at the macroscopic scale. The chemical nature of the entities imaged was demonstrated by the changes of the observed structure near the respective melting temperatures, see Fig. 7.

The electronic structure of these molecules is close to that of polyethylene: the energy states of the alkane electrons are constrained to remain within the limit of the polyethylene bands [30]. The upper polyethylene valence band formed by C(2p) H(1s) orbitals lies about 9 eV below the vacuum level, while the lowest conduction band lies just above the vacuum level. Since the Fermi level of tip and support will be located in this large energy gap [19], imaging of a monolayer of upright molecules cannot be explained simply in terms of tunneling through a passive molecular spacer of 20 Å. We have to consider the tip−molecule−support as a complete quantum system, where the molecule is not just a spacer but an active part of the system. An active participation of the long alkane chain for tunneling has to involve nonresonant through-bond tunneling processes, which in such a system are quantifiable. Such processes will be presented in detail in the lecture of C. Joachim.

To image large alkane moieties with many alkane chains in series, additional electron-transfer channels have to be considered because nonresonant tunneling is expected to fall off rapidly with the number of alkane chains in series. STM imaging of such mesoscopic entities has been explained [19,28] in terms of surface conductivity. The surface density of states, intrinsic as well as extrinsic, has been experimentally located in the middle of the energy gap, and thus it is

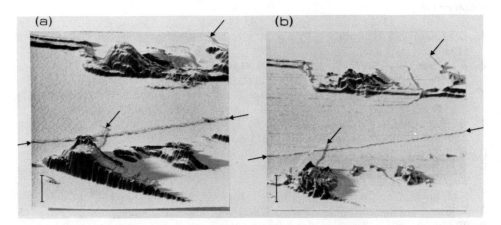

Figure 7. Three-dimensional images with simulated shading of uncoated n-heptadecane on graphite. (a) and (b) show the same area of 500 by 500 nm at temperatures of 18 and 20°C. Large structures of heptadecane (20 nm thick) disappear as soon as the temperature is raised above the melting temperature of 20°C. Smaller structures (8 nm) and layers recrystallize even at a temperature of 26°C. From [28], © Springer-Verlag 1989.

resonant with the Fermi levels of the tip or support. Such states could provide sufficient transfer paths on the surface to permit the imaging of these large alkane entities (but how "thick" is the surface?). STM experiments performed under UHV conditions with alkane entities (C_{34}-C_{44} chains) prepared in situ can elucidate the importance of extrinsic surface transfer for imaging, since no other extrinsic uncontrolled parameters will be involved in such experiments.

4. Conclusion

Pioneering STM experiments which substantially improved and finally established tunneling microscopy in biology have been reported. STM imaging of uncoated macromolecules is possible. Relief data of uncoated mesoscopic biological specimens have been obtained at submolecular level without averaging. The achieved resolution is better than that obtained for coated specimens. These results have to be judged as those of a recent, immature technique, for which sample preparation is still at an exploratory stage. This severely limits the resolution at present. The physical realization of STM in biology will provide information on local electronic structures of the macromolecule in addition to the structural information obtained so far.

The authors would like to thank R. Emch, A. Cricenti, C. Joachim, R. García and H. Siegenthaler for their active participation at the Erice School and for their very stimulating lectures.

References

1. G. Binnig and H. Rohrer *Trends in Physics, European Physical Society*, p. 38 (1984)
2. A.M. Baro, R. Miranda, J. Alaman, N. Garcia, G. Binnig, H. Rohrer, Ch. Gerber, and J.L. Carrascosa, *Nature*, **315**, 253 (1985)
3. G. Travaglini, H. Rohrer, M. Amrein and H. Gross, *Surface Sci.*, **181**, 380 (1987), presented at the *First Int'l STM Conference, Santiago, Spain (1986)*
4. A. Stemmer, R. Reichelt, A. Engel, M. Ringger, H.R. Hidber, J.P. Rosenbusch, and H.J. Güntherodt, *Surface Sci.*, **181**, 394 (1987)
5. H. Fuchs, W. Schrepp, and H. Rohrer, *Surface Sci.*, **181**, 391 (1987)
6. C.A Lang, J.K.H. Hörber, T.W. Hänsch, W.M. Heckl, and H. Möhwald, *J. Vac. Sci. Technol. A*, **6**, 368 (1987)
7. S.M. Lindsay and B. Barris, *J. Vac. Sci. Technol. A*, **6**, 544 (1987)
8. R. Emch, J. Nogami, M.M. Dovek, C.A. Lang, and C.F. Quate, *J. Appl. Phys*, **65**, 79 (1989)
9. J. Schneir, O. Marti, G. Remmers, D. Gläser, R. Sonnenfeld, B. Drake, P.K. Hansma, and V. Elings, *J. Vac. Sci. Technol. A*, **6**, 283 (1987)
10. H. Gross, T. Müller, I. Wildhaber, and H. Winkler, *Ultramicroscopy*, **16**, 287 (1985)
11. R. Wepf and H. Gross, (to be published)
12. M. Amrein, A. Stasiak, H. Gross, E. Stoll, and G. Travaglini, *Science*, **240**, 514 (1988), presented at the *Second Int'l STM Conference, Oxnard, USA (1987)*
13. J.A.N. Zasadzinski, J. Schneir, J. Gurley, V. Elings, and P.K. Hansma, *Science*, **239**, 1013 (1988)
14. R. Guckenberger, C. Kösslinger, R. Gatz, H. Breu, N. Levai, and W. Baumeister, *Ultramicroscopy*, **25**, 111 (1988)
15. M. Amrein, R. Wepf, R. Dürr, G. Travaglini, and H. Gross, *JUMSR*, (to be published)
16. G. Travaglini, H. Rohrer, E. Stoll, M. Amrein, A. Stasiak, and H. Gross, *Phys. Scr.*, **38**, 309 (1988)
17. G. Travaglini and H. Rohrer, *Spektrum d. Wissenschaft*, No. 8, p. 14 (August 1987); in English: *Scientific American*, **257**, No. 5, p. 30 (November 1987)
18. W.M. Heckl, K.M.R. Kallury, M. Thompson, Ch. Gerber, J.K.H. Hörber, and G. Binnig, *J. Am. Chem. Soc.* (to be published)
19. M. Amrein, A. Stasiak, R. Dürr, H. Gross, and G. Travaglini *Science*, **243**, 1708 (1989), presented at the *Third Int'l STM Conference, Oxford, England (1988)*
20. T.P. Beebe, T.E. Wilson, D.F. Ogletree, J.E. Katz, R. Balhorn, M.B. Salmeron, and W.J. Siekhaus, *Science*, **243**, 370 (1989)
21. A. Cricenti *et al.*, (to be published)
22. C. Bustamante *et al.*, (to be published)
23. G. Lee, P.G. Arscott, V.A. Bloomfield, and D. Evans, *Science*, **244**, 475 (1989)
24. P.G. Arscott, G. Lee, V.A. Bloomfield, and D. Evans, *Nature*, **339**, 484 (1989)
25. R. Guckenberger *et al.*, (to be published)
26. C.B. Duke, W.R. Salaneck, T.J. Fabish, J.J. Ritsko, H.R. Thomas, and A. Paton, *Phys. Rev. B*, **18**, 5717 (1978)
27. C.B. Duke and T. Fabish, *Phys. Rev. Lett.*, **16**, 1075 (1976)

28. B. Michel, G. Travaglini, H. Rohrer, C. Joachim, and M. Amrein, *Z. Physik B* (to be published)
29. B. Michel and G. Travaglini, *Proc. Int'l Engineering Conference "Molecular Electronics", Keautou Kona, Hawaii*, (to be published)
30. A. Karpfen, *J. Chem. Phys.*, **75**, 238 (1981)

STUDY OF THE BIOCOMPATIBILITY OF SURGICAL IMPLANT MATERIALS AT THE ATOMIC AND MOLECULAR LEVEL USING SCANNING TUNNELING MICROSCOPY

R. EMCH, X. CLIVAZ, C. TAYLOR-DENES, P. VAUDAUX[*], D. LEW[*], and
P. DESCOUTS

*Group of Applied Physics
University of Geneva
20 rue Ecole de Medecine
1211 Geneva 4, Switzerland
[*]Division of Infectious Diseases, University Hospital*

ABSTRACT. We have investigated some aspects of the interface between metallic biomedical implants and living tissue. To characterize this interface we have developed an STM coupled with a high resolution reflective optical microscope. The STM is used in connection with other microscopic techniques: TEM, SEM, and Auger spectroscopy. Initial biological measurements show that fibronectin (Fn), a glycoprotein which plays an important role in cell attachment and bacterial promotion, is attached on Ti and V substrates, but is not biologically active on V, a non-biocompatible substance. We present STM images of the Ti oxide layer and of single and multiple Fn molecules shadowed with a conductive layer.

1. Introduction

Many materials including metals, glasses, polymers, ceramics, carbons, and composites have become essential components of the medical practice as well as of medical and dental surgery. The tissue reactions to these biomaterials vary greatly depending on the type of material, the type of biological tissue and the surgical technique. However, even in the case of non-toxic materials one adverse effect of the implanted biomaterial is to increase the susceptibility of surrounding tissues to microbial infections.

When devices have to be implanted in the body for several months or years, they must have the specific mechanical properties to replace the functions of the defective body tissues or organs, and they must be biocompatible, i.e., composed of materials well accepted and integrated by the host in a controlled and predictable way. In the case of orthopedic prosthesis and dental implants it is necessary to have an intimate integration between the bone and the biomaterial to allow an efficient stress transfer from implant to bone. Branemark et al.[1] have observed such a close apposition of bone to titanium implant which they have called osteointegration.

The primary interactions between biomaterials and living tissues occur on the molecular level and in a very narrow interface zone of about 1 nm in width[2]. Therefore the surface properties of biomaterials on an atomic scale are extremely important. Early reactions include rapid adsorption of blood and connective tissue components which may act as bridges between artificial and biological surfaces. A relevant macromolecule in this context is fibronectin (Fn), a multifunctional glycoprotein found in both connective tissue and blood. Fn also appears to play an important role in promoting adherence of staphylococci which cause infection of metallic implants in orthopedic surgery[3].

The first objective of our study is to characterize the surface microstructure as well as oxide films of the metallic implant materials using Scanning Tunneling Microscopy together with Scanning Electron Microscopy, X-ray microprobe analysis, and Auger spectroscopy.

A second goal is to apply the aforementioned techniques together with Transmission Electron Microscopy and biological assays to study the interaction of biological macromolecules with metallic implant surfaces. We are particularly interested in analyzing the occurrence of conformation changes in the biological macromolecules adsorbed on metallic surfaces, in order to relate such changes to their pro-adhesive or anti-adhesive biological effects, tested against microbial or host cell lives. We now present our preliminary results of the biological and microscopic studies and the STM coupled with an optical microscope we have specifically designed for this project.

2. Pro-adhesive activity of fibronectin adsorbed on Titanium and Vanadium coverslips

We began our study with Titanium and Vanadium because they are at the two extremes on the scale of biocompatibility; Ti is inert and gives rise to osteointegration while V is toxic even though it appears in some Ti alloys used for implants[4].

P. Vaudaux et al.[5] have developed techniques for studying adsorption of macromolecules on the surface of biomedical polymers. Fn, either radiolabelled with ^{125}I or ^{3}H, was incubated in-vitro with polymer or metal coverslips and rinsed carefully allowing quantification of the surface-adsorbed protein. Fig. 1a shows the in-vitro adsorption of Fn on Ti, V, and Polymethylmetacrylate (PMMA) which was used as a reference. Adsorption of Fn showed a similar dose-response for each of the three different materials, in the presence of increasing concentration (1-16ug/ml) of Fn in the incubating solutions. The biological effects of Fn adsorbed in-vitro on various polymeric or metallic surfaces have been studied by using an in-vitro adherence assay developed by P. Vaudaux et al.[3] Figure 1b shows one of the various biological activities of Fn adsorbed on the various materials, namely the promotion of bacterial adherence. The number of adherent staphylococci (strain Wood 46 of Staphylococcus aureus) increased in proportion to the quantity of Fn adsorbed on the surface of either PMMA or Ti. In contrast, no promotion of bacterial adherence was ever recorded on V coverslips, whatever the concentration of surface-adsorbed Fn.

Fig.1

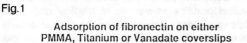

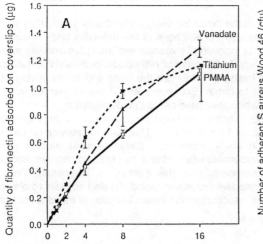
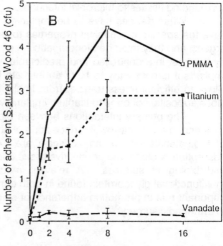

Concentration of fibronectin in incubating solution (μg/ml)

This latter observation suggested that fibronectin adsorbed on V coverslips could be inactivated by the conformation changes which would affect the majority of adsorbed macromolecules. Alternatively it was possible that the surface disposition of the adsorbed biomolecules was so irregular that it was inadequate to promote bacterial adherence. To discriminate between either possible explanation we attempted to view surface-adsorbed Fn on either Ti and V coverslips by electron microscopy. Control experiments ruled out the hypothesis that the reduced adhesion of staphylococci on Fn-adsorbed V was due to a toxic effect of the metal on the bacteria, since staphylococci remained fully viable during the assay. Immunological analyses are also in progress to obtain some information on possible antigenic alteration of the Fn molecules adsorbed on the various surfaces.

3. Microscopic analysis of Titanium and Vanadium coverslips, and observation of Fn with TEM

Titanium and Vanadium used for this study are commercially pure metals from Goodfellow: Ti (99.6%) and V (99.8%). Figure 2 shows an image of the Ti specimen obtained with the SEM where we can see the size of the grains and the roughness of the surface due to the etching. The surface of the two specimens is covered with a natural layer of oxide as we can see on Figure 3 which presents the depth profile obtained with Auger spectroscopy. We have observed that the thickness of the oxide layer is about 5 nm on the V and 10 nm on the Ti. The Ti oxide which is always present at the surface of Ti implants seems to play an important role in the process of osteointegration. We can also see Carbon at the surface of the specimen indicating the presence of a contamination layer of hydrocarbons we have to take into account when we observe at ambient atmosphere.

Fig.2   Fig.3

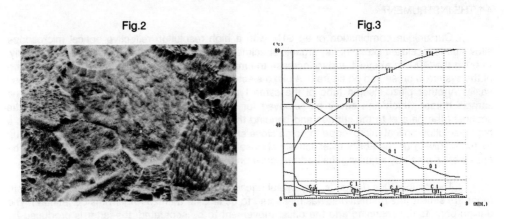

Fig.2 : SEM image of a cp Ti surface chemically etched (magn.: 1000x)
Fig.3 : Auger depth profile of the same surface presenting atomic composition versus sputtering time. The thickness of the oxyde is about 10 nm.

Fibronectin is a large glycoprotein with a molecular mass of about 440,000 daltons. It is composed of two subunits of about 60 nm in length and 2.5 nm in diameter linked with a disulfide bridge. Figure 4 shows a TEM observation of fibronectin adsorbed on freshly cleaved mica using the technique of rotary shadowing and of replica.

When the concentration of Fn becomes too large, the Fn forms very dense network where the bacteria can attach. Figure 5 represents a SEM image of bacteria attached to a network of Fn adsorbed on mica. It is clear that Fn is far too small to be precisely observed on metallic substrates using SEM techniques and to identify the conformation of that protein on different metals.

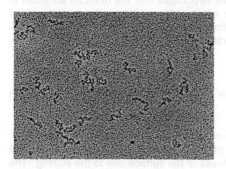

Fig.4 TEM image of Fn sprayed on mica (1 x 0.8 micrometer).

Fig.5 SEM image of bacteria attached to a network of Fn adsorbed on mica (15 x 12 micrometers).

4. STM combined with an optical microscope

4.1 THE INSTRUMENT

Our unique combination of an STM with a high resolution reflective optical microscope allows us to localize a site, e.g., an organic specimen deposited on a biomaterial surface, optically, to bring the STM tip above it and to increase the magnification up to atomic resolution. A drawing of the system is presented on figure 6. A piezo-electric tube supports the tip which can be changed easily. Another piezo-electric tube of the same type but with a different diameter supports the sample holder which can easily be removed for sample changing. This way of mounting the scanner tube parallel to the surface and bending the tip has previously been tested[6]. It enables a high resolution optical microscope to come close enough to take an image with a magnification up to 1000x, giving us the ability to effectively choose what we are about to scan with the STM. The approach is made by adjusting the bottom screw on the sample holder and with the z-electrodes.

The double tube design is very useful because it allows thermal compensation since both tubes are similar (the thermal drift was measured to be smaller than 20 Angstroms per minute). The design permits the scanning and the offset movement to be separated, the latter is produced by applying DC voltages on the tube supporting the sample.

Figure 6 shows the STM resting on a conventional damping system which can itself be positioned with micrometer precision to the optical microscope which is represented on the figure by an arrow situated above the STM tip. Despite the very large scanning and offset range (90 x 9 micrometers) this STM still enables atomic resolution which is shown on figure 7C. Figure 7B shows a photograph of our STM and figure 7A the complete system with the optical microscope. A CCD video camera located on top of the optical microscope transmits the optical image on a video monitor.

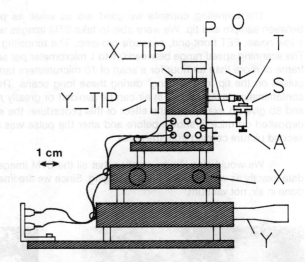

Fig.6 A schematic of the STM

A - approach screw
S - sample
T - Tip
P - Piezoelectric tube
O - Optical microscope

A very important point in imaging biological objects with an STM is that a high resolution image is needed immediately during the acquisition of data so that real and interesting features may be recognized. This capability is provided in our case by a video converter which digitalizes the signal coming from the feedback and the scanning units and transforms it in a video signal displayed on a conventional video monitor. The grayscale image represents a top view of the analyzed surface and the 512 x 512 bit resolution in X and Y is sufficient for our purposes.

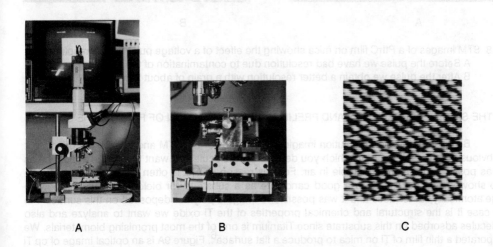

Fig.7A View of the setup including: the STM, optical microscope with video camera and TV monitor.
B View of the STM below the objective lens of the optical microscope.
C STM image obtained on HOPG with atomic resolution.

The tunneling currents we used are as small as possible to minimize risk of collision between sample and tip. We were able to take STM images with current down to 5 picoAmp using a low-noise, FET front-end, cascode pre-amp. The tunneling voltage was between 0.1 and 1 volt. The scanning speed range between 0.1 to 1 micrometer per second which gives a total time for one frame of more than an hour for a scan of 10 micrometers range. We found it useful to apply short pulses on the tip for "cleaning" during these long scans. The loss of resolution attributed to the contamination layer present in air can be avoided or greatly reduced by these pulses[7]. Figures 8a and 8b give an example of the effect of this procedure: the same surface composed of PtIrC film deposited on mica is shown before and after the pulse was applied, the hole in the center of the second picture coming from the pulse.

We would also like to mention that all the STM images presented in this paper are rough data exactly as we see them during imaging. Since we are imaging biological samples, all work was done in air, not vacuum.

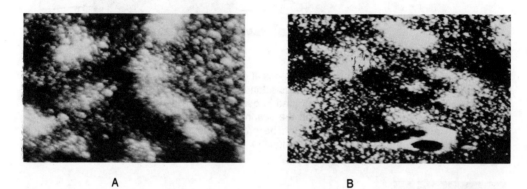

A   B

Fig.8 STM images of a PtIrC film on mica showing the effect of a voltage pulse on the resolution.
A Before the pulse we have bad resolution due to contamination of the tip.
B After the pulse we obtain a better resolution with a grain of about 2 nm.

4.2 THE SAMPLE PREPARATION AND PRELIMINARY OBSERVATION OF FN WITH THE STM:

Because of the high resolution imaging capabilities of the STM and its working principle it is obvious that the substrate on which you deposit the molecule you want to analyze should be as flat as possible and relatively stable in air. For this reason HOPG is often used in STM work. We also showed that gold (111) is a good candidate as a substrate for biological analysis showing large atomically flat terraces and it was possible to image a polymer deposited on this surface[8]. In our case it is the structural and chemical properties of the Ti oxide we want to analyze and also molecules adsorbed on this substrate since Titanium is one of the most promising biomaterials. We evaporated a thin film of Ti on mica to produce a flat surface[6]. Figure 9A is an optical image of cp Ti coverslip chemically etched. We can see the grains and the roughness due to etching. On the right we see the STM tip located exactly above the region imaged by the STM on figure 9B. Pulses on the tip were applied in this case at the beginning of each line in order to keep a good resolution (otherwise the image looks very smooth with no sharp structure).

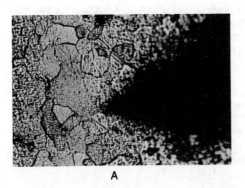 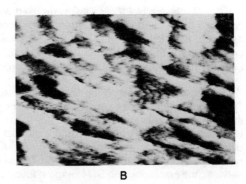

Fig.9A Optical image of the same Ti sample as on fig. 2 where we can see the tip in position for scanning (magn. 600x).
9B STM image with large scan on the same Ti sample (3.5 x 4.5 micrometers).

Tunneling through the natural 100 Angstrom thick Titanium oxide layer is not yet well understood and reveals some unexpected conducting mechanism. A threshold of about -200 mV was found to be necessary for electron tunneling out of the tip (unoccupied sample states) which could imply an empty state located above the Fermi level of the sample with a corresponding gap.

The contrast mechanism in STM imaging comes from a change in height in the topography or from a change of electronic characteristics of the sample, for example, the work function. In either case some conductivity is necessary to measure the tunneling current. For this reason we shadowed the fibronectin to exclude the problem of conductivity even if in the future we are interested in looking directly at bare molecules. The metals we used to coat the macro-molecule were PtC or PtIrC[9] evaporated on a cooled substrate to minimize the grain size.

Fig.10A STM image of a PtIrC film on mica (grain about 2 nm; 65 x 120 nm).
10B STM image of a sputtered Au film on mica (grain about 15 nm; 70 x 70 nm).

To avoid a denaturation of the fibronectin during this preparation procedure the molecule has to be dried in vacuum minimizing liquid superficial tension which could make the molecule collapse. The preparation can also be "freeze- dried" by rapid freezing and low-temperature sublimination of the vitrified ice in vacuum.

Thin layers of PtC or PtIrC were used with a grain size of about 30 Angstrom for the first case and down to about 20 Angstrom in the second case. An image obtained with the STM of Fn deposited on mica shadowed with PtIrC is shown on figure 10A. The usual technique of sample preparation for SEM is sputtering of gold on mica. The grains of gold shown in figure 10B range between 100 and 300 Angstrom which is too large for coating Fn samples.

Firstly we deposited the fibronectin by spraying it on freshly cleaved mica before coating. This technique was used to ensure a more homogeneous density of the fibronectin and to minimize the formation of clusters, since the molecules are predisposed to fix easily to other material but also to themselves. Images of PtC coated individual molecules at different places in the sample and different magnification are shown in figures 11A-C. The next step will be to image the molecule naturally adsorbed on mica and then on Titanium oxide. Figure 12A shows a SEM image of a Fn network; the corresponding STM image is found in figure 12B.

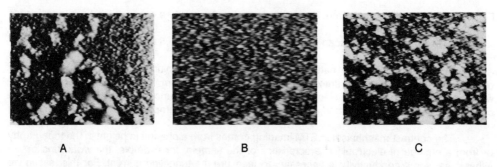

Fig.11 STM images of Fn deposited on mica and coated with PtC corresponding to different magnifications. We can see individual Fn molecules with various shapes.
Size of images: A: 360 x 360 nm; B: 200 x 250 nm; C: 100 x 150 nm.

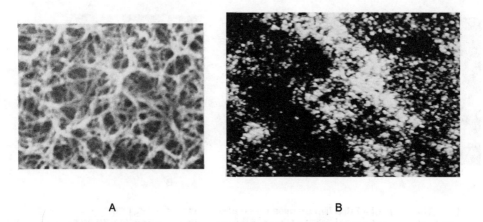

Fig.12A SEM image of Fn network on mica, gold labelled (2 x 1.5 micrometers).
12B STM image of Fn network on mica, PtIrC coated (130 x 250 nm).

5. Conclusion

We have shown that the adherence and biological activity of fibronectin, a glycoprotein playing an important role in the process of biocompatibility, is very different depending on the chemical composition of the material on which it is deposited. Using SEM and STM we showed the network formed by Fn attached to the substrate. With Auger spectroscopy and depth profile analysis we have investigated the chemical composition of the surface of chemically etched Titanium, showing the presence of a 10 nm thick natural oxide layer with hydrocarbon contamination on it, on which the reaction with the organic material will take place.

We reached our first goal in imaging Fn coated with a conductive layer by using an STM specially designed for biological analysis and coupled with a high resolution reflective optical microscope. Using this STM we showed images of individual Fn molecules and also of the network formed by the agglomeration of many molecules. Large STM images of chemically etched Titanium surface give rise to the question of the conducting mechanism through the Titanium oxide layer.

6. Acknowledgements

In particular we would like to thank Prof.Martin Peter and Dr.h.c.Fritz Straumann who have initiated this project, for their very helpful support and advice during this work. Sadly Fritz Straumann died last September and his personality is greatly missed.
We want to thank also Dr.H. Gross, M. Amrein from ETH Zurich and Dr.G. Travaglini from IBM Zurich for their help in the preparation of biological samples. Auger spectra have been made by Dr.H Mathieu (EPF Lausanne).

The studies described in this work have been performed with the support of Straumann Institute AG (Waldenburg) and of AO-ASIF Foundation.

References

1. P.I. Branemark, G.A. Zarb, and T.Albrektsson,(Eds.), 'Tissue Integrated Prostheses: Osseointegration in   Clinical Dentistry, Quintessence', Chicago (1985).

2. B. Kasemo and J Lausmaa, 'Biomaterials from a surface science perspective, in Surface Characterization of Biomaterials', (B.D. Ratner, Ed.) Elsevier (1988).

3. P. Vaudaux, R. Suzuki, F.A. Waldvogel, J.J. Morgenthaler, and V.E. Nydegger, *J. Infect. Dis.* **150**, 546-553 (1984).

4. S.G. Steinemann and S.M. Perren, 'Titanium alloys as metallic biomaterials', in Titanium, Science and Technology, vol. 1-4, (G. Lutjering,V. Zwicker and W. Bunk Eds.) Deutsche Gesellschaft fur Metallkunde, Oberursel (1985).

5. P. Vaudaux, F.A. Waldvogel, J.J. Morgenthaler, and V.E. Nydegger, *Infect. Immun.* **45**, 768-774 (1984).

6. R. Emch, P. Descouts, and Ph. Niedermann, 'A small scanning tunneling microscope with large scan range for biological studies', Proceedings of the STM 1988 Conference, Oxford, July 1988, to be published in *J. Micro*.

7. R. Emch, J. Nogami, M.M. Dovek, C.A. Lang, and C.F. Quate, 'Characterization and local modification of atomically flat gold surfaces by STM', Proceedings of the STM 1988 Conference, Oxford, July 1988 to be published in *J. Micro*.

8. R. Emch, J. Nogami, M.M. Dovek, C.A. Lang, and C.F. Quate, 'Characterization of gold surfaces for use as substrates in scanning tunneling microscopy studies', *J. Appl. Phys*. **65**, 79-84 (1989)

9. M. Amrein, A. Stasiak, H. Gross, E. Stoll, and G. Travaglini, 'Scanning tunneling microscopy of recA-DNA complexes coated with a conducting film', *Science* **240**, 514-516 (1988)

NAKED DNA HELICITY OBSERVED BY SCANNING TUNNELING MICROSCOPY

A.Cricenti, S.Selci*, A.C.Felici, R.Generosi, E.Gori, W.Djaczenko^ and G.Chiarotti*

Istituto di Struttura della Materia del Consiglio Nazionale delle Ricerche, via E.Fermi 38, I-00044 Frascati, Italy
^ Istituto di Medicina Sperimentale del Consiglio Nazionale delle Ricerche, I-00100 Roma, Italy
* Dipartimento di Fisica, Universita' di Roma "Tor Vergata", I-00073 Roma, Italy

ABSTRACT. Uncoated DNA molecules marked with an activated Tris 1-Aziridinyl Phosphine Oxide (TAPO) solution were deposited on gold substrates and imaged in air with high resolution Scanning Tunneling Microscope (STM). Constant current and barrier height STM images show a clear evidence of the helicity of the DNA structure: pitch periodicity ranges between 25 and 34 Å while the average diameter is 20 Å. Molecular structure within a single helix turn is also observed.

1. Introduction

Scanning Tunneling Microscopy has proved to be a unique tool for imaging with great detail structural and electronic properties of surfaces of metals and semiconductors [1-2]. Biological important systems have also been observed, for example biomembranes [3] bacteriophage particles [4], native circular DNA molecules [5], metal shadowed recA protein-DNA complexes [6], etc. . Recently STM observations of non-metal-shadowed DNA that clearly show the helicity of the molecule [7] greatly increased the interest of STM applications in biology. However all these STM observations did not furnish so far any image compatible with the structural model of DNA elaborated by X-ray crystallography and biochemical data.

In this paper we report on STM observations of native, non-metal-shadowed, TAPO marked DNA. Constant current and barrier height STM images show a clear evidence of the helicity of the DNA structure: the pitch periodicity ranges between 25 Å and 34 Å while the average diameter is 20 Å. The molecular structure within single helix turn is also shown with great detail.

2. Experimental apparatus

The mechanical assembly of our STM is shown in fig.1.

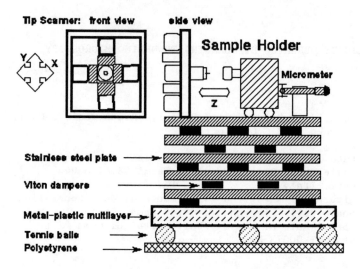

Fig. 1

2.1 SAMPLE HOLDER

The sample holder is a massive stainless steel single piece standing on three metallic balls on the antivibration stack. It is pushed by a precision micrometer to approach the tip until the gap sample-tip, observed by an optical microscope, is less than 5 μm. At this distance a piezoelectric ceramic disc, incorporated into the micrometer, pushes the sample holder until a tunneling current is flowing. The micrometer is then pulled back to avoid additional drift. The total time to be in tunneling operation is 30 seconds, a feature that is important for air operation in order to avoid excessive contamination of sample and tip [8].

2.2 TIP SCANNER

The tip scanner consists of a symmetric cross configuration of four piezoelectric ceramic discs for X and Y motion and one piezo for Z motion. The ceramics discs are pre-stressed with a force of several tens of N/cm^2 in order to increase the linearity of the motion and decrease creep and hysteresis. The cross configuration of the X and Y piezos results in thermal compensation against drift. The three movements are thus completely decoupled and the piezos provide motion with a sensitivity of about 5.5 Å/V.

2.3 VIBRATION ISOLATION

The vibration isolation system is obtained with the combined effect of a metal stack of discs with viton dampers and a metal-plastic multilayer isolated from the floor by nine tennis balls. On the top of the metal stack the tunneling unit is enclosed in a small stainless steel chamber that provides acoustic and electromagnetic screening.

2.4 TIPS SHAPING

Tungsten tips with atomic protrusions are essential for obtaining STM images with atomic resolution. Fig. 2 shows a method that permits to get very sharp tungsten tips in a very reproducible way.

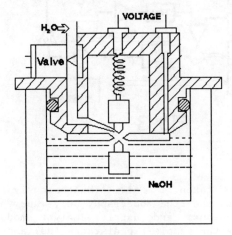

Fig.2

It consists of a Tungsten wire (diameter=1mm), notched by a lathe, immersed into an electrolytic solution of NaOH (5% molar). The current flows from a wedged ring electrode to the tungsten wire. The wire is loaded by a light weight (typically 5 g) and suspended by a spring that instantly moves the tip out of the solution when the load break the thinned wire. The electronic operated valve flushes then the tip with distilled water removing unwanted drops of the NaOH solution, that otherwise would smooth the tip.

2.5 ELECTRONICS

Fig. 3 gives the lay-out of the electronics used for controlling position of the tip, current flow, scanning movements and data acquisition. The black bold line shows the feedback loop used to maintain constant tunneling current. In this operation mode the tunneling current is revealed by an I-V converter, filtered and compared with a reference (fixed at the desired operating current) by a differential amplifier. This difference is digitalised and integrated by a Digital Integrator, where a fast and accurate integral response is obtained by an endless up-down counting and buffering accumulation. The digital output is continuously updated and converted by a 16-bit DAC to supply the high voltage amplifier of the Z-piezo.

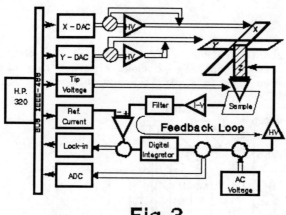

Fig.3

2.6 DATA ACQUISITION

STM images have been taken in air simultaneously in the constant current, which reflects the true topography only when the surface is electronically homogeneous, and barrier height modes. The latter consists in modulating the sample-tip distance (ds) and measuring the current variation (dI/ds) which is proportional to the square root of the local barrier height [1]. The barrier height depends on the local value of the work-function so that the method is sensitive to the chemical structure of the sample. Tip-sample voltage and tunneling current were set at 20 mV (sample positive) and 1.0 nA, respectively. The Z-piezo modulating frequency was set at 11 kHz with an amplitude (ds) of 0.5 Å. The figures shown in this paper are obtained from raw data without any post-image electronic treatment. Reproducible results have been obtained in several samples.

2.7 SAMPLE PREPARATION

Plasmid circular DNA in aqueous solution, concentration 1mg/ml, was used as base solution. DNA was marked with activated Tris 1-Aziridinyl Phosphine Oxide (TAPO) by mixing 20 l each of DNA and activated TAPO solutions for one hour at room temperature [9]. The solution was then deposited on a gold plated Aluminum stub and dried at room temperature for several hours . The possibility of preserving naked untreated DNA suitable for STM observation is still an open question. DNA molecules deposited on conducting substrates frequently collapse and a stable tunneling condition is not achieved (6). The success of the present experiment seems to suggest that naked DNA molecules are well preserved, presumably since ends of DNA segments are fixed by TAPO. Pure TAPO solution on a similar gold substrate can be imaged only in the constant current mode and appears as a crystalline structure of squares of 3 Å side. The absence of imaging in the barrier height mode is probably due to a negligible modification of the work function of gold by TAPO.

3. Experimental Results

Fig.4 is a dI/ds image in three-dimensional representation with simulated illumination. Two long straight and convoluted, partially overlapping segments of double-stranded DNA molecules may be clearly seen in the upper part of the image. The two segments of DNA molecules show a typical helical conformation. The periodicity of the helix ranges from 25 Å for the lower segment to 34 Å for the upper one. Barrier height value for DNA was $\sim$ 1 eV while the one of gold was $\sim$ 50 meV.

Fig.5 is a topographic image in amplitude traces representation, as it appears on the computer display during data acquisition. The image shows some DNA segments with similar periodicities of those observed in barrier height mode.

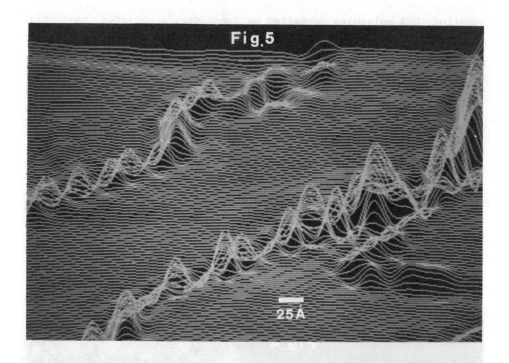

Fig.5

25Å

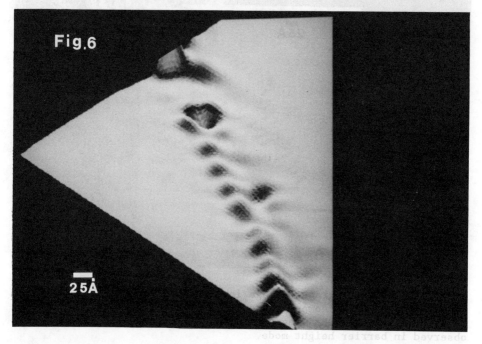

Fig.6

25Å

Fig.6 is a dI/ds image in three-dimensional representation with simulated illumination. The image is formed of regularly occurred bumps along the DNA molecule with a mean spatial periodicity of 30 Å and full width half maximum of 20 Å. Bumps correspond to ten turns of the DNA helix.

Further higher resolution images have been obtained in the barrier height mode and a fine structure within each bump of fig.6 showed up. The association of this fine structure to the different molecules or molecular groups is a hard task. A possible interpretation is that phosphate molecules of the DNA back-bones, which are negatively charged and thus increase the local work function, are associated with this fine structure.

The images of figs. 4-6 presumably show naked DNA, though the possibility of the presence of water molecules and TAPO bound to the DNA cannot be ruled out. Further experimental work is needed for a complete understanding of such a problem. However, the results reported here already show the great potentiality of the method for the characterization and possible sequentiation of the DNA.

4. Acknowledgments

We wish to thank L.Ferrari and A.Gavrilovich for helpful assistance.

5. References:

1. Binnig, G. and Rohrer, H., (1986) "Scanning tunneling microscopy", IBM J. Res. Develop. 30, 355-369.
2. Binnig, G., Rohrer, H., Gerber, Ch., Weibel, H., (1983) "7x7 reconstruction on Si(111) resolved in real space", Phys. Rev. Lett. 50, 120-123.
3. Zasadzinski, J.A.N., Schneir, J., Gurley, J., Elings, V., Hansma, P.K., (1988) "Scanning tunneling microscopy of freeze-fracture replicas of biomembranes", Science 239, 1013-1015.
4. Baro', A.M., et al., (1985) "Determination of surface topography of biological specimens at high resolution by scanning tunneling microscopy" Nature 315, 253-254.
5. Travaglini, G., Rohrer, H., Amrein, M., Gross, H., (1987) "Scanning tunneling microscopy on biological matter", Surf. Sci. 181, 380-390.
6. Amrein, M., Stasiak, A., Gross, H., Stoll, E., Travaglini, G., (1988) "Scanning tunneling microscopy of recA-DNA complexes coated with a conducting film" Science 240, 514-516.
7. Beebe, T.P. et al., (1989) "Direct observation of native DNA structures with the scanning tunneling microscope", Science 243, 370-372.
8. Schneir, J., Sonnerfeld, R., Hansma, P.K., Tersoff, J., (1986) "Tunneling microscopy study of the graphite surface in air and water", Phys. Rev. B 34, 4979-4984.
9. Djaczenko, W., Cimmino, C.C., (1973) "Visualization of polysaccharides in the cuticle of oligochaeta by the Tris 1-Aziridinyl Phosphine Oxide method", J. Cell Biol. 57, 859-867.

APPLICATIONS OF SCANNING TUNNELING MICROSCOPY TO LAYERED MATERIALS, ORGANIC CHARGE TRANSFER COMPLEXES AND CONDUCTIVE POLYMERS

S. N. MAGONOV * and H.-J. CANTOW
Freiburger Material-Forschungszentrum, F · M · F , and
Institut für Makromolekulare Chemie, Universität Freiburg
Stefan-Meier-Str. 31, D-7800 Freiburg, F. R. of Germany
* Permanent address: Institute of Chemical Physics
of the USSR Ac. Sci., Moscow, USSR

PROF. HANS ADAM SCHNEIDER DEDICATED CORDIALLY TO HIS 60. BIRTHDAY

ABSTRACT. The results of scanning tunneling microscopy (STM) at ambient conditions of several representatives of inorganic layered materials (α-RuCl$_3$, VS$_2$), conductive charge transfer complexes (tetracyanoquinodimethane with different donors: tetrathiofulvalene, 4-ethyl pyridine and triethylammonium) and conductive polymers (polypyrrole, polythiophene) are presented. In the case of α-RuCl$_3$ the threedimensional distortion of surface unit cell of chlorine atoms has been discovered. The STM imaging of atomic scale features of VS$_2$ surface had indicated the existence of charge density waves at room temperature. The STM images of crystal surfaces of the charge transfer complexes give detailed information concerning the surface charge distribution which has been analyzed in comparison with X-ray data. The arrangement of polymer chains in polypyrrole and polythiophene is discussed on the basis of the STM results.

1. Introduction

The creation of atomic scale resolution microscopy based on tunneling current effect is one of the latest scientific achievements [1]. The enormous capabilities of this new technique - scanning tunneling microscopy (STM) - are demonstrated in a rapidly increasing number of publications. The most advanced STM results were obtained during investigations of surfaces of semiconductors (expecially the technologically important Si and GaAs) [2], and of metals and adsorbates [3]. Among other attractive objects for STM studies there are inorganic layered materials, conductive organic charge transfer complexes and conductive polymers. Additionally to X-ray analysis and electron microscopic data important structural information may be obtained for inorganic layered materials. This is demonstrated below for α-RuCl$_3$ and VS$_2$. Detailed microscopic visualization of electronic features of surfaces of charge transfer complexes achieved by STM is of particular interest in the case of one-dimensional conductive complexes of tetracyanoquinodimethane (TCNQ). The STM images of such complexes will be discussed. The morphology of conductive polymers - polypyrrole and polythiophene - is far from being understood. The presented STM data of these polymers may add some further information.

2. Experimental

The STM studies were carried out at ambient conditions with the commercial instrument Nanoscope II (Digital Instruments Inc.). The tips were cut mechanically from Pt(70) / Ir(30) wire. The flat thin samples of α-RuCl$_3$ and VS$_2$ were prepared by sublimation and vapour transport techniques, correspondingly. The monocrystals of charge transfer complexes of TCNQ with tetrathiofulvalene (TTF), 4-ethyl pyridine (4EP) and triethyl ammonium (TEA) were prepared by a precipitation procedure. They exhibit needle-like shape, usually. The samples of layered materials and of organic complexes were fixed by silver glue on a metallic support and installed then on Nanoscope stage. The thin layers of polypyrrole and polythiophene were obtained by electrochemical polymerization on a conductive ITO glass electrode which directly is applied in STM experiments.

3. Results and Discussion

3. 1. LAYERED MATERIALS

The inorganic layered materials (dichalcogenides of transitions metals - TiS$_2$, TiSe$_2$, MoS$_2$, TaS$_2$, TaSe$_2$ - have been studied extensively by the STM. The typical structure of these compounds is represented in Fig.1. The three layered "sandwiches" with the

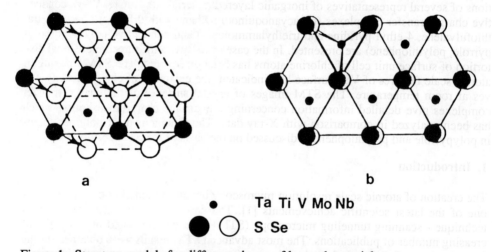

Figure 1: Structure models for different phases of layered materials
    a - 1T phase, cations are arranged in octahedral coordination
    b - 2H phase, coordination of cations is trigonal prismatic

central layer of metal atoms and two outer layers of chalcogenides atoms form bulk material via weak intermolecular forces. These materials are extremely stable at ambient conditions and exhibit semiconductive properties. The existence of temperature dependent structural phase transitions in dichalcogenides of transition metals which are correlated with conductivity changes was detected in bulk materials by electron beam diffraction experiments. The indicated structural transitions are accompanied by transformations of superlattices - charge density waves (CDW). The surface CDW were found by STM in several layered compounds (TaS$_2$, TaSe$_2$, TiS$_2$, TiSe$_2$ [4,5]) at diffe -

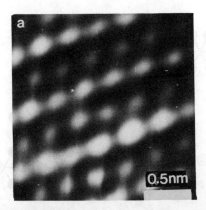
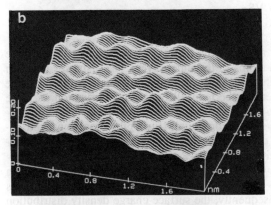
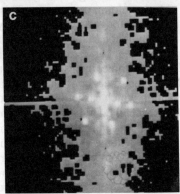

rent temperatures. Recently in the STM study of a less-common layered material, VS_2 [6], the existence of CDW was also visualized at room temperature. In Fig.2 it is evident - from the top view image and from the threedimensional line plots - that in addition to the close packed hexagonal symmetry of surface charge density "hills" - cell with a lattice parameter a = 3.3 Å - the periodically varied height of neighbouring rows represents a superlattice. The parameter of this superlattice is 2a. This type of CDW is different from the superstructure observed in TaS_2 and $TaSe_2$ but similiar to the one observed in TiS_2 [7]. It should be noted that surface area represented by Fig.2 is not easily to be detected. Generally chemical imperfections introduce different kinds of defects. In Fig. 3 the area (58 x 58 Å) includes various types of surface defects which may be caused by single atom missing or by steps in underlying atomic planes. Changes in the main direction of CDW were observed too, as well as crossing of CDW of different directions. This crossing causes formation of a (2a) x (2a) superlattice. Thus, it is evident that the STM technique yields unequivocal structural information which can be usefull in combination with electron diffraction data, when comparing bulk and surface properties. In the case of VS_2 the electron diffraction studies are in progress.

One of the important questions of the STM of layered materials still

Figure 2: The STM features of VS_2 measured at V = - 16.8 mV and I = 33.2 nA - <u>a</u> - top view image, <u>b</u> - threedimensional line plot <u>c</u> - twodimensional Fourier Transform o f <u>a</u>

has to be solved. It concerns the nature of the surface charge density "hills". Are they

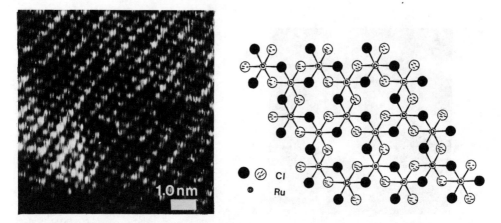

Figure 3: The STM top view image of VS$_2$ measured at V = 4.6 mV and I = 6.5 nA

Figure 4: Structure model of α-RuCl$_3$

representing sp orbitals of outer layer sulfur atoms (in case of disulfides), or are there contributions from d-orbitals of metal atoms in the inner layer? The distances between atoms in outer and inner layers of dichalcogenides of transition metals are in the same range, 3.2 - 3.4 Å. Thus, this factor cannot be helpfull for an indentification. A more simple situation exists in the case of α-RuCl$_3$ - another representative of layered materials. According to chemical structure of this compound (Fig.4) the closely packed hexagonal arrangement of the Cl atoms differs from the non-centered hexagonal symmetry of the Ru atom positions. The STM features of α-RuCl$_3$ as presented in Fig.5 clearly demonstrate a symmetry identical with that of the Cl atoms. The distances between hills in hexagonal cells correspond well with the distance between the Cl atoms, as known from X-ray data (3.25 Å). Consequently, the surface charge density distribution visualized by STM strongly correlates with the chlorine atom arrangement. This statement is the basis of very important structural information for α-RuCl$_3$. The analysis of the STM features of this compound at different tunneling conditions [8] shows that the threedimensional distortion of the hexagonal arrangement of surface charge density hills does not depend on experimental conditions. Thus, the threedimensional distortion of the unit cell of the Cl atoms is verified. This effect is the direct consequence of an inequivalence of Ru atoms. The Ru-Ru pairing interaction which earlier has been proposed [9] receives strong support from the STM data. Consequently, an impressive example of a principal advantage of the STM in structural analysis is presented, when compared with traditional diffraction techniques which usually present the averaged molecular information. The distortion of the unit cell gives rise to a commensurate superstructure (CDW). This is documented well by the twodimensional Fourier transform picture. The inner hexagon in Fourier space indicates the symmetry of superstructure with parameter $a\sqrt{3}$ (a - averaged parameter of hexagons of charge density hills) and an angle of 30° between the main directions of lattice and superlattice.

Figure 5: The STM features of α-RuCl$_3$ measured at V = - 241 mV, I = 2.6 nA (a, b, c), at V = - 785 mV, I = 27 nA (d, e, f) - a, d - top view images. b, e - threedimensional line plots. c, f - twodimensional Fourier Transform a and d

It should be noted that the STM information of the structure of layered materials is based on the electronic properties of their surfaces. The big corrugations of the charge density hills which in some cases (TaS_2, $TaSe_2$) are as big as 10 Å represent mainly the electronic effects. The corrugations in the STM images of a-$RuCl_3$ at different tunneling conditions are much smaller (0.2 - 0.5 Å).

In addition one more observation should be mentioned. It was found that shape and size of the surface charge density hills are depending on the distance between tip and surface. The maximum size of round shape patterns and the "flattening" of the hills as observed at closest tip approach to the surface (may be at point contact) have to be explained. A reasonable explanation may be based on the experimental results published earlier [10]. It was found that with decreasing tip-sample distance tunneling resistance decreases exponentially but reaches the plateau at very small distances. Other possible factors to be considered are the increased number of tip and surface orbitals taking part in current formation as well as tip-surface interatomic interactions.

3. 2. ORGANIC CHARGE TRANSFER COMPLEXES

The conductive organic charge transfer complexes appear to be good candidates for detailed STM studies. This idea is supported by the results for one of these complexes [11]. We have examined three different complexes TCNQ with TTF, 4EP and TEA by STM (Fig.6). In Fig. 7 the top view STM images of these complexes are presented. It is reasonable to assume that the observed surface charge density distribution belongs to one of the crystallographic planes. In case of TTF-TCNQ this correspondence was installed on the basis of crystallographic data and molecular orbitals calculations [11]. Similiar conclusions for 4EP-TCNQ are supported by preliminar X-ray analysis [12]. The main feature of the STM images of complexes are the stacking rows of TCNQ molecules. The three ball sets (Fig. 7) represent the visualization of lowest unoccupied molecular orbitals of TCNQ [11]. The biggest ball corresponds to benzene ring orbitals, and the smaller outer balls are centered near the N atoms of cyano groups. The separation between TCNQ molecules in row is 3.9± 0.1 Å. The distance between neighbouring rows is 12.2 ± 0.3 Å. The indicated parameters are similiar for all the complexes presented. The quality of the STM images of TTF-TCNQ and 4EP-TCNQ systems permits to visualize the molecular orbitals involved in tunneling current. The molecular orbital calculations using of X-ray data [11] showed that the highest occupied molecular orbitals of TTF molecule should be situated between and paralell to the rows of TCNQ molecules. In the image of TTF-TCNQ presented here the two balls occupy the positions perpendicular to rows. It seems that these balls represent the orbitals of TTF molecule which can be localized nearby sulfur atoms (perhaps p_z - orbitals). The reason of discrepancy between this image with one published before [12] is not clear.

Figure 6: Chemical structure of donor molecules, tetrathiofulvalene (TTF), 4-ethylpyridine (4EP), triethylamine (TEA) and of the acceptor tetracyanoquodimethane (TCNQ) used for the formation of conductive charge transfer complexes

Figure 7: The STM top view images of different charge transfer complexes
a - TTF-TCNQ (V = -34.5 mV, I = 1.1 nA) - b - TTF-TCNQ (V = 35.1 mV, I = 1.1 nA) - c - 4EP-$(TCNQ)_2$. (V = 37.5 mV, I = 4.9 nA)
d - TEA-$(TCNQ)_2$, (V = 26.6 mV, I = 1.2 nA)

4EP

TCNQ

TEA

TTF

In the STM image of 4EP-TCNQ the more intense patterns are observed between the TCNQ rows exactly. They may correspond to molecular orbital of the conjugated system of pyridine ring. This proposal correlates well with X-ray data of this complex [12]. Higher quality of STM images of TEA-TCNQ complex strongly is required for the determination of the molecular orbital arrangement in this compound.

In general STM observations of different crystallographic planes are possible. In some cases the obtained STM image of the TCNQ complex with 4EP shows up different from that presented in Fig. 7. The close packed hexagonal arrangement of more intense patterns is the main feature of these images. The parameter of the hexagonal lattice was approximately 7 Å. The comparison of such images with crystallographic data should clarify this situation.

A further observation supports the idea that the STM studies of organic charge transfer complexes are very informative, and very complex at the same time. In Fig. 7 STM images of TTF-TCNQ obtained at bias voltages of different polarity are presented also. It should be noted that the shape of molecular level features in these images are quite different. The round shape patterns are observed in the case of negative bias voltage, and less-regular patterns are characterictic for a STM image produced at positive bias voltage. This may be the manifestation of the fact that various molecular orbitals are involved in tunneling current at differing conditions. The corresponding effect is well known by STM studies on semiconductors [2].

3. 3. CONDUCTIVE POLYMERS

Due to non-conductivity of most of biological and synthetic polymers the first STM applications in the polymer field were done with a relatively restricted number of conductive polymers. Polypyrrole and polythiophene (Fig. 8) are well-known representatives of conductive polymers. Despite numerous physical studies the structure of these po-

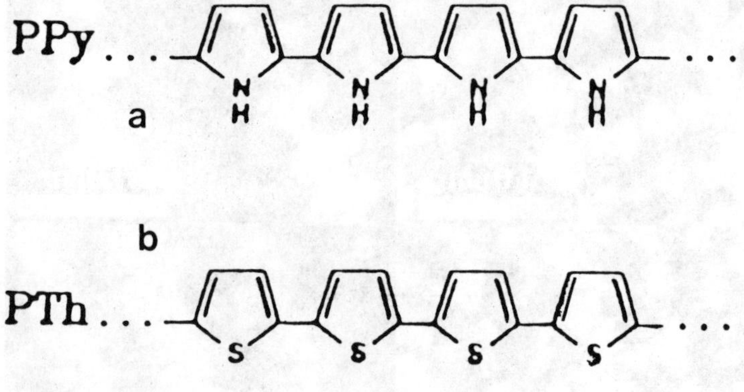

Figure 8: Chemical structure of conductive polymers
 a - polypyrrole, b - polythiophene

lymers still is poorly understood. For our STM experiments the indicated polymers were synthesized electrochemically on conductive ITO glass which was used as an electrode [13]. Material obtained during electrochemical polymerization is porous, and it exhibits highly developed surface even at the scale of hundreds of nanometers. Such

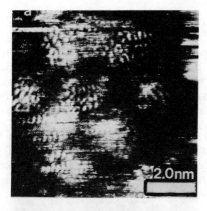

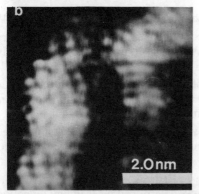

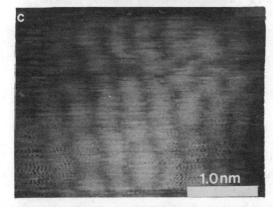

rough surfaces are very difficult for atomic scale STM investigations. Additional problems are connected with thermal motions which occur even on glassy and crystalline polymers at a molecular scale. Nevertheless in some parts of polypyrrole and polythiophene films we succeeded in observing molecular structural features. The corresponding images are presented in Fig. 9. These images are not ordered periodically so well as images of other compounds discussed above. They give definite information, however, concerning polymer chain organization. The collections of ordered white stripes and dots represent locally organized chain segments. This statement is based on the correlation between cross section sizes of the STM patterns and diameter of polypyrrole (polythiophene) chain. No long range orientation is observed in these macromolecular systems. The structural effects observed in the discussed conductive polymers also show the stacking phenomenon between chemical groups of neighbouring chains. This phenomenon should be expected an for explanation of conductivity. Definite similarity can be detected between structures of conductive polymers and conductive organic charge transfer complexes.

Figure 9: The STM top view image of conductive polymers - $\underline{a}$, $\underline{b}$ - polypyrrole (V = 131 mV, I = 1.9 nA) - $\underline{c}$ -polythiophene (V = 20.8 mV, I = 2.0 nA)

4. Conclusions

The scanning tunneling microscopy offers a new possibilities for atomic scale structure analysis of different compounds. Direct visualization of crystal surfaces can provide information which is difficult to receive by traditional crystallografic techniques. The advantage of STM was realized in discovery of threedimensional distortion of unit cell in α-RuCl$_3$. In case of VS$_2$ crystals which quality was not good enough for detailed X-ray studies application of STM was also succesfull. Atomic scale features of VS$_2$ surface indicate of existence of superlattice (charge density waves) at room temperature. The STM images of TCNQ complexes provide information about surface charge density distribution. Efforts to understand these images of surfaces composed of atoms of different nature in correlation with crystallografic data and molecular orbital considerations seems to be perspective. In not well ordered macromolecular systems such as conductive polymers STM can give structural information about morphology at polymer chain level. This was demonstrated in the studies of polypyrrole and polythiophene.

Acknowledgements. S.N.M. is pleased to express his gratitude to Deutsche Forschungsgemeinschaft for support of his research work in Freiburg. Thanks are to Dr. H. W. Rotter and H. Hillebrecht for samples of α-RuCl3 and VS_2, to S. Kempf for preparation of TCNQ complexes and to Prof. J. Heinze for samples of conductive polymers.

References

1. Binnig G., Rohrer H. 'Scanning tunneling microscopy' Surf. Sci. 152/153, 17 (1985)
2. Feenstra R. 'Scanning tunneling microscopy: semiconductor surfaces, adsorption and epitaxy' Lectures at the International Workshop 'Basic Concepts and Applications of Scanning Tunneling Microscopy and Related Techniques' Erice (Italy) (1989)
3. Behm J. 'Scanning tunneling microscopy of metals, adsorbates and reactions' Lectures at the International Workshop 'Basic Consepts and Applications of Scanning Tunneling Microscopy and Related Techniques' Erice (Italy) (1989)
4. Slough C.G., McNairy W.W., Coleman R.V., Drake B., Hansma P.K., 'Charge density waves studied with the use of a scanning tunneling microscopy' Phys. Rev. B 34, 994 (1986)
5. Thomson R.E., Walter U., Ganz E., Clarke J., Zette A., Rauch P., DiSalvo F.J. 'Local charge density waves structure in 1T-TaS$_2$ determined by scanning tunneling microscopy' Phys. Rev. B 38, 10734 (1988)
6. Magonov S., Cantow H.-J., Hillebrecht H., Drechsler M., Rotter W. 'The atomic structure and superstructure of vanadium disulfide by scanning tunneling microscopy' Optik 83, Suppl.4, 60 (1989)
7. Slough C.G., Giabattista B., Johnson A., McNairy W.W., Wang C., Coleman R.V. 'Scanning tunneling microscopy of 1T-TiSe$_2$ and 1T-TiS$_2$ at 77 and 4.2K' Phys. Rev. B 37, 6571 (1988)
8. Cantow H.-J., Hillebrecht H., Magonov S., Rotter H.W., Thiele G. 'Atomic structure and superstructure of a-Ruthenium trichloride by scanning tunneling microscopy' submitted to Angewandte Chemie (1989)
9. Brodersen K., Breitbach H.K., Thiele G. 'Die Feinstruktur des Ruthenium(III)-bromid', Z.Anorg.Allg.Chem. 357, 162 (1968)
10. Gimzewski J.K., Möller R. 'Transition from tunneling regime to point contact using STM' Phys. Rev. B 36, 284 (1987)
11. Sleator T., Tysko R. 'Observation of individual organic molecules at a crystal surface with use of a scanning tunneling microscope' Phys. Rev. Lett. 60, 1418 (1988)
12. Rotter H.W., preliminary data of X-ray analysis of 4EP-TCNQ complex
13. Bilger R., Cantow H.-J., Heinze, J., Magonov, S. 'Scanning tunneling microscope images of doped polypyrrole on ITO glass', EMSA Proc. (1989)

ELECTRON TUNNELING THROUGH A MOLECULE

C. JOACHIM[*] and P. SAUTET[+]
*Laboratoire de Chimie des Métaux de transition
4, place Jussieu
75230 Paris Cedex 05 (France)
+Laboratoire de Chimie Théorique
Ecole Normale Supérieure de Lyon
69364 Lyon Cedex 07 (France)

ABSTRACT. To interpret tunneling current through a molecule, it is important to distinguish between through bond and through space tunneling processes. A definition of these processes is proposed. For each one, the methods to calculate the tunneling current are briefly presented. A tight-binding example is analytically solved to illustrate them. Recent results on the STM imaging of Alkanes are discussed by considering through bond processes.

1. INTRODUCTION

The tunneling conductance of very thin film materials, insulator in the bulk, can be measured with a Scanning Tunneling Microscope (STM). For example, under an STM tip, organic monolayers [1] and more surprising multilayers [2] present a high conductance compared to the one measured with more traditionnal large scale metal - thin film - metal tunneling junctions. But no large discrepancies between theory and experiences has appeared since the Frenkel work [3] on the non-linear current-voltage characteristic of contact junctions between solid conductors. Therefore, to understand this apparently unexpected high conductance, it is important to study the electronic role of a molecule in a wire-molecule-wire tunneling junction. At which conditions a molecule is "passive", spacing only the two electrods like in Langmuir Blodgett films [4], or "active", increasing the tunneling current compared to a vaccum space between the electrods ?

A first element of answer was given by Schnupp [5]. He argued that interpretting the tunneling current through a very thin film from its bulk band structure may not be a good approximation: when the thickness of this film decreases, an overlap appears between the tunneling transmission through the valence and the conduction band in the region of the bulk energy gap [5]. This means that a molecule of the film may contribute indirectly to the tunneling current even if none of its molecular states are in resonance with the Fermi level of the electrods.

To analyse this contribution, tunneling processes can be classified using the well documented work on through space and through bond electron transfer processes.

The distinction between through space and through bond tunneling is first discussed in section 2. To study the competition between these two processes, an analytical tight-binding model is used in section 3. Tunneling through alkane chains is analyzed in section 4. In conclusion, the control of the tunneling current by a modulation of the through bond processes is discussed in section 5.

2. THROUGH SPACE AND THROUGH BOND TUNNELING

2.1. Definitions

Let us consider a wire-molecule-wire-bias voltage V circuit, for example the tunneling circuit of an STM with a molecule between the tip and the substrat. If the vibrations of the molecule are frozen, at least two processes may lead to a non zero tunneling current I in the circuit:
- electrons can tunnel from one wire to the other without any interaction with the molecule. It is a through space tunneling process.
- electrons can tunnel between the two wires using the specific tunneling path introduced by the molecule but with no direct tunneling between the wires. It is a through bond tunneling process.

This terminology is an extension of the distinction in intramolecular electron transfer processes between through space and through bond transfers [6]. It does not consider inelastic scattering processes. To justify this extension, a monoelectronic one dimension model for the wire-molecule-wire circuit is used here. The electrons are free in the wires far away from the molecule. They get a $W(z,V)$ potential energy on the molecule or when approaching a wire-molecule junction. The bias voltage V is applied far from these junctions. In this case, the I(V) current-voltage characteristic of the wire-molecule-wire unit is usually estimated, in the elastic scattering regime, from the average transmission coefficient [7]:

$$I(V) = 2eh^{-1} \int [f(E-u_1)-f(E-u_2)] \; T(E,V) \; dE \qquad (1)$$

Where f is the Fermi-Dirac distribution and $T(E,V)$ the transmission coefficient of an electron with an energy E through the $W(z,V)$ barrier. The quasi-chemical potential of the wires are respectively u_1 and u_2 with $eV=u_1-u_2$. Quite generally, $T(E,V)$ is obtained by solving the Schrödinger équation:

$$H \Psi(z) = E \Psi(z) \qquad (2)$$

after a decomposition of $\Psi(z)$ in plane wave on the left and right side of the molecule. H is the mono-electronic hamiltonian of the wire-molecule-wire unit. $W(z,V)$ is supposed to be zero out of the [-a,a] interval and $W(z,V)<W_{max}$. In the following, E is chosen lower than W_{max}.

To transpose the through band and space electron transfer terminology to a tunneling process, two wave functions must be constructed (the so called diabatic function of the quantum chemist [8]):

$$g_l(z) = \begin{cases} A\,e^{+ik(z+a)} + A'\,e^{-ik(z+a)} & z<-a \quad (3a) \\ B(z)\,e^{-s(z+a)} & z>-a \quad (3b) \end{cases}$$

$$g_r(z) = \begin{cases} B'(z)\,e^{+s(z-a)} & z<a \quad (4a) \\ C\,e^{+ik(z-a)} + C'\,e^{-ik(z-a)} & z>a \quad (4b) \end{cases}$$

with $s=[2mh^{-2}(W_{max}-E)]^{1/2}$ and k the electron wave vector.

The propagative parts (3a) and (4b) are defined in the wires far from the molecule and the tails (3b) and (4a) in the vicinity of the molecule. These functions are the eigenfunctions of H when the overlap between $e^{-s(z+a)}$ and $e^{+s(z-a)}$ is negligible and if W(z,V) does not introduce discret (intermediate) states. In this case, the electronic coupling between $g_l(z)$ and $g_r(z)$ is zero ($<g_l|H|g_r>=0$) and T(E,V)=0. Moreover, (3) and (4) are no more eigenfunctions of H and then T(E,V)≠0 in two cases:

- if there is an overlap between the two exponentials in (3b) and (4a) leading to a through space electronic coupling $<g_l|H|g_r>\neq 0$, with no intermediate state in W(z,V). It is a through space contribution to the tunneling current.

- if the potential W(z,V) introduces at least one intermediate state of wave function $h_i(z)$. In this case, T(E,V)≠0 due to the indirect coupling $<g_l|H|h_i>.<h_i|H|g_r>$ between $g_l(z)$ and $g_r(z)$ through this state, even if $<g_l|H|g_r>=0$. This is a through bond effect since only the molecule can produce such an intermediate state if the wires are considered free of defects on their surface and if the surface states are used only to define through space tunneling paths. Moreover, this intermediate state does not need to be in resonance with the conducting band of the wires to contribute to T(E,V) (see below).

2.2. Through space T(E,V)

To derive the I(V) law in the case of a through space tunneling, the most usefull and used approximation is the W.K.B expression for T(E,V) [9]:

$$T(E,V) = \exp - [4\pi h^{-1} \int_{-a}^{a} (2m(W(z,V) - E))^{1/2}\,dz] \quad (5)$$

Since the paper of Simmons [10], the conditions to apply this approximation were not often recalled. (5) is only valid for a thick and high barrier in such a way that W(z,V) does not vary too sharply in z. This is an important condition because when V increases, even without a molecule between the wires, the W(z,V) variations become too sharp in z for the W.K.B. approximation to be valid. In that case, more sophisticated techniques must be used to calculate T(E,V) [11]. The result is rarely analytical as compared with (5). Notice that the Bardeen transfer hamiltonian [12] does not work well either when V increases because it comes from a golden rule like approximation which is only valid in the weak coupling limit [13,14].

2.3. Through bond T(E,V)

Through bond tunneling like through bond electron transfer cover different processes: <u>double exchange</u> with simultaneous electron tunneling from one wire to the molecule and from the molecule to the other wire (or the reverse process), a <u>chemical</u> mecanism where the electrons reduce the molecule or a <u>super exchange</u> mecanism like in long range magnetic interaction. The second process does not occur when the LUMO and HOMO of the molecule are well separated and the LUMO is to far away from the wires Fermi level to participate to the tunneling current.

A way to get the I(V) law for tunneling through a molecule is again to use the W.K.B. approximation (4) with a W(z,V) taking into account the molecule. Since it is very difficult to get a good estimation of this W(z,V), an approximation is to remplace in (4) the work function of the wires by one taking the adsorbed molecule into account. The major advantage is to get an analytical expression for T(E,V). But this approximation is based on the hypothesis that W(z,V) varies slowly. This is clearly true without molecule and in the space between the two wires. This is questionable when a molecule is added because this molecule induces W(z,V) variations at the atomic scale. In STM imaging interpretations, the approximation to shift the work function of the surface using the molecular dipole moment density [15] intends to take the W(z,V) variations into account. But by construction, (4) accomodates only through space tunneling processes. Moreover, it is difficult to include all the molecular through bond tunneling processes described above in this modified work function which is an average property of the surface [16].

One possibility to overcome these problems is to use a W.K.B. approximation adapted for W(z,V) with intermediate states [9]. But again W.K.B. is only valid for slow varying potentials before and after the location of the intermediate state. This is incompatible with the range of W(z,V) variations on and near a molecule. Another possibility is the Bardeen transfer Hamiltonian method [12]. But like in the approximation (4), this method provides only information on through space tunneling. The matrix elements which have already been calculated with it for the wire-molecule-wire tunneling problem depend only on through space electronic coupling between wire orbitals like $g_l(z)$ and $g_r(z)$ ponderated by their mutual overlap [17].

Therefore, to include through bond coupling in T(E,V), one way is to remplace, in the Bardeen method, H by an effective hamiltonian constructed to take into account through bond processes [18] (see also section 3). The other way is to use the true scattering matrix S(E,V) of the electron-W(z,V) scattering process instead of its W.K.B approximation. S(E,V) is obtained by a non-unitary transformation of the spatial propagator associated with (2). This transformation is equivalent to decompose the H eigenfunctions on plane waves [11]. The spatial propagator takes into account through space and through bond coupling because the equation verified by this propagator is equivalent to (2). $T(E,V)=|S(E,V)|^2$ for one electronic channel per wire and $T(E,V) = N^{-1} tr[G(E,V).G^+(E,V)]$ for N channels per wire with G(E,V) the diagonal block of S(E,V) [19].

For example, S(E,V) can be calculated with the Elastic Scattering Quantum Chemistry (ESQC) procedure [20] using quantum chemistry hamiltonian. In this case, the limitation in the precision of T(E,V) is no more at the S(E,V) calculation step (which is exact in ESQC) but in the choice of an hamiltonian to describe properly the molecule and the wire-molecule interactions. The present version used the Extended Huckel Molecular Orbital method (EHMO). In section 4, the discussion on tunneling through alkane is based on ESQC-EHMO calculations.

3. A TIGHT-BINDING ANALYTICAL EXAMPLE

3.1. Calculation of the effective coupling

The model tight-binding wire-molecule-wire chain fig. 1 is composed of two semi-infinite chains (the wires) linked together by a single level w (a molecular level). The two chains are coupled through the space by the c interaction and to w by the b interactions. To calculate the effective electronic coupling $C_{eff}(E)$ between these two chains which results from a competition between the b and c interactions, an effective hamiltonian H_{eff} is needed [19]. For its construction, one can decompose the eigenfunctions $\Psi(z)$ of the system fig.1 on the basis set $f_n(z)$ where the matrix H of its hamiltonian is known:

$$\Psi(z) = \sum_n C_n(E) f_n(z) \qquad (6)$$

The $C_n(E)$ are solution of an infinite set of equations obtained from the projection of equation (2) on the functions $f_n(z)$ [21]. Since the tight-binding coupling in the chains extend only to the nearest neighbour, the coefficient $C_0(E)$ of $\Psi(z)$ on the level w can be eliminated by substitution. A new set of equations is obtained which can be attributed to a new hamiltonian H_{eff} constructed on H [18]. It is a Löwdin type hamiltonian because the elimination of $C_0(E)$ is equivalent to a distinction between two subspaces, one generated by all the semi-infinite chains $f_n(z)$ and one generated only by $f_0(z)$. Therefore, no perturbative development is needed to construct H_{eff} which is equal to:

$$H_{eff} = \begin{bmatrix} \ddots & \cdot & & & & \\ \cdot & e & d & & & \\ & d & e(E) & C_{eff}(E) & & \\ & & C_{eff}(E) & e(E) & d & \\ & & & d & e & \cdot \\ & & & & \cdot & \ddots \end{bmatrix} \qquad (7)$$

The interesting term is the effective coupling $C_{eff}(E)$ given by:

$$C_{eff}(E) = <f_{-1}| H_{eff} | f_1> = c + b^2.(E-w)^{-1} \qquad (8)$$

There are two contributions in $C_{eff}(E)$. The first one c, is the standard through space interaction which decreases generally exponentially as a function of the distance. The second one, $b^2.(E-w)^{-1}$, is an interaction between the chains through the level w. It is a through bond coupling since it does not depend on c and is function of the energy E of the incoming electron. It decreases as a function of the number of level w in series but much more slower than the exponential [6]. The through bond coupling induces also a shift in energy of the states of wave function $f_{-1}(z)$ and $f_1(z)$:

$$e(E) = <f_{-1} | H_{eff} | f_{-1}> = <f_1 | H_{eff} | f_1> = e + b^2.(E-w)^{-1} \qquad (9)$$

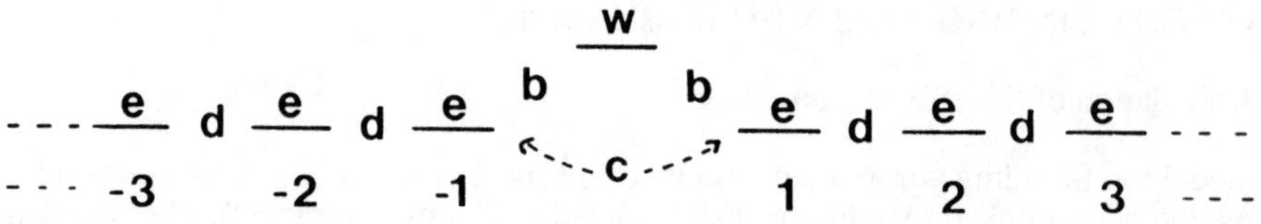

```
              w
            ╱   ╲
  e    e    e  b    b  e    e    e
---  d ─ d ─   ←--c--→   ─ d ─ d ─ ---
 -3    -2   -1            1    2    3
```

Figure 1: Tight binding chain to study the competition between through space and through bond tunneling processes: $b = \langle f_{-1}|H|f_0\rangle = \langle f_0|H|f_1\rangle$, $c = \langle f_{-1}|H|f_1\rangle$, $d = \langle f_i|H|f_i\rangle$, $e = \langle f_i|H|f_i\rangle$ for $i \neq 0$ and $w = \langle f_0|H|f_0\rangle$.

3.2. Calculation of T(E)

The electronic transmission coefficient $T(E)$ through the level w is a function of $C_{eff}(E)$. To calculate $T(E)$, the propagator $M(m,n)$ on the chain from the level n to the level m (see fig. 1) must be constructed to get the scattering matrix as discussed in section 2. Using (6), $M(m,n)$ can be defined by [21]:

$$\begin{bmatrix} C_{i+1}(E) \\ C_i(E) \end{bmatrix} = M(i+1,i-1) \begin{bmatrix} C_i(E) \\ C_{i-1}(E) \end{bmatrix} \quad (10)$$

With (7), the matrix element of this propagator from the left of w to the right of w are equal to [21]:

$$M_{11}(+2,-2) = [d\, C_{eff}(E)]^{-1} \cdot [(E - e(E))^2 - C_{eff}(E)^2] = X \cdot C_{eff}(E)^{-1} \quad (11a)$$
$$M_{12}(+2,-2) = -\, C_{eff}(E)^{-1} \cdot (E - e(E)) = -\, Y \cdot C_{eff}(E)^{-1} \quad (11b)$$
$$M_{21}(+2,-2) = C_{eff}(E)^{-1} \cdot (E - e(E)) = Y \cdot C_{eff}(E)^{-1} \quad (11c)$$
$$M_{22}(+2,-2) = -\, d \cdot C_{eff}(E)^{-1} \quad (11d)$$

After the decomposition of the $C_n(E)$ on plane waves, far from the level w, to get the scattering matrix, (10) and (11) give for $T(E)$:

$$T(E) = 4(1-q^2) \cdot C_{eff}(E)^2 \cdot [4Y^2 + (X-d)^2 - 4qY(X+d) + 4q^2Xd]^{-1} \quad (12)$$

with $q = (2d)^{-1}(E-e)$.

It is clear that $C_{eff}(E)$ controls $T(E)$. A detailed analysis of (12) shows that $T(E)$ depends on the three tunneling processes identified in section 2: through space tunneling controled by c, through bond tunneling controled by b and e-w and a mixing between these two processes which disappears when b=0 or c=0.

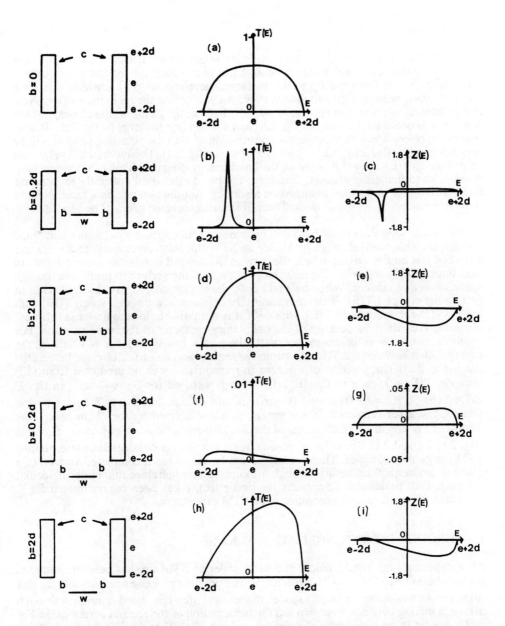

Figure 2: T(E) and Z(E) variations as a function of the tight binding matrix elements b and w. The relative position of the level w in the chain bands is represented on the left column. Each chain band extends between e-2d and e+2d. For 2.a, 2.c, 2.e, and 2.i, c=0.5d.

3.3. Discussion on T(E)

When c=0, there is no through space contribution to the current I. This is for example the case when a molecule under the tip of an STM is too long for the current to tunnel through the space from the tip to the surface. Therefore, to get a current, this long molecule must be more efficient than the vaccum in coupling the tip to the surface. As it is well known, when w is in resonance with the band of the semi-infinite chains, a sharp resonance is obtained for small b (fig. 2.b) and a broad one for large b (fig. 2.d). If now w is not in resonance with these bands, contribution of w to the tunneling process will be high for large b (fig. 2.h) and very low for small b (fig. 2.f). Therefore, the level w can contribute to T(E) even if it is out of the band energy domains. This is a non resonant through bond tunneling process. These are the basic rules used generally to interpret STM imaging of molecules adsorbed on a surface. But the non resonant through bond tunneling process is not often invoked even if it was discovered a long time ago [5].

Another interesting case is b≠0 and c≠0 i.e. when electrons can tunnel at the same time through the space and through w. Under an STM tip, this corresponds to a molecule adsorbed flat on the surface where the current has a small number of atomic layers to pass through [22] and when the tip is close enough to the surface to produce a through space tunneling current. When b≠0 and c≠0, the important property to deal with, in getting the complet T(E), is the non additivity between the through space T(E), with b=0, and the through bond T(E), with c=0. It is not possible to sum up these different tunneling contributions because of the ondulatory character of the wave function. For example, when w is in resonance with the chain bands and for a small b, the superposition between the T(E) is destructive near w and constructive elsewhere in the band (fig. 2.c). The position of the zero in transmission can be predicted from (12) because $T(E_o)=0$ leads to $C_{eff}(E_o)=0$ which is verified for $E_o=w-b^2c^{-1}$. In fig. 2, $Z(E)=T(E)-[T(E,b=0)+T(E,c=0)]$ is our indicator of non-additivity. T(E) for b=0 is given fig. 2.a for reference. In the energy window defined by the chain bands, the destructive or constructive character of the T(E) superposition depends on the E_o value (fig. 2.e, 2.g and 2.i). These interference effects were never considered in interpretting STM image of molecules. Therefore, fig.2 can be viewed as a completed version of a guide to understand tunneling through a molecule and therefore tunneling molecular imaging. But notice that to amplify the c≠0 effect, c has been overestimated fig. 2 compared for example to the condition of an STM experiment.

4. THROUGH BOND TUNNELING IN ALKANE

An unexpected high conductance, measured under an STM tip, has been attributed to alkane chains $CH_3(CH_2)_{n-2}CH_3$ with n>10. The possible processes responsible for this high conductance are: through space tunneling, through bond tunneling through alkane, a mixing between these two and the participation of surface and defect states. For n>10, the tip is far away from the surface, at least 20 A. Therefore, if the tip does not penetrate in the alkane layer, the through space tunneling between the tip and the surface, as defined in section 2, can be neglected. The participation of the surface states (or defects) will not be discussed here because such states have not been well identified so far even if there are some indications of their existence [2,23]. Concerning the through bond tunneling process, we have calculated T(E) for electrons able to tunnel through the gold wire-alkane-gold wire chain fig. 3 using the ESQC procedure [20].

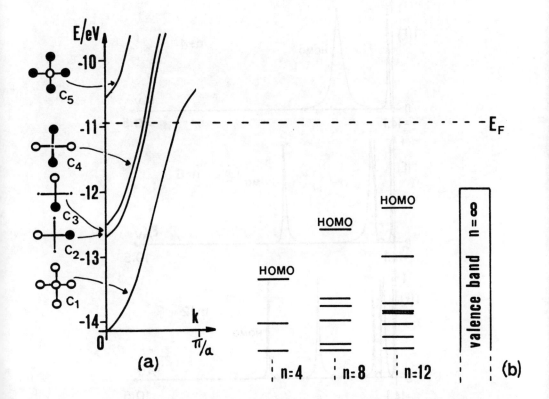

Figure 3: Structure of the gold wire-alkane chain-gold wire studied with ESQC. Notice that in ESQC, The two gold wires are taken semi-infinite respectivelly on the left and on the right side of the alkane.

Figure 4: Position of the $CH_3(CH_2)_{n-2}CH_3$ molecular levels in the gold wire fig. 3 conduction band. The 5 propagative channels C_i of the gold wire are defined in 4.a and the polyethylene chain valence band in 4.b for reference.

Before going into the details of the results, let us remark that high conductance for alkane means so far a resistance as high as 1 GigaOhms: with n>10, the current measured under an STM tip is of the order of 100 pA for V=100 mV [2]. Truly, this resitance is not the resistance of the alkane alone and takes into account the electrods to alkane tunneling barrier and the constriction effect [24]. But it is an indication of the T(E) magnitude near the electrods Fermi level which must be of the order of 10^{-4} to 10^{-5} since in (1), $(2e^2/h)^{-1}$=25.81 KOhms [24].

Because $CH_3(CH_2)_{n-2}CH_3$ appears to be only physiosorbed on usual metallic surface [25], a 3 A distance has been chosen between the last atoms of the gold wires and the first carbon atom of the alkane (fig. 3). To simplify the ESQC calculation, each gold atom is only described by its 6s orbital. Even with this simplification, there are 5 propagative electronic channels in the chosen gold wires (fig. 4.a). For n>3, the HOMO of the alkane is always in resonance with one the gold wire bands (figure 4.b).

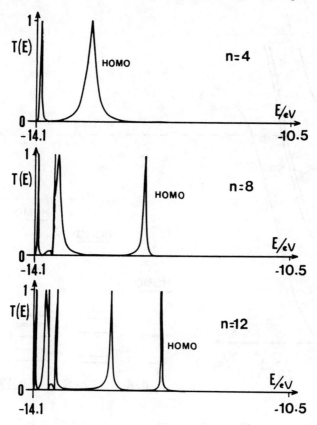

Figure 5: Transmission coefficient T(E) for an electron coming from one gold wire and tunneling through the $CH_3(CH_2)_{n-2}CH_3$ alkane chain fig. 3. It is an average T(E) taking all the propagative channels contributions into account. The calculations use an ESQC procedure with an EHMO hamiltonian.

The major contribution to T(E) comes from through bond coupling between each channel 1 of the gold wires using the alkane resonant molecular orbitals (fig. 5). Since the overlap between the 6s last gold atoms orbital and the first 2p carbon orbital is very low (0.03) compared with the gold to gold 6s atomic orbital overlap (0.135) inside a wire, T(E) for each resonant molecular level has the shape reported fig. 2.b. Constructive and destructive interference effects occur for example at the bottom of the channel 1 band between the different tunneling paths introduced by the alkane molecular orbitals. The 4 other channels of a wire are through bond coupled with the channel 1 of the other wire. These channels are independent in their respective wire but are coupled in the alkane. This contribution to T(E) is very small because channels 2,3,4 and 5 have not the same symmetry than the alkane molecular orbitals located in the same energy domain. The through bond coupling between channels 2,3,4 and 5 of each wire is negligible for the same reason.

Most interesting are the T(E) variations near the gold Fermi level, approximativelly located in the middle of the channel 1 band. The tail of the T(E) HOMO resonance extends far away in the region of the polyethylene gap and reaches 10^{-6} to 10^{-5} at the estimated gold Fermi level (fig. 6). Moreover, this tail does not disappear as fast as the exponential when the alkane chain length increases. This is due to a competition between the formation of the polyethylene valence band and the shift of the HOMO in direction of the top of this band. T(E) decreases in the region where the gap is under formation and the shift of the HOMO resonance increases T(E) in the same energy region.

In conclusion, tunneling through an alkane chain is an interesting example, at the Fermi level, of a through bond tunneling process in the absence of a resonant molecular level at this energy. The value of the T(E) tail at the Fermi level is a little lower than it would be expected from an STM experiment. The distance chosen between the gold wires and the alkane may be too long. But the order of magnitude of T(E) is the one expected.

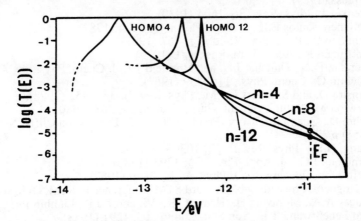

Figure 6: Log(T(E)) of the T(E) fig.5 taken after the HOMO tunneling resonance to follow the T(E) tail variations near the gold wire Fermi level E_F. Only the channel 1 contribution to T(E) is represented. Notice that T(E) is deformed at the top of the gold wire band due to a band edge effect.

5. CONTROL OF THROUGH BOND TUNNELING

If the molecule is long enough, through space tunneling does not exist. This open the way to experiment on purely intramolecular electronic phenomena produced and controlled by a single molecule. The same strategy was used to design active organic ligands which control electron transfer processes in valence mixte compounds [26]. One of these phenomena is the control of the through bond coupling by molecular conformation changes produced for example by a photo-isomerisation [26] or a TICT effect [27]. The STM would provide a way to follow electrically the change in conformation of one of these molecules depending, for example, on the intensity of the laser source.

Another interesting phenomena is due to the non superposition, discussed in section 3, between through bond and through space tunneling. This non additivity can be used to introduce controlable intramolecular interference effects between two through bond tunneling paths. As it is well known, interference patterns are controlled by the phase difference between the part of the wave travelling in each channel. In mesoscopic loops, this is done by electric or magnetic fields. At the molecular level, chemistry can do it for example by the attachment of a donor and or an acceptor group on a benzene ring [28]. This is an opportunity to develop intramolecular physics.

6. REFERENCES

[1] : Hörber J.K.H., Lang C.A., Hensch T.W., Heckl W.M. and Möhwald H., Chem. Phys. Lett., 145, 151 (1988).
[2] : Michel B., Travaglini G., Rohrer H., Joachim C. and Amrein M., Zeit. Phys. B, in press (1989).
[3] : Frenkel J., Phys. Rev., 36, 1604 (1930).
[4] : Polymeropoulos E.E., J. Appl. Phys., 48, 2404 (1987).
[5] : Schnupp P., Phys. Stat. Solid., 21, 567 (1967).
[6] : Hoffmann R., Acc. Chem. Res., 4, 1 (1973).
    Beratan D.N., Onuchic J.N. and Hopfield J.J., J. Chem. Phys., 86, 4488 (1987).
    Joachim C., Chem. Phys., 116, 339 (1987).
[7] : Lenstra D. and Smokers R.J.M., Phys. Rev. B, 38, 6452 (1988).
[8] : Newton M.D., Int. J. Quant. Chem. Symp., 14, 363 (1980).
[9] : Bohm D., Quantum Theory (Prentice-Hall Inc, Englewood Cliffs, New Jersey, 1951) p 275.
    Lang. N.D., Phys. Rev. B, 37, 10395 (1988).
[10]: Simmons J.G., J. Appl. Phys., 34, 1793 (1963).
[11]: Stein J. and Joachim C., J. Phys. A, 20, 2849 (1987).
    C. Noguera, Proceeding of the third STM conference in STM, Oxford 1988.
    Lucas A.A., Morawitz H., Henry G.R., Vigneron J.P., Lambin Ph, Cutler P.H. and Feuchtwang T.E., Sol. Stat. Comm., 65, 1291 (1988).
[12]: Bardeen J., Phys. Rev. Lett., 6, 57 (1961).
    Harrison W.A., Phys. Rev., 123, 85 (1961).
[13]: Feuchtwang T.E., Phys. Rev. B, 10, 4121 (1974).
[14]: Caroli C., Combescot R., Nozieres P. and Saint-James D., J. Phys. C, 4, 916 (1971).
[15]: Spong J.K., Mizes H.A., Lacomb Jr L.J., Dovek M.M., Frommer J.E. and Foster J.S., Nature, 338, 137 (1989).

[16]: Seitz F., The Modern Theory of Solids (Mc Graw Hill Book Compagny Inc, New York, 1940) p 395.
[17]: West P., Kramar J., Baxter D.V., Cave R.J. and Baldeschwieler J.D., IBM J. Res. Develop., 30, 485 (1986).
[18]: Joachim C., J. Mol. Elec., 4, 125 (1988).
[19]: Fisher D.S. and Lee P.A., Phys. Rev. B, 23, 6851 (1981).
[20]: Sautet P. and Joachim C., Chem. Phys. Lett., 153, 511 (1988).
[21]: Sautet P. and Joachim C., Phys. Rev. B, 38, 12238 (1988).
[22]: Ohtani H., Wilson R.J., Chiang S. and Mate C.M., Phys. Rev. Lett., 60, 2398 (1988).
Lippel P.H., Wilson R.J., Miller M.D., Wöll Ch. and Chiang S., Phys. Rev. Lett 62, 172 (1989).
[23]: Pireaux J.J., Private communication.
[24]: Gimzewski J.K. and Möller R., Phys. Rev. B, 36, 1284 (1987).
[25]: Chester M.A., Gardner P. and McCash E.M., Surf. Science, 209, 89 (1989).
[26]: C. Joachim and J.P. Launay, Chem. Phys., 109, 93 (1986).
[27]: Launay J.P. and Joachim C., J. Chim. Phys. Bio., 85, 1135 (1988).
[28]: Sautet P. and Joachim C., Chem. Phys. (1989) submitted for publication.

ELECTRONIC TRANSPORT IN DISORDERED ORGANIC CHAINS

Ricardo Garcia and N. Garcia
Departamento Fisica de la Materia Condensada
Universidad Autonoma de Madrid
28049 Madrid
Spain

ABSTRACT. We develop a model for calculating the conductance through disordered linear chains. The behaviour of the conductance for disordered systems shows a strong dependence with the length of the system, number of chains and energy of the incoming electrons. We suggest a link of these results and STM measurements on biologicals. Some of the features of those experiments could be explained in terms of properties of disordered systems.

Introduction

Pioneering attempts for imaging biological material with STM (1-2) opened the possibility for direct topographic images of biologicals at high resolution in air. So far STM experiments involve a wide range of samples: DNA (1,3-5), proteins (3,6-8), virus (2), and thin organic films (9-11). Those experiments have a two folded interest. First, provide a way for direct imaging of bare bilogical samples at high resolution. Second, pose the intriguing question of the electronic transport mechanism through a non-conducting sample.

The relevance of those experiments have been slightly blurred because imaging a biological seems to be a non well controlled event and because the lack of a theoretical model where the feasibility of those measurements could be undestood.

In this paper we show how the behaviour of the conductance in disordered systems could be applied for understanding some of the features of STM measurements on bare biological samples. Our model seems to be readily suitable for polypeptide chains.

A brief account of theoretical results together with its interpretation in terms of STM experiments follows.

1). The resistance of disordered systems has a local dependence. Its related to the molecular composition of the area illuminated by the incoming beam of electrons. It changes from place to place while scanning.

2). The ability of imaging biologicals depends on the molecular composition of the sample under study.

3). The resistance of the sample increases in an exponential way with increasing thickness. This implies the consistency of STM experiments with the known insulating properties of macroscopic biological samples.

4). From the results some limitations to the possibility of molecular or atomic resolution with the STM should be expected.

The connection of our results with STM measurements on biological samples stems mainly in two assumptions. i). The existence of energy relaxation. This means that there is a difference in the electronic properties of a system in a neutral state, and when the same system is brought into contact with a metal and an external field is applied. The injection of electrons can shift downwards the empty levels and upwards the occupied levels of the sample, i.e., inducing energy relaxation. ii). The biological system is view as a disordered system. The disorder can be *spatial*, in this the interatomic distance changes along the chain. The other possible source of disorder is known as *strength disorder*. This means that along a chain different atoms are present, i.e., different electronic levels appear. For a polypeptide chain the strength disorder is related to the presence of different amino acids.

Model and results

The magnitude of interest in the present work is the conductance G.

$$G = \frac{I}{V} \qquad 1$$

where I is the current and V the applied voltage. In a linear chain G can be expressed in terms of the transmission probability through the chain.

$$G = G_0 \times T \qquad 2$$

$G_0 = 2e^2/h$ (12). If we assume that the three dimensional system under study is composed of N linear chains without any kind of crosslinking, G will be G_0 times the average transmission probability $<T>$ by the number of chains N

$$G = G_0 \times <T> \times N \qquad 3$$

where

$$<T> = \sum_{i=1}^{N} \frac{T_i}{N} \qquad 4$$

T_i is the transmission probability for a single chain and is calculated by numerical solution of the time independent Schroedinger equation.

Our model has been choosen in a way that resembles some of the main features of a polypeptide chain.

```
  H    H    O         R₂        H    H    O
  |    |    ||        |         |    |    ||
  N----C----C----N----C----C----N----C----C —
       |         |    |    ||        |
       R₁        H    H    O         R₃
```

A----B_1----C---A----B_2----C----A----B_3----C —

A, B_i, C are the scattering centres along the chain that we call hereafter atoms. Each atom is allowed to have just one energy level that is monitored by a delta function potential. The strength disorder appears naturally in a polypepetide chain due to the presence of different side chains (R_1, R_2, R_3). The spatial disorder is related to the difference in N-C, CO-N, C-C bond's length.

We focus our work in the behaviour of G with increasing strength disorder. We also have performed calculations including spatial disorder but its effect on G is less important (13).

For comparison in figure 1 we calculate $<T>$ for an ordered system which unit cell has 3 atoms A,B,C. As a consequence each unit cell has three electronic levels E_1, E_2, E_3 associated to A, B, C respectively. The curve 1 shows the presence of 3 thin bands separtated by deep gaps. For an ordered insulator sandwhiched between two electrodes the Fermi level would lie somewhere within an energy gap. As a reference value we point out that for a standard STM measurement between two metals $T(E_F) \simeq 10^{-5}$.

The strength disorder appears along a polypeptide chain as a consequence of the presence of different amino acids and its enhanced by the following phenomenom: when one electron is injected in an organic chain this electron can induce intermolecular and intramolecular relaxation in the energy of the acceptor and/or donnor states. This idea was proposed for explaining contact-charge-exchange experiments between metals and polymers (14). The energies of the injected electrons lied in a region where the one-electron models of the electronic structure of polystyrene have an energy gap. In some aspects those experiments have some similarities with the STM experiments on biomacromolecules. As we stated above not all the chains are the same. This means that the inter and intramolecular relaxation will be different from amino acid to amino acid. In ref. 14 the energy relaxation is estimated as 4.2 eV for polystyrene and depends on the molecular composition. The inclusion of strength disorder shows a drastic improvement of G. T for energies close to the Fermi level increases about 10 orders of magnitude. This increase, $T \simeq 10^{-5}$, is the required for reaching nanoamp currents in STM experiments. Also besides

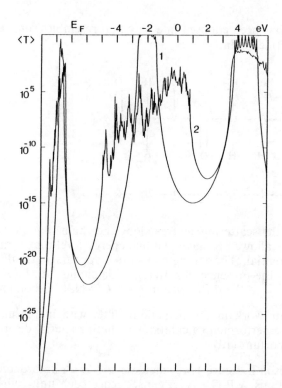

Figure 1. : $<T>$ (logarithmic plot) versus kinetic energy E of the incoming electrons. The origin of energies as well as the position of the Fermi level have been chosen arbitrarily. Curve 1 is for an ordered linear system with 3 atoms per unit cell. Characteristics, $a_0 = 1.4 \text{Å}$, $\Delta(E_2 - E_1) = 4.5 eV$, $\Delta(E_3 - E_2) = 4 eV$, $L=46.2 \text{Å}$. Curve 2 shows the influence of strength disorder. The atoms B_j get their energy eigenvalues from a random discrete distribution centered around E_2. $E = E_2 \pm (0.5, 1.5, 2.5, 3.5)eV$; $N = 400$. Notice that $<T>$ increases about 10 orders of magnitude for $E \simeq E_F$.

the increase of T(E), the curve presents noticeable fluctuations for small differences of energy. We note that only 1/3 of the atoms have modified their energy value with respect the ordered system.

The previous results have been done for a bunch of N = 400 linear chains. There is a proportionality between the number of chains an the area illuminated by the electrons. Larger N means more area and this, in turn, smaller resolution.

Figure 2 shows the curves for two systems with N = 10. There are some remarkable points.

1). For curve 1 we observe that the overall transmission probability over the entire range of energies is lesser for N = 10 than for N = 400. This is easily understood because the number of chains rise the probability for finding few chains with good transmittivity for a fixed energy. The decrease of G with decreasing N has another implication for STM measurements: it suggests a limitation for molecular resolution and/or a reduction of tip-sample distance. 2). Again for curve 1 (though the remark remains valid also for curve 2). We note that T(E) can change 2 orders of magnitude or more for small differences in the energy. This would imply a strong voltage dependence in STM experiments. 3). When comparing the results for two different systems (curves 1 and 2) we observe that for the same energy, T could fluctuate as much as 3 orders of magnitude. We understand this in terms of experimental features as follows. For a fixed applied voltage while the tip is scanning over the sample, the molecular sequence that electrons illuminate is changing from place to place. This

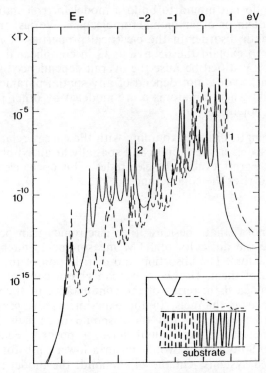

Figure 2. : For small systems (N small) $<T>$ shows strong fluctuations from system to system. Curves 1 and 2 are for two systems of N=10. The same parameters as in figure 1. The inset is a schematic picture of how those results could explain the sudden lost of contrast while scanning. Notice that the amplitude of fluctuations decreases with N decreasing (compare these curves with curve 2 of fig. 1).

would produce some local inestabilities and could be related to a sudden lost contrast in the images. Also it would imply the appearance of spikes in the current. Spike features have been observed scanning DNA samples (15). In fact, in most of the STM images only a small fragment of the whole biological structure have been recorded (see inset for a schematic picture).

Macroscopic samples of biological molecules are known for their insulating properties with respect electronic transport. So far we have shown the ability of disordered systems for carrying $10^7 A cm^{-2}$ currents through systems 50 Å length. Another fundamental consequence of the present model is that gives vanishing currents with increasing lengths. Figure 3 shows the behaviour for energies close to the Fermi level (small voltages) for systems of of different lengths L. The exponential decrease of G with L reflects the consistency of the model with the macroscopic limit.

We note that we do not intend to build a model for calculating the electronic structure of a protein or other biomacromolecule. We just try to monitor the influence of strength disorder in the electrical properties of linear chains and extrapolate them to explain the behaviour G in complicated systems as DNA or proteins. This is justified because the overall dependence of G with strength disorder is general, i). does not depend of any spatial parameter of the model, ii). it remains valid when the atoms a are modeled by other potential well different from delta potentials (16).

Summarizing, strength disorder coupling with the energy relaxation induced by the incoming electrons create a finite and spatially localized presence of allowed states around the Fermi level of the electrodes. This opens electron paths along the the chains and rises the conductance.

Conclusion

We do not ignore that besides the electronic transport through the biomacromolecule the difficulty of STM measurements on biologicals can be related to other factors. The absortion and distribution of the biological on the substrate are two of them. However, we can give a comprehensive picture of STM data on biologicals in terms of the conductance of the sample. We have considered only elastic channels. In the experiments the presence of inelastic processes should be expected and these should open new channels for the electrons. The inelastic effects could have a positive contribution to the conductance (17). So our work would mean a lower limit for the conductance through disordered systems. On the other hand, the model model seems specially suitable for polypeptide chains.

In short, we have developed and explored the consequences of a theoretical model for explaining the electronic transport through biological macromolecules. The proposed mechanism stemed on the enhancement of the conductance due to strength disorder shows the consistency with the experiments in the nanometer range as well as in the macroscopic limit. The main points are the

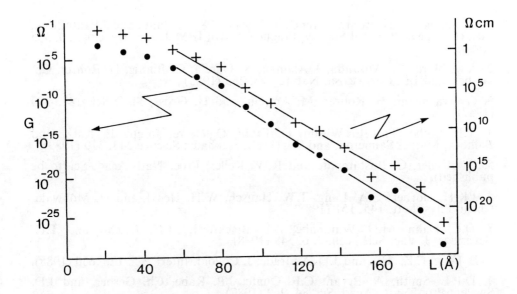

Figure 3. : Exponential decrease of the conductance with the thickness of the system. G versus L for small applied voltages. Deviations from a straight line are the very manifestations that we are dealing with disordered systems. On the right, we plot the resistivity. N = 400 and we estimate an equivalent area of 2000Å^2.

existence of spatial and energy fluctuations that can be related to the difficulty and lack of reproducibility of some STM experiments. The sharp dependence of G with E would imply some voltage dependence in the measurements. The behaviour of G for systems with N small suggests, on the other hand, a limitation for the spatial resolution when imaging biologicals.

Acknowledgement

We thank the IBM Zurich Research Laboratory for hospitality where part of this work was performed and for the finantial support through the Joint Agreement with the Universidad Autonoma de Madrid.

References

1. G. Binnig and H. Rohrer; Trends in Physics 1984, J. Janta and J.Pantoflicek, Eds. (European Physical Society, Prague (1984)); IBM J. Res. Develop. 30, 355 (1986).

2. A.M. Baro, R. Miranda, J. Alaman, N. Garcia, G. Binnig, H. Rohrer, Ch. Gerber, and J.L. Carrascosa; Nature 315, 253 (1985).

3. G. Travaglini, H. Rohrer, M. Amrein, and H. Gross; Surf. Sci. 181, 380 (1987).

4. T.P. Beebe, Trot E. Wilson, D. Frank Ogleterre, Joseph E. Katz, Rod Balhorn, Miquel Salmeron, and Wigbert J. Sickhaus; Science 243, 370 (1989).

5. D. Keller, C. Bustamante and R.W. Keller; Proc. Natl. Acad. Sci.(to be published).

6. J.K.H. Horber, C.A. Lang, T.W. Hausch, W.H. Heckl, and H. Mohwald; Chem. Phys. Lett. 145, 151 (1988).

7. D.C. Dahn, M.O. Watanabe, B.L. Blackford, M.H. Jericho, and T.H. Beveridge; J. Vac. Sci. Tech. A 6, 548 (1988).

8. L. Feng, C.Z. Hu, and J.D. Andrade; J. Colloid. Interf. Sci. 126, 650 (1988).

9. D.P.E. Smith, A. Bryant C.F. Quate, J.P. Rabe, Ch. Gerber, and J.D. Swalen; Proc. Natl. Acad. Sci. 84, 969 (1987).

10. H. Fuchs; Phys. Scr. 38, 264 (1988).

11. J.S. Foster and J.E. Frommer; Nature 333, 542 (1988).

12. R. Landauer; Phys. Lett. 85a, 91 (1981).

13. Ricardo Garcia and N. Garcia; to be published.

14. C.B. Duke and T.J. Fabish; Phys. Rev. Lett. 37, 1075 (1976).

15. Carlos Bustamante and David Keller; private communication.

16. Paul Erdos and R.C. Herndon; Advances in Physics, 31, 65, (1982).

17. M. Buttiker; IBM J. Res. Develop. 32, 63 (1988).

ELECTRON AND ION POINT SOURCES, PROPERTIES AND APPLICATIONS

HANS-WERNER FINK
IBM Research Division
Zurich Research Laboratory
CH-8803 Rüschlikon
Switzerland

ABSTRACT. This paper deals with point sources for charged particles. The importance of a well-characterized ensemble of particles in the framework of the quantum mechanical measuring process is illustrated. The engineering of these sources for electrons and noble gas ions and their emission properties are reviewed. Two practical microscopy applications, the resolution of which depends on the "quality" of the particle beam, are presented.

1. Introduction

Many experiments in solid state physics, including surface science and electron microscopy, can in their very basic design be described by a scheme as illustrated in Fig. 1. Three, in practice often complicated, parts make up the entire experiment. First, there is a "box" that prepares an ensemble of particles, quantum mechanical particles, photons, electrons, ions or atoms. This ensemble of particles impinges onto the second box, representing the system to be investigated. Depending on this sample, on the nature of the particles, their quantum states, position, energy, and spin, interactions have to take place between the particles and the sample. The sample must change the particle ensemble, otherwise no information on the sample is gained. The consequences of these interactions can be manifold: to name but a few, interference effects can occur after having passed the sample as

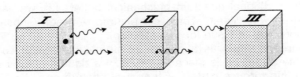

Fig. 1: Schematic separating an experiment with quantum mechanical particles into three physical systems. I: The system that prepares the particles, II: The system to be investigated, III: The system that detects an ensemble of particles like the result of the interaction between system I and II.

can elastic or inelastic scattering or emission of secondary particles from the sample. In order to deduce those interactions, a third box is needed, the detector. Detectors often detect the position of scattered primary particles (e.g. a diffraction or interference pattern caused by the object) or they may detect changes in the quantum state, energy losses or gains, or the yield of the secondary particles.

The above statements are certainly trivial in a sense. Nevertheless they help describe conceptually the structure of experiments with quantum mechanical particles. In looking at the design of various techniques one realizes that, for example in electron microscopy, major efforts have been undertaken to improve box 1, the preparation ensemble, an electron beam well localized in space. In other areas of solid state research, sample preparation or detector resolutions have been the center of attention in the experimental setup.

1.1 INTRINSIC POINT SOURCE PROPERTIES

I shall now concentrate on box 1, the system that prepares the particles with which to conduct the experiment. Intuitively it seems desirable that box 1 should have an opening as small as possible through which the particles can escape towards the system to be studied, because we wish to know as closely as possible where the particles are coming from. Unless searching for "new" particles, no one conducts experiments with particles from an arbitrary origin. On the other hand, small openings usually imply small currents, at least if one has classical dynamical models in mind. A reasonable current through the hole however is important for very pragmatic reasons: most experimentalists want results in a short time compared to the length of their career. Given a small opening and a reasonable flux in forward direction, since we mainly wish to put the particles on the system to be studied, the outcoming particles should also be as similar as possible. They should have the same mass and the same charge state; I shall ignore the spin because we are not set up to measure it. The quantum mechanical particles cannot be well localized in space and momentum at the same time; we therefore want the momentum spread to be small. What are the practical consequences of such a beam?

1.1.1 Interference, coherence of a fermion beam. To obtain a "good" interference pattern it is necessary to be able to observe a large number of interference fringes. In holography experiments with electrons, this will determine the resolution. Pioneering work in this area using a high-energy focussed electron beam, which is the demagnified image of a field emitter source, has been put forth by Lichte and his colleagues at the University of Tuebingen [1].

What determines the coherence of a quantum mechanical particle beam? In accordance with traditional quantum mechanical beliefs [2], we assume that every particle interferes with itself. Time correlation between subsequent emission events is therefore not an issue for this problem.

If a point source, defined as a source of atomic dimension comparable to the wavelength of the particles, is placed in front of a double slit, as illustrated at the top of Fig. 2, an interference pattern will arise if enough particles have reached the detector to distinguish a pattern from statistical noise. This is a matter of beam intensity per solid angle or measuring time. A macroscopic source, a standard field emitter with a curvature radius of 1000 angstroms or so, can be modelled by a set

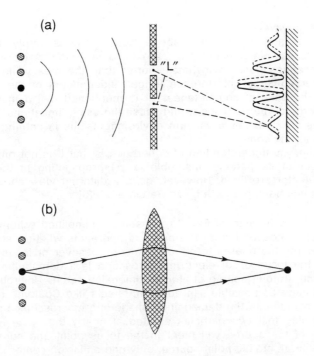

Fig. 2: Schematic indicating the situation of a particle beam originating from a physical point source (full cycle) versus an extended source, modelled by an array of point sources in (a) interference experiments, (b) an optical setup.

of point sources emitting electrons in a random fashion. Electrons originating from each of these emitters would give rise to individual interference patterns, displaced laterally to each other at the detector level. The stochastic nature of the emission events from each source precludes advance determination of which emitter is going to be active next. Otherwise one could, for example, compensate for an array of sources by displacing the screen at the proper time. However, since this is not feasible, only a superposition of all discrete patterns corresponding to individual emitters can be detected. Thus, the overall pattern becomes blurred and only a small number of fringes can be detected if the dimension of the source is too large in real space. The "spatial" coherence is poor. The nth maxima, still observable, defines a length L as indicated in Fig. 2. It is tempting to associate it with a length over which two wavepackets, having gone through each of the slits, still keep a phase relationship. However, this is in contradiction to the assumption of self-interference. Let us therefore associate $L(n)$ only with a number — the dimension of which is a length — that indicates the observability of n maxima for a given slit separation and wavelength.

The argument for the so-called "temporal" coherence is entirely analog: if the particles do not all have the same energy but exhibit a distribution of energies, this translates via the de Broglie relationship into a distribution of wavelengths. Accordingly, for every wavelength, different laterally spaced interference patterns arise. Even for a point emitter, the observability of interference fringes is now limited by the spread in wavelengths.

It should be added that the above considerations imply that the three boxes – source, sample and detector – are fixed in space. A detector, sample or source that changes position in time, shorter than the measuring time, will also limit the observability of interference fringes independent of the "quality" of the source. For the sake of completeness, it is mentioned that a medium that influences the energy of the particles would cause the same limitations. However, since in most cases we consider a vacuum environment, this is nothing to be too concerned about at this point.

In conclusion, my understanding of coherence is that it is not only a property of the source but of its particle ensemble $(\vec{x}, \vec{p})$ propagating in the space where object and detector are fixed. However, quite a different view about coherence of point sources has been put forth by N. Garcia et al. [3].

1.1.2 *Optics.* The lower part of Fig. 2 represents a simplified schematic of an electron microscope. Independent of the mode of operation, whether scanning or fixed beam microscopy, the goal is to obtain a small crossover or focus of the electron beam. Since the focus of an electron microscope is a demagnified image of the electron source, a point source of atomic dimension seems desirable for obtaining as small an image as possible. In addition, effects like Coulomb repulsion in the beam can be avoided. In the demagnified image of a macroscopic source there is a certain probability that two particles emitted at nearly the same time from separated regions of the source will meet in the focus point and consequently repel each other. With an atomic point source, only one emission center is present; particles are therefore generated sequentially, and reach the focus one after the other.

2. Physical realization of point sources

The scanning tunneling microscope [4] and in particular its feature that the information obtained by this technique is a convolution of the sample and tip electronic (and consequently atomic) structure has motivated the desire to be able to prepare and control probing tips on an atomic scale. The "obvious" approach for such a concept was to use field ion microscopy, a technique developed by Erwin Mueller in the fifties [5], which has since revealed significant information on individual atomic processes on surfaces [6]. With the above motivation in mind, it has been possible to prepare tips on an atomic scale that are terminated by a selected individual atom [7].

Both the rapid development of STM as well as my belief in the significance of a well-characterized charged particle beam have led me to depart from the goal described above towards the concept of viewing single atom tips as point sources for ions and electrons in a framework as outlined in the introduction [8].

2.1 ENGINEERING OF POINT SOURCES

The techniques for producing point sources for ions or electrons have been described elsewhere [7,8]; they shall not be repeated here in detail. The important steps towards an electron or ion source of atomic dimension are sketched in Fig. 3.

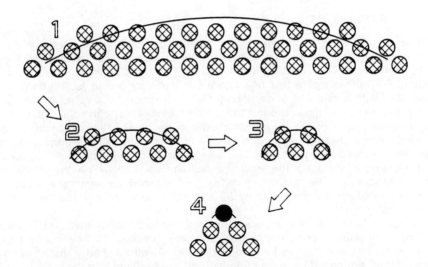

Fig. 3: Schematic showing the steps leading to a point source for electrons or ions. 1: Blunt and facetted tip after high-temperature treatment in UHV. 2: Tip after in situ sharpening by neon ion bombardment followed by short annealing to heal out crystal defects. 3: By controlled field evaporation, an apex of only three atoms remains. 4: A foreign atom from an evaporator has been deposited onto the trimer which leads to the single atom tip. The decrease of the overall radius of curvature is also indicated schematically in the figure. In reality it is about three orders of magnitude.

(1) A single tungsten crystal wire of [111] orientation is electrochemically etched to a tip and placed in an UHV chamber, where it is cleaned by heating above 2000 K. At this high temperature, the tip tries to reach its equilibrium shape, forms facets and becomes blunt. (2) The sharpening of the tip, as well as the removal of segregated carbon to the surface, is achieved by in situ bombardment with neon ions. They are produced by collisions with electrons which are field-emitted from the tip. In order to heal out defects produced during sputtering, the tip is annealed at around 1000 K to re-establish a crystallographic order. (3) In the field ion mode, individual atoms are removed in a controlled manner. While observing the FIM pattern, atoms are field evaporated one by one until a tip terminated in its upper terrace by a cluster made up of three atoms remains. At this preparation stage, well-controlled heating alone also leads to a well-structured apex without the need for further shaping by field evaporation. This produces tips terminated by an apex made up of 7,6,4,3 [7,8] or even a single atom, as demonstrated by Roger Morin and Heinz Schmid [9]. (4) With a trimer tip, as routinely achieved by field evaporation, the last step towards a single-atom tip is the deposition of an additional atom from the gas phase. This can be another tungsten atom or one of a different chemical nature; we also deposited a silicon atom for example. The chemisorption of this foreign atom leads to a field enhancement above this last atom of the pyramidic atomic arrangement. Thus this structure fulfills the requirement for the definition of a particle point source as discussed in the introduction.

2.2 EMISSION PROPERTIES

2.2.1 *Electrons*. Owing to the extremely small curvature radius of single-atom or trimer tips, extraction potentials of only a few hundred volts are needed for field electron emission. The resulting electron beam is directed rather well in forward direction. For currents of a few nA, where the intensity profile is still conveniently measurable with a channel plate detector, the emission is directed into a cone of about 2 degrees half-angle around the emitter axis. Total currents that still leave the atomic structure of the tip undisturbed can be higher than 10 micro-amperes. The width of the energy distribution has been measured to be around 250 mV for single atom as well as trimer tips. The deposition of a few cesium atoms, however, has a significant effect on the emission voltage as well as on the width of the energy distribution. The emission voltage is reduced by almost a factor of 4, leading to an electron beam of only 50 eV, which exhibits a spread in energy of less than 100 mV [10].

A complete and detailed understanding of the processes involved in electron emission from point source tips has not yet been reached. However a first step in this direction has been made by Norton Lang et al. with a theory that models field emission from an individual atom chemisorbed on a jellium surface [11].

2.2.2 *Ions*. The single-atom tip leads to field ionization of noble gases, usually helium or neon, exclusively above the last tungsten atom that is chemisorbed on a trimer, the second layer of the tip. The emission area is confined to the size of one atom or even smaller if one considers the virtual size of the source. This is the region defined by the crossing of the tangentials to the trajectories of the ions at some distance from the emitter. This distance is determined by the location of the object, that "sees" the source; it might be a lens, for example. The cone of the ion beam originating from this atomic or even sub-atomic region in space exhibits an opening of only 0.5 degrees.

The rate of ionization or the resulting current is limited by the supply of atoms from the gas phase. In a normal UHV chamber the partial pressure must not exceed a few 10^{-4} mbar in order to avoid discharges and to keep detectors operable. Heinz Schmid from our laboratory has circumvented this limitation by placing the tip into an only locally high helium pressure environment. This leads to the generation of a helium ion every few nano-seconds above a tip terminated by a trimer apex. Given the small emission region, the brightness of this helium ion source amounts to 10^7 A/cm^{2}sr, as has recently been demonstrated in preliminary experiments [12].

3. Applications

In the introduction, where interference and optics were discussed, applications of point sources in holography as well as in microscopy are implied in the sense of using an electrostatic or magnetic lens or lens system to obtain as small an image of the source as possible (a fine electron or ion beam) with which to experiment. From a conceptional point of view this might indeed be the most straightforward application. A small beam of low-energy electrons or ions certainly also has important implications in science and technology. However, there are a number of technical problems yet to be solved before such a concept can be realized, for example the manufacturing of sub-micron size lenses with a uniform surface potential to influence the low-energy charged particle beam in a well-controlled way. Recent electron optical calculations on these microlenses appear promising [13] and confirm that the efforts involved in actually building such a lens are worthwhile.

At this point I would like to discuss two applications that in a sense are precursory to such a concept. Both, however, are high-resolution techniques with a promising range of applications in their own right, particularly the second one, the Low Energy Projection Microscope, developed by Werner Stocker et al. [14]. The principles of these techniques are illustrated in Fig. 4. Both techniques combine

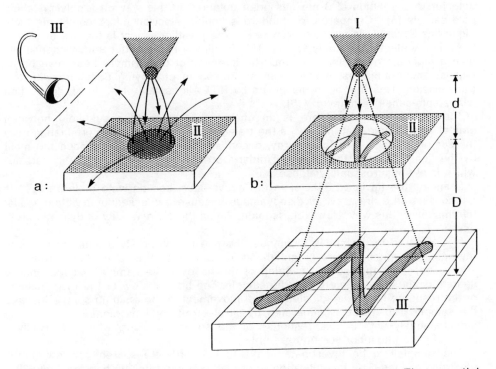

Fig. 4: Schematic of two applications of point source particle beams. The essential parts of the setups, in accordance with Fig. 1 are the source of particles (I), the sample (II) and the detector (III). Both techniques rely on the close proximity between source and sample and their resolution is determined by the source size. a: Low-energy proximity imaging with backscattered or secondary electrons. The tip biased at -15 V or higher with respect to the sample generates free electrons that impinge on the sample. While the tip is scanned over the surface, the count rate at a channeltron detector is monitored. This rate is brought about by elastic reflected or secondary electrons from the sample. b: Low-energy projection microscopy. The emitter is positioned close to (d) a partly transparent film. At a macroscopic distance (D) a projection image of the sample is generated simultaneously on a channel plate detector. With $d = 1$ micron, an image of carbon fibers magnified 60000 times is generated with an electron beam of 30 eV.

the knowledge of point sources with precise mechanical manipulation that has now become routine with the development of the STM.

a) The proximity imaging with secondary electrons, as sketched in Fig. 4a, has in its principle design already been demonstrated by Russel Young [15], however with a much inferior resolution due to a standard field emitter that was employed by Young in the seventies. Analog to STM, the tip is a short distance from the sample to be investigated. In order to produce free electrons the tip/sample separation has to be larger than the few nm used for STM operation. In our setup, a minimum voltage of -15 V was required in order to obtain a signal. The tip is scanned over the surface at a constant emission current of 0.1 nA and the signal to be measured is the yield of backscattered or secondary electrons as detected by a channeltron detector. A resolution of 3 nm has been obtained in this way on a polycrystalline gold sample [8]. Compared to standard scanning electron microscopy, this technique has advantages but also drawbacks. The positive aspect is the low energy of the primary electrons of only 15 eV. This implies a much better surface sensitivity than a high-energy focussed beam. The intensity of the primary beam can be as high as micro-amperes, which is an advantage in employing detectors with low transmission. This has been discussed by Rolf Allenspach et al. in an application with a spin-sensitive detector [16].

The drawback is that there is an unavoidable inhomogeneous field between emitter and sample that influences the trajectories of the electrons originating from the sample surface. The low-energy, secondary electrons are influenced the most by the field; they will therefore be under-represented at the level of the detector which is at a macroscopic distance.

Scanning the tip in an STM or in the above-described derivate or in an SEM is not considered a drawback or disadvantage because it is essential to obtain spatial information in this way. However, scanning limits the observability of dynamic processes.

b) Compared to the above technique, the one I shall briefly discuss now has in my opinion a much greater potential. In the Low Energy Electron (Ion) Projection Microscope [14], the lateral scanning of the tip over the sample is used only to search for the area of interest of the sample. The field of view or the magnification of the resulting image is adjusted by the movement of the emitting source towards the sample. The image is generated by all electrons or ions emitted from the tip simultaneously with a time response of the spatial detector, which is usually a channel plate phosphor screen assembly.

The samples to be investigated have to be partly transparent for low-energy electrons or ions. The magnification in preliminary experiments has reached 60000 with a tip sample separation of 1 micron. With a perforated carbon film, fibers exhibiting a width of 30 nm have been imaged with a low-energy electron beam with an energy of only 30 eV.

4. Conclusion

The concept of a well-characterized point source beam of electrons or ions brings about a variety of scientific problems and applications, that, at least by those involved in such efforts, are considered a worthwhile endeavor.

Acknowledgement

I would like to thank my friends and colleagues at our laboratory in Rüschlikon, in particular Werner Stocker and Heinz Schmid for fruitful discussions and pleasant collaboration. The same gratitude is due my friends outside of Rüschlikon, in particular to Roger Morin, CNRS Marseille, and to Nobert Ernst, FHI Berlin. I am grateful to Heini Rohrer and Eric Courtens, who supported this project, and I would also like to thank them for their comments on the manuscript.

References

[1] Lichte, H. (1986) Ultramicroscopy 20, 293-304.
[2] Feynman, R. (1985) The Character of Physical Law, MIT Press.
[3] Garcia, N. and Rohrer, H. (1989) J. Phys. C 1, 3737-3742.
[4] Binnig, G. and Rohrer, H. (1982) Helv. Phys. Acta 55, 726. See also contributions within this proceedings book.
[5] Müller, E.W. and Tsong, T.T. (1969) Field Ion Microscopy, Principles and Applications, American Elsevier Publishing Co., Inc. New York.
[6] For recent reviews, see: Ernst, N. and Ehrlich, G. (1986) in Microscopy Methods in Metals, ed. by U. Gonser, Topics Current Phys. 40, Springer, Berlin, Heidelberg, 75-115: Fink, H.-W. (1988) Diffusion at Interfaces: Microscopic Concepts, Springer Series in Surface Science 12, ed. by M. Grunze, H.J. Kreuzer and J.J. Weimer, Springer, Berlin, Heidelberg, 75-91.
[7] Fink, H.-W., (1986) IBM J. Res. Develop. 30, 460.
[8] Fink, H.-W., (1988) Physica Scripta 38, 260.
[9] Morin, R. and Schmid, H. unpublished data.
[10] Morin, R. and Fink, H.-W., in preparation.
[11] Lang, N.D., Yacoby, A. and Imry, Y., (1989) Phys. Rev. Lett. 63, 1499.
[12] Schmid, H. and Fink, H.-W., in preparation.
[13] Chang, T.H.P., Kern, D.P. and McCord, M.A., J. Vac. Sci. Technol., in press.
[14] Stocker, W., Fink, H.-W. and Morin, R., Ultramicroscopy, in press; (see also references therein for related earlier work).
[15] Young, R., Ward, J. and Scire, F., (1972) Rev. Sci. Instrum. 43, 999.
[16] Allenspach, R. and Bischof, A., (1989) Appl. Phys. Lett. 54, 587.

FIELD ELECTRON EMISSION FROM ATOMIC-SIZE MICROTIPS

J.J. SAENZ [1], N. GARCIA [1,2], VU THIEN BINH [3] and H. DE RAEDT [4,5]

[1] Dept. Materia Condensada, Univ. Autónoma de Madrid, Cantoblanco,
28049 Madrid, Spain
[2] Dept. Applied Mathematics, Univ. of Waterloo, Waterloo,
Ontario, Canada N21 3G1
[3] Dept. de Physique des Materiaux (UA CNRS), Univ. Claude Bernard - Lyon 1,
F-69622 Villeurbanne, France
[4] Physics Dept., Univ. of Antwerp, Universiteitsplein 1,
B-2610 Wilrijk, Belgium
[5] Natuurkunding Laboratorium, Univ. of Amsterdam, Valckenierstraat 65,
1018 XE Amsterdam, The Netherlands

ABSTRACT. The influence of quantum size effects and atomic geometry on the field emission characteristics (energy distribution, intensity-voltage, angular spread of the emitted beam, resolution) of atomic-size microtips is analyzed. Theoretical models are propose to analyze the influence of atomic-size protrusions on the properties of the emitted beam. It is shown that a tip with a small protrusion can be used as a source of very collimated electron beams. The conditions under which is it possible to obtain atomic resolution are discussed. Simple formulas relating the angular spread and the resolution with the experimental parameters are presented. Field emission experiments on "build up" and "teton" tips are presented together with the basic principles of the fabrication technique.

1. Introduction

Electron emission from very small sources, i.e. sources which have a size comparable to the wavelength of the emitted electrons is a very interesting problem related to the observation of mesoscopic quantum effects.

Recent theoretical work /1,2/ on coherent electron emission from small sources shows that the properties of the emitted beams differ markedly from those emitted by macroscopic sources. The development of coherent field-emission electron beams has facilitated some practical apllications of electron holography /3-6/ and interferometry /7/. Although there have been advances in this direction by using conventional field-emission tips, the funcionality of these devices is still very limited compared with optical systems. It has been suggested /1,2/ that the properties of electron beams emitted from an atomic-size microtip could lead to the practical realization of three-dimensional visualization of atomic objects.

Another point of practical interest in the study of atomic-size microtips is the relevance of the tip structure on the image interpretation in Scanning Tunneling

Microscopy (STM) /8/. Experiments with combined Field Ion Microscope (FIM) and STM /9-11/ have shown the relation between atomic resolution in STM and atomic geometry of the tip apex. It is then important to have at our disposition simple fabrication methods of microtips with controlled geometry at the atomic level /12-13/.

The aim of the present paper is to study the general properties of the electron field-emission from small (atomic-size) microtips as well as discuss some simple techniques of its fabrication. There are several aspects of the problem which lead to electron beam properties different from those obtained with "macroscopic" tips. As the source size is comparable to the electron wavelength inside the tip, quantum effects play an important role in determining the properties of the emitted beam. Strong quantum effects have been observed in recent experiments on point contacts in 2D electron gas /14,15/. By fabricating a device having two electron reservoirs connected by a narrow constriction, it was shown that the conductance is quantized, the quantization being a function of the diameter of the constriction /14-16/. A fundamental difference between the atomic tips and the GaAs devices is that in the latter case the electrons has to move through a constriction whereas in the former it also has to tunnel through the metal vacuum potential. Theoretical calculations /17/ have shown that, in the field emission case, the properties of the emitted beam are almost independent of the source size and mainly controlled by the shape of the tunnel barrier. Only if the electron source has a plane emitting surface (i.e. a plane triangular metal-vacuum tunnel barrier) it is possible to obtain a strong collimation of the electron beam. The influence of the quantum difraction on the field emission characteristics (energy distribution, angular spread ...) is discussed in Chapter 2.

The tip geometry at microscopic scale has also a great influence on the field-emission characteristics of small microtips. For example, it is well known that the presence of protuberances or asperities on the tip leads to anisotropies in the field emission pattern of classical tips /18/. In fact, the experimental shapes of buid-up and teton tips /13/ can be considered as having a nanometer size protrusion erected on top of a large support tip. In Chapter 3 we study the local field and tunnel barrier around a small bump on the tip apex. The strong electron beam collimation and the peculiar behavior of the intensity-voltage (I-V) characteristics of a protruded tip is also discussed. Another remarkable property of these tips appears when the protrusion size is large compared with the atomic dimensions, but still in the few nanometer range. Within the context of classical electrodynamics, Rose /18/ showed that these protrusions can provide areas for which the magnification (resolution) is much larger than that computed for a smooth tip. Localized emission centers (atoms) on the protrusion surface could lead to separated spots in the field-emission pattern, provided the electron wave packets emitted from these centers do not overlap /19/. Some conditions under which it is possible are also studied in Chapter 3.

Teton tips /13/ provide an ideal scenario to study the increased magnification and to check the possibility of observe some of their atomic structure in field-emission experiments. In Chapter 4 we will discuss on the basic principles of the pseudo-stationary profile technique used to produce teat-like shaped tips. We will discuss some recent experiments /19,20/ on electron field emission from "build up" and "teton" tips, both with a single atom and with a trimer placed at the tip appex. These results, still to be confirmed by field ion microscopy, indicate the possibility of atomic resolution with electron field emission.

2. Quantum effects in Electron Field Emission

When the size of the emitting region is of the order of the wave length of the electrons inside the tip, quantum effects can play an important role in determining the nature of the emitted beam. It is an open question to what extent the level quantization, observed in GaAs systems /14-16/, will affect the properties of the emitted electron waves and if it does so, whether the focusing of the beam is governerd by the geometry of the tip, the tunnel barrier or both.

Our main objetive here is to study general features of the effects of the size of the source on the angular spread and energy distribution of the emitted electrons. A full self-consistent solution of this problem is extremely difficult. In order to tackle the problem, drastic approximations in describing the metal tip have to be made. As a first step we can model the emission tip by a free electron gas. The simplest model that captures the basic physics can taken to be two-dimensional (2D) /17,21/. In fig. 1 we present the geometry of the emitting source together with an energy diagram. In order to simulate the possible level quantization in the tip, we will model the tip as a semi-infinite constriction of width W. The potential, $V(\vec{r})$, is taken to be $V = \infty$ at the source boundary except at the exit plane of the constriction. At this plane we introduce a triangular tunnel barrier, characterized by the applied field F, which would simulate the tunnel barrier at the tip apex.

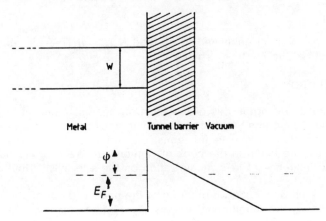

Figure 1. Geometry of the model used in the calculation of the electron emission from a 2D source (a), and energy diagrams for the tunnel barrier.

From a general point of view, in a 2-D field emission model, the total intensity emitted from the tip can be written as

$$I = \frac{e}{\pi \hbar} \int_0^{E_F} dE \sum_i \sum_f T_{f,i}(E) = \frac{e}{\pi \hbar} \int_0^{E_F} dE \sum_i T_i(E) \qquad (2.1)$$

where E is the total electron energy inside the tip, $T_i(E)$ is the transmission probability for an angle of incidence θ_i, and the summation runs over all angles of incidence θ_i and over all possible outgoing angles θ_f. Note that if only electrons at the Fermi level contribute to the current, (2.1) leads to the approximate quantum conductance formula for narrow constrictions /14,16/. In the case of a semi-infinite constriction, the electron energy is fully quantized and given by

$$E(k_z, n) = \frac{\hbar^2}{2m}[(\frac{n\pi}{W})^2 + k_z^2] \tag{2.2}$$

where k_z is the longitudinal wave vector (parallel to the constriction) and n is the quantum number that determines the wave vector $k_n = \frac{n\pi}{W}$ in the transverse direction. Then (2.1) reads

$$I = \frac{e}{\pi\hbar} \int_0^{E_F} dE \sum_{n=1}^{N_{max}} T_n(E) \tag{2.3}$$

where the sum runs over all quantized levels and $N_{max} = INT[\frac{2W}{\lambda_F}]$, being λ_F the electron wave length at the Fermi energy E_F.

2.1. ENERGY DISTRIBUTION

One of the most important properties of the electron beams is the energy distribution of the emitted electrons. A discussion about the total energy distribution, $J_T(E)$, and the distribution normal to the emission surface, $J_N(E_z)$, for a standard field emission tip can be found in the book of Gomer /22/. We can use our simple model to study the influence of level quantization on the energy distribution of the emitted beam.

Assuming there is no diffraction on entering of the barrier, the tunnel probability, $T_n(E)$, only depends on the longitudinal energy, $E_z = E_z(E,n) = E - (\hbar^2/2m)(n\pi/W)^2$, and is given by

$$T_n(E) = T(E_z)\Theta(E_z) \tag{2.4}$$

where $\Theta(x)$ is the Heaviside step-function. For simplicity $T(E_z)$ is taken to be the WKB expression

$$T(E_z) \simeq \exp[-\frac{4}{3}\sqrt{\frac{2m}{\hbar^2}}\frac{(\phi + E_F - E_z)^{\frac{3}{2}}}{F}] \tag{2.5}$$

where ϕ is the work function of the metal and F the applied field.

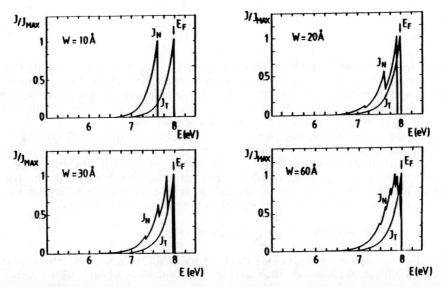

Figure 2. Total (J_T) and normal (J_N) energy distribution of the current emitted from a planar source for different sizes W for a 2D electron gas with a Fermi energy $E_F = 8eV$ and work function $\phi = 4.5eV$. The tunnel barrier is assumed to be triangular with a field $F = 0.5V/\text{Å}$.

From (2.3) it is clear that the total energy distribution for a "finite-size" 2D source, is given by

$$J_T(E) = \frac{e}{\pi\hbar} \sum_{n=1}^{N_{max}} T_n(E) \tag{2.6}$$

whereas the normal energy distribution is

$$J_N(E_z) = \frac{e}{\pi\hbar} \, INT\left[\frac{W}{\pi}\sqrt{\frac{2m}{\hbar^2}(E_F - E_z)}\right] T(E_z) \tag{2.7}$$

In fig. 2 we show the energy distribution of the emitted electrons at zero temperature for constrictions of different width and for a constant field $F = 0.5V/\text{Å}$. As it can be seen, the normal energy distribution, $J_N(E_z)$, clearly reflects the level quantization of the source, showing peaks corresponding to the energy levels present in the source. However, the total energy distribution, $J_T(E)$, is almost a constant function, independent of the constriction width W, i.e. $J_T(E)$ is mainly determined by the filtering effect of the tunneling barrier. We can estimate the value of ΔE from (2.5) and the condition $T(E_F - \Delta E) \simeq 1/e$,

$$\Delta E \simeq \frac{F}{\sqrt{\phi}} \qquad (2.8)$$

where we have assumed $\Delta E \ll E_F$, and ϕ is in eV, F in V/Å. The generalization of these results to a three dimensional system (with two different set of quantum levels) is straigthforward leading to the same qualitative result. Experimental observation of level quantization from energy measurements would be difficult unless one is able to discriminate between normal and total energies.

For small constrictions, the diffraction at the tunnel barrier (as discussed in the next section) can modifies the normal (not the total) energy distribution, simply because the tunnel barrier favours scattered waves which are normal to the barrier. This focusing effect results in a smoothing of the peaks in the normal energy distribution, $J_N(E_z)$, and in a shift of its possition towards higher normal energies E_z (i.e. towards directions closer to the normal to the barrier).

2.2. DIFFRACTION THROUGH A TUNNEL BARRIER

When the electron emission comes from a very small region, the electron focusing could be affected by the difraction and level quantization effects. In a normal diffraction problem it is possible to get an estimation of these effects by using the Heisenberg principle. However, in field emission we have to deal with a problem of diffraction throughout a tunnel barrier. The influence of a tunnel barrier on the diffraction process at the exit of a constriction have been studied recently /17/. Here we will discuss some of these results for the case of the approximate triangular tunneling barrier typical in field emission.

The relevant function to describe the angular spread of the beam is the intensity as a function of the outgoing angles θ_f. From (2.1), in analogy with the energy distribution, we can define the angular distribution, $J(\theta_f)$, as

$$J(\theta_f) = \frac{e}{\pi \hbar} \int_0^{E_F} dE \sum_i T_{f,i}(E) \qquad (2.9)$$

Then, the angular spread can be defined by the angle $\Delta \theta_t$ at which $J(\Delta \theta_t) = \frac{1}{e} J(0)$ (i.e. half of the total angular width).

To calculate the angular distribution (2.9), we have to deal with a very complicated quantum mechanical problem. Different approaches have been used to solve this problem /17/. The first approach relies on the basic assumption that the waves incident on the barrier can be thought of as being (incoherent combinations of) plane waves. In this case, the $T_{f,i}$ can be calculated by solving the appropiate stationary Schroedinger equation (SSE). This has been solved by using a "matching scattering technique" /16/. The second approach assumes that the electrons may behave like a wave packet, i.e. as a particular combination of plane waves. Under such circumstances the problem requires the solution of the time-dependent Schroedinger equation (TDSE). The computer simulation consists of preparing a gaussian wavepacket moving towards the constriction and letting the wavepacket evolve in time according to the TDSE. The TDSE was solved numerically by means of an algorithm based on a fourth-order Trotter formula /23/. A detailed discussion of the results of extensive calculations for different constriction and barrier geometries, using the methods described above, is given elsewhere /17,21/.

One of the most interesting results from these calculations is that the angular spread of the emitted electrons is almost independent of the constriction width W for a given tunnel barrier geometry. For example, for a planar triangular barrier similar to that of Fig. 1, and for an applied field $F = 0.5V/Å$ the angular spread is constant, $\Delta\theta_t \lesssim 10°$, for the range of widths calculated, $2\lambda_F \lesssim W \lesssim 5\lambda_F$ (i.e. $8Å \lesssim W \lesssim 20Å$ for $E_F = 8eV$). The angular spread is calculated for the same electron wavelength inside and outside the tip, i.e. is measured when the potential has drop $\phi + E_F$ (see Fig. 1).

It is interesting to compare these scattering calculations with the results of a simple semi-classical diffraction approximation (SDA) /17/. Although, the results of the latter are only qualitative, it will give insight to the importance of the different parameters involved. Let us consider again the simple geometry of Fig. 1. To analyze the angular spread for this case, we now take into account the scattering of electrons at the end of the constriction. As a first approximation to this problem, we will envisage each scattered plane wave to be filtered incoherently by the tunneling barrier, i.e. we will assume that, for electrons having a total energy E and a transverse momentum $k_n = \frac{n\pi}{W}$, the $T_{f,i}$ in (2.8), are given by

$$T_{f,n}(E) \simeq |f_n(k \sin \theta_f)|^2 \; T(E \cos^2\theta) \qquad (2.10)$$

where $k = \sqrt{(2m/\hbar^2)E}$ and $f_n(k \sin \theta_f)$ is the diffraction function of the constriction or, in other words, the Fourier transform of the slit function. In this case the sum in (2.8) runs over all quantized levels.

In Fig. 3 we have plotted the angular distribution, $J(\theta_f)$, for different widths W and an applied field $F = 0.5V/Å$. Also shown in Fig. 2 are the contribution of the two first levels of the source (n = 1,2). For small W, the emitted current consists mainly of electrons coming from only one level, yielding a wave which, in this case, has an angular spread $\Delta\theta_t \simeq 10°$. For wide constrictions, a different picture emerges. In this case, diffraction effects are very weak, and each electron wave has an angular width which is very small. However, the current density now is given by the non-coherent sum of a large number of electrons coming from different levels. In the end, the total current has an angular spread $\simeq 10°$ which is much larger than that of the individual electron waves. These results are in full agreement with the full quantum mechanical approaches SSE and TDSE discussed above.

These results can be understood as follows. Because of the filtering effect of the barrier, only those electrons having energies close to the Fermi energy will contribute to the current. For narrow constrictions the main contribution comes from the first level and $|f_1(k \sin \theta_f)|^2$ is a smooth function of the angle. As the transmission probability decays exponentially, the angular width will be determined by the tunneling properties of the barrier. The angular spread in this case can be estimated from

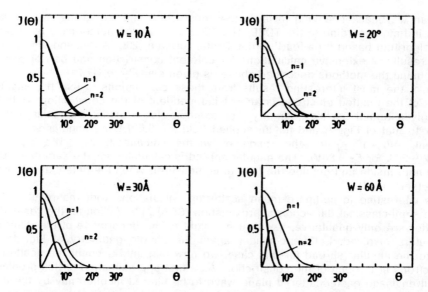

Figure 3. Angular distribution of the emitted current $J(\theta)$ for the same parameters as in the previous figure. Dashed lines correspond to the contribution of the first (n=1) and the second (n=2) energy level of the source. The angular width at 1/e intensity remains approximately constant ($\simeq 10°$), independent of the source width W, whereas the angular width of the contribution of each level becomes smaller as W increases.

$$T(\frac{\hbar^2 k_F^2 \cos^2 \Delta\theta_t}{2m}) \simeq \frac{1}{e} \qquad (2.11)$$

From (2.5) and assuming small angular spreads we obtain the approximate formula

$$\Delta\theta_t \simeq \sqrt{\frac{F}{E_F \phi^{1/2}}} \simeq \sqrt{\frac{\Delta E}{E_F}} \qquad (2.12)$$

where, here and in what follows, $\Delta\theta$ is in radians, E_F and ϕ are given in eV, F in V/Å and ΔE is given by (2.8). For large constrictions, the diffraction is very weak ($|f_n(k \sin\theta_f)|^2 \propto \delta(k \sin\theta_f - n\pi/W)$) and the angle of incidence θ_i is conserved ($\theta_i = \theta_f$). In this case the angular spread is controlled by the statistical distribution of momenta transverse to the emission direction, rather than by diffractions effects. The condition (2.11) could be written now as $T(E_F \cos^2\theta_i) \simeq 1/e$ leading to the same estimation for $\Delta\theta_t$. Although the physical mechanim associated with the angular spread is quite different, the final value of $\Delta\theta$ is the same, independent of the size of the emitting source. For typical values $E_F = 8eV, \phi = 4.5eV, F = 0.5V/Å$ the angular spread given by (2.12) is 9.8° in agreement with the quantum mechanical calculations /xx/.

2.3. THE EFFECT OF THE TUNNELING BARRIER GEOMETRY

Semi-classical calculations /2/, neglecting electron diffraction, have shown that the emission from smooth hyperboloidal tips having a small radius of curvature do not show strong beam collimation. The $\Delta\theta_t$ obtained from a planar emitting barrier is much lower than that obtained from tips with a finite radius of curvature. In this case the spread of the beam is mainly controlled by the curved geometry of the barrier, rather than by the diffraction or level quantization in the tip. This have been shown by SSE and TDSE calculations for 2D paraboloidal tips. For radius of curvature $2\lambda_F \lesssim R_0 \lesssim 5\lambda_F$ and applied field $F = 0.5 V/Å$ at the tip apex (i.e. for the same applied field and constrictions size as in the planar case), $\Delta\theta$ is of the order of 25°, compared with the 10° obtained for a planar source. These results are of the same order of those obtained with semiclassical calculations on hyperboloidal tips: for radius of curvature $R_0 \lesssim 50Å$, $\Delta\theta$ is between 20° and 30° for total intensities $10^{-10} - 10^{-6}$ A /2/.

In a smooth curved tip, the tunnel barrier (the field) increases (decreases) as we move out of the tip apex. From each point of the equipotential surface defining the tunnel barrier outside the tip, the emission is directed around the direction normal to it. Then, as the emission angle, θ_e (defined as the angle between the normal to the equipotential surface and the normal to the apex), increases, the field decreases and the total emission is collimated within a cone of angles around the normal to the apex (see fig. 3). The angular opening at the tip surface, $\Delta\theta_e$, is given by the half angle of this cone (see fig. 3). If the radius of curvature of this equipotential surface is R_c then it can be seen that for small θ_e, the field, $F(\theta)$, goes as

$$F(\theta) \simeq F_0 [1 - \frac{(R_c - \Delta s - R_0)(R_c - \Delta s)}{2\Delta s R_0} \theta_e^2] \equiv F_0 [1 - \frac{\alpha}{2}\theta_e^2] \tag{2.13}$$

where $\Delta s = \phi/F_0$ is the barrier length for $\theta_e = 0$ and $R_c \geq (R_0 + \Delta s)$. Then $\Delta\theta_e$ is obtained approximately by setting $T(E_F, F(\Delta\theta_e)) = 1/e$ in (2.5), i.e.

$$\Delta\theta_e \simeq 2\sqrt{\frac{F}{\alpha\phi^{3/2}}} \tag{2.14}$$

The angular spread of the beam near the tip surface would be given mainly by the above expression unless R_c would be large enough $(R_c \gg (R_0 + \Delta s))$. In this case, for an almost flat equipotential, $\Delta\theta$ would be given by (2.12) as discussed in the previous section.

2.3.1. Hyperboloidal tips.

It is interesting to compare this approximate expression for $\Delta\theta_e$ with the results obtained for hyperboloidal tips /2,17/. Let us consider an hyperboloid with radius of curvature R_0 at a distance L from a flat screen. In this case the solution of the Laplace equation $\nabla^2 V_h = 0$ can be separated in terms of spheroidal coordinates /24/, and the potential V is given by

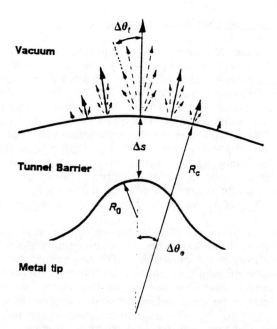

Figure 4. Influence of the tunneling barrier geometry on the angular spread of the emitted current. Thick arrows indicate the current density distribution on the emitting surface. Dashed arrows show the spread in the transversal velocities of the emitted electrons.

$$V_h = -V_0 \times [1 - \ln(\frac{1+\xi}{1-\xi})\{\ln(\frac{1+\xi_0}{1-\xi_0})\}^{-1}] \qquad (2.15)$$

The equipotential surfaces are characterized by the coordinate ξ, and, in cylindrical coordinates (r,z), each equipotential surface is defined by the equation of the hyperboloid

$$\frac{(z-L)^2}{\xi^2} - \frac{r^2}{1-\xi^2} = \frac{L^2}{\xi_0^2} \qquad (2.16)$$

where the origin is taken at the tip apex, $0 \leq \xi \leq \xi_0$ and $\xi_0 = (1 + R_0/L)^{-1/2}$. The tip is defined by the equipotential $\xi = \xi_0$, $(V_h = 0)$, while the flat screen corresponds to $\xi = 0$, $(V_h = -V_0)$. The field at the tip apex F_0 is given by

$$F_0 = \frac{V_0}{kR_0} \quad ; \quad k \simeq \frac{1}{2}\ln(4\frac{L}{R_0}) \qquad (2.17)$$

We are interested on the radius of curvature R_c of the equipotential surface defining the tunneling barrier outside the tip, i.e. the equipotential corresponding to $V_h = -\phi$. Assuming that $\phi/F_0 = \Delta s \ll R_0 \ll L$ it is easy to see that R_c is $\simeq R_0 + 2\Delta s$. Then α defined in (2.13) is $\simeq 1$ and the beam opening becomes

$$\Delta \theta_e \simeq 2 \left(\frac{F_0}{\phi^{3/2}} \right)^{1/2}$$

For $F = 0.5 V/\text{Å}$, $\phi = 4.5 eV$ we have $\Delta\theta_e \simeq 26°$ in agreement with the full quantum mechanical result of $\simeq 25°$ /17/.

3. Influence of the microtip geometry on the Field Emission characteristics

3.1. ATOMIC-SIZE PROTUBERANCES ON A TIP SURFACE

In the previous chapter we have seen that the curvature of the tunneling barrier have a strong influence on the properties of the emitted beam. Of course, in a real tip, this curvature is determined by the tip geometry at an atomic scale. The experiments on "buid-up" and "teton" tips discussed below (Chapter 4) indicate that the atomic-size microtips can be considered as having a nanometer size protrusion erected on top of a larger support tip. In this case, the presence of these protuberances would distort and compress the equipotential lines in the vicinity of the apex, causing a local field enhancement and an increase of localized emission /18,22/. In order to study the main characteristics of the electron emission from such tips, we construct a simple model to calculate the field distribution and tunneling barrier around a protruded tip.

For simplicity, we will consider in our calculations a hyperbolic shape (see 2.3.1.) for the macroscopic support tip. In order to simulate a protuberance at the tip apex we can use the following trick, based on the superposition principle. Let us superpose the potential V_h of the hyperboloid-screen system, given by (2.15), and the potential of a point charge placed in the symmetry axis at a distance δ from the tip apex /19/. The total electrostatic potential will be

$$V(r,z) = v_h(r,z) + V_0 \frac{\gamma}{\sqrt{r^2 + (z-\delta)^2}} \qquad (3.1)$$

where, as before, all the distances are measured from the tip apex. This new potential lead to a set of equipotential lines which, near the tip apex, resembles a teat-like geometry. Provide the tip-screen distance, L, is large enough, the new potential is still a solution of the Laplace equation and fulfit our boundary conditions, i.e., the equipotential which defines the screen is flat and, far from the tip apex, the tip equipotential has a hyperbolic shape. In fig. 5a we show schematically the basic arrangement associated to equation 3.1. The position, δ, and strength, γ, of the point charge can be adjusted to match an equipotential line to the protrusion/support-tip geometry. Figure 5b illustrate the equipotential lines corresponding to the superposition of the potentials of fig. 5a. The field enhance-

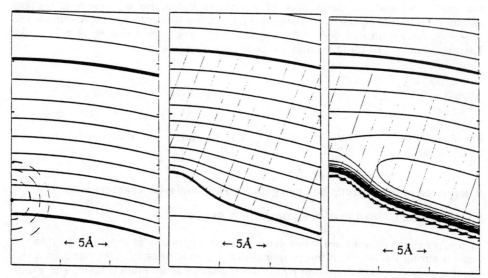

Figure 5. (a) Equipotential lines for a hyperbolic tip ($R_0 = 50$Å and applied voltage $V_0 = 175V$) and a point charge ($\gamma = 0.02$, delta = 2 angstrom). (b) Equipotential lines corresponding to the superposition of the potentials given in (a). The dashed region corresponds to the tunneling region. (c) the same as (b) including the image force correction. In this case, the tunnel barrier is lower and the equipotentials near the protrusion are almost flat.

ment near the apex can be estimated as follows. Assuming $\delta \simeq 0$, and if R_0 is large compared with the bump height ρ_0, the potential (3.1) near the apex can be written as

$$V \simeq F_0 \times (z - \frac{\rho_0^2}{\rho}) \tag{3.2}$$

where F_0 is given by (2.17), $\rho = \sqrt{r^2 + z^2}$, and we have taken $V = 0$ as the equipotential defining the tip surface. The field at the tip apex is then $\simeq 2F_0$, independent of ρ_0 (If instead of a point charge, the protrusion is simulated by a dipole on the surface the field is $3F_0$ /18/). However, the potential near the apex is not linear and the effective field enhancement factor lies between 1 and 2. If we define Δs_p as the barrier width at the tip apex, the effective field should be close to $F_p \simeq \phi/\Delta s_p$ and

$$\Delta s_p = \rho_0(a - 1 + \sqrt{1 + a^2}) \tag{3.3}$$

with $a = \phi/(2F_0\rho_0)$. Then, for large protrusions or large fields ($a \ll 1$) the field increases a factor of 2 with respect to the field on the support tip in the vicinity of the bump. On the other hand, for ρ_0 or F_0 small there is no significant field enhancement.

The influence of the image force on the equipotential lines is shown in figure 5c. We have calculated the image force correction in an approximated way. At every point we calculate the shortest distance, d, to the tip. Then, the image potential at this point is calculated as if it were at a distance d from a flat surface, i.e. $V_{im} \simeq -3.6/d$ with V in volts and d in angstroms.

3.2. INTENSITY-VOLTAGE CHARACTERISTICS

Once we know the potential distribution, we calculate the total current density within a semiclassical approximation similar to that used by Dyke and Dolan /25/ and Serena et al. /2/. We assume that the current density, at every point of the equipotential line defining the tunneling region (see Fig. 5), is the same as that obtained from a flat surface, with a tunneling barrier defined by a straight line from the equipotential to the nearest point on the tip. In all the cases analyzed, we will take a fermi energy $E_F = 8eV$ and a work function $\phi = 4.5eV$ which corresponds to a typical tungsten tip.

Vu Thien Binh and Marien /12/ have measured the I-V characteristics of build up and teton tips. For build up tips, as in the case of standard tips, the I-V curve follows the Fowler-Nordheim (F-N) law, as indicated by a straight line in the $\log(I/V^2)$ versus $1/V$ plot (In the F-N approach $I \propto V^2 \exp(-C/V)$). However for the teton tips deviations from this behavior are found, as discussed in the next chapter. The deviation from the F-N behavior had been interpreted /12/ as a space-charge effect /25/. However, it can be intrerpreted as a result of the particular geometry of the teton tip.

In order to understand the experiments, we have performed calculations for different tip geometries. In Fig. 6a we show the results obtained for a support tip of $R_0 = 400Å$, with three different protusion sizes ρ_0. As it can be seen, for a smooth tip (or if ρ_0 is very small like in the case of the atomic corrugation of a build up tip) the F-N plot is almost a straight line. However, as ρ_0 increases the deviation from the F-N behavior is clear. Even for a smooth tip, it is possible to observe similar deviations provide the radius of the tip is small (see fig. 5b). This deviation is related with the small radius of curvature of the emitting regions. When this radius is of the order or smaller than the typical tunneling distances the potential drop is not linear and the effective field can be much lower than that calculated just at the tip surface. Then, as the voltage decreases the current decreases faster than predicted from the Fowler-Nordheim law (for large fields ($\gtrsim 1V/Å$) there is also a saturation effect because of the influence of the image force on the height of the barrier).

In the case of a protruded tip and for low fields, the main contribution to the total current comes from the protrusion at the tip apex. As the field increases, there is an increasing contribution comming from the support tip (with a lower field but larger emitting area). Eventually the support tip contribution dominates and there is a crossover in the F-N plot (fig. 6b). This crossover is qualitatively similar to that observed in the experiments. For a given support tip curvature, the crossover in the F-N plot would be observable only within a small window of protrusion sizes. For build up tips, in which ρ_0 is of the order of the atomic size, the current intensity comming from the protrusion is of the order of the intensity comming from the support tip. On the other hand, for large ρ_0 the current is always comming

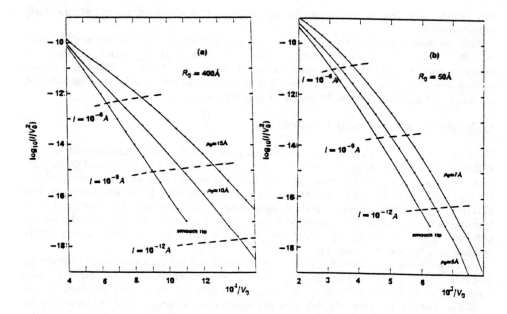

Figure 6. (a) Calculated current-voltage characteristics for a support tip with $R_0 = 400$Å and different protrusion sizes. V is in Volts and I in Amperes. (c) The same as (b) but for $R_0 = 50$Å. Dashed lines indicate constant intensity.

from the protrusion, and before to get any crossover, the intensity is high enough to destroy the tip. When the radius of curvature of the tip is smaller we do not found any anomalous behavior other than that associated to the saturation of the intensity. In this case, the emitting area of the support tip is not large enough to compensate the lower field at the small protrusion at the tip apex. As an example, the results corresponding to a support tip of $R_0 = 50$Å are shown in Fig. 6c.

3.3. ELECTRON BEAM FOCALIZATION

For practical applications, one of the most important parameters of an electron source is the degree of focalization of the emitted beam. Experimentally, it is possible to measure the radius of the spot observed at the screen, d_S and then, define the angular spread of the beam at the screen as $\Delta\theta_S \simeq d_S/L$. This angular spread, $\Delta\theta_S$ is different from the $\Delta\theta_t$ and $\Delta\theta_e$ discussed in 2.2. and 2.3.. $\Delta\theta_S$ is determined not only by the barrier geometry or by the transverse velocity of electrons, but also by the electric field in the free region between tip and screen.

3.3.1. Angular Spread and Transverse Velocity

The angular spread $\Delta\theta_t$ discussed in 2.2, is related with the average transverse velocity, v_t, by

$$\Delta\theta_t \simeq \frac{v_t}{v_F} = \left(\frac{m}{2E_F}\right)^{1/2} v_t \qquad (3.4)$$

with $\Delta\theta_t$ given by (2.12). v_t defines an spot diameter at the screen /22/

$$d_S \simeq (2/\beta_1) v_t t \qquad (3.5)$$

where t is almost the time required by the electrons of energy eV_0 to traverse the distance L ($t \simeq L(2eV_0/m)^{-1/2}$), and β_1 is a constant factor which lies between 2 (for a planar condenser geometry) and 1 (for a spherical geometry). Substituying (2.12) and (3.5) in (3.4) we have

$$\Delta\theta_{St} \simeq \frac{2}{\beta_1} \sqrt{\frac{F}{V_0 \phi^{1/2}}} \qquad (3.6)$$

In general F is proportional to the applied voltage V_0 and $\Delta\theta_{St}$ given by (3.6) will be determined by the tip geometry (independent of the applied voltage). For a smooth hyperbolic tip the field F is given by (2.17) and

$$\Delta\theta_S \simeq \frac{2}{\beta_1} \frac{1}{\{kR_0\phi^{1/2}\}^{1/2}} \propto \frac{1}{\sqrt{R_0}} \qquad (3.7)$$

Assuming $\beta_1 \simeq 1$ and taking $k \simeq 8$ (see (2.17)) we have $0.7° \lesssim \Delta\theta_S \lesssim 2°$ for $50\text{Å} \lesssim R_0 \lesssim 400\text{Å}$. As we have seen, the field enhancement factor in the vicinity of a bump on the tip surface is lower than 2. This implies that if the emission comes from an atomic-size bump the angular spread increases, at most, a factor $\leq \sqrt{2}$ larger.

3.3.2. Angular Spread and Magnification

Let us consider that the emission comes from a small spot of radius d_0 on the tip. The electron trajectories obtained by classical electrodynamics define a spot radius at the screen d_S and a magnification factor $M \equiv d_S/d_0$. For a spherical tip, the magnification would be L/R_0, where R_0 is the tip radius. However, for a paraboloidal tip, because of the compression of the lines of force towards the tip apex, the magnification M is reduced by a factor $\beta \simeq 2$ /18,22/. The angular spread associated with the magnification would be then,

$$\Delta\theta_{Se} \simeq M \frac{d_0}{L} = \frac{1}{\beta} \frac{d_0}{R_0} \simeq \frac{1}{\beta} \Delta\theta_e \qquad (3.8)$$

where $\Delta\theta_e$ is the angle for the apex, defined in Fig. 4 and eq. (2.14). Then, for a smooth hyperboloid, (3.8) becomes

$$\Delta\theta_{Se} \simeq \sqrt{\frac{F_0}{\phi^{3/2}}} \qquad (3.9)$$

with F_0 given by (2.17). In this case $\Delta\theta_{Se}$ is a factor $\simeq (V_0/\phi)^{1/2}$ larger than that associated with the transverse velocities, $\Delta\theta_{St}$. For example, for $F_0 = 0.5 V/\text{Å}$ we have $\Delta\theta_S \simeq 13°$. Then for a smooth tip the experimental angular spread should be given mainly by (3.9) and would increase with the applied voltage as $\propto V_0^{1/2}$. This cualitative estimation is in agreement with more detailed calculations by Serena et al. /2/.

It is generally assumed that small protrusions on the surface of the tip provide areas for which the magnification M is much greater than that of a smooth tip. The additional magnification was calculated by Rose /18/ within the context of classical electrodynamics. Assuming a radius of the support tip R_0 much larger than the protrusion size ρ_0 and simulating the protrusion as a dipole on top of a smooth tip, the potential may be written as

$$V \simeq -F_0 z [1 - (\frac{\rho_0}{\rho})^3] \qquad (3.10)$$

A calculation of the classical trajectories gives, for distances $z \gg \rho_0$, $r \simeq 2 d_0 (z/\rho_0)^{1/2}$, where r_0 is the initial displacement from the axis on the protrusion surface. If instead of use (3.10) for the potential we use our point charge model (equation (3.2)) the solution of the equation of the trajectories near the axis is much simpler than in the former case and gives

$$r \simeq d_0 (\frac{z}{\rho_0})^{1/2} \qquad (3.11)$$

for any distance $z \geq \rho_0$. The factor of $\simeq 2$ between (3.11) and Rose's result arises from the different radius of curvature of the protrusion surface (in our case the radius of curvature is $2\rho_0$). As pointed out by Rose /18/, the slope of the trajectories at any distance z from the origin and projected backward intersects the axis at (-z,0). At a distance $z = R_0$ the electron trajectories are approximately normal to a surface of radius $2R_0$ (see fig. 7). Then, one may consider the electrons as being emitted from a spot of radius $\simeq d_0 (R_0/\rho_0)^{1/2}$ on a virtual tip with radius of curvature $2R_0$. The new magnification M_p is therefore

$$M_p \simeq 0.5 (\frac{R_0}{\rho_0})^{1/2} M \qquad (3.12)$$

This analysis would be valid if ρ_0 is large compare with the width of the tunneling barrier, Δs_p (but still small compared with R_0). Thus, the equipotential defining the tunneling barrier outside the tip (i.e. the classical starting point for the electron trajectories) follows more or less the shape of the protrusion and the the radius of curvature, R_c, will be close to $2\rho_0$. Then α, defined in (2.13), will be $\simeq 1$ and

$d_0/R_c \simeq \Delta\theta_e \simeq 2(F_p\phi^{-3/2})^{1/2}$ (see section 2.3.) Substituying (3.12) in (3.8) we will have

$$\Delta\theta_{Se} \simeq (\frac{\rho_0}{R_0})^{1/2} \sqrt{\frac{F_p}{\phi^{3/2}}} \quad ; \quad (\rho_0 \gg \Delta s_p) \tag{3.13}$$

i.e., the angular spread is reduced a factor $\simeq(\rho_0/R_0)^{1/2}$ with respect to the case of the smooth tip (eq. (3.9)). For fields $F_p \simeq 0.5 V/Å$ the above expression would be valid provide $\rho_0 \gg 10Å \simeq \Delta s_p$. For $R_0 \simeq 400Å$ and $\rho_0 \simeq 20Å - 50Å$ we have $\Delta\theta_{Se} \simeq 3° - 4°$.

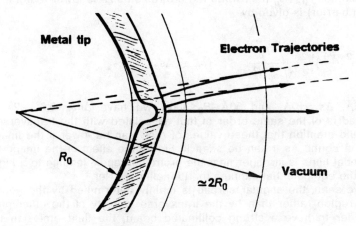

Figure 7. Typical trajectories of electrons leaving a small protrusion on a tip. To a first approximation, the trajectories are normal to a surface of radius $2R_0$, which therefore forms a virtual tip (After D.J. Rose /18/).

For small protrusion sizes, the field contribution of the protrusion itself falls off very fast with the distance from the tip surface. Then, the curvature R_c can be much larger than that of the protrusion itself, $2\rho_0$ (see Fig. 5). The distance from the origin to the classical turning point is $z_p = \rho_0 + \Delta s_p$, with Δs_p given by (3.3). The trajectories in this case can be approximate by

$$r \simeq 2d_0(\frac{z_p}{R_c})(\frac{z}{z_p})^{1/2} + d_0(1 - \frac{2z_p}{R_c}) \tag{3.14}$$

with $R_c = z_p(1 + (z_p/\rho_0)^2)$. If we applied now the same arguments leading to (3.12) we obtain

$$M_p \simeq \frac{z_p}{R_c}(\frac{R_0}{z_p})^{1/2} M \qquad (3.15)$$

and the angular spread will be now

$$\Delta\theta_{Se} \simeq M_p \frac{d_0}{L} = \frac{1}{2}(\frac{z_p}{R_0})^{1/2} \frac{d_0}{R_c} \simeq [(\frac{z_p}{R_0})^{1/2}(\frac{1}{\alpha})^{1/2}] \sqrt{\frac{F_p}{\phi^{3/2}}} \qquad (3.16)$$

where we have used the equation (2.14) with $d_0/R_c \simeq \Delta\theta_e$. It can be seen that the angular spread, for $\rho_0 \lesssim \Delta s_p$ (i.e., when the protrusion size is lower than the width of the tunneling barrier) is given by

$$\Delta\theta_{Se} \simeq 2(\frac{\rho_0}{\Delta s_p})^2 (\frac{\rho_0}{R_0})^{1/2} \sqrt{\frac{F_p}{\phi^{3/2}}} \quad ; \quad (\rho_0 \lesssim \Delta s_p) \qquad (3.17)$$

Taking $\rho_0 \simeq 5\text{Å}$, $\Delta s_p \simeq 10\text{Å}$, and $50\text{Å} \lesssim R_0 \lesssim 400\text{Å}$ we have $\Delta\theta_{Se} \simeq 0.6° - 2°$ i.e., the angular spread is of the same order of that associated with the transverse velocities. We should mention that these values of $\Delta\theta_{Se}$ can be lower if the image force is taken into account. As it can be seen in Fig. 5, the effect of the image force on the equipotential lines is stronger near the protrusion apex, leading to a higher curvature R_c of the equipotential defining the tunneling barrier.

As we have seen, the angular spread is mainly determined by the geometry of the emitting region rather than by the transverse velocity of the outcoming electrons. In order to have a strong collimated beam, the field emission tip must consist on a small protrusion placed on top of a larger support tip. The focusing is better smaller is the height of the bump ρ_0. Of course, this is true if the emitted current comes mainly from the bump. The smallest size ρ_0 which lead to strong focusing would be limited by the typical size of the irregularities (steps, defects ...) on the surface of the support tip in the vicinity of the bump.

As a ressume of the problem of collimation of the electron beam we can say that for tungsten tips ($\phi = 4.5eV$, $E_F = 8eV$) with small radius of curvature ($\simeq 50\text{Å}$) it is possible to focus the electron beam to $\lesssim 2°$ provide there is an atomic-size bump ($\simeq 5\text{Å}$ height) placed on top of the tip apex. For larger R_0 ($\simeq 400\text{Å}$) it is possible to go bellow $2° - 3°$ even for protrusions of the order of few nanometers.

3.4. RESOLUTION

3.4.1. *Semiclassical Approach*. In the previous subsection we have discussed the emission properties of a microtip assuming a smooth surface both for the protrusion and the support tip. However, because of the atomic corrugation, the field should be stronger near the apex of the atoms forming the protrusion. Depending on the exact geometrical arrangement of the atoms the emission could be localized around some of the atoms at the apex (i.e., the atoms would be like atomic bumps on the protrusion). Whithin the context of classical electrodynamics Rose showed /18/ that, because of the enhanced magnification M_p (see 3.3.2.), some of the atomic structure of the protrusion should be observable. For example, if we have an electron emission localized on two atoms separated at a distance D and placed symmetrically with respect to the tip axis, they will be resolved if the distance between spots at the screen ($\simeq DM_p$) is larger than the size of each spot ($\simeq L\Delta\theta_{St}$), i.e. if

$$D \gtrsim (\frac{\rho_0}{k\phi^{1/2}})^{1/2} 2\frac{R_c}{z_p} \qquad (3.18)$$

where we have use for M_p and $\Delta\theta_{St}$ the expressions given in (3.15) and (3.7). The distance between spots at the classical turning point is assumed to be $\simeq (z_p/\rho_0)^{1/2} D$. If we define the emission angle θ_e as the angle between the normal to the barrier at the emitting points and the normal to the apex (see Fig. 3), the condition (3.18) can be rewritten as

$$\theta_e \gtrsim = (\frac{R_0}{\rho_0})^{1/2}(\frac{\rho_0}{z_p})\Delta\theta_{St} \qquad (3.19)$$

The resolution given by (3.18) or (3.19) is the generalization of the Rose's result when the shape of the equipotential defining the tunneling barrier is taken into account. Notice, that the resolution is not limited because of diffraction effects (see chapter 2), but by the curvature of the equipotential surfaces R_c near the protrusion apex. As the applied voltage increases, the equipotential lines become closer to the the tip shape, and R_c decreases leading to a small improvement in the resolution. For smooth protrusions like those simulated by our point charge model (see 3.1.) the magnification factor, as well as the resolution, are maxima when the protrusion size ρ_0 is of the order of $\simeq 2$ times the width of the tunnel barrier, $\Delta s_p = (z_p - \rho_0)$, i.e. typically when $\rho_0 \simeq 10\text{Å} - 30\text{Å}$. For this protrusion sizes the resolution would be between $\simeq 5\text{Å} - 8\text{Å}$ depending on the applied field. Because of the linear dependence of (3.18) with R_c the final resolution will depend on the exact geometry of the apex of the protrusion. For example, for a hemispherical protrusion the resolution will be improved by a factor of 2 (i.e., $D \gtrsim 3\text{Å}$).

Using the arguments given above, the emission angle θ_e (eq. 3.19) can be as small as $\simeq 3°$. From this small angles the semiclassical electron beams would overlapp very close to the tip surface, although far from the tip, because of the field focusing, they would be well separated. Then, the discussion about resolution would be valid if the emitting spots are independent. In other words, if there is no interference phenomena between the different emitted beams. If we want to have atomic resolution, in view of the small distance between the atoms ($\simeq \lambda_F$) it is not clear whether we can apply the previous arguments.

3.4.2. *Simulation of Electron Wave Packets* . If there is some field enhancement in the vicinity of the atoms forming the protrusion apex, the electrons inside the tip would observe a tunneling barrier which has a minimum on top of each apex atom. The distance between the different minima would be of the order of the electron wave length. Moreover, the size of each emitting spot on the surface would be lower than the electron wave length. In this situation it is not possible to consider

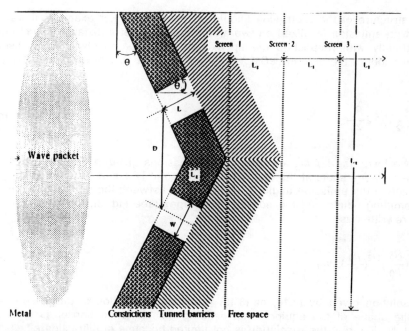

Figure 8. Geometry of the 2D model used in TDSE simulations.

the atoms as independent emission centers, and the problem will be more like a problem of electron diffraction throughout several slits. The conditions under which the diffracted beams (i.e., the diffracted electron wave packets) do not overlap are not known. To investigate this conditions theoretically a model is constructed that contains the most prominent features of the problem. This model is solved by exact numerical integration of the corresponding time-dependent Schroedinger equation (TDSE) /17/. Analysis of the time development of the wave packets should then reveal which physical mechanism(s) govern the emission of electron beam.

The geometry of the model is depicted in Fig.8. To construct a model that contains the basic ingredients it is sufficient to work in 2 spatial dimensions. The metal-vacuum tunnel barrier is assumed to have a plane triangular shape. Let us consider in our 2D model two atoms separated by a distance D on the surface of a protrusion and placed symmetrically around the tip axis. The localized emission on top of each atom can be simulated by two slits of the size of an atom (see Fig. 8). As we have seen in the previous sections, the emission from each point on the tip can be considered to be directed towards the normal to the equipotential defining the tunnel barrier outside the tip. If the radius of curvature of the equipotential around the protrusion is R_c, the normal to the surface of the tunneling barrier around each atom will be tilted an angle $\theta_e \simeq d_0/R_c$ ($d_0 = D/2$), with respect to the tip axis. Then the metal-vacuum tunnel barrier can be assumed to have a plane triangular shape, properly tilted as indicated in Fig. 8. In all our TDSE calculations the length scale is expressed in units of λ_F the relevant wavelength of the electron, wave numbers are measured in units of $k_F = 2\pi/\lambda_F$, energies in units of E_F and times in units of h/E_F. The dimensions of the slits are $W = L = 1/2$ their separation $D = 1.5$, the work function $\phi = 0.5$, and the potential slope corresponds to a field of $0.33V/\text{Å}$, all reasonable values from experimental viewpoint.

A convenient choice for the initial wave packet is

$$\psi(x, z, t = 0) \propto \sin(\frac{n\pi x}{L_x}) \exp(iK_z) \exp(-\frac{1}{2}(\frac{x - x_0}{\sigma_x})^2 - \frac{1}{2}(\frac{z - z_0}{\sigma_z})^2) \qquad (3.20)$$

i.e. of Gaussian packet moving in the z-direction. The angle of incidence θ_i is fixed by value of the "mode" index n as $\sin \theta_i = n/(2L_x)$ and $k_z^2 + (n/2L_x)^2 = 1$. In practice $x_0 = 0$, $\sigma_x = 5$, $\sigma_z = 2$ and z_0 is chosen such that the initial wave packet does not overlap with the constrictions. The size of the simulation box used is $40\lambda_F \times 23\lambda_F$ corresponding to a grid of 801×467 lattice points. Other technical details about the simulation itself are given elsewhere[17,23].

As the transmitted wave packet is propagating, probability is accumulated at each of the ideal "transparent" screens labeled 1,2,... (see Fig.8) in order to gain additional information about the lateral distribution of intensity. At regular time intervals, snap-shots of the probability distribution are taken.

Figure 9 shows the TDSE simulation results for $\theta_e = 10°$. Both from the intensities at the screens 2,3,4 and 5 and a snap-shot of transmitted wave, it is clear that only one spot will be observed at a screen far away from the tip. For $\theta_e = 30°$ the behavior has changed tremendously as is clear from Fig. 10. Now well-separated spots are observed. At first sight this might be explained quite trivially by purely geometrical reasoning. Tilting the slits more and more has to result in a situation where the incident wave splits in two outgoing waves. However if this reasoning is correct, turning off the tunnel barrier should not alter the pictures on a qualitative level. As demonstrated in Fig.11 this is not what happens as without tunnel barrier the transmitted wave looks as one big packet with a lot of structure. From the intensities at the different screens it is also obvious that the main part of the wave packet is centered around $x = 0$. On a genuine screen, such a packet would produce one instead of two spots. To understand this behavior in simple terms first recall that that the two slits are very close ($\simeq \lambda_F$) to each other. Secondly, it has been demonstrated [17] that in the absence of a tunnel barrier, there is a lot of diffraction at the exit plane of a narrow constriction, i.e. the angular spread of a wave emitted by a constriction is large. Without tunnel barrier the two waves join to form one object.

Simulations with other initial conditions, triangular potentials of different height and slope have been carried out and all lead to the same conclusion: The tilt angle θ_e necessary to observe well-separated electron beams is in the range of $20° - 30°$. This angles are much larger than those predicted by the semiclassical arguments assuming no interference. From the above results we can conclude that in order to obtain a resolution of the order of the electron wave length λ_F the radius of curvature at the tip apex should be $R_c \simeq \lambda_F/(2 \tan \theta_e) \lesssim \lambda_F$.

Provided we have a localized emission comming from every atom on the protrusion apex, our calculation imposes severe constrictions to the possible geometries of the tip which are able to give separated spots at the screen. For example, in the case of a tungsten tip the wave length is of the order of 4.3Å. For a <111>

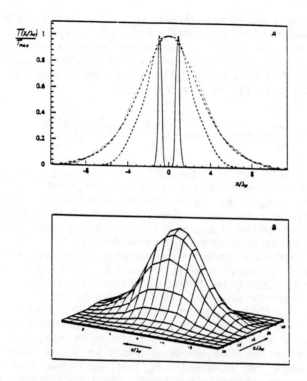

Figure 9. TDSE simulation of the 2D teton model with $\theta_e = 10°$ A) Accumulated intenstity at the screens (see Fig.9). Screen 1: solid line. Screen 2: Dashed line. Screen 3: Dashed-dotted line. Screen 4: dotted line. B) Real-space probability distribution of the transmitted electron wave.

oriented tip the interatomic distance is $\simeq 4.5$Å. In order to observe individiual spots coming from neighbouring atoms at the apex of a protrusion, the radius of curvature of the equipotential around the tip apex should be of the order of the interatomic distance. This implies that the protrusion must ended in a trimer (otherwise R_c would be too large and the different spots would merge). On the other hand, the radius of curvature R_c is always larger than the actual radius of the protrusion apex ρ_0 except if the protrusion height, h, is much larger than the typical width of the barrier ($h \gg 10$Å). Only in this case we would have $R_c \simeq \rho_0$. The shape of the protrusion must be such that $h \gg \rho_0$ (notice that this is not the case of the approximate hemispherical protrusions). As discussed bellow, a "teton" tip with a trimer at the apex fulfil these geometry criteria. In fact, recent field emission experiments employing "teton tips", suggested that atomic resolution might have been obtained /19,20/ (see section 4).

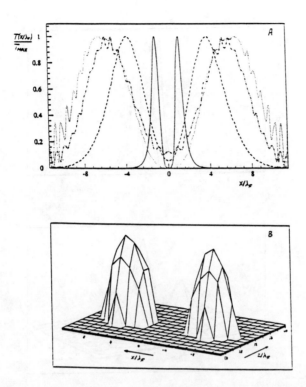

Figure 10. TDSE simulation of the 2D teton model with $\theta_e = 30°$ A) Accumulated intenstity at the screens (see Fig.1). Screen 1: solid line. Screen 2: Dashed line. Screen 3: Dashed-dotted line. Screen 4: dotted line. B) Real-space probability distribution of the transmitted electron wave. Wiggles in the screen signals are due to reflections from the simulation box boundary.

4. Field Emission experiments on atomic-size microtips

4.1. FABRICATION OF TIPS WITH A SINGLE ATOM APEX

We have seen that the tip geometry at a microscopic scale plays an important role in determining the properties of the emitted beam. In this chapter we will discuss the basic principles of the production of tips with controlled geometry at the atomic level based in the pseudo-stationary profile (PSP) technique /13/ together with some experimental results.

In the PSP technique /13/, the tip profile is shaped up until the atomic scale by using surface self-diffusion under different driving forces (capillarity and gradient of electric field), and evaporation of the tip atoms. The specificity of this technique is that the end form of the tip is an equilibrium profile for a given set of parameters (temperature, pressure, applied field...), thus the regeneration and production of

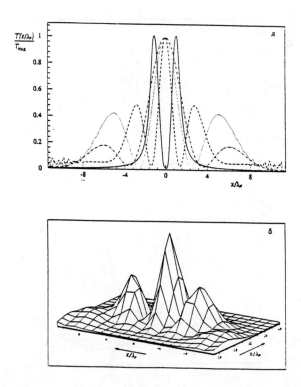

Figure 11. TDSE simulation of the 2D teton model with $\theta_e = 30°$ but *without* triangular tunnel barrier. A) Accumulated intenstity at the screens (see Fig.1). Screen 1: solid line. Screen 2: Dashed line. Screen 3: Dashed-dotted line. Screen 4: dotted line. B) Real-space probability distribution of the transmitted electron wave. Wiggles in the screen signals are due to reflections from the simulation box boundary.

reproducible tip profiles could be done on this basis if the experimental conditions could be fixed and with the use of only simple monitoring devices, i.e. is a "blind" technique.

4.1.1. *Basic Principles* . Sub-micron tip profiles undergo modifications under heat treatments in vacuum. The two principal mechanisms which are responsible for the tip profile variations are surface self-diffusion and evaporation. The evaporation, which means a loss of matter, leads to a decrease of the dimensions of the tip, and in particular the apex radius. The mass transport of surface atoms by surface self-diffusion is according to Nernst-Einstein relation proportional to the gradient of the chemical potential $\overline{\nabla}_s \mu$,

$$\bar{J} = -\frac{N_0 D_s}{kT} \bar{\nabla}_s \mu \qquad (4.1)$$

where $\bar{J}$ is the flux of surface atoms, N_0 is the number of diffusing atoms per unit area, D_s is the surface diffusion coefficient, k is the Boltzman constant, $\bar{\nabla}_s$ is the gradient along the surface, and T the temperature.

One driving force for surface diffusion is due to capillarity and exists for curved surfaces /26,27/. The chemical potential gradient could then be expressed in function of the gradient of the curvature of a surface, and the surface atoms will migrate from regions of a certain curvature to those of lower curvature. For a tip, this corresponds to a surface diffusion flux from the apex of the tip towards the shank with consequence a blunting of the tip /28,29/.

The other driving force which is meaningfull for shaping up the tip is the electric field strength gradient. When a surface is exposed to an electric field, a surface diffusion flux exists from regions of a certain field strength to neibouring regions of higher field strength /30/. Considering the tip geometry, two effects will be present, one at the macroscopic level and the other at the microscopic scale. At macroscopic level, the shape of a tip could be approximated by a cone, a geometry which will induced a surface diffusion flux under field gradient from the shank to the apex. This flux is in the opposite direction of surface diffusion flux due to capillarity, it will then counter-balance the blunting of the tip, and could also in principle lead to a sharpening under certain conditions. On the microscopic scale, an other field gradient exists across a plane surface and in particular across the flat facets present at the apex of the tip and which are due to surface energy anisotropy /31/. It has been studied that the electric field across a plane varies with the radial distance from the plane center (see for example /32/) with consequences a migration of adsorbed atoms towards the edge of the facets (for example the review paper by Tsong /33/) , or the growth of certain crystal planes leading to polyhedral-like shape of the tip apex /30,34-37/. This last phenomenon is called tip "facetting" or "build-up".

Each combination the upper phenomena yields tips having a different geometry and consequently a judicious choice of experimental conditions will allow the set up of recipes to fabricate tips with a controlled apex geometry at the atomic level /13/.

4.1.2. *Fabrication and Regeneration of Microtips* . Two methods were proposed for the controlled fabrication of two different microtip geometries, the build-up tips and the teton tips /13/. They are a two-step methods, the first step is devoted to the macroscopic sharpening of the tip until a fixed radius smaller than 100 nm it is the same for the two methods, and the second one consists essentially of local atomic rearrangement in order to shape it up until an atomic size apex.

The first step consists essentially of heating the tip under vacuum at very high temperature and in presence of an electric field. It has been studied that at very high temperature annealings without field, the tip profile evolves from an initial geometry to reach a constant geometry which is called the pseudo-stationary profile (PSP) /38,39/, the dimensions of the PSP is function exclusively of the blunting rate by surface diffusion and the evaporation rate. For tungsten the limiting radii obtained experimentally are in the interval of 200 nm to 250 nm. To obtain smaller values the blunting rate has to be lowered by counter-balancing the surface diffusion flux due to capillarity by an opposite direction surface diffusion

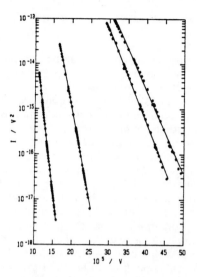

Figure 12.: Successive Fowler-Nordheim plots of the same W tip heated at 3000K with an applied voltage of 3kV in UHV. The displacement of the plots from left side to the right side is correlated to the tip sharpening.

flux under field gradient driving force. Experimental results give limiting radii in the interval of 70 nm to 100 nm, in presence of an electric field. This tip sharpening could be followed by monitoring its intensity-voltage field emission characteristics and the tip radius could be estimated from these values /22/, in fig.12 we show in example the successive Fowler-Nordheim plots during the sharpening.

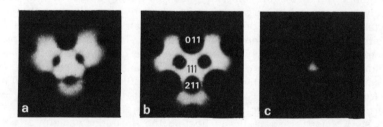

Figure 13. FEM patterns for a W (111) tip during the build-up process (1750K, 3kV). (a) Clean thermal end form with near hemispherical apex with {011} and {112} facets. (b) Enlargement of these facets by a build-up process. (c) Equilibrium build-up apex.

4.1.3. *Build-up tips* . However, and because the heating temperature is very high, the tip apex obtained after the above first step is still an hemisphere, this can be controlled by FEM (fig.13a). Atomic size protrusion at the apex is then shaped up by a build-up process which leads to the enlargement of the {011} and {112} planes (fig.13) and will end when the three {112} facets meet ach other with a monoatomic boundary. The microtip obtained is the protrusion formed by the corner resulting of the meeting of the three {112} build-up planes, it ended with three atoms at the apex. Atomic structure of such tips could be controlled by FIM, as well as the underneath geometry by field desorption technique (fig.14). showing that the height of the apex protrusion is only of atomic size, a few angtroms.

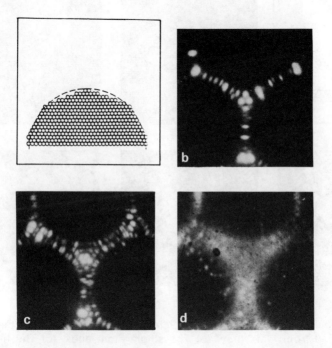

Figure 14. FIM patterns of a build-up W (111) tip (a) Schematic representation of the polyhedral-like equilibrium end form of the apex. (b) The three atom apex and the mono-atomic boundaries between two build-up facets. (c) The same tip after field desorption of the top single atom boundaries and the three atom apex. (d) The blurred regions represented local bumps relative to the surrounding regions. (Vu Thien Binh and J. Marien, unpublished).

4.1.4. *Teton tips* . In this second technique the surface diffusion was hindered and the loss of matter was increased by adequate adsorption species. Oxygen was used for this effect with W tips, and the shaping-up of the microtip was obtained after thermal annealing in presence of an electric field and at temperatures necessary to form volatile oxides. The parameters we used were 10^{-6} Torr to 10^{-4} Torr of oxygen, 2kV to 3kV, and 1500K to 1800K. The 3-D geometry of the microtip were analyzed by FIM and by using the field evaporation technique to peel it (fig.15), and the study have shown that the equilibrium emitter profile is made up of a single atom microtip erected on top of a support tip (teat-like geometry), a schematic draw is exemplified in fig.15a. The triangular base and the height of the microtip are around 5 nm a value which is estimated from the FI patterns. This tip is called teton-tip owing to its special geometry.

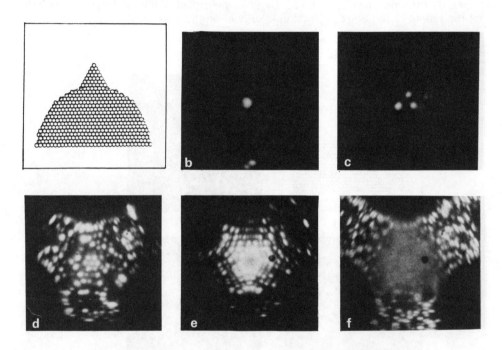

Figure 15. FIM patterns of a teton tip. (a) Schematic representation of the teat-like geometry deduced from FIM study. (b) The single atom apex. (c) The three atom apex obtained after field evaporation of the top atom at the apex of fig. (b). (d). The underneath structure obtained after field evaporation of several layers. (e).Structure of the base of the microtip with a triangular form obtained after further field evaporation from (d). (f). Structure of the support tip, the blurred region, where the field is greater than the best image field, indicates the location of the base of the microtip on the support tip. (FI images are made with Ne, except for (e) where He is used as imaging gaz).

4.2. INFLUENCE OF THE MICROTIP GEOMETRY ON THE FIM PATTERNS

The surface local curvature is a fundamental parameter for the resolution and magnification of field ion patterns /40/ The field ion pattern of the build-up tip (fig.14b) shows the three atoms at the apex, but also the atoms of the monoatomic boundaries between the facets. On the contrary, the field ion patterns of the teton tip do show only the ultimate atoms at the microtip apex location, fig 15b for one atom apex and fig.15c for three-atom apex. Even when the microtip is field evaporated, which means that its height is lowered and the radius is increased compared to the initial geometry, FIM images only the bump which constitutes the base of the microtip (fig.15d and 15f). Imaging the support tip with atomic resolution leads to the blurring of the region of the base of the microtip (fig.15e). This is an indication that the field over the bump region of the microtip base is still much greater than the best image field of the surrounding regions of the support tip. These observations indicate that the field enhancement is so great over the apex of the teton tip, that field ions (or field emission) come exclusively from this region, and confirm the teat-like geometry which has been deduced from the FI patterns after successive field evaporation sequences.

An other remark on the influence of the microtip geometry could be set out when we compare the FI patterns of the three atom apex of the build-up tip (fig.14b) on one hand, and of the teton tip (fig.15c) on the other hand. The distances between the three FI spots are not the same for the two tip geometries. Measured values give a ratio of about 2 between the two distances. Assuming, at first approximation, that the real distances between each atoms of the three atom apex are the same for the build-up tip and for the teton tip, differences between the two FI patterns could be imputed mainly to the local curvature. The same phenomenon is observed if we compared the distances between atoms of fig 15c and of fig.15d (the seven atoms at the center of the FI pattern), the ratio is even greater and is now between 2 and 3. These results are consistent with the magnification concept discussed in the previous chapter, but it is not straitforward to deduce the ratio between the two local radii of curvature because the image compression factor should not be the same. However, these results are an indication that the ionization zones over the three atoms at the apex are not at the same position for the two geometries, and there is an outward tilting of these zones when we pass from the build-up to the teton tip geometry.

4.3. INFLUENCE OF THE MICROTIP GEOMETRY ON THE FIELD EMISSION CHARACTERISTICS

The first difference in field emission characteristics between build-up and teton tip is observed with the current-voltage characteristics. For build-up tips, the variation of the field emission current I in function of the applied voltage V follows strictly the Fowler-Nordheim law, which is represented by a straight line in the plot of $\log(I/V^2)$ versus $1/V$ (fig.16a). Teton tips present a different behaviour, in particular the field emission current increase in function of the applied voltage is smaller than under the Fowler-Nordheim behaviour. This is exemplifed in fig.16b, and is related to the very high field enhancement over the microtip apex due to the teat-like geometry of the teton tips (see section 3.2.)

The second difference is the stability of the field emission current between the two tips /20/. The stability of the current as a function of time mainly depends on

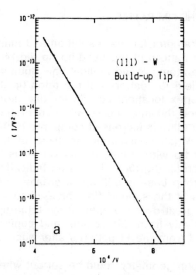 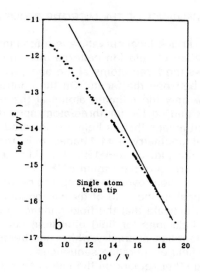

Figure 16. Fig.5: Current-voltage characteristics of the microtips. (a) Build-up tip. (b) Single atom teton tip.

the adsorption at the tip surface. For build-up tips, the current decrease shows mainly the same behaviour as for hemispherical apex tips but with a greater ratio between the minimum current and the initial current. For teton tips, in view of the small emitting area, such atomic-size source are extremely stable over a period of few hours, occasionally interrupted by reversible jumps caused by atom adsorbed on the apex which is signalled by the appearance of extra spot.

Recent electron field emission experiments have been carried out, employing both build-up and teton tips, suggesting that with the latter atomic resolution might have been obtained /19,20/. Fig.17 shows the FE patterns of the three-atom apex of a teton tip. Its shows clearly three circular, well separated spots (fig.17a), the size of which increases with the applied field (for one-atom teton tip only one spot is observed). By further increase of the voltage the three-fold symmetry of the <111> direction , characteristics for electron emission of the atoms of the shank underneath the three top atoms becomes faintly visible (fig 17b). In combination with (1) the experimental observation of the deviation of the I-V charavteristics of this three-atom teton tip from the Fowler-Nordheim law (fig.17c), and (2) the current stability of these three spots over a period of five hours, occasionally interrupted by small reversible jumps which are signalled by the concomitant appearance of extra spots as exempified in fig.17c, this strongly suggests that each of the three main spots is an image of an atom and not of some small facets. As discussed in the previous chapter, this property of the teton tips can be imputed also to the teat-like geometry which increases the magnification factor, and must be set in parallel with the above FI magnification observations between build-up and teton tips.

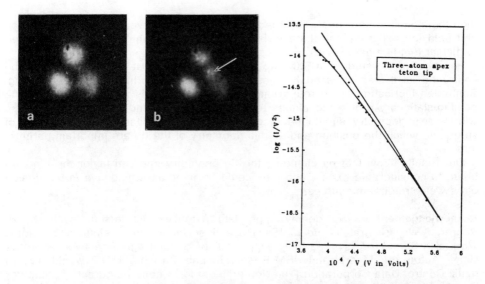

Figure 17. (a) Field emission pattern of a three-atom teton tip. (b) Same as (a) but in addition an adsorbed atom, indicated by an arrow, is present. (c) Current-voltage characteristics of the three-atom apex teton tip

5. Conclusions

We have seen that the tip geometry plays a fundamental role to understand the remarkable properties of the electron beams emitted from the atomic-size tips.

Small protrusions on a smooth support tip can lead to a very collimated electron beam. The origin of the beam focusing can be found in the geometry of the equipotentials near the tip apex. Because of the field distribution in the vicinity of the protrusion the electrons are emitted from a region in which the equipotentials are almost flat. The filtering effect of the barrier and the low diverging field are the main mechanisms which control the angular spread of the beam. The focalization increases as the radius of the underlying tip increases. However, if the radius increases the applied voltage needed to get the same current increases. Therefore in practical applications, if both focussing and low electron energy are required, a compromise has to be found. (It has been suggested /41/ that a constriction of adiabatically changing shape, formed around the tip apex, could also focus the electron beam. However, this adiabatic approximation only holds, if the width of the constriction changes in a region large compare with the electron wave length which is not the case of the emission from an atomic-size region)

We have shown that the usual estimations of resolution can not be applied when there are interference effects between the localized emission centers on the surface of the tip. The results of our quantum mechanical simulations show that atomic resolution is only possible for teat-like tip geometries, provide the emission is localized on each atom at the apex and the size of the protrusion is much larger than the atomic dimensions. The special behaviours observed in the field emission and field ion experimental characteristics of teton tips can be associated to their particular teat-like geometry.

For most of the studies in SXM, tip-surface interactions at atomic scale must be controlled as it was stressed by most of the authors of this NATO-ASI meeting. Controlled fabrication of microtips until the atomic scale is possible by using the pseudo-stationary profile techniques to produce build-up and teton tips. These tips with known geometry should be of great help for quantitative interpretations of studies in which the position and atomic geometry of the tip are important parameters.

Our results show that by changing the tip geometry one can taylor the electron beam to considerable extent, and they could be used as a guidance in the design of new electron beam sources.

Acknowledgement . We deeply appreciate extensive interactions with Dr. H. Rohrer. We are grateful to L. Escapa and A. Jacoby for helpful discussions. Work of N.G. and J.J.S. is partially supported by a joint agreement between the Univ. Auton. de Madrid and IBM Research Lab. Zurich. H.D.R would like to thank Control Data Corporation (The Netherlands) for a generous grant of computer time on the CYBER 205, the University of Leuven for providing unrestricted access to their IBM 3090-300/VF, and the Belgian National Science Foundation for financial support.

References

1. García N. and Rohrer H. (1989) J. Phys Cond. Matt. 1, 3737
2. Serena P., Escapa L., Sáenz J.J., García N., and Rohrer H. (1988) J. Microscopy, 152, 43
3. Gabor D. (1949) Proc. Roy. Soc. London A54, 197
4. Gabor D. (1951) Proc. Roy. Soc. London B64, 449
5. Lichte H. (1986) Ultramicroscopy 20, 293
6. Tonomura A. (1987) Rev. Mod. Phys. 59, 639
7. Missiroli F.G., Pozzi G. and Valdre U. (1981) J. Phys. E 14, 649
8. Binnig G. and Rohrer H. (1986) IBM J. Res. and Develop. 30, 355
9. Kuk Y. and Silverman P.J. (1986) Appl. Phys. Lett. 48, 1597-1600
10. Hashizume T., Kamiya I., Hasegawa Y., Sano N., Sakurai T., and Pickering H.W. (1988) J. Microscopy 152, 347
11. Nishikawa O., Tomitori M. and Katsuki F. (1988) J. Microscopy 152, 347
12. Vu Thien Binh and Marien J. (1988) Surface Science 102, L539
13. Vu Thien Binh (1988) J. Microscopy 152, 355
14. Van Wees B.J., Van Houten H., Beenakker J., Williamson J.G. Kouwenhoven L.P., van der Maret D. and Foxon C.T. (1988) Phys. Rev. Lett. 60, 848
15. Wharam D.A., Thorton T.J., Newbury R., Pepper M., Ahmed H. , Frost J.E.F., Hasko D.G., Peacock D.C., Ritchie D.A. and Jones, G.A.G. (1988) J. Phys. C 21 L209
16. Escapa L. and Garcia N. (1989) in this Proceedings, and references therein
17. García N., Sáenz J.J. and H. De Raedt (1989) J. Phys. Condens. Matt. (in press)
18. Rose D.J. (1956) J. Appl. Phys. 27, 215

19. Sáenz J.J., García N., De Raedt H. and Vu Thien Binh (1989) in "Proceedings of the 36th Field Emission Symposium", Oxford (to appear in J. de Physique).
20. Vu Thien Binh, Saenz J.J., De Raedt H., and Garcia H. (1989) to be published.
21. A general discussion of the emission characteristics of 2D systems including applications to GaAs devices and light experiments can be found in ref. 17
22. Gomer R. (1961) "Field Emission and Field Ionization", (Harvard University, Cambridge Mass.)
23. De Raedt H. (1987) Comp. Phys. Rep. 7, 1
24. Smythe W.R. (1952) "Static and Dynamic Electricity", McGraw Hill, New York
25. Dyke W.P. and Dolan W.W. (1956) in "Advances in Electronics and Electron Physics", 8 (Academic Press, New York)
26. Herring C. (1953) in "Structure and Properties of Solid surfaces" Editors R. Gomer and C.S. Smith, Univ. of Chicago Press.
27. Mullins W.W. (1957) J.Appl.Phys. 28, 333.
28. Nichols F.A. and Mullins W.W. (1965) Trans.Met.Soc. AIME 233, 1840.
29. Vu Thien Binh, Piquet A., Roux H., Uzan R., and Drechsler M. (1971) Surface Sci. 25, 348.
30. Sokolovskaia I.L. (1956) Sov.Phys.Techn. 1, 1147.
31. Drechsler M. (1983) in "Surface Mobilities on Solid Materials: Fundamental Concepts and Applications", Editor Vu Thien Binh, Plenum Press, NATO ASI Series B 86, pp.405).
32. Plummer E.W. and Rhodin T.N. (1969) J.Chem.Phys. 49, 3473.
33. Tsong T.T. (1983) in "Surface Mobilities on Solid Materials: Fundamental Concepts and Applications" Editor Vu Thien Binh, Plenum Press, NATO ASI Series B 86, pp.109).
34. Drechsler M. (1957) Z.Elektrochemie 61, 48.
35. J.P. Barbour, Charbonnier F.M., Dolan W.W., Dyke W.P., Martin E.E., and Trolan J.K. (1960) Phys.Rev. 117, 1452.
36. P.C. Bettler P.C. and Charbonnier F.M. (1960) Phys.Rev. 119, 85.
37. Neddermeyer H. and Drechsler M. (1988) J. of Microscopy 152, 459.
38. Vu Thien Binh, Piquet A., Roux H., Uzan R., and Drechsler M. (1974) Surface Sci. 44, 598.
39. Vu Thien Binh and Uzan R. (1987) Surface Sci. 179, 540.
40. Muller E.W. and Tsong T.T. (1969) in "Field Ion Microscopy: Principles and Applications", American Elsevier Publ. Co.
41. Lang N.D., Jacoby A. and Imry Y. (1989) to be published

FORCE MICROSCOPY

H. HEINZELMANN, E. MEYER, H. RUDIN, and H.-J. GÜNTHERODT
Institute of Physics, University of Basel
Klingelbergstrasse 82
4056 Basel, Switzerland

ABSTRACT. Force microscopy is a new technique which allows the investigation of minute interactions on a micrometer down to an atomic scale. We will give a short overview of the most important experimental aspects and describe recent trends in instrumentation and force probe design. Different forces which have been studied by force microscopy are summarized. The emphasis is put on the imaging mode of force microscopy in the regime of repulsive contact forces, commonly called *atomic force microscopy* (AFM). Numerous examples from our group as well as from other laboratories illustrate the high resolution capability of AFM, and the wide variety of samples studied so far show the applicability of AFM to different fields of actual interest, such as surface science, biology and technology.

1. Introduction

Force is an old concept in human science and philosophy. For hundreds of years man has observed and studied omnipresent interactions such as gravitation and electromagnetism. Modern science allows precise investigation of various types of forces. Often specially designed instruments are used, such as the surface forces apparatus developed by the schools of Tabor and Israelachvili [1, 2, 3]. These methods give a detailed description of the force *vs.* distance law, but are mostly limited in their lateral resolution.

The use of point probes is a well known way of studying physical properties of a sample on a local scale. Scanning a probe with respect to the surface thus gives a highly resolved map of a specific physical property. A breakthrough of this concept was the development of the scanning tunneling microscope (STM) in 1981 [4, 5]. Using a metallic point probe in close distance to conducting samples, one could trace the electron density of states near the Fermi energy of the sample surface. This method allowed atomic resolution imaging of a whole variety of systems, e. g. layered materials such as graphite or transition metal dichalcogenides, semiconductors such as Si or GaAs, or metals like Au or Al. In addition to its imaging capability, the STM proved to be useful in I-V spectroscopy and for surface modifications on an atomic scale. The STM has opened a new era in physics in that the observation and accessibility of individual atoms has become possible.

The success of STM triggered the development of a variety of other scanned probe microscopes. The principle behind all these microscopes is similar, but they differ in the physical property which they probe (figure 1). For these new techniques the name SXM has been proposed, where X stands for tunneling, force, near field optical, thermal etc.. For a recent review on scanned probe microscopes see reference [6]. The most prominent among these is probably the scanning force microscope (SFM), originally called *atomic* force microscope (AFM) and introduced in 1986 by Binnig, Quate and Gerber [7].

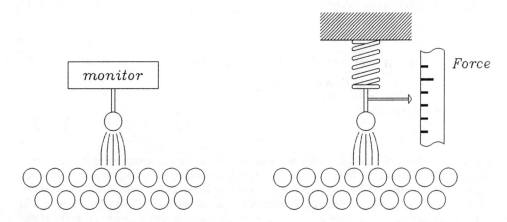

Figure 1: Principle of a point probe. In force microscopy (right) the property sensed is the interaction force between tip and sample.

The force microscope connects earlier force measurements with the high resolution capability of a point probe. Detailed studies of interactions such as interatomic, frictional, magnetic or electrostatic forces, are now possible on a very fine scale. Probably even more important is its potential as an imaging device. The topography of insulating surfaces can not directly be imaged by other high resolution techniques such as electron microscopy. In this respect the force microscope follows the stylus profilometer [8], and indeed the first AFM design was a hybrid of the stylus and the STM.

There are many questions of great scientific and technological interest connected to insulating samples, and there is much hope that the force microscope can contribute to solve open problems in surface physics, biology (a subject covered by Paul Hansma in these proceedings), and technology. Some workers have observed considerable forces [9] and even severe sample damage [10] during STM operation in air, suggesting that for some applications AFM might be advantageous even on conducting surfaces.

2. Force Microscopy

The principle of force microscopy is rather simple. A fine probe tip mounted on a small spring is tracked across the sample surface. The spring deflects according to the force between tip and surface, and this deflection is monitored as a function of the lateral displacement of the tip. This scheme resembles that of a record player, although the displacements and forces have to be controlled on a much finer scale.

2.1. THE INSTRUMENT

The tunneling current between two electrodes depends exponentially on their separation. It varies by a factor of up to 10 (depending on the effective barrier height) when the separation is changed by 1 Å, giving the STM its extremely high sensitivity. The first AFM by Binnig, Quate and Gerber [7] was derived from a tunneling microscope. They positioned a tunneling probe behind the force sensing lever, providing a highly sensitive displacement sensor. With some modifications of the setup a STM can be turned into an AFM. Most of the groups who have started working with AFM have chosen this tunneling detection scheme [11, 12, 13, 14, 15].

Optical interferometry allows the detection of ac displacements down to 10^{-4} Å. Some workers used either the homodyne [16, 17, 18, 19, 20] or the more sophisticated heterodyne [21] technique to build a force microscope. In the homodyne setup drifts of the optical path show up in the signal. The use of fiberoptics has allowed the reduction of the length of the interferometer cavity to some microns and has thus led to a minimization of the drift problems [18]. Further, the setup of the microscope could be made simpler and more robust, and it has even been possible to achieve atomic resolution in dc operation [22]. The heterodyne interferometry [23] is insensitive to optical path length drifts, but needs more optical and electronic components.

An alternative interferometric method is laser diode interferometry [24]. The light reflected off the lever is fed back into the laser diode. The diode is extremely sensitive to this optical feedback, and the displacement signal is contained in the diode current or detected by an integrated photodiode. No other optical components are needed, but the analysis of the signal is rather complex [25]. Other optical methods are the use of focusing error detection based optical heads [26] and the detection of a laser beam reflected off the lever with a position sensitive detector [27, 28]. Both setups are relatively simple and the signal analysis is straightforward. The latter method is very attractive since it readily yields a vertical resolution of roughly 0.2 Å, sufficient for many imaging applications of force microscopy.

Optical detection schemes have several advantages compared to tunneling detection. Since the laser spot averages over some μm^2, the signal is less dependent on the roughness of the lever. This reduces the sensitivity to thermal drift. The optical signal is less sensitive to contamination of the lever backside, which can be a severe problem when using tunneling detection. Further, when working at ambient conditions the additional force exerted on the cantilever by a laser beam is probably smaller than that by a tunneling tip. This makes the microscope operation more reliable and gentle.

Electron tunneling and optical methods are the widest used techniques in force microscopy, and atomic resolution has been achieved with both. An alternative detection scheme is the measurement of the capacitance between the lever and a reference electrode [29]. It is also possible to measure forces in an STM-like setup where the sample is mounted on a flexible support. The force information is then contained in the tunneling current signal [30, 31] or taken from a bimorph element connected with the sample holder [32, 33].

So far, most of the force microscopes are operated in air. Very soon after its first demonstration the technique was extended to liquid-covered surfaces [13], an important step for its application to biological samples. More recent developments have produced first prototypes for operation in ultra-high vacuum [27] or at low temperatures [34]. A great deal is expected in particular from UHV force microscopy, since only few alternative methods exist to access the surface of insulators or to study minute forces under controlled environmental conditions.

2.1.1. *Setup of an actual force microscope:* The schematics of an actual force microscope used in our laboratory is shown in figure 2. It uses electron tunneling for detection of the lever displacements.

The tunneling tip and force sensor are mounted on the same block of glass ceramics (Macor). This block can be rotated around a vertical axis. The force sensor is glued to a metal plate fixed to a leaf spring with the help of a magnet. It can be laterally positioned relative to the tunneling tip by moving the holding magnet with an external xyz-micropositioner. Tip and lever are approached by bending the leaf spring with two screws to within the dynamic range of the tunneling tip piezo (1μm). These components are thus connected by a very short thermal path. Once a certain force is preset by the appropriate bending of the leaf spring, it is nearly drift-free during the experiment. The sample is fixed on a single tube scanner on a second block made from ceramic which can be rotated around another vertical axis.

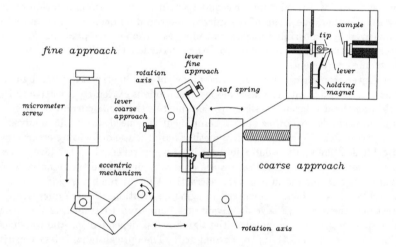

Figure 2: An AFM based on electron tunneling. The white parts visible in the photograph are Macor blocks carrying tip and lever, and sample, respectively.

The approach of the whole tip-lever assembly to the sample is achieved by two screws which press both blocks together against the force of strong stabilizing springs (not shown in the schematics). The coarse approach screw is for rough positioning under optical control. The fine approach is more elaborate: a micrometer screw drives an eccentric mechanism which moves the tip-lever assembly, with an additional geometrical reduction, towards the sample. This allows controlled movements with a step size as small as 5 nm. Although this microscope proved to be rather insensitive to external vibrations, we mainly operate it on a commercial air-damped table (Newport Corp.). For high-resolution work, a bell jar is used to shield the instrument from acoustic noise.

2.2. THE FORCE SENSOR

The force sensor is probably the most crucial part of a force microscope. It consists of a probing tip and a flexible system which acts as a spring. In most experiments reported so far this spring is in the form of a rectangularly or triangularly shaped cantilever beam or a circular rod. In initial experiments pieces of metal foil were used, with [7] or without [12] an additionally attached diamond. Other groups have used wires [17, 21] which were bent and etched to a tip similar to tunneling tips in STM. Alternative lever designs have also been developed [13, 35].

Common to all different designs is the need to combine a spring constant of 0.1 - 10 N/m with a resonant frequency as high as possible, which would make the instrument less sensitive to low frequency noise and would allow high scanning rates. Since the relation $f_{res} \propto \sqrt{C/m}$ holds for every geometry used so far, the mass of the force sensor must be kept small. Very soon the advantages of microfabrication techniques [36] were recognized, which led to the production of AFM cantilevers made from thin films of silicon oxide and nitride [37, 38].

The shape of the probing tip is one of the factors determining the spatial resolution of the force microscope. The SiO_2 microcantilevers provide enough inherent surface roughness due to protruding microtips to allow atomic resolution imaging on very flat surfaces. The cantilever is simply tilted until an edge points towards the sample, no additional tip is needed. For more corrugated surfaces, however, the overall shape of the probe tip becomes more important. Some groups have tried the use of fragments of a shattered diamond

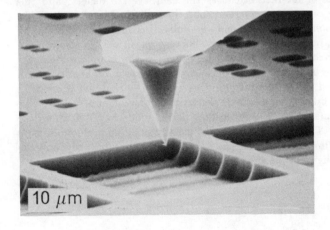

Figure 3: Silicon force sensor hovering over a microfabricated test pattern.
(Courtesy of O. Wolter)

[13, 39] which gave good results also on rougher surfaces. However, the mounting of such a tip on the lever is a delicate and time consuming job, and the success not only depends on the experimentalists' skill, but also on his luck in choosing a sharp fragment. As with the bare microcantilevers, not every force probe gives atomic resolution.

The need for reproducibility and batch fabrication of sharp tips already attached to the cantilevers led to the development of a microfabrication process for integrated force sensors. Levers with integrated tips have been produced from SiO_2 and Si_3N_4 [40]. Tips prepared in this way are sharp down to a scale of perhaps hundred Å. A probably even better way is the fabrication of tips from a single crystal. Olaf Wolter from the IBM manufacturing center in Sindelfingen succeeded in producing integrated levers with tips made entirely from crystalline silicon [41]. There is a good chance that, due to its crystallographic orientation, such a tip may end in a single atomic protrusion, and the atomic species at its apex is known in principle. This is very important in order to compare force microscope experiments with theory. Figure 3 shows such a force sensor hovering over a microfabricated test pattern. The lever deflection is sensed optically in this case.

2.3. MODES OF OPERATION

There are several ways to operate a force microscope. Force microscopes based on interferometric detection often use *dynamic* measuring modes where either the force sensor or the sample is vibrated at a high frequency. In the first case the lever is driven slightly off its natural resonance frequency, usually by a piezoelectric element. The interaction force gradient F' felt by the tip causes a shift in the lever resonant frequency, $\Delta f = (f_{res}/2C) \cdot F'$, where C is the spring constant of the lever. This leads to a measurable change in its vibration amplitude. The absolute force is calculated from the gradient by integration. The vibration amplitude of the lever is usually of the order of some nanometers, and consequently the average tip to sample distance has to be rather large, leading to a natural limit for the lateral resolution. This method is predominantly used for work on magnetic and electrostatic forces, and for non-contact imaging. The second dynamic method is seldom used. The sample is vibrated by a piezoelectric element, and a transfer function has to be evaluated to calculate the force [42].

Quasistatic modes are best suited for microscopes with tunneling or direct optical beam deflection feedback. The force of interaction causes the cantilever to deflect, and this deflection is measured directly. In most cases the probe tip is in close contact with the sample. This allows a high resolution laterally and vertically. Depending on the kind of feedback operation, one can basically distinguish two different quasistatic modes. In the constant force mode, the feedback is connected to the z-piezo of the sample so that the lever deflection and thus the average loading force between tip and sample is kept constant. The feedback signal is monitored. Scan lines taken in this mode show contours of constant force above the surface. The mode of variable lever deflection is more common. The force felt by the tip is monitored, analogous to the constant height mode in STM where the tunneling current is registered. The variation in force is usually small and seldom exceeds a few percent. In larger scale images, however, the force may vary significantly, and the influence of the local force gradient on the signal has to be taken into account.

3. Forces and their Relevance to Force Microscopy

When two bodies come into close distance there is a whole variety of interactions arising between them, depending on the nature of the two bodies. For example van der Waals forces are always present and are sensible over hundreds of Ångstroms, whereas metallic adhesion as a quantum mechanical exchange interaction only acts if the surfaces are not more than some Ångstroms apart.

In force microscopy one of the two bodies is the sample, the other is the force probe. In the limit of high contact forces ionic repulsion between the ions of the sample and the ions of the tip is the dominant interaction. Scanning in this regime reveals the surface topography of the sample. But it is clear that all the other interactions influence the force signal depending on their magnitude and range. Furthermore, they give rise to a different image contrast and thus allow the imaging of such features as magnetic domains or isolated charges, or they are interesting from a more principle point of view. Figure 4 summarizes those most relevant to force microscopy.

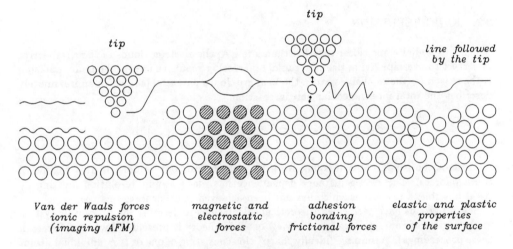

Figure 4: Some interactions relevant to force microscopy.
The probe tip can follow lines of constant force. Van der Waals and contact repulsion forces in general reveal the topography of the sample, magnetic and electrostatic forces lead to an additional attraction or repulsion, adhesion and binding forces play a role in friction and can lead to an atomic-scale stick-and-slip behaviour of the tip, and the elastic and plastic properties of the sample can be studied with indentation-type experiments and change the image contrast in the contact profiling mode.

3.1. VAN DER WAALS INTERACTION

Van der Waals forces are present between all types of bodies. They include induction forces (the interaction of a dipole with an induced dipole), orientation forces (the interaction between two orientated permanent dipoles) and dispersion forces. Dispersion forces arise from induced dipole - induced dipole interaction and are quantum mechanical in origin.

They are long-ranged (2 Å up to > 100 Å) and can be attractive or repulsive. They are always present and are the origin for such macroscopic phenomena as adhesion, surface tension, physisorption, the magnitude of the strength of solids, etc.. Some solids such as hexane are solely held together by dispersion forces and are therefore called van der Waals solids. In general, dispersion forces do not follow a simple power law. Their potential is approximately proportional to $1/r^6$ (London forces). In most cases they represent the dominant term of the van der Waals interaction.

In dynamic measuring modes the probe tip is typically some nanometers above the surface. At this distance the van der Waals interaction is the dominant force (except for the case of charged or magnetized surfaces). Contours of constant force or constant force derivative reflect the sample topography and, since the van der Waals force is material dependent, can reveal chemical inhomogeneities of the sample surface. Imaging in the regime of van der Waals forces is non-contacting, saving the tip from damage. Moiseev et al. [43] have calculated the distance dependence of van der Waals forces and derived an expression for the lateral resolution of the force microscope as a function of the tip to sample distance.

3.2. IONIC REPULSION

When two bodies come closer together than a few Å, the electron clouds of their respective ions start to overlap, giving rise to a rapidly increasing repulsive force. The ionic repulsion is the reason for the repulsive part in the Lennard-Jones potential [44] which approximately describes the total intermolecular pair interaction

$$w(r) = C_1/r^{12} - C_2/r^6 = 4\varepsilon \left[\left(\frac{\sigma}{r}\right)^{12} - \left(\frac{\sigma}{r}\right)^6 \right]$$

Most of the topographic imaging work by force microscopy is done in this regime of contact forces. The strong distance dependence has made atomic resolution imaging by AFM possible, a resolution which has not been achieved in other force regimes.

Several groups have performed theoretical calculations to describe the tip-sample system in the regime of contact force (an overview of AFM theory is presented by Salim Ciraci in these proceedings). Assuming "ideally sharp" tips consisting of one or two individual atoms and realistic interaction potentials they were able to reproduce the high lateral resolution observed in the experiment [45, 46, 47, 48]. However, the maximum force for tip-stability and for damage-free imaging of the surface, calculated to be 10^{-8} and 10^{-9} N for the C-C [46] and Si-Si [47] tip-sample systems, is much lower than the tracking force used in most AFM experiments. There are some indications that the effective contact area is much larger than a single atom diameter, and it has been shown that an atomic contrast in the force signal can also be observed in this case [49]. To clear this point of actual debate, more theoretical and experimental work is needed.

3.3. MAGNETIC FORCES

Magnetic microstructure is of both scientific importance and technological interest [50]. Domain wall structure, formation and movement, or various magnetic ordering phenomena are a field of intense experimental and theoretical research. Many of these questions are related to equally important technological problems. The increased understanding of surface magnetism has led to the development of new materials such as thin film recording media which allow today's highest information storage densities. In both fields of research, fundamental as well as applied, high resolution microscopies are necessary to access the length scale of interesting magnetic phenomena.

There are already a number of techniques which allow the study of surface domain structures with high resolution, such as Bitter pattern, Lorentz microscopy, scanning electron microscopy with polarization analysis, or Kerr microscopy.

The need for higher resolution and easier operation makes the force microscope a promising candidate for the investigation of magnetic surfaces. The experimental procedure is similar to the topographic imaging. A magnetized probe tip is scanned across the sample and responds to the sample surface magnetic stray field, thus tracing lines of equal force or equal force gradient. Usually the distance between tip and sample is larger than in the surface profiling mode (usually some nanometers) in order to be sensitive to surface magnetic properties instead of sample topography.

Besides the topographic imaging of various surfaces the most effort in force microscopy has been put into its adaption to the study of magnetic materials [51, 52]. Laser-written bits in TbFe thin film recording media [53, 54], magnetization patterns in Co alloys [55, 56], the fine structure of Bloch walls in Fe crystals [32] and the domain structure of permalloy films [57] have been made visible by magnetic force microscopy. A detection scheme was developed recently which allows simultaneous recording of magnetic forces and sample topography [58], an important step towards the understanding of surface magnetic properties. Considerable theoretical effort is needed to get quantitative information from such experiments (see for example references [59, 60]).

3.4. ELECTROSTATIC FORCES

The sensitivity of force microscopy to electrostatic forces has led to different applications. Surface dielectric properties can be mapped by a potentiometric scheme. Martin et al. [61] were able to map simultaneously topography and capacity of a silicon substrate partly covered with a photoresist, thus revealing the distribution of the dielectric resist on the substrate. Stern et al. [62] deposited localized charge on an insulating surface and could observe its decay due to surface charge mobility. In a refined scheme it is possible to simultaneously determine the amount and sign of surface charges as small as 6 electron charges, along with the surface topography [63]. The lateral resolution of 0.2 μm is not achieved by any other charge detection method. This allows the study of tribocharging effects on a submicron scale [64].

In many force microscopy experiments an additional voltage applied to a back-electrode behind the sample is used to create an additional electrostatic force. This can be tuned to shift the total force to an experimentally appropriate range.

3.5. FRICTIONAL FORCES

The friction between two surfaces is strongly related to their adhesion. It is currently a field of rapidly increasing interest for many reasons:

- the miniaturization of mechanical elements and its operating distances makes friction and wear one of the largest problems for future technology
- a microscopic understanding of friction is still lacking

- the increasing possibilities of computers allow the calculation of more complex problems, such as the sliding of two surfaces under load
- the force microscope represents a promising microscopic tool which may allow the study of friction on a true atomic scale

So far friction and wear had to be studied by indirect means such as wear tests and subsequent investigation of the surfaces in contact. Mate et al. [16] showed that the force microscope is suited for the observation of friction on an atomic scale. Scanning a tungsten tip across a surface of highly oriented pyrolytic graphite they could detect an atomic periodicity in the frictional force. A similar behaviour was found on mica [65]. There is much hope that the force microscope will help to understand important frictional mechanisms from a microscopical point of view, e.g. as the energy dissipation in the contact zone [66].

3.6. ADHESION

The strength of adhesive forces between two bodies depends on the fundamental properties of the two solids, both bulk and surface. Surface properties correlated with adhesion include reconstructions, segregation and the chemistry of the surface species. Practical models of adhesion do not consider the details of the interatomic forces, but include them in the more general term of surface energy of adhesion $\Delta\gamma = \gamma_1 + \gamma_2 - \gamma_{12}$, where γ_1 and γ_2 are the surface energies of the two free surfaces and γ_{12} is the interfacial energy. Experimental adhesion studies involve the testing of not only bond strengths at interfaces, but also the effects of elastic and plastic deformation. They are further complicated by surface impurities, defects and by the short timescale of the adhesion process. A good review of the experimental work is given in reference [67].

So far, detailed theoretical descriptions have only considered the most ideal surfaces such as the bimetallic interface [68]. By calculating self-consistent electron densities and minimizing the total energy they could derive a universal curve for the basic adhesive energy E_{ad}. This type of adhesion has been studied by Dürig et al. by analyzing the interaction between the tunneling tip and a sample mounted on a cantilevered support [30, 69]. They were able to measure this force with the high resolution offered by the STM and thus distinguish the metallic adhesion force from the repulsion felt over oxidized parts of an Al sample.

The force microscope allows the study of surface forces on a local scale. When approached to the sample, the lever suddenly jumps into contact with the surface. On retraction it has to be pulled a larger distance away from the surface until it snaps back. This phenomenon is well known from various force measurement techniques: when the gradient of the measured force becomes larger than the spring constant of the force probing spring, the situation becomes unstable and the spring jumps into a stable position. This mechanism has been exploited in the surface forces apparatus to determine force vs. distance laws [1, 2]. Several groups have reported similar observations with a force microscope and obtained comparable values for the maximum attractive force [15, 21, 39, 42, 70]. Martin et al. [21] recorded simultaneously sample topography and peak attractive force of a silicon sample. They could distinguish between regions of bare silicon and other regions which were covered with a photoresist.

Under atmospheric conditions this attraction can have several origins. We have tried to describe this attraction with a van der Waals potential and to fit it to our experimental data [39]. Figure 5 shows the calculated (left) and experimental (right) curves for a diamond tip and a graphite surface. The tip was modelled as a sphere, the C-C interaction parameters used for the calculation were $\varepsilon = 3.79 \times 10^{-22}$ J and $\sigma = 3.41$ Å [71]. The force constant for the lever used in this experiment was $C = 0.7$ N/m. The total approach and attraction cycle took several seconds. The tip radius was used as a fit parameter, giving a value of $R = 250$ Å, in reasonable agreement with scanning electron micrographs. On LiF surfaces, however, the experiment could not be described in this way. Differences in the surface forces between a tungsten tip and highly oriented pyrolytic graphite and gold surfaces were observed with the same technique by Burnham et al. [72], however, the maximum applied loading force was much higher.

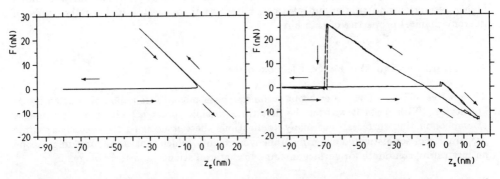

Figure 5: Force vs. distance curve, calculation (left) and experiment (right). The point of zero distance is defined where the tip is in contact with the sample at zero force. A positive force means attraction.

A second way to account for the observations is the assumption of the presence of a liquid film. Such films, preferentially from water, are present on most surfaces exposed to ambient conditions. When the liquid films covering sample and probe tip touch they will build a liquid bridge, and a sudden attraction due to the capillary binding force will bring the tip into contact with the surface. When separating the tip from the surface, the liquid bridge between the surfaces contracts and is finally destroyed, and the lever suddenly snaps back. Mate et al. have used the distance the lever must retract as a measure for the thickness of polymeric liquid films deposited on a solid substrate [73]. Weisenhorn et al. have shown that on mica the peak attractive force is reduced from the 10^{-7} N range to the order of 10^{-9} N when the same experiment is done in water, where no liquid capillary forms between tip and sample [74]. These experiments show the strong influence liquid surface films may have. AFM operation in water is a useful way to minimize the loading force of the probe tip and thus very important when imaging delicate samples, such as biomolecules.

3.7. ELASTIC AND PLASTIC RESPONSE OF THE SAMPLE

Elastic and plastic behaviour is an important property of a material. Indentation measurements [75] are a widespread method for obtaining these magnitudes. In these techniques the indentation is measured in-situ or the penetration depths of a sharp point probe loaded with different forces are analyzed using the electron microscope. The indentation has to be small compared to the total sample thickness in order to exclude size effects. For the characterization of thin films this sets limitations to the applied load. The influence of the substrate on the load-penetration depth curve generally begins to show up at penetration depths of about one tenth of the film thickness, and refined setups are necessary [76]. With the force microscope it is possible to apply much smaller loads than commonly used in conventional indentation meters, and the technique certainly has a great potential for hardness measurements.

Also in imaging-AFM the response of the sample to the probing tip plays a role and contributes to the image contrast. Since reproducible imaging is possible on most samples with loading forces of 10^{-8} to 10^{-7} N, the surface deformation is then elastic.

Burnham et al. have used a force microscope to probe the elastic and plastic properties of various samples by extending the loading force to the μN range [72]. This method allows higher resolution and gentler indentation and seems to be best suited for the study of micromechanical properties of thin films.

4. Imaging with the Force Microscope

The greatest effort in force microscopy so far has been put into its application as an imaging device. With AFM it was for the first time possible to get atomic resolution images of non-conducting surfaces. For many applications there is no need for excessive sample preparation or sophisticated experimental environments, which makes the force microscope an interesting candidate for surface imaging in material science and technology.

4.1. ATOMIC-SCALE IMAGES

Materials with a layered structure showed to be the easiest to resolve atomically by AFM. Three examples are briefly described. It is experimentally evident that it is much more difficult to obtain atomic resolution images of non-layered materials. The reason is a point of actual debate, but the different elastic behaviour of these surfaces probably plays an important role. There is much confidence that routine atomic-resolution imaging also on non-layered samples will be possible in the near future.

4.1.1. *Graphite:* Highly oriented pyrolitic graphite (HOPG) has proven an ideal sample for easy atomic resolution imaging by STM. Soon after the first paper on AFM, Binnig et al. demonstrated atomic resolution by AFM [37]. An actual example from our laboratory is shown in figure 6. The image was taken in the variable deflection mode, bright spots correspond to a larger repulsive force. The scan direction is from left to right, with the opposite direction blanked. The graphite surface consists of two inequivalent atoms, one with a neighbour (type A) and one with no neighbour (type B) in the layer beneath. These inequivalent atoms respond to the AFM tip in different ways, so that only one set of atoms is imaged in most experiments. Like in STM, atomic resolution AFM images of graphite may have many different appearances [38], depending on various controllable and arbitrary conditions, such as loading force between lever and sample, cleanliness of the surface, and state of the tip.

Figure 6: AFM image of highly oriented pyrolitic graphite. Every second atom of the surface is visible. Friction at the beginning of each scan line distorts the image at the left margin.

|——————|
10 Å

The distorted region at the left margin is attributed to frictional effects. At the beginning of each scan line, the tip stays at a certain location on the surface until the lateral force in the scan direction is large enough to overcome the sticking force. This feature is commonly seen on most AFM data of HOPG.

4.1.2. *Boron Nitride:* Highly oriented pyrolytic nitride (HOPBN) was the first insulator resolved *atomically* by any technique [11]. The observed lattice spacing indicates that only one atomic species is visible by AFM. By approximating the interaction between their SiO_2 probe and the sample surface with a Gordon-Kim potential, Albrecht *et al.* found a lever deflection which should be 0.2 Å larger on the N sites than on the B sites. The height maxima should therefore show the N atoms.

4.1.3. *Transition Metal Dichalcogenides:* These materials have attracted much interest in the STM community since some of them develop charge density waves (CDW) at relatively high temperatures. The CDW state is invoked by a small shift of the atomic positions (Peierls transition) of the crystal lattice, causing a redistribution of the charge at the Fermi energy. In the case of 1T-TaS_2 and 1T-$TaSe_2$ a two-dimensional $\sqrt{13} \times \sqrt{13}$ electronic superstructure is formed at room temperature.

The STM is sensitive to the local density of states at the Fermi level, and it is well established that CDW's can be studied by STM (see for example references [77, 78]). STM images often show the atomic lattice together with the electronic superstructure. The AFM responds to the force which is related to the total density of states. The charge transfer associated with the CDW can be up to 0.9 electron charges for the atoms located at the minimum and the maximum of the CDW [79], and it is of interest to see to what extent AFM responds to the Fermi level effect.

Figure 7 shows a typical STM image (a) and a typical AFM image (b) on 1T-TaSe$_2$ at room temperature [80]. In both images the atomic lattice is visible. The experimental value of 3.5 Å ± 0.1 Å agrees well with the known lattice constant of 3.477 Å. The CDW is only visible in the STM image where it actually dominates the image. The AFM corrugation varies between 0.2 and 0.4 Å for the atomic lattice, and no additional corrugation due to a CDW could be resolved down to approximately 0.01 Å. The absence of a superstructure in the AFM images is surprising as compared to the results from He scattering experiments, which also probe the total charge density and do show the CDW modulation [81].

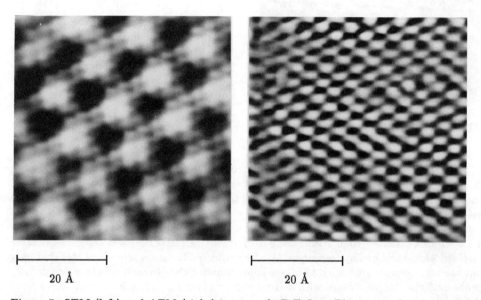

|⊢―――――⊣| |⊢―――――⊣|
20 Å   20 Å

Figure 7: STM (left) and AFM (right) images of 1T-TaSe$_2$. The atomic lattice is visible with both techniques, but STM additionally shows the CDW modulation.
(STM image courtesy of D. Anselmetti)

The loading force for the AFM experiments was 3×10^{-8} N which is close to the smallest values possible with our setup. At these forces a surface distortion caused by the probing tip cannot be ruled out, which might suppress the CDW state. However, the same argument would hold for the STM experiments where considerable forces may also build up, but in this case the CDW is still visible. This discrepancy is not yet solved.

4.1.4. *Silicon Oxide:* One of the first non-layered materials studied in our laboratory was non-crystalline quartz (SiO_2) [12, 70]. Figure 8 shows an AFM image of 27 x 27 Å^2. Distinct bumps $\sim$ 0.5 Å $\pm$ 0.2 Å are resolved which are separated by an average distance of $\sim$ 5 Å. Since quartz glass is built up from SiO_4 tetrahedra with a bonding length of 1.5 Å, this image does not reflect individual atoms, but more probably individual SiO_4 units. However, this result indicates that a resolution of 2.5 Å is possible by AFM, even on non-layered materials. A similar resolution was found by Marti *et al.* on the native oxide of silicon [82]. Atomic steps, without lateral atomic resolution, could be imaged on a variety of non-layered materials, among them lithium fluoride [39] and sapphire [83].

Figure 8: AFM image of non-crystalline quartz. The area is 27 x 27 Å^2, the height of the bumps is 0.5 $\pm$ 0.2 Å. The data are raw data.

4.2. MOLECULAR IMAGING

During the last years a great deal of experimental data have accumulated on molecular imaging of organic materials with the STM. Little is known about the electronic transport through organic molecules, and the contrast mechanism is not well understood. The actual situation is that there is a contrast in STM experiments, but the interpretation is very problematic. Therefore, much expectation has been put into the AFM. Since it relies on force, there is no need to explain a certain sample conductivity. It has been shown to work in liquids, allowing to image biological material in its natural environment [13].

Many molecules are soft, leading to deformations or even destruction under the probe tip. Calculations suggest that forces as low as 10^{-11} N are necessary to image soft biological materials [84]. The tracking force of the AFM becomes an important parameter, and a proper interpretation of the images must consider the elastic properties of the sample.

The first reported molecular scale images were obtained on a monolayer of N-(2-aminoethyl)-10,12-tricosadiyanamide deposited on a glass substrate [85]. Although the forces used for imaging were as high as 10^{-8} N, the film was only damaged after some dozens of scans. A similar resolution of 5 Å was obtained on amino acid crystals of DL-leucine on a glass substrate. Fibrils of submonolayers of poly(octadecylacrylate) on graphite could be imaged even with a force of 10^{-7} N [86].

One of the most exciting series of images of biological materials was recently reported by Paul Hansma's group [87]. Operating their force microscope in water they were able to observe the thrombin-catalyzed polymerization of the protein fibrin over more than half an hour. The fact that fibrin and thrombin are responsible for blood clotting shows the potential of AFM for observing important biological processes in natural environments.

AFM on biomaterials is still at an initial stage, and it is difficult to state how important this method will become in the future. More development is needed to make even gentler force microscopes and better defined probe tips. With increasing knowledge on the elastic properties of macromolecules it may become possible to reveal their internal structure.

4.3. TECHNOLOGICAL APPLICATIONS

Although force microscopy is quite new, there are already a number of technological applications. It is an ideal method to image microfabricated patterns such as finished microcircuits. A resolution of the order of 100 Å is sufficient in most cases, and the possibility to look at the samples directly, without excessive sample preparation, is very attractive. A better understanding of friction on a microscopic scale or the elastic properties of thin film materials are important issues for future technology. The study of magnetic properties of recording media or man-written bits is important for the development of more powerful recording devices. Further, the force microscope may be successfully applied to electrochemistry. In contrast to a STM tip, the force sensor would represent an electrically neutral probe and thus not influence the electrochemical system. In the following some imaging applications are presented.

4.3.1. *Microfabricated Structures:* The imaging aspect of force microscopy has attracted special attention for the control of microfabricated structures. Martin *et al.* have demonstrated this by imaging V-shaped profiles on a silicon wafer with a resolution of 50 Å [21]. They were even able to distinguish between different materials on the surface, bare silicon and a photoresist.

An example from our laboratory is shown in figure 9. Three tracks of a magneto-optical storage disk are shown in the left image. The tracks are 1.6 μm apart. The right image shows a zoom on one individual track. The fine structure is mainly due to a gold film which was deposited for subsequent STM imaging. These images were acquired with force sensors

 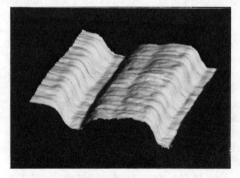

Figure 9: Magneto-optical disk imaged by AFM. The tracks are 1.6 μm apart. The zoom on one track shown in the right image reveals fine details on top of the ridge.

(Courtesy of H. Hug and Th. Jung)

made from a piece of metal foil. The grooves are flattened due to limited sharpness of the probe tip. Much smaller man-made structures, a tilted superlattice fabricated by molecular beam epitaxy, where recently imaged by Chalmers et al. [88]. Investigations of this type are necessary to optimize the fabrication process.

The surface roughness of microfabricated three-dimensional structures is an important factor. As typical length scales shrink to hundreds and tens of nanometers, a surface roughness in the order of nanometers can be unacceptable. In a combined electron microscope and force microscope study Robrock et al. investigated the planarization of cobalt silicide lines on silicon [89]. Various reacting conditions for the $CoSi_2$ formation were tested and the roughness of the composite silicide-silicon structure was determined. The high contrast vertical to the sample surface makes the force microscope well suited for this application.

4.3.2. *Thin Films:* Another investigation of surface roughness was conducted in our laboratory [90, 91]. Amorphous carbon layers [92] fulfill many of the requirements for modern tribological overcoats, e.g. for the protection of magnetic storage disks. Of particular interest is the decomposition of hydrocarbon gases in a glow discharge, leading to films of hydrogenated amorphous carbon (a-C:H). To increase the performance of computer disks it is necessary to learn more about the properties of such tribological coatings, such as wear-resistance, friction, adhesion to the substrate, surface roughness, as well as the correlation between these magnitudes.

Today's thin film technology is able to deposit coatings with a height corrugation of less than 200 Å peak-to-peak. A value which is very close to the resolution limit of stylus profilometry. This method can thus no longer be applied to determine the surface roughness of state-of-the-art coatings. Figure 10 shows two AFM images, out of four different samples of magnetic recording thin films, identically coated with 300 Å thick a-C:H layers. After imaging, the samples were submitted to a wear test in a conventional slider-on-disk device. Whereas the starting friction coefficients of all four samples were comparable, the final friction coefficient (measured after 2000 revolutions of the slider on the same track of the disk) separated the samples into two classes. A comparison with the root mean square values of the AFM data revealed that the samples with higher roughness (such as sample A) show an abrasive wear behaviour. This kind of measurements may help to optimize the deposition parameters of the a-C:H film.

 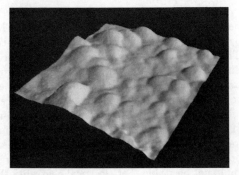

Figure 10: AFM images of two magnetic recording disks covered with a-C:H protection layers of different roughness. The area shown is approximately 500 x 500 nm^2.

4.3.3. Single Crystals:
STM has proven to be very useful for the study of crystal surfaces of a variety of conducting materials. Atomic resolution images of metals and semiconductors revealed detailed information on surface reconstructions on a local scale.

There is much hope that the AFM will allow the study of crystals of insulating samples in the same way. Even with a resolution in the nanometer range, AFM can give valuable evidence difficult to obtain with other techniques. On surfaces of high-T_c superconductors the AFM has already provided information on the topography complementary to STM [93, 94]. While STM is insensitive to an eventual non-conducting surface layer, AFM traces the surface.

Other samples which have been studied intensively in our laboratory are single crystals of silver bromide [95, 96]. Two representative images of the AgBr(100) and AgBr(111) surfaces are shown in figure 11. The AgBr(100) exhibits terraces which are several thousand Å wide. These regions are separated by steps with a height varying from 10 to 80 Å. The steps cross each other, leaving no continuous connection between the terraces. The minimum step height found was 9 ± 0.8 Å. The AgBr(111) surface looks characteristically different. Several steps run almost parallel across the surface and do not cross each other, in striking difference to the (100) surface. This leads to a very large lateral extension of the different terraces, without intersection by a step. On both surfaces the steps could be attributed to low index crystal directions.

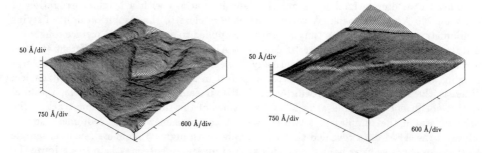

Figure 11: AFM images of the AgBr(100) (left) and the AgBr(111) (right) surfaces. On the (100) surface steps cross each other, while extended terraces occur on the (111) surface.

Surfaces of silver halides are of particular interest in photography [97]. The difference between the two crystal faces found by AFM may explain the observation that certain dye chromophores may cover the AgBr(111) completely, but leave holes on the AgBr(100). These dyes are known to form highly stable two-dimensional crystals which may attain dimensions up to 100 μm [98]. By the appropriate application of these site-directors the overgrowth of AgCl can be conducted to the edge of twinned tabular AgBr grains, leading to a better performance of the photographic emulsion [99].

The AgBr(111) surface is also interesting from a more fundamental point of view. The (111) plane of a NaCl-type ionic crystal is composed entirely either of positive or negative ions. A crystal face of a perfect (111) plane would have a high prohibitive electrostatic potential of effectively one half electronic charge for every ion on the surface. Such a surface could not be stable. For electrical neutrality the surface must consist of a layer with half as many ions as are in a bulk (111) plane. Hamilton and Brady [100] studied the (111) surface of epitaxial AgBr films by low-angle electron diffraction on vacuum deposited metal nuclei. They proposed three different models for this surface. However, an exact determination of the structure of AgBr(111) is still missing. With a better lateral resolution AFM will hopefully be able to reveal the atomic arrangement on this surface.

5. Outlook

Force microscopy is still a very young technique, and many workers have new ideas and find new applications. Force Microscopy has not only proven to be sensitive to many different types of forces, but also to reveal the high spatial resolution of a point probe. Besides these demonstrations it has already been successfully used in obtaining new information on technologically interesting surfaces. Since the method is relatively easy and many technological samples are insulating, force microscopy can have a large impact in materials control and process optimization.

There is much hope that this rapid initial progress will continue. Force microscopy makes magnetic, electrostatic, frictional and adhesive forces accessible on a scale not possible before. AFM can fill the gap as an ultra-high vacuum compatible high resolution microscopy suited for insulators. It seems that very soon routine atomic resolution imaging also of non-layered materials will be possible, opening the door to a whole class of samples. First results on soft organic systems further demonstrate that biological samples can be imaged by force microscopy, and even the visualization of biological processes is possible. To access very delicate samples it is necessary to develop even gentler and more sophisticated microscopes. Much will depend on the microfabrication of appropriate force sensors.

Our understanding of the contrast mechanism in AFM is poor. There is much speculation on the size of the true contact area or the effective tracking force. First theoretical descriptions treat ideal cases of clean and well defined tips, and more work is needed to describe more realistic situations. On the other hand the experimentalists are challenged to come close to the model cases, where the experiment can really be compared with theory. With these developments we may arrive at a better understanding of various interactions on an atomic level.

Acknowlegdments

We want to thank D. Anselmetti, H. Hug, Th. Jung and O. Wolter who kindly provided illustrations of their work for including them in this article. For the microfabricated force sensors used in many of our experiments we are grateful to T.R. Albrecht, R. Buser, N. de Rooij, and C.F. Quate. Ch. Gerber, P. Grütter, H.-R. Hidber, P. Junod, J. Ketterer, C. Schmidt and R. Steiger had many useful comments and provided well characterized samples. T. Richmond and A. Wadas carefully proofread the manuscript. This work was financially supported by the National Science Foundation and the Kommission zur Förderung der wissenschaftlichen Forschung.

References

[1] D. Tabor and F. R. S. Winterton. The Direct Measurement of Normal and Retarded Van der Waals Forces. *Proc. R. Soc. A* **312**, 435 (1969).

[2] J. N. Israelachvili and D. Tabor. The Measurement of Van der Waals Dispersion Forces in the Range of 1.5 to 130 nm. *Proc. R. Soc. Lond. A* **331**, 19 (1972).

[3] J. N. Israelachvili. *Intermolecular and Surface Forces*. Academic Press, (1985).

[4] G. Binnig and H. Rohrer. Scanning Tunneling Microscopy. *Helv. Phys. Acta* **55**, 726 (1982).

[5] G. Binnig and H. Rohrer. Scanning Tunneling Microscopy. *IBM J. Res. Dev.* **30**, 355 (1986).

[6] H. K. Wickramasinghe. Scanned-Probe Microscopes. *Scientific American*, October 1989.

[7] G. Binnig, C. F. Quate, and Ch. Gerber. Atomic Force Microscope. *Phys. Rev. Lett.* **56**, 930 (1986).

[8] E. C. Teague, F. E. Scire, S. M. Baker, and S. W. Jensen. Three-dimensional Stylus Profilometry. *Wear* **83**, 1 (1982).

[9] C. M. Mate, R. Erlandsson, G. M. McClelland, and S. Chiang. Direct Measurement of Forces during Scanning Tunneling Microscope Imaging of Graphite. *Surf. Sci.* **208**, 473 (1989).

[10] J. B. Pethica and W. C. Oliver. Tip Surface Interactions in STM and AFM. *Physica Scripta* **T19**, 61 (1987).

[11] T. R. Albrecht and C. F. Quate. Atomic Resolution Imaging of a Nonconductor by Atomic Force Microscopy. *J. Appl. Phys.* **62**, 2599 (1987).

[12] H. Heinzelmann, P. Grütter, E. Meyer, H.-R. Hidber, L. Rosenthaler, M. Ringger, and H.-J. Güntherodt. Design of an Atomic Force Microscope and First Results. *Surf. Sci.* **189/190**, 29 (1987).

[13] O. Marti, B. Drake, and P. K. Hansma. Atomic Force Microscopy of Liquid-Covered Surfaces: Atomic Resolution Images. *Appl. Phys. Lett.* **51**, 484 (1987).

[14] R. Yang, R. Miller, and P. J. Bryant. Atomic Force Profiling by Utilizing Contact Forces. *J. Appl. Phys.* **63**, 570 (1988).

[15] H. Yamada, T. Fujii, and K. Nakayama. Experimental Study of Forces between a Tunnel Tip and the Graphite Surface. *J. Vac. Sci. Technol. A* **6**, 293 (1988).

[16] C. M. Mate, G. M. McClelland, R. Erlandsson, and S. Chiang. Atomic-Scale Friction of a Tungsten Tip on a Graphite Surface. *Phys. Rev. Lett.* **59**, 1942 (1987).

[17] G. M. McClelland, R. Erlandsson, and S. Chiang. Atomic Force Microscopy: General Principles and a New Implementation. *Rev. Progr. Qual. Non-Destr. Eval.* **6** (1987). Plenum, New York.

[18] D. Rugar, H. J. Mamin, R. Erlandsson, J. E. Stern, and B. D. Terris. Force Microscope Using a Fiber-Optic Displacement Sensor. *Rev. Sci. Instr.* **59**, 2337 (1988).

[19] A. J. den Boef. Scanning Force Microscope Using a Simple Low-Noise Interferometer. *Appl. Phys. Lett.* **55**, 439 (1989).

[20] C. Schönenberger and S. F. Alvarado. A Differential Interferometer for Force Microscopy. Submitted to *Rev. Sci. Instr.*

[21] Y. Martin, C. C. Williams, and H. K. Wickramasinghe. Atomic Force Microscope - Force Mapping and Profiling on a Sub 100 Å Scale. *J. Appl. Phys.* **61**, 4723 (1987).

[22] D. Rugar, H. J. Mamin, and P. Guethner. Improved Fiber Optic Interferometer for Atomic Force Microscopy. To be published in *Appl. Phys. Lett.*

[23] D. Royer, E. Dieulesaint, and Y. Martin. Improved Version of a Polarized Beam Heterodyne Interferometer. *Ultrasonics Symposium* page 432 (1985).

[24] D. Sarid, D. A. Iams, V. Weissenberger, and L. S. Bell. Compact Scanning-Force Microscope Using a Laser Diode. *Opt. Lett.* **13**, 1057 (1988).

[25] D. Sarid, D. A. Iams, and J. T. Ingle. Performance of a Scanning Force Microscope. To be published in *J. Vac.Sci. Technol. A*.

[26] R. Kaneko, K. Nonaka, and K. Yasuda. Summary Abstract: Scanning Tunneling Microscopy and Atomic Force Microscopy for Microtribology. *J. Vac. Sci. Technol. A* **6**, 291 (1988).

[27] G. Meyer and N. M. Amer. Novel Optical Approach to Atomic Force Microscopy. *Appl. Phys. Lett.* **53**, 1044 (1988).

[28] S. Alexander, L. Hellemans, O. Marti, J. Schneir, V. Elings, P. K. Hansma, M. Longmire, and J. Gurley. An Atomic-Resolution AFM implemented using an Optical Lever. *J. Appl. Phys.* **65**, 164 (1989).

[29] T. Göddenhenrich, H. Lemke, U. Hartmann, and C. Heiden. Force Microscope with Capacitive Displacement Sensor. To be published in *J. Vac. Sci. Technol. A*.

[30] U. Dürig, J. K. Gimzewski, and D. W. Pohl. Experimental Observation of Forces Acting during STM. *Phys. Rev. Lett.* **57**, 2403 (1986).

[31] S. L. Tang, J. Bokor, and R. H. Storz. Direct Force Measurement in Scanning Tunneling Microscopy. *Appl. Phys. Lett.* **52**, 188 (1988).

[32] T. Göddenhenrich, U. Hartmann, M. Anders, and C. Heiden. Investigations of Bloch Wall Fine Structures by Magnetic Force Microscopy. *J. Microsc.* **152**, 527 (1988).

[33] M. Anders and C. Heiden. Imaging of Tip-Sample Compliance in STM. *J. Microsc.* **152**, 643 (1988).

[34] M. D. Kirk, T. R. Albrecht, and C. F. Quate. Low Temperature Atomic Force Microscopy. *Rev. Sci. Instr.* **59**, 833 (1988).

[35] P. J. Bryant, R. G. Miller, and R. Yang. Scanning Tunneling Microscopy and Atomic Force Microscopy Combined. *Appl. Phys. Lett.* **52**, 2233 (1988).

[36] K. E. Petersen. Silicon as a Mechanical Material. *Proc. IEEE* **70**, 420 (1982).

[37] G. Binnig, Ch. Gerber, E. Stoll, T. R. Albrecht, and C. F. Quate. Atomic Resolution with Atomic Force Microscope. *Europhys. Lett.* **3**, 1281 (1987).

[38] T. R. Albrecht and C. F. Quate. Atomic Resolution with the Atomic Force Microscope on Conductors and Nonconductors. *J. Vac. Sci. Technol. A* **6**, 271 (1988).

[39] E. Meyer, H. Heinzelmann, P. Grütter, Th. Jung, Th. Weisskopf, H.-R. Hidber, R. Lapka, H. Rudin, and H.-J. Güntherodt. Comparative Study of Lithium Fluoride and Graphite by Atomic Force Microscopy (AFM). *J. Microsc.* **151**, 269 (1988).

[40] T. R. Albrecht, S. Akamine, T. E. Carver, and C. F. Quate. Microfabricated Cantilever Stylus for Atomic Force Microscopy. Manuscript in preparation.

[41] O. Wolter. Micromechanics: Overview and Applications to SXM. Presented at the Seminar "SXM": Ultramicroscopy, Physics and Chemistry on the Nanometer Scale. IBM Europe Institute, Garmisch-Partenkirchen, August 14-18, 1989.

[42] R. Erlandsson, G. M. McClelland, C. M. Mate, and S. Chiang. Atomic Force Microscopy using Optical Interferometry. *J. Vac. Sci. Technol.* A **6**, 266 (1988).

[43] Yu. N. Moiseev, V. M. Mostepanenko, V. I. Panov, and I. Yu. Sokolov. Force Dependences for the Definition of the Atomic Force Microscopy Spatial Resolution. *Phys. Lett.* A **132**, 354 (1988).

[44] J. E. Lennard-Jones. Processes of Adsorption and Diffusion on Solid Particles. *Trans. Faraday Soc.* **28**, 334 (1932).

[45] I. P. Batra and S. Ciraci. Theoretical STM and AFM Study of Graphite Including Tip-Surface Interaction. *J. Vac. Sci. Technol.* A **6**, 313 (1988).

[46] F. F. Abraham and I. P. Batra. Theoretical Interpretation of Atomic-Force-Microscope Images of Graphite. *Surf. Sci.* **209**, L125 (1989).

[47] F. F. Abraham, I. P. Batra, and S. Ciraci. Effect of Tip Profile on Atomic-Force Microscope Images: A Model Study. *Phys. Rev. Lett.* **60**, 1314 (1988).

[48] D. Tománek, G. Overney, H. Miyazaki, and S. D. Mahanti. (in preparation).

[49] U. Landman, W. D. Luedtke, and A. Nitzan. Dynamics of Tip-Substrate Interactions in AFM. *Surf. Sci.* **210**, L177 (1989).

[50] R. J. Celotta and D. T. Pierce. Polarized Electron Probes of Magnetic Surfaces. *Science* **234**, 333 (1986).

[51] Y. Martin and H. K. Wickramasinghe. Magnetic Imaging by "Force Microscopy" with 1000 ÅResolution. *Appl. Phys. Lett.* **50**, 1455 (1987).

[52] J. J. Sáenz, N. García, P. Grütter, E. Meyer, H. Heinzelmann, R. Wiesendanger, L. Rosenthaler, H.-R. Hidber, and H.-J. Güntherodt. Observation of Magnetic Forces by the Atomic Force Microscope. *J. Appl. Phys.* **62**, 4293 (1987).

[53] Y. Martin, D. Rugar, and H. K. Wickramasinghe. High Resolution Magnetic Imaging of Domains in TbFe by Force Microscopy. *Appl. Phys. Lett.* **52**, 244 (1988).

[54] D. W. Abraham, C. C. Williams, and H. Wickramasinghe. Measurement of in-plane Magnetization by Force Microscopy. *Appl. Phys. Lett.* **53**, 1446 (1988).

[55] H. J. Mamin, D. Rugar, J. E. Stern, B. D. Terris, and S. E. Lambert. Force Microscopy of Magnetization Patterns in Longitudinal Recording Media. *Appl. Phys. Lett.* **53**, 1563 (1988).

[56] P. Grütter, E. Meyer, H. Heinzelmann, L. Rosenthaler, H.-R. Hidber, and H.-J. Güntherodt. Application of Atomic Force Microscopy to Magnetic Materials. *J. Vac. Sci. Technol.* A **6**, 279 (1988).

[57] H. J. Mamin, D. Rugar, J. E. Stern, R. E. Fontana, and P. Kasiraj. Magnetic Force Microscopy of Thin Permalloy Films. *Appl. Phys. Lett.* **55**, 318 (1989).

[58] C. Schönenberger, S. F. Alvarado, S. E. Lambert, and I. L. Sanders. Separation of Magnetic and Topographic Effects in Force Microscopy. Submitted to *J. Appl. Phys.*

[59] U. Hartmann and C. Heiden. Calculation of the Bloch Wall Contrast in Magnetic Force Microsocpy. *J. Microsc.* **152**, 281 (1988).

[60] A. Wadas and P. Grütter. Theoretical Approach to Magnetic Force Microscopy. *Phys. Rev. B* **39**, 12013 (1989).

[61] Y. Martin, D. W. Abraham, and H. K. Wickramasinghe. High-Resolution Capacitance Measurement and Potentiometry by Force Microscopy. *Appl. Phys. Lett.* **52**, 1103 (1988).

[62] J. E. Stern, B. D. Terris, H. J. Mamin, and D. Rugar. Deposition and Imaging of Localized Charge on Insulator Surfaces using AFM. *Appl. Phys. Lett.* **53**, 2717 (1988).

[63] B. D. Terris, J. E. Stern, D. Rugar, and H. J. Mamin. Localized Charge Force Microscopy. To be published in *J. Vac. Sci. Technol. A*.

[64] B. D. Terris, J. E. Stern, D. Rugar, and H. J. Mamin. Novel Study of Contact Electrification Using Force Microscopy. Submitted to *Phys. Rev. Lett.*

[65] R. Erlandsson, G. Hadzioannou, C. M. Mate, G. M. McClelland, and S. Chiang. Atomic Scale Friction Between the Muscovite Mica Cleavage Plane an a Tungsten Tip. *J. Chem. Phys.* **89**, 5190 (1988).

[66] G. M. McClelland. Friction between Weakly Interacting Surfaces. In *Adhesion and Friction*, eds. H. J. Kreuzer and M. Grunze, Springer Series in Surface Science, Springer Verlag, Berlin, (1989).

[67] D. H. Buckley. *Surface Effects in Adhesion, Friction, Wear and Lubrication.* Elsevier, Amsterdam, (1982).

[68] J. Ferrante and J. R. Smith. Theory of the Bimetallic Interface. *Phys. Rev. B* **31**, 3427 (1985).

[69] U. Dürig, O. Züger, and D. W. Pohl. Force Sensing in Scanning Tunneling Microscopy: Observation of Adhesion Forces on Clean Metal Surfaces. *J. Microsc.* **152**, 259 (1988).

[70] H. Heinzelmann, E. Meyer, P. Grütter, H.-R. Hidber, L. Rosenthaler, and H.-J. Güntherodt. Atomic Force Microscopy: General Aspects and Application to Insulators. *J. Vac. Sci. Technol. A* **6**, 275 (1988).

[71] L. A. Girifalco and R. A. Lad. Energy of Cohesion, Compressibility, and the Potential Energy Functions of the Graphite System. *J. Chem. Phys.* **25**, 693 (1956).

[72] N. A. Burnham and R. J. Colton. Measuring the Nanomechanical Properties and Surface Forces of Materials using an Atomic Force Microscope. *J. Vac. Sci. Technol. A* **7**, 2906 (1989).

[73] C. M. Mate, M. R. Lorenz, and V. J. Novotny. Atomic Force Microscopy of Polymeric Liquid Films. *J. Chem. Phys.* **90**, 7550 (1989).

[74] A. L. Weisenhorn, P. K. Hansma, T. R. Albrecht, and C. F. Quate. Forces in Atomic Force Microscopy in Air and Water. *Appl. Phys. Lett.* **54**, 2651 (1989).

[75] J. B. Pethica, R. Hutchings, and W. C. Oliver. Hardness Measurement at Penetration Depths as small as 20 nm. *Phil. Mag.* **48**, 593 (1983).

[76] M. Yanagisawa and Y. Motomura. An Ultramicro Indentation Hardness Tester and Its Application to Thin Films. *Lubrication Engineering*, page 52, October 1989.

[77] R. V. Coleman, B. Drake, P. K. Hansma, and G. Slough. Charge-Density Waves Observed with a Tunneling Microscope. *Phys. Rev. Lett.* **55**, 394 (1985).

[78] R. E. Thomson, U. Walter, E. Ganz, J. Clarke, and A. Zettl. Local Charge-Density-Wave Structure in 1T-TaS$_2$ Determined by Scanning Tunneling Microscopy. *Phys. Rev. B* **38**, 10734 (1988).

[79] N. V. Smith, S. D. Kevan, and F. J. DiSalvo. Band Structures of the Layer Compounds 1T-TaS$_2$ and 2H-TaSe$_2$ in the Presence of Commensurate Charge-Density Waves. *J. Phys. C: Solid State Phys.* **18**, 3175 (1985).

[80] E. Meyer, D. Anselmetti, R. Wiesendanger, H.-J.Güntherodt, F. Lévy, and H. Berger. Different Response of Atomic Force Microscopy and Scanning Tunneling Microscopy to Charge Density Waves. *Europhys. Lett.* **9**, 695 (1989).

[81] P. Cantini, G. Boato, and R. Colella. Surface Charge Density Waves Observed by Atomic Beam Diffraction. *Physica B* **99**, 59 (1980).

[82] O. Marti, B. Drake, S. Gould, and P. K. Hansma. Atomic Resolution Atomic Force Microscopy of Graphite and the "native Oxide" on Silicon. *J. Vac. Sci. Technol. A* **6**, 287 (1988).

[83] R. C. Barrett and C. F. Quate. Imaging Polished Sapphire with Atomic Force Microscopy. To be published in *J. Vac. Sci. Technol. A*.

[84] B. N. J. Persson. The Atomic Force Microscope: Can it be Used to Study Biological Molecules. *Chem. Phys. Lett.* **41**, 366 (1987).

[85] O. Marti, H. O. Ribi, B. Drake, T. R. Albrecht, C. F. Quate, and P. K. Hansma. Atomic Force Microscopy of an Organic Monolayer. *Science* **239**, 50 (1988).

[86] T. R. Albrecht, M. M. Dovek, C. A. Lang, P. Grütter, C. F. Quate, S. W. J. Kuan, C. W. Frank, and R. F. W. Pease. Imaging and Modification of Polymers by Scanning Tunneling Microscopy and Atomic Force Microscopy. *J. Appl. Phys.* **64**, 1178 (1988).

[87] B. Drake, C. B. Prater, A. L. Weisenhorn, S. A. C. Gould, T. R. Albrecht, C. F. Quate, H. G. Hansma D. S. Cannell, and P. K. Hansma. Imaging Crystals, Polymers, and Processes in Water with the Atomic Force Microscope. *Science* **243**, 1586 (1989).

[88] S. A. Chalmers, A. C. Gossard, A. L. Weisenhorn, S. A. C. Gould, B. Drake, and P. K. Hansma. The Determination of Tilted Superlattice Structure by Atomic Force Microscopy. To be published in *Appl. Phys. Lett.*

[89] K.-H. Robrock, K. N. Tu, D. W. Abraham, and J. B. Clabes. Study of Planarization of Cobalt Silicide Lines and Silicon by Scanning Force Microscopy and Scanning Electron Microscopy. *Appl. Phys. Lett.* **54**, 1543 (1989).

[90] H. Heinzelmann, E. Meyer, L. Scandella, P. Grütter, Th. Jung, H. Hug, H.-R. Hidber, and H.-J. Güntherodt. Topography and Correlation to Wear of Hydrogenated Amorphous Carbon Coatings: An Atomic Force Microscope Study. *Wear* **135**, 107 (1989).

[91] E. Meyer, H. Heinzelmann, P. Grütter, Th. Jung, H.-R. Hidber, H. Rudin, and H.-J. Güntherodt. Investigations of Hydrogenated Amorphous Carbon Coatings for Magnetic Data Storage Media by Atomic Force Microscopy (AFM). To be published in *Appl. Phys. Lett.*

[92] H. Tsai and D. B. Bogy. Critical Review: Characterization of Diamond-Like Carbon Films and their Application as Overcoats on Thin Film Media for Magnetic Recording. *J. Vac. Sci. Technol. A* **5**, 3287 (1987).

[93] H. Heinzelmann, D. Anselmetti, R. Wiesendanger, H.-R. Hidber, H.-J. Güntherodt, M. Düggelin, R. Guggenheim, H. Schmidt, and G. Güntherodt. STM and AFM Investigations of High-T_c Superconductors. *J. Microsc.* **152**, 399 (1988).

[94] H. Heinzelmann, D. Anselmetti, R. Wiesendanger, H.-J. Güntherodt, E. Kaldis, and A. Wisard. Topography and Local Modification of the $HoBa_2Cu_3O_{7-x}$ (001) Surface using Scanning Tunneling Microscopy. *Appl. Phys. Lett.* **53**, 2447 (1988).

[95] E. Meyer, H. Heinzelmann, P. Grütter, and H.-J. Güntherodt. A Study of the AgBr(111) and AgBr(111) Surfaces by Means of Atomic Force Microscopy. To be published in *J. Appl. Phys.*

[96] H. Heinzelmann, E. Meyer, H.-J. Güntherodt, and R. Steiger. Local Step Structure of the AgBr(100) and (111) Surfaces using Atomic Force Microscopy. *Surf. Sci.* **221**, 1 (1989).

[97] J. F. Hamilton. The Silver Halide Photographic Process. *Advan. Phys.* **37**, 359 (1988).

[98] C. Duschl, W. Frey, and W. Knoll. The Crystalline Structure of Two-Dimensional Cyanine Dye Single Crystals as Revealed by Electron Diffraction. *Thin Solid Films* **160**, 251 (1988).

[99] J. E. Maskasky. Epitaxial Selective Site Sensitization of Tabular Grain Emulsions. *J. Imaging Sci.* **32**, 160 (1988).

[100] J. F. Hamilton and L. E. Brady. A Model for the AgBr(111) Surface, based on the Symmetry of Nucleation Sites for Evaporated Metal. *Surf. Sci.* **23**, 389 (1970).

Electret-Condensor-Microphone used as a very sensitive Force Sensor

E.Schreck, J.Knittel and K.Dransfeld

Universität Konstanz, Fakultät für Physik, Postfach 5560
D-7750 Konstanz, FRG.

ABSTRACT:

The pressure sensitivity of commercially available electret microphones is typically about 1 mV/µbar. A pressure of 1 µbar corresponds to a force of 10^{-6} N acting on the diaphragma.
We show that, with suitable electronics, it is possible to detect - at the center of the diaphragm - even forces in the range of 10^{-9} N.
This sensitivity is in an interesting range, because the interaction forces occuring in standard tunneling microscopy between the tip and sample are of the same order of magnitude.

We present data of STM experiments under ambient air conditions on samples of graphite (HOPG) and of gold mounted at the center of the diaphragm.
During the tunnel process the height of the tip was modulated (z-direction) with an amplitude of about 1 nm at a frequency of 2 kHz. We were able to measure the vibrations of the diaphragm caused by the periodic interaction forces between tip and sample.

In summary, we present a new method of simultaneous tunneling and force-microscopy.

Introduction

In the early beginning of investigations with the scanning tunneling microscope (STM) [1,2], this method was described as a very soft method that avoids modification of the surface by the tip.
Lateron the advent of atomic force microscopy (AFM) [3,4] started a new area of microscopy. For the AFM it is no longer the tunneling effect between two close conductive materials which is most important, but also the strongly distance dependant Van der Waals forces between two, even nonconducting, materials.
Now it was possible to learn more about the nature and strength of the interaction forces between tip and surface [5], and subsequently different methods have been developed to achieve this goal [6,7,8].
These AFM experiments showed, that the forces are in the range of 10^{-7} N up to 10^{-9} N and that these forces are high enough to produce deformations and modifications of the surface or of the tip.

From this point of view it seemed most interesting to us to carry out *combined* STM/AFM measurements. Thus we tried - during an STM investigation - to measure simultaneously the interaction forces (for example: Van der Waals forces) between tip and sample surface. Since the AFM itself is a very sensitive and sophisticated instrument it is not so easy to combine both instruments.
Clearly there is a need to find a simple and small force sensor that could be placed and used just below a normal STM.

In this contribution we describe the test of a commercial electret microphone [9], which in view of its high pressure sensitivity and small mechanical dimensions is a very suitable sensor for a combined operation together with a STM.

1. The microphone

The microphone we used for the following experiments was a common electret microphone, as used in most tape-recorders. For the amplification of the microphone signal we used the set-up explained in fig.1.

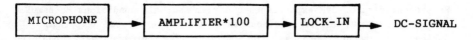

Fig. 1: This diagram shows the simple electronic circuitry that is sufficient for the measurement of intermolecular forces.

The housing of the electret microphone also contains a field effect transistor (FET) as an impedance converter. The output of the FET-follower is connected to the input of a two stage audio preamplifier having a gain of about 100. The signal - after passing a lock-in amplifier - produces a dc voltage which is proportional to the absolute amplitude of the diaphragma.

2. Force calibration of the microphone.

In typical applications the microphone has to be sensitive only to the pressure fluctuations, i.e. the resulting force is acting homogenously on the whole membrane. In our application, however, we like to measure only a point force due to the tip surface interaction forces acting only at the center of the membrane. Since these two modes of operation are quite different, it is necessary to calibrate the microphone response under the influence of a *point* force.
For this purpose we used a cantilever carrying a small tip at its free end in order to exert a periodic force on the membrane. The cantilever, prepared from an iron wire (diameter of 0.1 mm) had an calculated spring constant of 24 N/m and was fixed to a bimorph (see also fig.2).

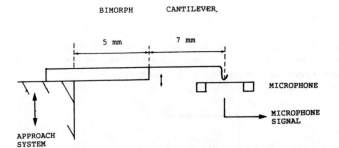

Fig. 2: Experimental set-up to calibrate the response of the microphone. A bimorph driven cantilever exerts a force on the microphone membrane.

Using a micrometer approach system the cantilever was carefully placed onto the membrane. The bimorph was then vibrated sinusoidally and the microphone signal was recorded by means of the set-up explained in fig.1. Fig.3 shows the result of this calibration for different sites near the center of the membrane.

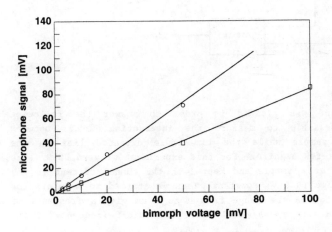

Fig. 3: Calibration experiment. The two curves were recorded on different sites around the center of the membrane.

Interpretation

As the deflection of the membrane (having a dynamic force constant of about 1000 N/m) is much smaller than that of the cantilever it is possible to neglect the displacement of the membrane. Applying a voltage amplitude of 100 mV at a frequency of about 2 kHz to the bimorph led to a mechanical modulation amplitude of about 6 nm. Using the known spring constant of the cantilever (24 N/m) we can calculate the force sensitivity of our microphone to be about $2*10^{-9}$ N/mV.

3. Tunneling Experiments

For the tunneling experiments we used a commercially available scanning tunneling microscope (Nanoscope I). The experimental set-up is shown in Fig.4. The sample was glued to the center of the membrane.

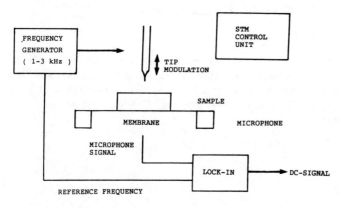

Fig. 4: Experimental set-up for the STM experiments.

Preliminary Results

The first experiment was planned in order to answer the following question: Is it possible to measure the interaction forces between tunneling tip and sample while the tip-sample distance lies in the tunneling range of a few Angstrom? For this experiment we used HOPG (size about 1*2*0.05 mm²) as a sample and kept both the tunnel current (1 nA) and the tunnel voltage (0.5 V) constant. Then we started to modulate the tip in the z-direction in the range from 0 to 4 nm with a frequency of about 2 kHz. (This frequency was well above the cutoff frequency of the control loop.)

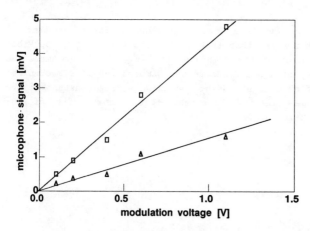

Fig. 5: Dependency of the microphone signal on the STM-tip modulation amplitude (1 V corresponds 4 nm modulation amplitude). The two curves correspond to different sites on the graphite sample

Fig.5 shows the result of two experiments, carried out on different sites on the sample. In both curves the linear relationship between the microphone signal and the modulation amplitude is clearly visible.

In the following experiments we investigated the dependence of the interaction force on the tip-sample distance. By keeping the tunnel bias voltage constant an increase of the tunnel current is caused only by a decrease of gap distance.

For very low tunnel currents we find a microphone signal of at least 5 mV in all curves of fig.6. Following the calibration experiment this corresponds to interaction forces of about 10^{-7} N. Now, while increasing the tunnel current up to 1 nA we observe also an increase of the microphone signal. With increasing tunnel current the microphone signal reaches - in two curves - a minimum located at about 2 nA. After a further increase of the tunnel current the microphone output signal increases again and shows a saturation behavior between 3 and 10 nA.

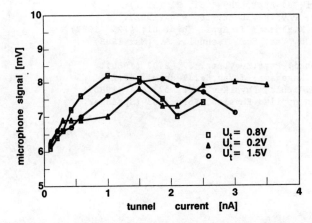

Fig. 6: Measurement of the microphone signal during standard tunneling on graphite while the tip was oscillating in the vertical z-direction with an amplitude of 4 nm at a frequency of 1.2 kHz.

In fig.7 we show the results of a similar experiment, using a microphone, whose membrane was evaporated with 100 nm of gold.

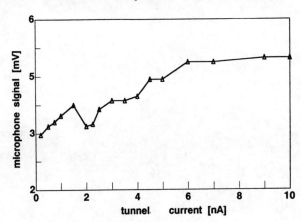

Fig. 7: Measurement of the microphone signal during standard tunneling on a gold sample (z-modulation of the tip: 2 nm with a frequency of 2.6 kHz).

Conclusion

The experimental results presented above demonstrate that an electret microphone is a very suitable sensor to measure forces in the range between 10^{-7} and 10^{-9} N. Furthermore, we demonstrate that it is possible to get a stable tunneling current for samples glued on to a microphone membrane, thereby measuring the interaction forces between tip and sample during the normal tunnel process.

Acknowledgments

We would like to thank H.Birk and S.Akari for their effective help. This work was supported by the SFB 306.

REFERENCES

1. G.Binnig, H.Rohrer & E.Weibel; Phys.Rev.Lett. 49, 57 (1982).
2. P.K.Hansma & J.Tersoff; J. Appl. Phys. 61, R1 (1987).
3. G.Binning & C.F.Quate; Phys.Rev.Lett. 56, 930 (1982).
4. Y.Martin, B.Drake & P.K.Hansma; J. Appl. Phys. 61, 4723 (1987).
5. I.P.Batra & S.Ciraci; J. Vac. Sci. Technol., A. (Mar 1988) v. 6(2) p. 313-318.
6. U.Düring & J.K.Gimzewski; Phys.Rev.Lett. 57, 2403 (1986).
7. H.Yamada, T.Fuiji & K.Nakayama; J.Vac.Sci.Technol. A6(2).
8. J.M.Soler, M.Baro & N.Garcia; Phys.Rev.Lett. 57, 444 (1986).
9. G.M.Sessler, Topics in Applied Physics, Volume 33 (Springer Verlag New York 1980).

RESOLUTION AND CONTRAST GENERATION IN SCANNING NEAR FIELD OPTICAL MICROSCOPY

U. CH. FISCHER
Nanotec GmbH
D 6330 Wetzlar
FRG

ABSTRACT. In scanning near field optical microscopy (SNOM), an antenna of subwavelength size is scanned accross the object in close proximity. The modulated light emission due to near field interactions with the object is recorded for image formation. With 50 - 100 nm apertures in a metal film or protrusions serving as antennas a resolution of 30 - 100 nm - limited by the antenna size - is demonstrated. Phase contrast depending on electric excitation and contrast enhancement depending on local plasmon excitation of the antenna are demonstrated in two types of reflection mode SNOM, one of which - scanning optical tunneling microscopy - is closely analogous to STM.

INTRODUCTION

Scanning near field optical microscopy (SNOM) is a new type of optical microscopy without lenses, which, in its concept, is closely related to scanning tunneling microscopy (1). In SNOM, an antenna of dimensions small as compared to the wavelength is scanned across the surface of the object in close proximity. The mutual near field interactions of the antenna and the object lead to a modulation of light emission of the antenna, which is used as a signal for image formation. The resolution of this kind of microscopy is not limited by the wavelength but only by the dimensions of the antenna and the distance to the object. SNOM opens a way to optical microscopy beyond the diffraction limit with similar contrast as in different modes of conventional optical microscopy. It therefore can be considered as a true extension of light microscopy of surfaces to the submicroscopic domain.

The concept of SNOM can be traced back to O'Keefe (2) as already mentioned by Betzig et al (3) and thus it has earlier origins than the concept of STM. The first concepts of SNOM (4 - 6) use small apertures in a metal screen as antennas. Ash, Nichols (7) and Husain (8) devised an aperture scanning microscope in the microwave domain. Their investigations have important implications for SNOM as well, as they demonstrate the important properties of this kind of microscopy:

1) The limits of resolution are determined by the size of the aperture and not by the wavelength.

2) Contrast generation is determined by the type of aperture exitation. Phase contrast for non absorbing pure phase objects can only be obtained with electrical exitation of the aperture.

3) Coupling of the aperture to a resonator leads to enhanced sensitivity.

A plausible probe for aperture scanning microscopy would be a transparent pointed body covered with an opaque metal film with a tiny hole at its apex. The aperture is irradiated, and light transmitted through the aperture and a semitransparent object is recorded, as the object is scanned laterally in close proximity to the surface of the object. With such a design optical images of perforations in an opaque metal film were obtained for the first time at a resolution in the range of 20 - 150 nm using visible light by Pohl et al (9).

The fabrication of submicron apertures is a critical step in the realisation of aperture scanning microscopy. An elegant and reliable method of their formation was described by Betzig et al (10). They used sharply pointed pipettes which were drawn from capillaries in a similar way as microelectrodes as employed for the patch clamp technique (11). The micropipettes are subsequently covered with an opaque metal film. In this way very smooth metallised tips with apertures of a diameter as small as 30 nm can be formed. With these probes scanning aperture microscopy in transmission was also demonstrated at a resolution of 70 nm. Betzig et al also realised an alternative transmission mode of SNOM, the collection mode (10), where by irradiation of the object, evanescent waves are generated at the surface of the object. When the pipette is approached to the surface, light which is picked up by the aperture is recorded. In addition, they demonstrated imaging of a fluorescent structure (12).

Here we are mainly concerned with SNOM in a reflection mode. With SNOM in reflection it is also possible to demonstrate the aspect of superresolution of SNOM, and in addition it is possible to demonstrate near field phenomena which are closely connected to the mechanisms of contrast formation. With SNOM in reflection the concept of near field microscopy is generalised to other types of antennas than apertures such as a small pin.

A CONCEPT OF NEAR FIELD MICROSCOPY

Our concept of scanning near field optical microscopy was partly inspired by a scheme of forming contact copies of planar structures with visible light at a resolution in the order of 10 nm as proposed by H. Kuhn (13) and demonstrated experimentally with a resolution of about 70 nm (14). Although not bgeing a scheme of a scanning microscopy it illustrates the concept of near field microscopy in a very pure way. Single fluorescent molecules serve as antennas for imaging in the following sense: A planar, weakly absorbing pattern, as e.g. a thin patterned metal film, is brought into contact with a monomolecular layer of a fluorescent dye. The fluorescent dye molecules can be regarded to a good approximation as small dipole antennas. Due to the near field interaction of the dye molecules and the slightly absorbing metal, energy is transfered from the electronically exited dye molecules to the metal over distances in the order of 10 nm which is related to the near field range of the dye molecules (16). Since energy transfer competes with light emission, the fluorescence of the dye is quenched at distances from the metal within a range of about 10 nm (16) and a complementary fluorescent image of the metal pattern is formed at a resolution of 10 nm. This scheme can be extended at least conceptually to a scheme of scanning microscopy. Imagine a single fluorescent molecule being scanned at a constant distance of less than 10 nm over the planar pattern, and that the fluorescence is being recorded. The modulation of the fluorescence will generate a scanned image of the pattern at a lateral resolution of about 10 nm. Our experiments aim at a realisation of such a concept with other means which are determined mainly by practical considerations such as antenna fabrication, and signal detection. The transfer of energy , which is stored in the near field of the antenna to a receiving body accross a vacuum gap is analogous to vacuum electron tunneling of bound electrons from a tip to an object in close proximity, and therefore the process of optical energy transfer may be called photon tunneling and near field microscopy may be called scanning optical tunneling microscopy (SOTM). In this special kind of SOTM the "optical tunnel current" is, however recorded indirectly by its competition to radiative energy transfer.

THE SCANNING NEAR FIELD OPTICAL MICROSCOPE

A schematic view of our instrument is shown in fig. 1. In scanning aperture microscopy the antennas are holes of 70 - 100 nm in diameter in a 20 - 30 nm thick aluminum film on a

Fig.1

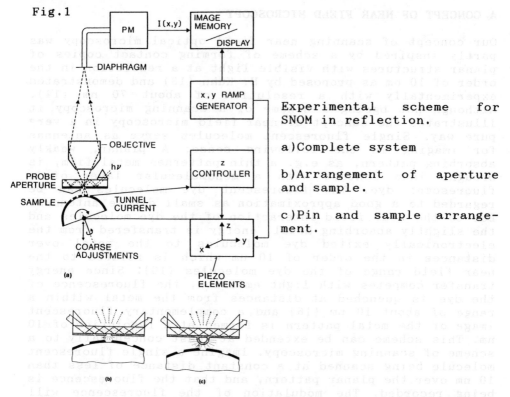

Experimental scheme for SNOM in reflection.

a) Complete system

b) Arrangement of aperture and sample.

c) Pin and sample arrangement.

glas slide (6,15,17) (fig.1b). They are formed by metal shadowing of latex spheres of 90 nm nominal diameter which are adsorbed on the glas slide and subsequent removal of the spheres. In scanning pin microscopy (18,19,33) the antennas are small protrusions of about 70 to 100 nm from a 30 - 40 nm thick film of gold on a glas slide (fig. 1c). They are formed by rotary shadowing of the latex sheres with a 1.5 nm thick film of Ta/W and subsequently with a 30 - 40 nm thick film of gold, such that the spheres are embedded in a thin film of gold. One of the antennas is selected by appropriate focusing and centering of a conventional microscope (fig 1a). A He- Ne laser beam is coupled into the slide such that the antennas appear in dark field illumination, as bright points on a dark background. The object is mounted on a piezoelectric scanning device. The test object is a convex glass surface covered with a pattern of patches only 5 nm thick. The curvature facilitates approach of the antenna while being sufficiently small to make the object appear flat within the scan range. The light scattered from the antenna into the microscope is guided onto a photomultiplier; the signal is stored, together with positional information in an image memory device. The scattered intensity varies when the object is brought into the proximity of the antenna or moved along it. Two methods of image retrieval are demonstrated. For an electrically conducting object, a tunnel distance regulation is used to scan the object at a constant distance from the antenna (9). The variation of the optical signal is used for image formation. With the pin microscope the optical signal itself can be used as a measure of distance between antenna and object. Using a feedback loop similar as in scanning tunneling microscopy (1), the antenna can be scanned at a constant height accross the surface and the feedback signal be recorded for image formation (18,19,33). This mode of SNOM is therefore also in an operational sense analogous to STM, and may be called scanning optical tunneling microscopy (SOTM).

THEORETICAL CONSIDERATIONS

Near fields arise, when the dimensions of a radiating object are small compared to the wavelength. A few general properties of near field patterns, which are strikingly different from the corresponding far field patterns, can be deduced from a simple example. Consider a sphere of radius a, which is irradiated by a plane, electromagnetic wave $\vec{E}_o \exp(i\vec{k}\vec{r})$, where E_o is the amplitude of the electric field, k the wavevector. The electric field scattered by the small sphere (ka << 1), corresponds to the electric field of an oscillating electrical dipole (22) of a dipole moment p:

$$\vec{p} = ga^3 \vec{E}_o; \quad (g = (\epsilon - 1)/(\epsilon + 2)) \tag{1}$$

where $\epsilon = \epsilon_s/\epsilon_o$ is the ratio of the complex dielectric constant of the sphere ϵ_s and the one of the embedding medium ϵ_o. For the near field $E_N(r)$ with (kr << 1) one obtains:

$$\vec{E}_N = (3\vec{n}(\vec{n}\cdot\vec{p}) - \vec{p})/r^3 \tag{3}$$

where n is the unit vector in the direction of r. For the corresponding far field $E_F(r)$ with kr >> 1 one obtains:

$$\vec{E}_F = k^2(\vec{p} - \vec{n}(\vec{n}\cdot\vec{p}))\exp(i\vec{k}\vec{r})/r \tag{4}$$

In order to compare the relative contributions of the near and far field in the proximity of the sphere the ratio of their maximal intensity for r = a is determined:

$$E_N\max(a)^2/E_F\max(a)^2 = 4/(a^4 k^4) \tag{5}$$

The ratio of intensities of the near field and the incoming plane wave is independent of the size of the sphere

$$E_N max \ (a)^2 / E_0^2 = 4g^2 \tag{6}$$

The range d of the near field is determined by the relation

$$E_N max \ (d + a)^2 = 1/2 \ E_N max(a)^2 \tag{7}$$

to be: $d = 0.12a$. From these relations we draw the following conclusions, which are valid qualitatively also for other geometries than a sphere, and which are relevant for scanning near field optical microscopy:

1) The ratio of intensities of the near and far field at the rim of the sphere ($r = a$) increases with the fourth power of the inverse radius $1/a$ (eq 5).

2) The ratio of the field intensities of the near field of a dielectric sphere of a typical refractive index of $n = 1.5$ and of the incoming wave assumes a value of 0.35 (eq. 6) irrespective of the size of the sphere. The distortion of the incoming field in the proximity of the sphere is thus comparable to the magnitude of the incoming field.

3) For special materials such as silver or gold the near field may have a much larger intensity than the incoming wave, at conditions where the value of g according to eq. 2 is nearly divergent. This is the case when ϵ assumes a value close to -2. Using the optical constants of silver at 354 nm (21) one obtains $g^2 = 100$. At these conditions plasma resonances are excited, which lead to an enhanced near field and to enhanced light scattering.

4) The range of the near field is very small. It is only a fraction of the dimension of the scattering obstacle (eq.7).

5) The angular dependence of the near field is very different from the one of the far field. Whereas the near field assumes a maximal value in the direction parallel to the dipole orientation, the far field is maximal at the perpendicular direction (eq. 3,4).

The notion of the near field is not restricted to electromagnetic fields but exists also in acoustics. A longitudinally vibrating small sphere can be regarded as an acoustic dipole with the characteristic nearfield decaying with the third power of distance and a radiation field. This analogy led to a concept of scanning near field acoustic microscopy (23). Near field phenomena can be demonstrated in acoustics much more easily than in optics, using 30 kHz quarz tuning fork resonators as a source of acoustic waves, because the wavelength is in the cm range and the phenomena can be demonstrated at a much coarser scale than in optics.

SUPERRESOLUTION IN SNOM

In order to understand the resolution capability of SNOM, consider the model of an aperture in an opaque screen, irradiated from one side by light of wavelength λ. The electromagnetic field directly behind the screen is localised to the area of the aperture. The field in the plane at a distance z behind the screen may be decomposed in terms of a scalar wave theory into its 2-D Fourier components $g(k_x, k_y, z)$ (24):

$$f(x,y,z) = \int_{k_x} \int_{k_y} g(k_x, k_y, z) \exp(i(k_x x + k_y y)) dk_x dk_y \quad (8)$$

$$g(k_x, k_y, z) = g(k_x, k_y, 0) \exp(i (k^2 - k_x^2 - k_y^2)^{1/2} z)$$

for: $k_x^2 + k_y^2 < k^2$

$$g(k_x, k_y, z) = g(k_x, k_y, 0) \exp(-(k_x^2 + k_y^2 - k^2)^{1/2} z)$$

for: $k_x^2 + k_y^2 > k^2$

where $k = 2\pi/\lambda$. k_x and k_y are the components of the wavevector in the plane parallel to the screen. Therefore, one has to distinguish between the components propagating as plane waves and those decaying exponentially with distance from the screen as evanescent waves according to whether k^2 is larger or smaller than $k_x^2 + k_y^2$. The dominant components for small apertures with a diameter $a \ll \lambda$ are centered around values of $(k_x^2 + k_y^2)^{1/2} = 2\pi/a$. Therefore, the localisation of the electromagnetic field to the dimension of the aperture is lost at distances larger than $a/2\pi$ from the aperture. If a sample is scanned within this very small distance, the aperture will sense its optical properties at a resolution, which is limited by the size of the aperture. Contrast generation of near field microscopy depends on the interaction of the electromagnetic field created by the aperture with the sample and its influence on the near field distribution in the aperture plane. This near field distribution also determines the radiation field, which is detected as a signal for SNOM.

MODES OF CONTRAST AND CONTRAST ENHANCEMENT IN SNOM

Contrast generation is an important aspect in any kind of microscopy, because it determines the modes of imaging which are possible. In SNOM, the design of the instrument strongly determines the kind of contrast which can be obtained. With SNOM in reflection it is possible to demonstrate different modes of contrast generation and to demonstrate a design with contrast enhancement.

Electric and Magnetic Excitation of Apertures. A distinction between electric and magnetic types of aperture excitation is important for an understanding of contrast generation in SNOM. Such a distinction, well known for dipole radiators is useful for near fields in general. One can distinguish between two types of evanescent waves (25):

1) TEE, transverse electric evanescent wave. Let the electric field vector point in the y- direction parallel to the screen:

$$E_y = E_o \exp(ik_x x) \exp(-(k_x^2 - k^2)^{1/2} z); \quad (z>0) \quad (9)$$

The wave decays exponentially in the z - direction normal to the screen and propagates in the x- direction with a wavenumber $k_x > k$

2) TME, transverse magnetic evanescent wave:

$$H_y = H_o \exp(ik_x x) \exp(-(k_x^2 - k^2)^{1/2} z); \quad (z>0) \quad (10)$$

For TME, the conditions

$$\text{rot } E = -kH \text{ and div } E = 0 \quad (11)$$

yield the relations

$$E_y = 0, \quad E_z = (k_x/k)H_y, \quad E_x = -i(1-k^2/k_x^2)^{1/2} E_z \quad (12)$$

For TEE, one obtains similar relations, e.g., $H_z = (k_x/k)E_y$. Thus the differences in the electric and magnetic field of a near field are the more pronounced, the larger k_x, i.e., the smaller the sources are. At optical frequencies, the interaction of electromagnetic fields with matter occurs mainly via the electric polarisability. It is therefore expected, that contrast generation in SNOM for purely dielectric phase objects is more pronounced for aperture excitations with large electric fields. According to Bethe (26), the radiation field of an aperture corresponds to the superposition of an electric dipole with a dipole moment $P = -1/3 \, a^3 \, E_o$ and a magnetic dipole with dipole moment $M = -2/3 \, a^3 \, H_o$, where a is the radius of the aperture, small compared to the wavelength. E_o and H_o are the unperturbed electric and magnetic fields on the screen, which would exist in the absence of the aperture. For an infinitely conducting film, these are the normal component E_z of the electric and the tangential components H_x and H_y of the magnetic field. Thus a TE or s- polarised plane wave can only excite a magnetic dipole, whereas a p -polarised wave will excite both magnetic and electric dipoles. The significance of the polarisation of aperture excitation can be demonstrated in SNOM with apertures in reflection as was shown previously (6,30).

Contrast Enhancement. As a crude model of our pin antenna we assume a dielectric sphere suspended freely above the object (fig. 2a).

Fig.2 a)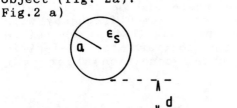

b)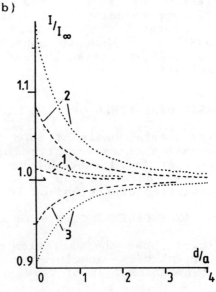

a) sphere above halfspace.

b) Calculated normalised intensity according to Eq. 13 for a silver sphere as a function of relative distance d/a to a halfspace of $\epsilon_o = 2.25$.
··· p, ---- s-polarised
1) 630nm, $\epsilon = -18 + i0.48$
2) 350nm, $\epsilon = -1.7 + i0.3$
3) 360 nm, $\epsilon = -2.3 + i0.27$
ϵ values were taken from ref 21.

The radius a of the sphere and the distance d to the object is assumed to be small compared to the wavelength. The effective dipole moment p of the sphere depends on the distance due to the polarisation of the object. In a quasistatic dipole approximation one obtains (27):

$$p = (\epsilon_s - 1)/ (4\pi + (\epsilon_s - 1) F_{p/s}) \qquad (13)$$
$$F_p = 4\pi/3 (1 - 1/4 (a/(a+d))^3 (\epsilon_o-1)/(\epsilon_o+1))$$
$$F_s = 4\pi/3 (1 - 1/8 (a/(a+d))^3 (\epsilon_o-1)/(\epsilon_o+1))$$

$\epsilon_s = \epsilon_s' + i \epsilon_s''$ is the complex dielectric constant of the sphere. F is a parameter which is different for p-polarisation (F_p) and s polarisation (F_s) respectively. Fig 2 shows the calculated dependence of the scattering intensity I ~ pp* of a silver sphere as a function of relative distance d/a to a halfspace of dielectric constant 2.25, normalised to the intensity at infinite distance. At 630 nm the change in relative intensity is only about 0.02. This change is smaller for s - polarisation than for p-plarisation. As shown in fig. 2b this change is, however, by a factor of about 10 larger at a wavelength where the real part of ϵ is close to -2, which is the condition for surface plasmon excitation of the sphere. For these conditions, the absolute value for the intensity is by a factor of about 100 larger. The intensity increases with the approach of the halfspace for a frequency above the plasmon resonance of the free sphere and decreases below it, corresponding to a decrease of plasmon frequency with the approach. These considerations show that it should be possible in SNOM to enhance the contrast for dielectric objects by selecting resonance conditions for pin excitation.

BASIC NEAR FIELD OPTICAL EXPERIMENTS

Experimental basis of SNOM is the variation of the antenna radiation characteristics upon approach of an object into the near field range of the antenna and also the variation of the radiation due to local differences of the optical properties of the object in close proximity.

DISTANCE DEPENDENCIES WITH APERTURE ANTENNAS.

Fig. 3 shows the variation of scattering intensity for the approach of an aperture towards a glas surface covered with a 5 nm thick film of Ta/W (15).

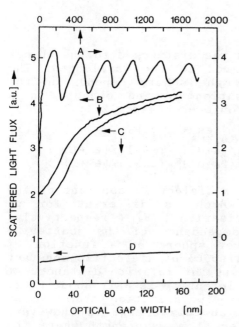

Fig.3

Scattering signal of a 0.09 µm aperture as a function of distance between aperture and sample. Note different scales for curves A and curves B,C, and D. C is a repetition of B, slightly displaced. D. was obtained with the microscope focused away from the aperture. (from (15)).

The approach was stopped in this case with the onset of the tunnel current. Interference undulations of $\lambda/2$ periodicity are observed with a last maximum at a separation of ca $\lambda/4$. We focus the attention to the range between the last maximum and tunnel contact. The signal decreases rather abruptly within the last 60 nm of the final approach. This abrupt decrease in scattering intensity is interpreted as a near field phenomenon similar to the quenching of fluorescence of dye molecules, which was mentioned above. The range of this steep decay scales with the diameter of the aperture (6,30) as is also expected for the range of the near field according to eq.7. The extent of quenching depends sensitively on the nature of the surface and on the polarisation of the exciting beam (6). For non - absorbing dielectric interfaces it is significant only for p - polarisation. The difference between s- and p- polarisation was attributed to the electric type of excitation for p - polarised and the magnetic type of excitation for the s - polarised incident light respectively (30). In the latter case no significant influence of a dielectric object on scattering is to be expected.

ENHANCED SPECTROSCOPY WITH APERTURES

The same experimental scheme used in SNOM was used to detect small changes in optical properties in very small volumes (17). The modified experimental scheme is shown in fig. 4a. The glas slide with the metal film containing small apertures is sealed to a compartment which can be filled with a liquid. Changes in scattering intensity of an aperture indicate changes in liquid composition. For apertures in a nearly opaque aluminum film, a decrease in scattering is always observed, when the compartment is filled with a liquid. This decrease is stronger for p- than for s- polarised incident light (17). Enhanced sensitivity of the apertures was acheived by choosing silver or gold as a surrounding metal. For apertures in a silver film, irradiation conditions can be found using p-polarised irradiation such that the scattering intensity is strongly enhanced for liquids of refractive index around 1.37 as shown in fig. 4b. The scattering maximum nearly coincides with a slight minimum in the reflectivity of the silver film, which indicates the occurrence of thin film plasmons in the silver film (31). The enhanced scattering of the apertures and its sensitivity to small changes in refractive index of the liquid is obviously due to surface plasmon exitation of the metal film and to localised surface plasmons at the rim of the apertures, similar as in the case of a metal sphere mentioned above. The scattering maximum strongly depends on the wavelength and the angle of incidence of the irradiating beam (17). Close to resonant conditions , a change of 10^{-4} in refractive index and a concentration of 10^{-4} molar of a dye can easily be detected

Fig. 4 a)

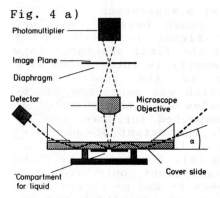

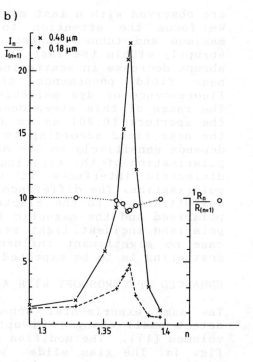

a) Experimental scheme

b) The ratio I(n)/I(1) of scattering intensities I(n) as a function of refractive index n for apertures of a diameter of 0.48 µm (x) and of 0.18 µm (+) in a silver film. The scattering intensity I(1) was about an order of magnitude larger for the 0.48 µm as compared to 0.18 µm apertures.) In comparison, the relative intensity R(n)/R(1) (o) is plotted. Methanol, water, ethanol, propanol and butanol and mixtures of these liquids were chosen to vary the refractive index of the liquid. The wavelength of the irradiating light was 568 nm. (from 16)

by a change of the scattering signal of a 0.3 µm aperture (17). Enhanced fluorescence of a 10^{-4} molar solution of rhodamine b can also be observed from a single 0.3µm aperture (17). These results demonstrate a remarkable sensitivity of the scattering and fluorescence signals of an aperture to small changes in the aperture environment. Similar, but less pronounced effects are obtained by using gold as a metal.

Let us assume, that the volume probed by a 0.3 µm aperture is determined by its diameter and a range of 100 nm of the near field of the aperture. into the liquid corresponding to a volume of $10^{-18} l$. For a 10^{-4} molar dye solution, this would correspond to 500 molecules being probed. It should be interesting to find a way to combine this kind of an enhanced sensitivity of an antenna with SNOM.

DISTANCE DEPENDENCIES WITH PIN ANTENNAS.

Fig. 5

Scattering signal of pin antennas as a function of distance to a glas surface.

a) Gold embedded 0.09 μm latex sphere as short optical antenna. Angle of incidence of exciting beam 26 deg, p-polarised,

b) as a but s-polarised incident beam

c) as a but 14 deg,

d) as c but s - polarised (from 31).

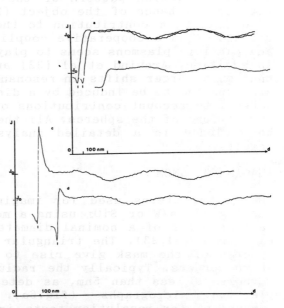

Contrast enhancement in SNOM can be demonstrated with pin antennas choosing gold as a metal. Fig. 5 shows the variation of radiation intensity for the approach of a pin antenna toward a dielectric interface. In this case a much steeper slope of the final intensity decay over a range of only about 20 nm is obtained. Again the decrease is much larger for p - polarised incident radiation. In addition a very pronounced change in the approach characteristic is obtained for a more shallow angle of incidence of the exciting beam, especially pronounced for the p - polarised incident beam. A narrow scattering maximum is observed before the final quenching of emission. In other cases even more pronounced resonances in the scattering intensity were observed as a function of distance (18). With gold as a material, particle plasma resonances of the antenna may be expected to arise in the visible spectral range. The appearance of a scattering maximum during approach of a dielectric surface is interpreted as a tuning through resonance of the pin antenna (18). From the simple model of an isolated sphere above a dielectric object no such large changes in scattering intensity as found experimentally are to be expected, as is shown above for the model of a silver sphere. A more refined model may lead to an explanation of the larger effect observed in the experiments. In our experiments we don't measure the total radiated power but only the one accepted into an angle which is determined by the numerical aperture of the objective lens. Since the angular radiation pattern of the pin may change strongly with the distance of the object (28,29), this change will also lead to a contribution to the change in the measured signals. Also a cooperative coupling of thin film plasmons and particle plasmons seems to play an important role (18). In addition, Aravind et al (32) and Ruppin (41) point out that much larger shifts in resonance frequency of a sphere are expected to be induced by a dielectric substrate if one takes into account contributions of higher order multipole excitations of the spheres. All these effects would have to be included in a detailed analysis of the experimental results.

IMAGING BY SNOM

The test patterns used for imaging are formed by shadow casting of Ta/W or SiO_2 using a monolayer of close packed latex spheres of a nominal diameter of 0.31, 0.49 or 1 µm as a mask (34,33). The triangular interstices between the spheres of the mask give rise to triangular patches with sharp corners. Typically the radius of curvature of their corners is less than 5nm, as determined from transmission electron micrographs (fig.6a). These patterns are convenient for resolution tests in the 5 - 1000 nm range. Fig 6b-d show high resolution optical micrographs obtained with a conventional light microscope using an oil immersion objective of a numerical aperture of 1.4. The triangular interstices are clearly resolved for the 1 µm pattern with an indication of their triangular form. In the 0.49µm pattern the triangular interestices are resolved, but their triangular form is not recognised. In the 0.31 µm pattern only the 0.28µm periodicity of the grating is resolved. Fig 8e shows an image of the 1 µm pattern obtained with scanning aperture microscopy, using tunnel distance regulation. The triangular metal patches are clearly resolved at a resolution of about 50 nm corresponding approximately to the diameter of the apertures used. Fig. 6e shows an image of a 0.31 µm Ta/W pattern obtained with a pin antenna and tunnel distance regulation. In some parts of the image the form of the triangular interstices is resolved indicating a resolution of about 30 nm. Fig 6f shows a line scan image obtained with scanning pin microscopy in the optical tunneling mode. Conditions of antenna excitation were chosen similar as in fig 4b (p - polarised). The object were only 5 nm thick patches of SiO_2 using 0.49 µm close packed spheres as a mask.

Fig.6 a)

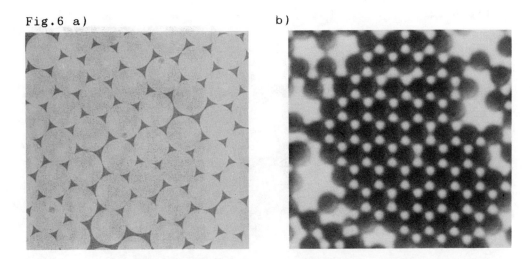

Images of test structures a) Transmission electron micrograph of a 0.31 μ metal test pattern. The test pattern are 5 nm thick Ta/W patches obtained by metal shadowing of a mask made of patches of close packed latex spheres of a nominal diameter of 0.31 μm. b) Optical micrograph of a 1 μm latex metal test pattern of a thickness of 20 nm Ta/W obtained with an oil immersion objective of numerical aperture of 1.4.

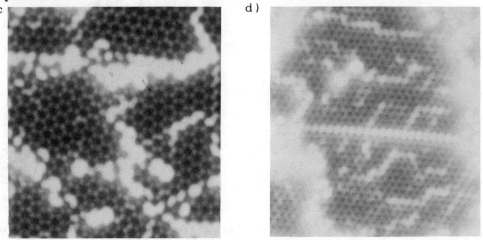

c) The same as b with a 0.49 μm test pattern; d) the same as b and c with a 0.31 μm latex test pattern.

Fig.6 e) f)

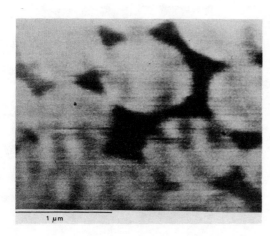
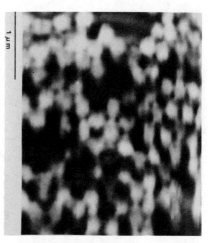

e) SNOM image of a 1 μm latex metal test pattern of 5 nm thickness using a 90 nm aperture and electronic tunnel current distance control (from 15). f) SNOM image of a 5 nm thick 0.31 μm latex metal pattern using a 90 nm pin and electronic tunnel current distance control (from 18).

g) h)

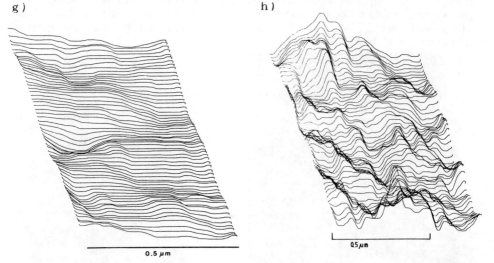

g) SNOM image of a 0.49 μm latex test pattern consisting of 5 nm thick SiO_2 patches using a 90 nm pin as an antenna in the SOTM mode (from 33). h) The same as g but of a 0.31 μm SiO_2 test pattern.

packed sheres as a mask. In this case one can resolve the patches but their form is not clearly revealed. Fig 4g shows an image obtained with the same method of a 0.31 µm test pattern. Again there is an indication of the pattern, but its form cannot be clearly recognised, although there are details in the image revealing a resolution of about 30 nm. We assume that the image reveals less clearly the evaporated pattern in the SOTM case, because the roughness of the substrate and remainders of latex spheres not completely removed from the substrate are also contributing to the recorded profile and prevent us from recognising the superimposed pattern. This is not to be expected in the tunnel distance control mode to such a large extent, because there the feedback signal reflecting the topography is not used for image formation but only the optical signal reflecting mainly material contrast.

SNOM WITH TIP GEOMETRIES.

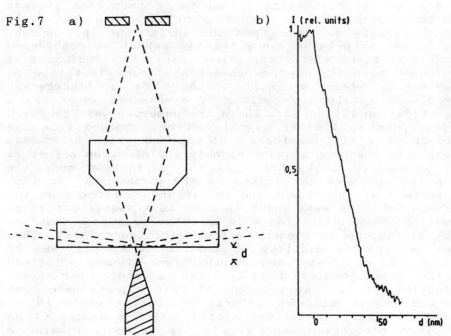

a) setup for the measurement of the approach of a metal tip towards a transparent surface in an SNOM. b) Light intensity scattered by the tip as a function of distance d between tip and surface. Irradiation was at an angle of 10^0 to the surface using 633nm light. The approach of the tip was stopped with the onset of an electronic tunneling current

A severe drawback of the SNOM mode discussed here is the flat sensor head. With this geometry microscopy is limited to very flat surfaces with a local roughness which doesn't exceed 20 nm. Therefore the kind of objects which can be investigated is very limited. It is therefore desirable to have a sensor head with a tip geometry similar as in STM. As shown in fig 7b it is indeed possible to extract an optical signal of the approach of a regular electrochemically etched metal tip, as they are standardly used in STM, towards a semitransparent object indicating the approach of the tip about 30 nm before the onset of an electronic tunnel current with a set up shown schematically in fig.7a. The transparent object is irradiated such, that evanescent waves are generated at its surface. As during the approach the tip penetrates into the evanescent field, the tip is excited and scatters light, a part of which can be collected by means of a light microscope. The result of such an approach experiment is shown in fig.7b. The increase of the scattering signal is in this case steeper than the increase in the intensity of the evanescent field with decreasing distance from the surface of the object. This is not surprising since the tip cannot be considered as an ideal point scatterer, which would be an indicator of the local intensity of the evanescent field. This type of experiments, where a small obstacle or a fluorescent molecule is excited to radiate by evanescent waves has a long tradition (35 36 37). There are numerous ways by which in principle an optical signal can be obtained from the approach of a tip towards an object and there have been attempts to make use of such signals for SNOM. Betzig et al (10) introduced what they call the collection mode, in which the aperture is excited by the evanescent field of the surface of the object and the light scattered from the aperture into the wave guide is used as a signal for image formation. This collection mode is rather simple because no light microscope is needed in the detection pathway. This mode was recently modified by Reddick et al (39) and by Courjon (38) et al who used a transparent glass tip instead of the metal coated tip with an aperture for light collection. In all cases images of test objects have been produced demonstrating a lateral resolution close to or beyond the limit of optical microscopy. The design of an SNOM in a reflection mode with a tip geometry remains a challenging task.

APPLICATIONS OF SNOM

Until now SNOM in reflection has been developed only so far, that the principle could be demonstrated on test objects. There have been no applications of the method to other fields of science. One reason is the restrictive conditions posed on sample preparation with the present instrument. The following suggestions give an idea of the kind of problems, SNOM may be applied to in the future:

1) Submicroscopic mapping of spectroscopic properties such as luminescence of surfaces or thin films.

2) SNOM may turn out to be a unique tool to deposit information on - and read it out from - thin electrooptically active layers at a submicron scale. By applying a small voltage between the antenna and an electrically conducting surface as a support of the electrooptic layer, a strong electric field can be applied locally across the layer, leading to some electric field induced change in the layer. This change can be read out with the same antenna from the same spot.

3) An SNOM may not only be used as a scanning microscope. There are also interesting possibilities of using SNOM in a stationary arrangement in order to make use of near field optical phenomena to monitor time dependent fluctuations in very small areas using the approach of surface enhanced spectroscopy of single apertures described above, rather than surface enhanced spectroscopy from large ensembles of scattering centers (40). Apart from apertures in a thin film of gold or silver, the configuration of the pin antenna described above may be a useful probe in this context due to the large field enhancements which can be expected in the gap between the pin and an appropriate reference surface (32,41). Aravind et al find that the calculated field intensity in the gap between a gold sphere and a gold substrate can be enhanced by a factor of 10^4 as compared to the intensity of the light irradiating the sphere. This is an enhancement which is by two orders of magnitude larger than the enhancement in the vicinity of an isolated sphere of gold. Considering our estimate, that 500 dye molecules can be detected by surface enhanced spectroscopy with apertures as mentioned above, it may well be possible to detect single dye molecules which are trapped in a gap between a pin antenna and a gold surface.

ACKNOWLEDGEMENTS

Previously unpublished results shown in fig.6h and 7 were obtained in collaboration with D.W. Pohl from the IBM Zürich Research Laboratory. Test objects (Ta/W) were made in collaboration with W. Jahn at the department. of the kinetics of phase formation, Max Planck Institute of biophysical Chemistry, Göttingen, and with C. Gunkel ($Si O_2$), at the Wild Leitz company, Wetzlar.

REFERENCES

1) Binnig G, Rohrer H. (1984) Scanning Tunneling Microscopy. Physica 127 B, 37 - 45.
2) O'Keefe J. A. (1956) Resolving power of visible light. J. opt. Soc. Am. 46, 359.
3) Betzig E., M. Isaacson, H. Barshatzky, A. Lewis, K. Lin (1988). Near field scanning optical microscopy (NSOM). Proc. SPIE (Los Angeles)
4) Pohl D.W., W. Denk, M. Lanz (1984). Optical Stethoscopy: Image recording with resolution \/20. Appl. Phys. Lett. 44, 651 - 653.
5) Lewis,A., M. Isaacson, A. Harootunian, A. Muray (1984). development of a 500 Å resolution light microscope. Ultramicroscopy 13, 227 - 232.
6) Fischer U.Ch. (1985). Optical characteristics of a 0.1 µm circular apertures in a metal film as light sources for scanning ultramicroscopy. J. Vac. Sci. Technol.B3, 386 - 390.
7) Ash E.K, N. Nichols (1972). Superresolution aperture scanning microscope. Nature 237, 510 - 513
8) Husain A. (1976). Super resolution aperture scanning microwave microscope. Thesis, Univ. College, Dept. Electronic and Electric Engineering, London UK.
9) Dürig U., D.W. Pohl, F. Rohner (1986). Near - Field optical scanning microscopy. J. Appl. Phys. 59, 3318 - 3327.
10) Betzig E., M. Isaacson, A. Lewis (1987). Collection mode near field scanning optical microscopy. Appl. Phys. Lett. 51, 2088 - 2090.
11) Neher E., B. Sakmann (1976) Single channel currents recorded from membrane of denervated frog muscle fibres. Nature 260. 7909 - 802.
12) Harootunian A., E. Betzig, M. Isaacson, A. Lewis (1986). Super resolution fluorescence near - field scanning optical microscopy. Appl. Phys. Lett. 49, 674 - 676.
13) Kuhn H. (1968). On possible ways of assembling simple organised systems of molecules. in: Structural Chemistry and Molecular Biology, A. Rich and N. Davidson (eds.), Freeman, San francisco, 566 - 571.
14) Fischer U.Ch, H.P. Zingsheim (1982). Submicroscopic contact imaging with visible light by energy transfer. Appl. Phys. Lett. 40, 195.
15) Fischer U.Ch, U. Dürig , D.W. Pohl (1988). Near field optical scanning microscopy in reflection. Appl Phys. Lett. 52 , 249 - 251.
16) Kuhn H, D. Moebius, H. Buecher (1972) Spectroscopy of monolayer assemblies. In: Physical Methods of Chemistry.Vol. 1, Part 3B, A. Weissberger B. Rossiter (eds.), Wiley, NY, 577.

17) Fischer U.Ch. (1986). Submicrometer aperture in a thin metal film as a probe of its microenvironment through enhanced light scattering and fluorescence. J. Opt. Soc. Am B 3, 1239 - 1244.
18) Fischer, U.Ch., D.W. Pohl (1989). Observation of single particle plasmons by near field microscopy. Phys. Rev. Lett. 62, 458 - 461.
19) Pohl D.W., Fischer UCh, Dürig UT. (1988). Scanning near - field microscopy (SNOM): basic principles and recent developments. SPIE Vol 897, 84 - 90.
20) Jackson JD (1975). Classical Electrodynamics. Wiley NY. Chapter 9.2.
21) Johnson, P.B., R.W. Christy (1972). Optical constants of the noble metals. Phys. Rev. B 6, 4370 - 4379.
22) Ruppin,R. (1976). Infrared active modes of dielectric crystallites on a substrate. Surface Science 58, 550-556.
23) Güthner P., U.Ch. Fischer, K. Dransfeld (1989). Scanning near-field acoustic microscopy. Appl. Phys. B48, 89-92.
24) Massey G.A, (1984). Microscopy and pattern generation with scanned evanescent waves. Appl. Optics 23, 658-660.
25) Lukosz W., R.E. Kunz (1977). Light emission by magnetic and electric dipoles close to a plane interface. I. Total radiated power. J. Opt. Soc. Am 67, 1607 - 1614.
26) Bethe H.A. (1944). Theory of diffraction by small holes. Phys. Rev. 66, 163-182.
27) Ruppin R. (1983). Surface modes and optical absorption of a small sphere above a substrate. Surface Science 127, 108 - 118.
28) Lukosz,W., R.E. Kunz (1978). Light emission by magnetic and electric dipoles close to a plane dielectric interface II. Radiation patterns of perpendicular oriented dipoles. J. opt. Soc. Am. 67, 1615 - 19.
29) Lukosz,W. (1979). Light emission by magnetic and electric dipoles close to a plane dielectric interface III. radiation patterns odf dipoles with arbitrary orientation. J. Opt. Soc. Am. 69, 1495 - 1503.
30) Fischer U.Ch, Dürig U, Pohl DW. (1987) Near - field optical Scanning microscopy and enhanced spectroscopy with submicron apertures. Scanning Microscopy Supplement 1, 47 - 52.
31) Raether H. (1988). Surface Plasmons an smooth and rough surfaces and on gratings. Springer tracts in modern physics Vol 111, G. Höhler ed.

32) Aravind P.K., R.W. Rendell, H. Methiu (1982). A new geometry for field enhancement in surface enhanced spectroscopy. Chem Phys. Lett. 85, 396 - 403.
33) Fischer, U. Ch. , U. T. Dürig, D. W. pohl. (1989). Scanning near field optical microscopy (SNOM) in Reflection or scanning optical tunneling microscopy (SOTM). Scanning. Microscopy.3, 1-7.
34) Fischer U.Ch., H.P. Zingsheim (1981) Submicroscopic pattern replication with visible light. J. Vac. Sci. Technol., 19, 881 - 885.
35) Selenyi,P. (1913). Sur l'éxistence et l'observation des ondes lumineuses sphériques inhomogènes. Comptes Rend. 157, 1408-1410.
36) Fröhlich,P. (1921). Die Gültigkeitsgrenze des geometrischen Gesetzes der Lichtbrechung. Ann. Phys. (Leip) 65, 577-592.
37) Carniglia C.K., L. Mandel, K.H. Drexhage. (1972) Absorption and emission of evanescent photons. J. Opt. Soc. Am. 62, 479 - 486.
38) Courjon D., K Sarayeddine, M. Spajer (1989) Scanning tunneling optical microscopy.To be publ. in Opt. Commun.
39) Reddick R.C., R.J. Warmack , T.L. Ferrell (1989). New form of scanning optical microscopy. Phys Rev. B 39, 767-770.
40) Metiu,H. (1984). Surface enhanced spectroscopy. Progr. Surf. Sci. 17, 153.
41) Ruppin, R. (1983). Surface modes and optical absorption of a small sphere above a substrate. Surface Science 127, 108 - 118.

SCANNING TUNNELING OPTICAL MICROSCOPY

Daniel Courjon
Laboratoire d'Optique P.M. Duffieux
Université de Franche-Comté-Besançon
25030 Besançon cédex France

The study of the photon tunneling effect is a part of the programme connected to the near field optics developed in the following group :

- Daniel Courjon
- Michel Spajer
- Khaled Sarayeddine
- Daniel Van Labeke

Laboratoire d'Optique-Besançon

- Jean Marie Vigoureux
- Christian Girard

Laboratoire de Physique Moléculaire-Besançon

Abstract

For overpassing the classical limit of resolution in optical microscopy, it is necessary to detect the light diffracted from small objects in the near field and not in the far field as in classical microscopy. A particular case is the detection of the evanescent field lying on the surface of a guiding structure. These surface waves interact with the object details and then can be used for determining the topography of the object. The main problem is the detection because the light beam is confined on the object surface. A solution consists of frustrating the evanescent field by means of a dielectric probe. The conversion of the inhomogeneous waves into homogeneous ones is fundamentally similar to the electronic tunneling effect. Subwavelength resolution can be obtained by placing a suitable optical stylus connected to an optical fibre near the surface. A xyz piezo-electric micropositioning system allows then to scan the object surface under test.

I - HISTORY OF THE OPTICAL TUNNELING EFFECT

Newton's experiments

The optical tunneling effect was discovered by Isaac Newton at the end of the 17th century. Newton noticed that a light beam apparently totally reflected <u>inside</u> a prism must necessarily propagate outside in order to explain the following experiment. Let us consider a first prism illuminated in total reflection, when a second prism whose basis is slightly convex is pressed against the first one, the light beam can pass from one prism to the other one even if the close contact between the two surfaces is not ensured. In order to explain this surprising effect, Newton assumed the light beam in total reflection was coming out of the prism along a curved trajectory before coming back into the prism by "attraction".

This effect in fact can be explained in term of field continuity conditions along the surface : in each point the incident field is the sum of the reflected and the transmitted fields. Since the latter cannot be classically detected we have to assume that it propagates along the surface and vanishes perpendicularly. If a second prism or any refracting or diffracting center is placed inside this peculiar field, the latter will be converted into a propagating one. This property is called optical frustration or optical tunneling effect and the field on the prism surface, the evanescent field.

The Evanescent Microscopy

We have to wait for the middle of the 20th century to see the first applications of the optical tunneling effect in microscopy. An object whose flatness is unknown is placed in the evanescent field generated by a prism working in total reflection. The parts of the object very close to the prism will frustrate the beam and a small amount of light will be then reflected towards the observer while the object points far from the prism will not affect the reflected beam. A kind of object topography will be imaged by this method.

Recently, because of the stimulation caused by the development of the scanning tunneling microscopy, the evanescent optical microscopy is improved and renamed "photon tunneling microscopy" in comparison with the "electron tunneling microscopy". In the experiment of Guerra a peculiar objective is substituted for Newton's prism. This objective works in total reflection and images by frustration the object surface. A digital image processing provides a 3D visualization of the object topography.

* J. Guerra SPIE Vol. 1009. P 254 september 1988.

Scanning Tunneling Optical Microscopy (STOM)
Photon Scanning Tunneling Microscopy (PSTM)

At the same time the transposition of the electron STM is proposed simultaneously in several laboratories. This technique involves a scanning system, a tip and a detection by tunnel effect.

* R. Reddick et al Phys. Rev. B Rapid Com. p. 767, 1989
* D. Courjon et al Optics Com. Spring 1989, SPIE (Paris) april 1989

II - SPATIAL RESOLUTION IN FAR FIELD OPTICS

Let us consider a usual imaging system and a plane wave impinging onto a diffracting object. The emerging light is then collected by a lens and focused in the image plane. Let us analyze the diffraction mechanism. The large object details (in comparison with the wavelength) will diffract the light beam slightly whereas the small details will deviate the rays strongly. It is clear that there is a physical limit of diffraction corresponding to an angle of diffraction of 90 degrees. From a certain value of the spatial frequency of the object, the beam is no more classically diffracted as a propagating one but converted into an evanescent field confined in the vicinity of the object surface.

Such a field vanishes in the direction of observation and thus cannot participate to the image formation.

Finally, a light beam intercepted by a limited object will be diffracted both as a propagating homogeneous field and as an evanescent (non homogeneous) one confined near the object surface. Because 1) the evanescent field does not propagate towards the detector, and 2) it is generated by the small details of the object, such details cannot be retrieved in the far field image. The limit of resolution is given by the formula :

$$R \sim \frac{1.2\lambda}{2n.\sin\theta}$$

where λ is the wavelength, n the index of the medium between object and lens, and θ the semi angle of aperture (diffraction angle).

Example : optical system with $\sin\theta = 1$, $\lambda = 500$nm $\underline{R \sim 300 \text{ nm}}$.

Let us recall that the longitudinal resolution (along the propagation direction) can reach 10^{-3} A by using interferometric techniques.

III - SOLUTIONS FOR INCREASING THE RESOLUTION

* Classical Solutions (far field microscopy). It consists of decreasing the value of R by
 - increasing θ (limited to $\pi/2$)
 - increasing n (by immersion)
 - decreasing the wavelength (UV or X microscopy)

* Exotic Solutions (Near field microscopy)
It is possible to increase fictitiously the index n by "immersion" in an evanescent field. (Photon tunneling microscopy)
The best way consists of circumventing the classical limit by detecting the diffracted field before propagation.

Example : Near Field Scanning Optical Microscopy (NSOM)
Scanning Tunneling Optical Microscopy (STOM)
In this case Rnear field < R

IV - OPTICAL METHODS IN METROLOGY AND TOPOGRAPHY

It is often assumed that optics is unable to provide high resolution methods for metrology and topography analysis. In fact such an assumption is questionable since the near field contains the information missing in the far field. It is thus possible by analyzing the near field to obtain superresolved images in experimental conditions rarely met by electron microscopy :

* The interaction light-matter is elastic (soft interaction). It implies :
 - non destructive testing
 - weak sensitivity to material properties
 - no sensitivity to electromagnetic environment
 - air propagation

* Real time non destructive testing methods can be developed in the range : 1µm - 10 nm leading to

- in situ, in line testing (metrology)
- in vivo analysis (biology)

V - THE SCANNING TUNNELING OPTICAL MICROSCOPY (STOM)

The scanning tunneling microscopy is the equivalent in optics of the electron tunneling effect.

Principle

The basic idea is to detect the evanescent field generated by total reflection or by diffraction at the surface of a transparent object by means of a dielectric stylus placed in the vicinity of the interface object-air.

Why "Tunneling" Microscopy ?

Because of the fundamental analogy between the properties of the evanescent wave and the tunneling current.

In order to prove these similarities of behavior, let us compare the electron and photon properties.

ELECTRONS	PHOTONS
energy $E = \hbar\omega$	energy $= \hbar\omega$
momentum $p = \hbar k$	propagation vector $k = p/\hbar$
mass $\neq 0$	mass $= 0$
wave function $= \Psi(x,y,z)$	electric field $= F(x,y,z)$
Wave equation :	Wave equation :
Schrödinger equation :	Helmholtz equation :
$\Delta\Psi + (2mh^2)(E-V_o)\Psi = 0$	$\Delta F + k^2 F = 0$
+ boundary conditions	+ boundary conditions
Solutions	Solutions
$\Psi = \Psi_o \exp\left[(i/\hbar)p.r\right]$	$F = F_o \exp(ik.r)$

PROPAGATION CONDITIONS

The parameter connected to the spatial behavior of the wave is the momentum p or the k vector.

$p = (p_x, p_y, p_z) = p_{\|\|}, p_z$	$k = (k_x, k_y, k_z) = k_{\|\|}, k_z$
$p_z = \left[2m(E-V_o) - p^2_{\|\|}\right]^{1/2}$	$k_z = \left[(n^2\omega^2/c^2) - k^2_{\|\|}\right]^{1/2}$
if $\left[2m(E-V_o) - p^2_{\|\|}\right] > 0$	if $(n^2\omega^2/c^2) - k^2_{\|\|} > 0$
p_z is real	k_z is real

then electrons and photons propagate as homogeneous waves

if $[2m(E-V_o) - p^2_{||}] < 0$

p_z is imaginary
the wave function becomes

$\Psi = \Psi_{xy} \exp(-p_z \cdot 2/h)$

if $(n^2\omega^2/c^2) - k^2_{||} < 0$

k_z is imaginary
the electric field becomes

$F = F_{xy} \exp(-k_z \cdot z)$

it is a damped function called evanescent wave in optics.

Interaction of an electron with a potential barrier

Interaction of a wave with a double optical interface

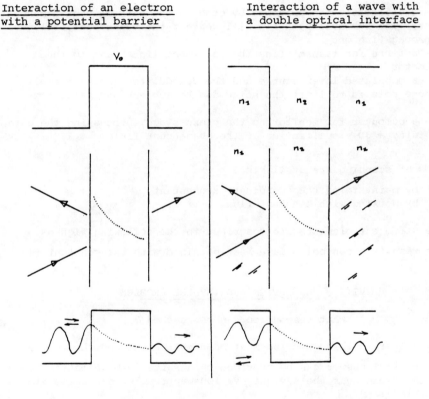

VI - PROPERTIES OF THE OPTICAL TUNNELING EFFECT

A beam falling onto a surface characterized by subwavelength details is partially converted into surface waves which cannot be detected by far field analysis.

An interesting property of the optical tunneling detection is connected to the k vector which is directly related to the spatial resolution. It is easy to show that the propagating component of the K vector in the evanescent wave, is larger than its modulus. This astonishing property is due to the fact that the z component of K is imaginary.

Practically, a wave characterized by a wavelength in vacuum = 500 nm will be "seen" in evanescent mode with a new wavelength of about 380 nm for an incident angle of 60 degrees although the evanescent wave propagates in air. This property is similar to the principle of immersion microscopy which allows to increase the spatial resolution.

VII - EXPERIMENT

The experimental device is composed of :

- a xyz piezo-positioning stage allowing the object surface scanning
- a dielectric tip obtained by etching the extremity of a monomode fibre. The subwavelength tip will convert the evanescent field into a propagating one
- a fibre for transmitting the resulting light beam to the photodetector
- a modulated laser source and the demodulator (lock' in amplifier) whose role is to limit the noise due to ambient and parasitic light
- a computer for monitoring the piezo stage, processing the data and providing a 3D visualization of the evanescent field.

Two modes of scanning are available :

-1) by maintaining the coordinate z constant
-2) by working with constant flux.

The different elements are described in the following figures.

The results shown below have been obtained with the constant flux mode.

VIII - RESOLUTION AND INVERSE PROBLEM

What resolution can be reached ?

The answer to this question must be experimental and theoretical.

- A partial answer can be provided by a suitable mathematical modeling of the field near the surface. Various approaches can be developed by exploiting :

* the Maxwell equations,
* the molecular physics,
* the antenna problems in microwave technology,
* the analogy with surface acoustic wave propagation.

- As suggested by physicists working in near field microscopy, the resolution seems to depend directly on the size of the stylus. A suitable modeling of the tip in interaction with the light field has to be developed.

- From our first results, it seems that two cases of interactions light-object have to be considered ; if the object size is of the same order of the wavelength, the interaction can be analyzed in terms of resonance phenomenon and the detected field is strongly different from the object. This effect clearly appears in the case of gratings. Now if the object is much smaller than the wavelength, it acts as a perturbation of the light beam which is confined very close to the object. Such conclusions are verified in the figure associated to the lecture.

- Finally, details of about 20 nm in the xy plane have been detected with this method corresponding to a resolution of $\lambda/30$. The resolution in the z direction is better than in xy plane ; it attains 3nm i.e. $\lambda/200$. Such results have been obtained without any image processing. By implementing suitable filtering and deconvolution procedures higher resolutions can be expected.

IX - BIBLIOGRAPHY

- E. Wolf, M. Nieto-Vesperinas", "Analicity of the angular spectrum amplitude of scattered fields and some of its consequences", J. Opt. Soc. Am., vol. 2, 1985, p. 886-889.
- E. Ash, Nicholls, "Super-resolution Aperture Scanning Microscope", Nature, vol. 237, 1972, p. 510-512.
- G. Massey, "Microscopy and pattern generation with scanned evanescent waves", Appl. Opt. vol. 23, 1984, p. 658-660.
- D. Pohl, W. Denk, M. Lanz, "Optical stethoscopy : Image recording with resolution/20", Appl. Phys. Lett., vol. 44, 1984, p. 651-653.
- U. Fischer, "Optical Characteristics of 0.1µm circular apertures in a metal film as light sources for scanning ultramicroscopy" J. Vac. Sci. Technol., B3(1), 1985, p. 386-390.
- E. Betzig, A. Lewis, A. Harootunian, M. Isaacson, E. Kratschmer, "Near-field scanning optical microscopy (NSOM) Development and Biological Applications", Biophys. J., vol. 49, 1986, p. 269-279.
- U. Fischer, D. Pohl, "Observation of Single-Particle Plasmons by Near-field Optical Microscopy", to be published.
- E. Betzig, M. Isaacson, A. Lewis, "Collection mode near-field scanning optical microscopy", Appl. Phys. Lett., vol. 51, 1987, p. 2088-2090.
- R. Reddick, R. Warmack, T. Ferrell, "New form of scanning optical microscopy", Phys. Rev. B, vol. 39, 1989, p. 767-770.
- D. Courjon, K. Sarayeddine, M. Spajer, "Scanning Tunneling Optical Microscopy", Optics Commun., vol. 71, 1989, p. 23-27.
- J. Guerra, "Photon tunneling microscopy", SPIE, vol. 1009, 1988.
- J.M. Vigoureux, C. Girard, D. Courjon, "General Principle of Scanning Tunneling Microscopy", Opt. Lett., to be published Oct. 1989.

APPLICATIONS

Tunneling optical microscopy could be used for transparent or metallic objects (by plasmons generation).
The field of applications could be :

* metrology in microelectronics
* biology
* integrated optics (beam propagation analysis inside the guide)
* infra-red optics
* acoustic wave analysis
* etc..

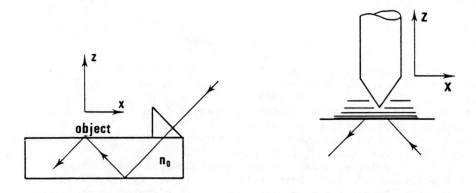

Technique of evanescent field generation

The tip is placed in the evanescent field

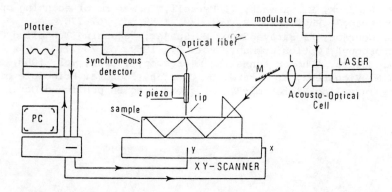

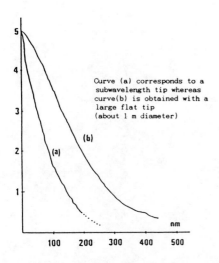

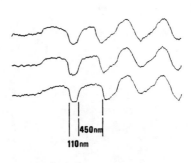

GLASS GRATING OBTAINED BY CHEMICAL ETCHING
(thickness about 150 nm)

QUASI EXPONENTIAL ATTENUATION OF THE
EVANESCENT FIELD VERSUS THE DISTANCE
SURFACE OBJECT–TIP

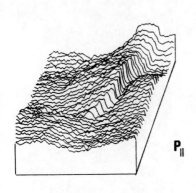

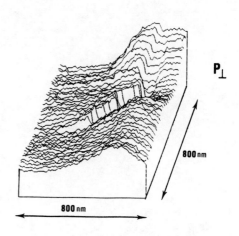

a SMALLEST DETECTED DETAIL : 20 nm ~ λ/30

Analysis of a defect on a glass grating
in parallel and orthogonal polarization

506 = blanko

SCANNING NEARFIELD ACOUSTIC MICROSCOPY

P. GUETHNER*, E. SCHRECK*, K. DRANSFELD
Univerity of Konstanz, Physics Departement, D-7750 Konstanz, W. Germany
)present address: IBM Almaden Research Center, San Jose CA 95120, USA

U. CH. FISCHER
Nanotec GmbH, D-6330 Wetzlar, W. Germany

ABSTRACT. Scanning Nearfield Acoustic Microscopy (SNAM) is a new method for imaging the topography of non-conducting surfaces using a vibrating tip of high Q as distance sensor. A 32 kHz quartz tuning fork is used as an oscillator driven at resonance by a feedback loop. The feedback signal is used both to excite and to detect the oscillation. An edge of one leg serves as a tip for imaging. If the tip is approached to the surface, the resonance frequency and the oscillation amplitude both decrease. The dependence of this decrease on the gas pressure shows that hydrodynamic forces in the gas and adsorbed vapor films are responsible for the interaction. For imaging, the sensor is piezoelectrically scanned accross the surface. The distance is feedback controlled such that the vibration amplitude stays constant. Using the control signal for imaging, a resolution of 1.5 μm laterally - limited by the tip geometry - and of 10 nm vertically are demonstrated.

INTRODUCTION

The achievement of G. Binnig and H. Rohrer to image the topography of electrically conducting surfaces at an atomic scale by scanning tunneling microscopy (1) has stimulated the invention of new types of microscopes such as the atomic force microscope (AFM) (2) for imaging the topography of insulating surfaces as well. The AFM detects the local interaction between a tip and a sample by measuring the minute deflections or the change of compliance of a small cantilever to which the tip is attached. The deflections of the cantilever have been measured by a tunneling gap (2), optical interferometry

(3,4), laser beam deflection (5) or piezoelectrically (6). The force sensitivity increases, if the cantilever has a lower compliance, a higher Q-value, a higher resonance frequency, and if driven close to its resonance frequency.

We chose a 32.7 kHz quartz tuning fork - such as they are used in many types of watches - having a quality factor of 14000 (in air) and a compliance of 3200 N/m (7,8) as a force sensing element. This quartz resonator of a very high quality factor is, however, also very stiff. It is therefore not an ideal force sensor, but the sensitivity should be sufficient for force microscopy. An advantage is its piezoelectricity which allows us to exite and detect the oscillation by one pair of electrodes deposited on the quartz.

THE ACOUSTIC NEARFIELD EFFECT AS A DISTANCE SENSOR

In order to test the function of the tuning fork as a force sensor, a few experiments were performed using a rather coarse geometry (8,9). The tuning fork was mounted such as to oscillate in a direction perpendicular to the sample which had a slightly spherical shape as shown in Fig. 1. In this way the effective area of interaction between the surface of the leg and the sample is estimated to be a few 10^{-2} mm^2. The tuning fork was excited by a feedback loop to oscillate at its resonance frequency at an amplitude of approximately 1 μm. The high quality factor allows us to measure the resonance frequency at a resolution of a fraction of one Hz. Fig. 1 shows the frequency shift as a function of relative distance between the two surfaces. A decrease of the resonance frequency starts at a distance of about 200 microns and the maximum frequency shift we could measure was a few Hz. For very small distances the oscillation stops presumably by mechanical contact.

We have also measured the piezoelectric signal of the oscillating tuning fork after passing through the amplifier of the oscillator feedback circuit and a rectifier. This signal will briefly be called the amplitude of vibration. This amplitude increases with the quality factor of the resonator and it strongly decreases when the distance between quartz and sample is reduced. For large distances of the order of 100 μm the amplitude decreases by a few percent. At small distances of less than 10 μm the amplitude decreases by more than 50 % (8,9)!

These changes of frequency and amplitude versus distance strongly depend on the pressure of the coupling gas (fig.1). A detailed analysis of these dependencies led us to conclude that a hydrodynamic damping force in the air gap is responsible for the observed effects and thus the dominant interaction between the vibrating leg and the sample surface in this case is not the van-der Waals force as usually assumed in AFM. The interaction arises at distances in the acoustic near field range of the vibrating leg of less than 200 µm well below the acoustic wavelength of 1 cm in air at 32 kHz.

Fig. 1

Frequency shift as a function of relative distance for different gas pressures.

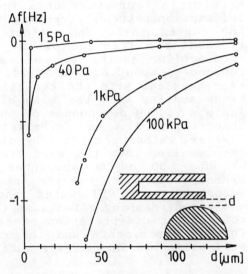

For distances larger than 100 nm the damping of the resonator is caused by the air streaming in and out of the gap between leg and sample. For dimensions which are small compared to the acoustic wavelength gases can be considered as incompressible viscous media (10). If the quartz moves up and down the air is pressed out and sucked in periodically. This streaming causes a hydrodynamic friction damping the resonator.

The decrease of the vibrational amplitude is, however, surprisingly large and we assume that it may not only b/ due to the hydrodynamic damping. The high quality factor the tuning fork, which is by two orders larger than the of simple cantilevers (11) of similar low loss materi is mainly due to the compensating opposed vibration c two legs of the tuning fork which inhibits the trans kinetic energy to and dissipation within the sup structure. By the interaction of only one leg wi surface, the symmetry of the oscillation is p{s leading to an onset of dissipation of energy support, as the leg is approached to the surfa phenomenon is expected to add to the sensitivity sensing as compared to a simple cantilever and represent an additional damping being responsibl large observed change in amplitude.

Our main interest was to investigate the interaction between a vibrating tip and a sample. Since it was not possible for us to attach a tip to the tuning fork without an unacceptable decrease of the quality factor we reversed the set up of fig 1a, replacing the sample by a tip having a radius of curvature of a few microns and measuring the interaction between a resting tip and the vibrating leg of the tuning fork. The effective area of interaction is reduced to a fraction of ca. 1 µm^2 and the range of the interaction leading to an observable frequency shift is reduced to about 0.2 µm (8,9). This reduction of the range is consistent with the typical property of the near field range scaling with the size of the radiation source (12). Again a strong dependence of the effects on gas pressure is observed (8,9). An interpretation of the effects is, however, rather complex. The distances are comparable to or even smaller than the mean free path of the gas molecules of 60 - 80 nm at atmospheric pressure and the simple hydrodynamic model for the interaction fails. In addition, adsorbed layers of water and other vapours may cause additional damping effects at these very small distances. A conclusive interpretation of the observed effects requires further experiments and these near field effects may be well suited to investigate such layers.

IMAGING OF TEST-SAMPLES

Irrespective of the exact nature of the dissipative force responsible for the observed effects, the strong distance dependance of the interaction stimulated us to build a scanning nearfield acoustic microscope. We simply used one edge of the quartz as an imaging tip by inclining the tuning fork with respect to the surface of the sample such that it oscillates at an angle of 45° (fig. 2). Also in this case the range of the interaction is reduced to distances of less than one micron.

Fig. 2
arrangement of the tuning
fork for imaging

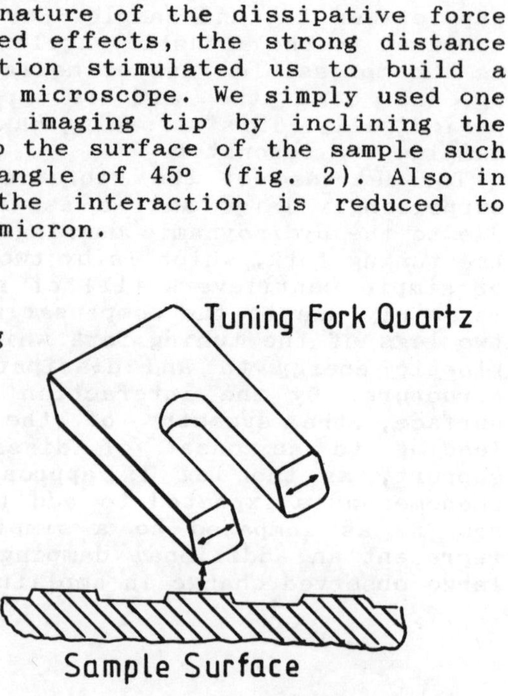

The first experiments were one-dimensional linescans of a linear grating. For simplicity we scanned the quartz across the sample without any distance adjustment and measured the vibration amplitude of the quartz as a function of the lateral displacement (9). It was possible to clearly image a line grating of a periodicity of 8 µm consisting of only 30 nm thick stripes of chromium evaporated on glass.

We now implemented an imaging mode as shown in fig.3, in which the sample is scanned at a constant vibration amplitude corresponding to the constant current mode of the STM. We measure the electrical vibration amplitude with a rectifier and compare this voltage to a reference voltage. The difference voltage is amplified and controls the z-piezo. The scan in x- and y-direction is made by commercial piezo translators and is controlled by a computer. The voltage at the z-piezo is registered at each point by the computer and used as a signal for image formation. A three dimensional image of the surface topography is finally obtained by image processing.

Fig.3

Experimental scheme for imaging in the constant amplitude mode

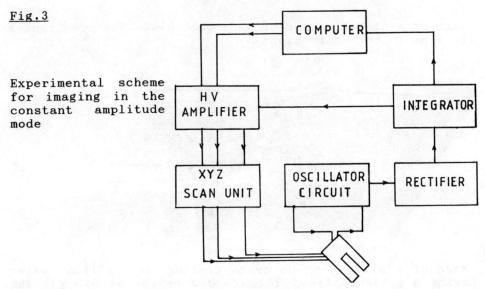

Fig. 4 shows an image of a nickel mask of a compact disc taken with the SNAM. In this image the well known track distance of 1.6 microns is revealed, but the height of 160 nm and the form of the hills is not imaged correctly because the radius of curvature the "tip" of a few microns is too large. It is also possible to image a PMMA replica obtained by stripping off the aluminum coating from a compact disk. The image of a silicon-oxide grating on a silicon wafer with a periodicity of 10 microns and a height of 0.3 µm is shown in fig. 5. It is easy to measure the lateral structure of this sample with an optical microscope or a SEM but it is impossible to measure the height and form of the silicon- oxide grating with such a microscope without destroying the sample (13). These two images demonstrate a lateral resolution of our microscope of about 1.5 µm and a height resolution of about 10 nm.

Fig. 4

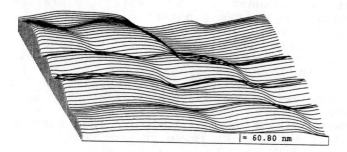

Image of a nickel master of a compact disk. The scan ranges are 7 µm in both directions

Fig. 5

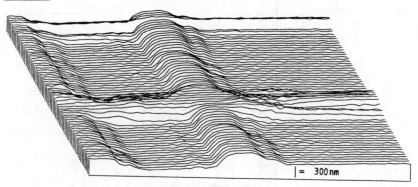

Image of a silicon oxide cross grating on a silicon wafer having a periodicity of 10 µm and a height of 0.3 µm. The scan range was 20 µm in both directions.

CONCLUSION

The Scanning Nearfield Acoustic Microscope (SNAM) uses the friction of the gas and the adsorbed vapor film in the gap between a vibrating tip and the surface of the sample in close proximity as a measure of distance. The lateral resolution is presently limited by the radius of curvature of the imaging tip to about 1.5 µm and it should be possible to increase this resolution by designing special high-Q resonators with finer tips. This microscope is a new type of force microscope capable of imaging the topography of non-conducting surfaces. Whether the forces discussed here are also important for other AFM's using a vibrating tip (3) has to be further clarified. The most interesting features of SNAM are the simplicity of detecting the vibration amplitude and the small interaction forces. Imaging at moderate lateral resolution the topography of non-conducting surfaces with an unknown variation of the refractive index, where optical profilometry fails, and of soft surfaces such as polymers, where other mechanical profilometers cannot be used, are a promising field of technically interesting applications of SNAM.

REFERENCES

1) Binnig, G., H. Rohrer, C. Gerber, E. Weibel. (1982). 'Surface studies by scanning tunneling microscopy' Phys. Rev. Lett. 441, 57 - 61.
2) Binnig, G., C. F. Quate, Ch. Gerber (1986). 'The atomic force microscope'. Phys. Rev. Lett. 56, 930.
3) Martin Y., C. C. Williams, H. K. Wickramasinghe. (1987). 'Atomic force microscope - force mapping and profiling on a sub 100-A scale.' J.Appl. Phys. 61,4723
4) Rugar, D., H. J. Mamin, R. Erlandsson, J. E. Stern, B. D. Terris (1988). 'Force microscope using a fibre - optic displacement sensor.' Rev. Sci. Instrum. 59, 2337 - 2340.
5) Alexander, S., L. Hellemans, O. Marti., J. Schneir, V. Elings, P.K. Hansma (1989). 'An atomic-resolution atomic-force microscope implemented using an optical lever.' J. Appl. Phys. 65, 164 - 167
6) Anders, M., C. Heiden. (1988). 'Imaging of tip - sample compliance in STM.' J. Microscopy 152, 643 - 650.
7) Christen,M. (1983). 'Air and gas damping of quartz tuning forks.' Sensors and Actuators 4, 555 - 564.
8) Guethner, P., Fischer U. Ch., Dransfeld, K. (1988) 'Experimente zur akustischen Nahfeldmikroskopie.' Beitr. elektronenmikroskop. Direktabb. Oberfl. 21, 27-32.
9) Guethner, P., Fischer U. Ch., Dransfeld, K. (1989) 'Scanning Near-field acoustic Microscopy.' Appl. Phys. B48, 89-92.
10) Landau, L. D., Lifschitz E. M. (1966) Lehrbuch der theoretischen Physik, Bd. 6 (Hydrodynamik), Akademie Verlag, Berlin

11) Mc Clelland, G. M., R. Erlandsson, S. Chiang. (1987). 'Atomic force microscopy: general principles and a new implementation.' Rev. prog quant. non destr. eval.6B, 1307 -1312.
12) Fischer, U. Ch. (1985) 'Optical characteristics of a 0.1 µm circular aperture in a metal film as light source for scanning ultramicroscopy.' J. Vac. Sci. Technol. B3 (1), 386 - 390.
13) Hersener, J. (AEG Forschungsinstitut Ulm), priv. comm.

INDEX

Abrikosov flux line pattern 261
adhesion 452
adsorbates
- adsorbate structures 187, 188
- electric field effects 197
- local bond effects 190
- local observation of adsorbates 193
- mobility of adsorbates 194, 195
- spectroscopy of adsorbates 192

AFM - atomic force microscopy 443
Ag
- Ag(100) 326
- Ag(111) 323
- Ag/AgCl electrode 321
- electrodeposition 288, 304, 319
- imaging by SNOM 484
- photon emission 269, 273

AgBr 461
Al
- Al(111) atomic corrugation 103, 107, 109
- Al(111) atomic resolution imaging 178
- Al(111) barrier height 105
- Al(111)-C 188
- Al(111)-O 194
- atomic corrugation 119
- corrosion 306

AlF_2 282
alkanes 343
- electron structure 344
- through bond tunneling 384

Andreev reflection 252, 256, 257, 258
angular spread
- in electron field emission 414, 415, 425
atomic corrugation 179, 103, 107, 109, 454
atomic force microscopy (AFM) 443
- atomic resolution 454
- imaging 454
- modes of operations 448

Au
- Au(100), atomic resolution imaging 178
- Au(110), (1x2) reconstruction 183
- Au(110), terrace topography 184
- Au(111) 324
- Au(111), atomic resolution imaging 181, 182
- Au(111), reconstruction 182
- Au(111), STM spectroscopy 177
- Au(111)-Ag 196
- Au(111)-Au 196
- Au(111)-Cl 329
- electrodeposition 304, 319
- plasmon observation by SNOM 270

bacteriophage φ29 336
bacteriophage T4 338
ballistic electron transport 136, 143, 151
band gap

- wide band gap materials 234
Bardeen, Cooper, Schriever (BCS) model 252, 253
Bardeen formula 80, 82, 97
BEEM - scanning electron emission microscopy 16
Bi
- band gap of Bi covered surfaces 234
biomaterials 349
blooming 282
branching ratio
- in photon emission 274
Büttiker-Landauer expression 65
carbon
- amorphous hydrogenated carbon 459
charging effects 255, 260
- single-electron charging 27
charge density waves (CDW) 369, 455
charge transfer complexes 372
CITS - current imaging tunneling spectroscopy 229, 235
coherence
- of a fermion beam 400
coherence length in tunneling 252, 256
coherent electron emission 409
collimation of electron beams 426
conductance through a constriction 157, 159
conductive polymers 374
conductivity
- of biological matter 343
corrosion 302, 306, 319,
- photocorrosion 330
Coulomb blockade 12, 17, 252, 253, 260
Coulomb gap 252
Coulomb staircase 255
crystal field levels 246
Cu
- Cu(111), STM spectroscopy 177
- Cu(110)-O 188, 201
deformation 453
- elastic deformation 115, 118
- plastic deformation 116
- plastic deformation region 137, 139
density of states 249, 252
diffraction
- semi-classical approximation 415
- through a tunnel barrier 414
diffusion controlled steady-state current 318
diffusive coupling 318
direct interaction model
- in inelastic tunneling spectroscopy 263
DNA 306, 336, 338, 342, 363
dwell time 59
Ehrenfest theorem 103
elastic conductance 143, 145
electric field 97
- effects in STM imaging of adsorbates 197
electrochemical current 316
electrochemical deposition 288, 304, 323, 326, 327, 330
electrochemical reactions 317

electrolytic transport 318
electron beam
- angular spread 414, 415, 417, 422, 425
- classical trajectories 424, 425
- collimated electron beam 409, 426
- electron wave packet 427
- focalization 422
- magnification 423, 426, 438
- resolution 426
- transverse velocity of electrons 422, 423
electron charge distribution 165
electron focusing
- in electron field emission 414
electron focusing mirror 248, 259
electron-hole pair 277
electron-impurity scattering 242
electron mean free path 144, 149, 155, 241
electron microscopy 399
electron-phonon interaction
- in alkali metals 243, 245
- in noble metals 243
- in normal metals 27
- in superconductors 27
- in zinc 243
- scattering 242, 258
electron-surface interaction 246
electron thermalization 270
electron transfer
- in disordered organic chains 391
- in organic molecules 335
electron traps 17
electron tunneling spectroscopy 27
electron wave packet 427
Eliashberg function 243, 248
emission properties 269, 270, 404, 409, 419
epitaxial growth 196
faradaic current 322
faradaic reactions 321
equipotential surface 417, 418, 428
FEM - field emission microscopy 434, 435
ferroelectric films 287
Feynman path integral approach 61
fibrinogen 304
fibronectin 350
field electron emission 163, 409, 419
- angular spread 414, 415, 417, 422, 425
- classical trajectories 424, 425
- coherent electron emission 409
- current-voltage characteristics 421, 422, 438
- electron focusing 414
- energy distribution 412
- level quantization 411
- magnification 423, 426, 438
- protrusions on the tip 424, 425
- quantum size effects 409, 411
- Schrödinger equation 414, 428
field emission microscopy (FEM) 434, 435

field ion microscopy (FIM)   169, 170, 402, 435, 436, 437
fluorescence spectra   273
focussing
- of an electron beam   149
focussing field   246, 248
force measurements   444
forces
- adhesion   452
- electrostatic forces   451
- frictional forces   451
- ionic repulsion   450
- loading force   453
- magnetic forces   450
- van der Waals forces   449
force sensor   447, 469
Fowler-Nordheim plot   166, 168, 169, 421, 434, 438
friction   459
Friedel oscillations   150, 151, 154
GaAs
- GaAs(110)   85, 226
- GaAs(110)-Au   222
- GaAs(110)-O   228
- GaAs(110)-Sn   221
- GaAs(110)-Sb   223
- electrodes   303
- photocorrosion   330
Ge
- Ge(100)   221
- Ge(111)   220, 228
GeSi
- GeSi(111)   235
giant corrugation   117
graphite   301, 319
- giant corrugation   117
- polypyrrole   331
- reaction with water vapor   293
Gundlach oscillations   9, 31, 177
Helmholtz layer   343
hydrodynamic friction   509
IETS - inelastic tunneling spectroscopy   49, 261
image force   164, 426
incremental charging   252, 254
inelastic excitation probability   272
inelastic mean free path   279
inelastic tunneling   263, 269
Inelastic Tunneling Spectroscopy (IETS)   49, 261
- benzoic acid   262
- C-O stretching vibration   262
- direct interaction model   263
- sorbic acid   263
information storing   281
instrumentation
- AFM   445, 470
- SNAM   507
- STM   215, 352, 360
integrated photon yield
- in photon emission   270

interference
- of a fermion beam   400
interrupted feedback   230, 235
ion enhanced etching   293
Ir
- tip-sample interaction   119
isochromat photon spectra   269
KNO_3 287
Knudsen limit   242, 243
Langmuir-Blodgett film   336
Larmor clock approach   63, 74
layered materials   367
leakage current   343
liquid crystals   300
liquid metal ion source   284
liquid-solid interface   299
lithography   281, 300
loading force   139, 453
local density of states (LDOS)   279
local probe methods   1
low energy projection microscope   405
magnetic imaging   286
magneto-optical devices   285
Matthiesen's rule   242
mesoscopic systems   270
metal surfaces
- adsorbates   187, 188
- d-states   84
- image effects   177
- sample preparation   174
- STM imaging of metal surfaces   176, 178
- STM spectroscopy   177
- superposition of atomic charge densities   84
- surface contaminations   187
- terrace topography   183, 184
- tunnel current   188
metallic glasses   289
microfabrication   447
microfacets   183, 185
Mo(100)-S   188
MoS_2 368
Morse potential   118
nanolithography   282
$NbSe_2$
- superconducting vortices in $2H-NbSe_2$ 12
negative resistance   288
Ni
- Ni(100)-O   188, 197
- Ni(110)-CO   187
- Ni(110)-O   17, 201
- electrodes   303
non-ballistic transport   152, 154, 155
nonlinear effects   241
nucleation and growth   197, 199
organic molecules
- conductivity   343
- electron transfer   335

organic chains   391
- electron transport   393, 396
- inelastic processes
- transmission   395
osteointegration   351
oxidation
- of metal surfaces   202
patch fields   165
Pauli exclusion energy   114
Pb
- electrochemical deposition   323, 326, 327
Pd
- Pd(111) STM spectroscopy   177
phonon assisted tunneling   52
phonon spectroscopy
- normal state phonon spectroscopy   46
- of superconductors   44
photon emission   269, 270
- integrated photon yield   270
- spectral distribution of photons   273
- topography effects   273
PMMA   282, 284, 350
point contact   116, 134, 147, 241
- contact aperture   136
- electronic contact regime   138
- excess current   257
- spectroscopy   1, 241
point probes   443
point source beams   23, 399
Poison equation   242
polarizable electrode   321
polypeptides   343, 396
potential
- energy profile for tunneling   28
- in the barrier region   82
potentiometry   14
potentiostatic concept   320
pressure effects in tunneling   252, 255
- dependence in T_c 256
- $La_{1.85}Sr_{0.15}CuO_4$ 255
proteins   343
PSP - pseudo-stationary profile technique   431
PSTM - photon scanning tunneling microscopy   498
Pt
- Pt(100), 'hex' reconstruction   181
- Pt(100)-C_2H_2 187, 198
- Pt(100)-CO   187, 198
- Pt(100)-NO   187, 198
- Pt(110), (1x2) reconstruction   183
- Pt(110), terrace topography   184
- Pt(110)-CO   198, 200
- Pt-C films   335, 336
- Pt-Ir-C films   355
- electrochemical deposition   319
- electrodes   303
purple membrane   341
quantized conductance   143

quantum elastic resistance   145
quantum of resistance   144
quasi-bound states 37
radiative damping of surface plasmons   277
reconstruction of images   11
reference elctrode   319
resist   282
- contamination resist   283
- PMMA   282, 284, 350
resolution   86
- far field optics   498, 502
- SNOM   476, 481
- STM, lateral resolution   2, 3, 89, 90
- STM, spatial/energetical resolution of electronic states 211
- STM, vertical resolution   2, 3, 89, 90
resonant structure   143, 147
resonant tunneling   27, 31
- through a constriction   157, 159
Rh
- Rh(111)-C_6H_6/CO   188
Ru
- Ru(0001), subsurface oxygen   202
- Ru(0001)-Cu   196
- Ru(0001)-O   193
$RuCl_3$ 368, 370
Sb
- band gap of Sb covered surfaces   234
scanning electrochemical microscope   304
scanning ion-conductance microscope (SICM)   306
scanning near field acoustic microscope (SNAM)   270
scanning near field optical microscope (SNOM)   270, 475
scanning tunneling optical microscope (STOM) 497
scattering
- efficiency factor scattering   243
- electron-excitation scattering   249
- electron impurity scattering   242
- electron-phonon scattering   242
- stationary-state scattering   27
scattering time
- electron-phonon scattering   243
screw dislocations   186
semiconductor surfaces
- buckling   225
- imaging energy   85
- imaging of band-edge states   86
- resolution   90
- STM imaging   85, 225
Sharvin contact   143, 144, 242, 243, 247, 257
Si
- Si(100)   221, 226
- Si(111)   85, 220, 228, 235
- Si(111) - dimer adatom stacking fault (DAS) model   227
- Si(111)-Ag   223
- Si(111)-Al   221
- Si(111)-Au   230
- Si(111)-Ga   221

- Si(111)-In   221
- Si(111)-NH$_3$ 236
SICM - scanning ion conductance microscope   306
Si$_3$N$_4$ 282
single atom tip   404
single electron charging   27, 53
single electron tunneling (SET)   252, 253, 254
small contacts   143, 144
- constrictions in small contacts   143
- elastic resistance   143
- quantized conductance   143
Smoluchowski effect   165, 176
SNAM - scanning near field acoustic microscope   270
SnGe
- SnGe(111)   235
SNOM - scanning near field optical microscope   270, 475
- imaging   488
- resolution   476
sorbic acid   300
space charge   164
spectroscopy
- electron tunneling spectroscopy   27
- gap spectroscopy of superconductors   43
- inelastic tunneling spectroscopy   261
- phonon spectroscopy of superconductors   44
spherical tip model   80
stationary-state scattering approach   29
STOM - scanning tunneling optical microscopy   497
STM imaging
- capacitance imaging   22
- conductivity imaging   229
- constant current imaging   220, 230
- constant height mode   8, 83
- constant interaction mode   8, 83
- contact   9
- magnetic imaging   20, 286
- measurement mode   7
- mixed interactions   19
- of biological matter   336, 342, 396
- of electrode surfaces   316
- of support material   336
- principles   212
- probe object distance   9
- reconstruction of images   11
- secondary electron imaging   22
- smoothening of state density   218
- technical features   214
- thermal imaging   22
- tip shape effects   114
- tracking mode   8
- voltage dependent imaging   223
- work function imaging   229, 339, 363
STM spectroscopy   114
- background current   236
- ballistic electron emission microscopy (BEEM)   16
- conductivity imaging   229
- constant separation spectroscopy   230

- current imaging tunneling spectroscopy (CITS) 12, 229, 235
- normalized conductivity 231
- qualitative theoretical description 91
- quanitative theoretical description 92
- scanning tunneling potentiometry 14
- spatially resolved spectroscopy 234
- spectroscopy at many points 235
- variable separation spectroscopy 233
- voltage dependent imaging 223
- work function imaging 13, 229, 339, 363

structural transformations
- of metal surfaces 198

subsurface oxygen 202
superconductivity gap 43, 253
superconductor
- gap energy 257, 260
- gap spectroscopy of superconductors 43
- heavy Fermion superconductors, tunneling 252
- high-T_c superconductors, tunneling 252, 258
- insulator-normal metal (SIN) junction 253
- Pb 258
- phonon spectroscopy of superconductors 44
- vortices 12

surface diffusion 17, 195, 196, 330, 431
surface modification 281, 283
- ablative systems 284, 285
- atomic scale modifications 283
- attachment of molecules 288
- deposition of atoms 284
- hole melting 285
- metallic glasses 289
- modification of molecular films 288, 290
- phase change systems 285
- surface charging 289

surface plasmons 275, 483
- dipole-active modes 275
- electromagnetic coupling 275
- excitation and decay 269, 276
- multiple modes 275
- nonradiative damping 277
- radiative damping 277
- short-wavelength surface plasmons 275

surface reconstruction 180
surface resonances 168
surface roughness 459
surface states 216, 217, 230, 231
- dangling bonds 224
- effects in tunneling 82
- smoothening of state density 218
- two-band model 227

Ta-W imaging by SNOM 484
TaS_2 368
$TaSe_2$ 300, 368
Taylor cone 284
Te 285
teton tips 409, 421, 430, 433, 435, 439
thermionic emission 163

Ti 350
TiO$_2$ 330
TiS$_2$ 369
TILS - tip induced localized states 130
tip
- atomic size microtips 409
- blunt tip 139
- build-up tips 433, 435
- coating 319, 322, 337
- elastic deformation of the tip 104, 107
- electronic structure 98
- etching 337, 361
- fabrication of tips 431
- hyperboloidal tips 417
- multiple tips 83
- pseudo-stationary profile (PSP) technique 431
- single atom tip 404
- spectral resolution of tip orbitals 101
- spherical tip model 80
- teton tip 409, 433, 435, 439
- transition metal tips 97
tip-sample interactions 92, 116, 118, 472
- Al(tip)-Al(sample) 131
- Al(tip)-graphite(sample) 125
- C(tip)-graphite(sample) 123
- forces 472
- local density approximation 121
- SCF pseudopotential modeling 122
- tip induced localized states (TILS) 130
- tip induced modifications in the atomic and electronic structure 120, 129
topografiner 22
Transfer Hamiltonian 27, 40, 78, 80, 82, 97, 114
transverse velocity of electrons 422, 423
tribology 459
trumpets 144, 148, 428
tunnel barrier
- barrier collaps 115, 119
- barrier lowering 129
- barrier height 105, 169
- diffraction through a tunnel barrier 414
- distance dependence 107
- effective barrier 279
- geometry 417
tunnel current 102, 215
tunneling
- Andreev reflection 252, 256, 257
- Bardeen, Cooper, Schriever (BCS) model 252
- Bardeen formula 80, 82
- between semi-infinite electrodes 27
- coherence length 252, 256
- Coulomb blockade 252, 253
- Coulomb gap 252
- description by model Hamiltonian 97
- effect of surface states 82
- elastic deformation of the tip 97, 104, 105
- electronic structure of the tip 98

- electron thermalization   270
- energy diagram   216
- heavy Fermion superconductors   252
- high-T_c superconductors   252, 258
- inclusion of d-states on the tip   97, 104, 105
- incremental charging   252, 254
- local orbital basis set   81
- local work function   79
- Monte Carlo simulations   252, 253
- potential energy profile   28
- pressure effects   252, 255
- resonant tunneling   27, 31
- single electron tunneling (SET)   252, 253, 254
- spherical tip model   80
- stationary-state scattering approach   29
- s-wave approximation   80
- through electrolyte junctions   331
- through-bond effective coupling   381, 388
- through-bond transmission   382
- through-bond tunneling   344, 378, 381, 384
- through-space tunneling   378, 379
- wave-vector conservation   82

tunneling time   27
- Büttiker-Landauer expression   65
- dwell time   59
- Feynman path integral approach   61
- for one-dimensional barriers   58
- Larmor clock approach   63, 74
- wave packet approach   70

tunnel junction
- capacitance   252
- charging energy   252
- Coulomb blockade   252
- electrolyte junction   331
- evaporated junction   252
- metal-oxide-metal junction   270
- point contact   249
- single electron tunneling   252
- superconductor-insulator-normal metal (SIN) junction   253
- TTF-TCNQ   252
- vacuum junction   252
- $YBa_2Cu_3O_{7-\delta}$ 253, 258

V   350
VS_2 368
vacuum level   164
van der Waals energy   114
vortex   12, 261

wave packet approach
- for tunneling times   70

WKB approximation   28, 216
- transmission in WKB approximation   219

work function   13, 163
- image effects   13
- of adsorbates   166
- work function imaging   229

zero bias anomaly   245